Diatom Morphogenesis

Scrivener Publishing
100 Cummings Center, Suite 541J
Beverly, MA 01915-6106

Diatoms: Biology and Applications

Series Editors: Richard Gordon (dickgordoncan@xplornet.com) and Joseph Seckbach (Joseph.Seckbach@mail.huji.ac.il)

Scope: The diatoms are a single-cell algal group, with each cell surrounded by a silica shell. The shells have beautiful attractive shapes with multiscalar structure at 8 orders of magnitude, and have several uses. 20% of the oxygen we breathe is produced by diatom photosynthesis, and they feed most of the aquatic food chain in freshwaters and the oceans. Diatoms serve as sources of biofuel and electrical solar energy production and are impacting on nanotechnology and photonics. They are important ecological and paleoclimate indicators. Some of them are extremophiles, living at high temperatures or in ice, at extremes of pH, at high or low light levels, and surviving desiccation. There are about 100,000 species and as many papers written about them since their discovery over three hundred years ago. The literature on diatoms is currently doubling every ten years, with 50,000 papers during the last decade (2006-2016). In this context, it is timely to review the progress to date, highlight cutting-edge discoveries, and discuss exciting future perspectives. To fulfill this objective, this new Diatom Series is being launched under the leadership of two experts in diatoms and related disciplines. The aim is to provide a comprehensive and reliable source of information on diatom biology and applications and enhance interdisciplinary collaborations required to advance knowledge and applications of diatoms.

Publishers at Scrivener
Martin Scrivener (martin@scrivenerpublishing.com)
Phillip Carmical (pcarmical@scrivenerpublishing.com)

Diatom Morphogenesis

Edited by
Vadim Annenkov,
Limnological Institute, Siberian Branch of Russian Academy of Sciences
Joseph Seckback
The Hebrew University of Jerusalem, Israel
and
Richard Gordon
Gulf Specimen Marine Laboratory & Aquarium, Panacea, FL, USA
and Wayne State University, Detroit, MI, USA

Scrivener Publishing

WILEY

This edition first published 2022 by John Wiley & Sons, Inc., 111 River Street, Hoboken, NJ 07030, USA and Scrivener Publishing LLC, 100 Cummings Center, Suite 541J, Beverly, MA 01915, USA

For more information about Scrivener publications please visit www.scrivenerpublishing.com.

Wiley Global Headquarters
111 River Street, Hoboken, NJ 07030, USA

For details of our global editorial offices, customer services, and more information about Wiley products visit us at www.wiley.com.

Library of Congress Cataloging-in-Publication Data

ISBN 978-1-119-487951

Cover image: Colored scanning electron micrographs (SEMs) of a morphogenetic sequence of the diatom *Fragillaria capucina var. mesolepta* by Dr. Mary Ann Tiffany, Biology Department, San Diego State University, USA
Sample taken from Lake Murray (a freshwater San Diego Reservoir) on 3/18/2000.
Cover design by Russell Richardson

Set in size of 11pt and Minion Pro by Manila Typesetting Company, Makati, Philippines

10 9 8 7 6 5 4 3 2 1

FSC
www.fsc.org
MIX
Paper from responsible sources
FSC® C013604

Topics Addressed in This Volume

Topics include computer simulation of morphogenesis, silicic acid to silica frustules, inhibition in valve morphogenesis, pores within frustules, mesopores of pennate diatoms, frustule photonics and light harvesting, clonal chains, silica cell wall, geometric models of centric diatoms, morphology, surface features, buckling of valve morphogenesis, on mantle profiles, genetic-biochemical approaches, modeling silicon pools, valve morphogenesis, diatom teratology in taxonomy, phenotypic plasticity, geometric and morphometric analysis, silica morphogenesis in sister algae, and the uncanny symmetry of some diatoms.

This volume is the third book in the series Diatoms: Biology and Applications. The first book, *Diatoms: Fundamentals and Applications* appeared in 2019, and was edited by Joseph Seckbach and Richard Gordon. The second book, *Diatom Gliding Motility*, was published in September 2021 and is edited by Stanley A. Cohn, Kalina M. Manoylov and Richard Gordon.

We would like to thank the authors, the reviewers, the guest editor (Vadim V. Annenkov), and our publisher Martin Scrivener of Massachusetts, USA.

Joseph Seckbach
Hebrew University Jerusalem, Israel
September 2021

Part 1
GENERAL ISSUES

1

Introduction for a Tutorial on Diatom Morphology

Kalina Manoylov[1]* and Mohamed Ghobara[2]

[1]*Dept. of Biological & Environmental Sciences, Georgia College and State University, Milledgeville, GA, United States*

[2]*Department of Physics, Freie Universitat Berlin, Berlin, Germany*

Abstract

Diatoms are an exceptionally successful group of unicellular microalgae with a large contribution of global primary production in aquatic environments and contributing a significant amount of oxygen to both hydro- and atmospheres. They are fascinating throughout their life and even after death, thanks to their unique cell walls made from ornamented silica. The diatoms include centric species, which may have radial or polar symmetry, and pennates, which include araphid, monoraphid, and biraphid species. Several applications have utilized diatomite, i.e., the fossil form of diatom frustules. To date, many diatoms' secrets have been understood; however, there are still more hidden. Thus, there is a need for more research on diatom basic biology and applications. Seeking this goal, more people should be encouraged to work on diatoms. Often novice researchers are overwhelmed by the terminology associated with the diverse morphology, the discrepancy between expected features for published descriptions, and the actual observation of those complex 3D organisms, which can be a barrier for more progress. Here, we provide a brief introduction to the beginners with a guide to approach the complex diatom morphology focusing on the tools that can be used for its study.

Keywords: Diatom morphology, tutorial, LM and SEM, frustule morphology

1.1 Diatoms in Brief

Diatoms are unicellular, eukaryotic, microscopic algae (range from 1.5 μm to 5 mm in length, or diameter [1.9]), which maintain large population numbers and contribute considerably to the carbon and oxygen cycle on a global scale [1.8]. This ecologically successful group of algae is present in all aquatic habitats e.g. [1.1, 1.2] and even extends to humid terrestrial places. In aquatic habitats, diatoms are present in the photic zone, i.e., the region of water that light strongly penetrates, as well as in the benthic zone, i.e., the lowest level of water adjacent to the bottom with dim light conditions, depending on water column height and water's turbidity. Diatoms can exist as planktonic

**Corresponding author*: kalina.manoylov@gcsu.edu

Vadim Annenkov, Joseph Seckback and Richard Gordon (eds.) Diatom Morphogenesis, (3–18) © 2022 Scrivener Publishing LLC

(i.e., suspended in the water column), benthic (i.e., living near the bottom), epiphytic (i.e., adhered to aquatic plants [1.19], Figures 1.2c–d), or epizoic (i.e., adhered to a wide range of marine organisms such as crustaceans, mollusks, and vertebrates [1.19, 1.38]), or epilithic (i.e., attached completely or partially to submerged rocks). The adhesion ability of some diatoms is related to their mucilage secretion from specialized areas within their rigid cell walls (such as examples shown in Figures 1.1d and 1.2c–d). Some diatoms can form colonies in different arrangements such as chains and ribbons (examples shown in Figures 1.1 and 1.2).

Diatoms are a unique group of microalgae for several reasons, but one of the most notable and unique differences is the glass cell walls they possess [1.45]. This cell wall is called the "frustule" and is composed of amorphous hydrated silica that gives it unique properties. In general, the frustule is composed of two pieces that fit together like a petri-dish, meaning that the lower part of the frustule, called the hypotheca, sits inside of the upper part of the frustule, called the epitheca. The frustule volume extends by adding strips of silica called girdle bands (cingulum) to the mantle, i.e., the curved edge of the valve. It should be noted that there are plenty of frustule morphologies that vary between taxa.

Diatoms reproduce both asexually (visible in Figure 1.6) and sexually. Most of the time, they reproduce asexually via binary fission through adding new hypovalves to the parent valves. Those new hypovalves are synthesized inside the silica deposition vesicle (SDV). Only after the new hypovalves have completely synthesized and the protoplast cleavage, as well as the exocytosis of siliceous parts, has occurred, the final splitting apart will occur, leaving two daughter diatoms in place. Because the SDV forms inside of each new cell

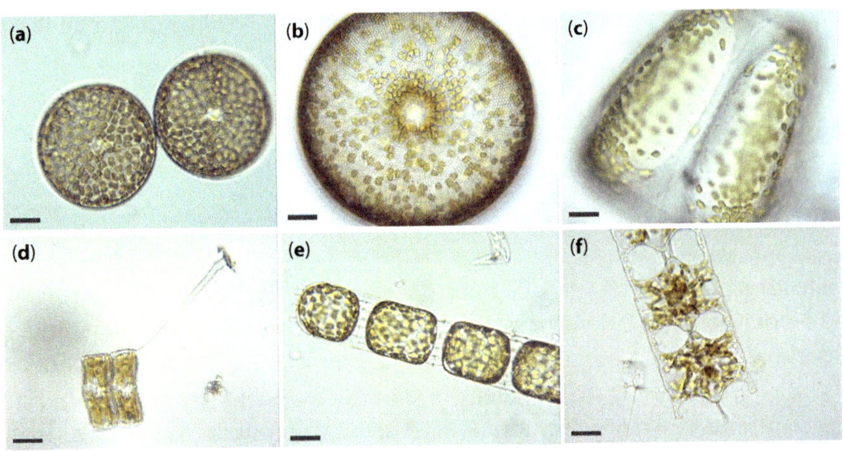

Figure 1.1 Living diatoms as observed under LM, brightfield. (a) Two living cells of *Actinoptychus senarius* (Ehrenberg) Ehrenberg at the valve view. (b) The valve view of a single living cell of *Coscinodiscus wailesii* Gran and Angst. (c) The girdle view of a single living cell of *Coscinodiscus granii* L.F. Gough. (d) Two living cells of *Achnanthes brevipes* C. Agardh at the girdle view attached to each other with a prolonged stalk for the attachment to the substrate. (e) A living colony of *Stephanopyxis turris* (Greville) Ralfs with visible linking spines. (f) A living colony of *Odontella longicruris* (Greville) M.A. Hoban with discoid chloroplasts. Copyright reserved Mary Ann Tiffany, used with her permission. The identification was carried out by Mary Ann Tiffany. All the scale bars are 50 μm.

cell. Xing *et al.* [1.42] is an inspiring reference for the 3D reconstruction of diatom frustule using the 2D image series resulting from FIB-SEM.

1.3.1 Recommended Steps to Understand the Complex Diatom Morphology: A Guide for Beginners

1. Fresh samples, if available, which means the sample is not diatomite (i.e., the fossil form of diatom frustules) [1.31], allow evaluation of the physiological state of the population at the time of collection. The features of living cells might help in the identification process, through the descriptions of colonial forms, cell attachment, extruded materials, plastids, and nucleus position (for instance see Figures 1.1 and 1.2). Since diatom frustules are rigid silica, they appear from two different viewpoints, the girdle view (from the side) and the valve view (from the front) under LM Figure 1.4. Some genera are easily identifiable in girdle view, such as *Mastogloia* and *Amphora pediculus* (Kütz.) Grunow, others are not. Lately, confocal microscopy uses are very promising [1.3, 1.13], also combining microscopic and molecular information allowed reclassification of a population in culture annotated as a radial centric species related to *Leptocylindrus danicus* Cleve, as an araphid pennate species in the staurosiroid lineage, within the genus *Plagiostriata* [1.14].
2. The selection of the suitable cleaning procedure, depending on the source of diatoms, is a crucial step to extract the siliceous frustule parts. Routinely, diatoms are identified and enumerated without their protoplasts where striae morphology & number, raphe shape, and other morphological characteristics could be verified for identification (details in Figures 1.3b and 1.4). Among the best guides for the cleaning process could be Wang *et al.* [1.43]. More gentle cleaning methods (e.g., H_2O_2 method) should be considered in case we need to preserve mesopores (i.e., fine pores with diameters ranging from 50 to 3 nm that present on pore occlusions, which cover the areolae from inside or outside [1.48]) especially within genera, such as raphid pennate diatoms (Figure 1.5b).
3. Preparation of permanent slides should be carried out for LM observations using a suitable high refractive index mounting medium such as Naphrax and Hyrax [1.44]. Figure 1.4 considers a good example of observing cleaned frustules fixed in a permanent slide under LM.
4. Using LM to observe the 2D morphology of the cleaned valve and girdle bands to assign the most similar genus and, if possible, the species. The overall morphology, striae, valve symmetry, ribs, and the presence of raphe will be clear under LM; however, the pores, the spines, and special structures within the valve and girdle band might be not observable due to their smaller size. Though, in other few cases, the areolae might be observable even before the frustule cleaning procedure, especially for large taxa (Figure 1.1b). Valve outline, valve ends, shapes of sturdy, and hyaline silica (not penetrated with striae or other openings) should be recorded. Classical diameter in centric diatoms from valve view or length and frustule depth form girdle view allows verifications of descriptions from literature. Moreover, the frustule silicification rate can be observed under LM reflects information about the environmental condition that surrounds the living cells.

5. Using SEM, after suitable sample preparation and coating with a conductive material [1.38], to observe the ultrastructure of the valve and girdle bands and define the species accordingly. Under SEM, the valve will have two different viewpoints, the internal and external view. Some important criteria should be recorded using SEM including:

- the areolae structure from the internal and external valve view,
- striae (the periodic rows of areolae) shape,
- striae count in 10 microns or fibulae (in Nitzschioid diatoms [1.11], folds in Surirelloid diatoms [1.18]),
- sternum size and shape (in Fragilaroid and Naviculoid pennate diatoms),
- nodule zone or annulus shape and its number (if more than one present),
- raphe structure and shape of raphe at the end and in the middle of a valve,
- the pore occlusions (if observable),
- the presence of spine or other projections like fultoportulae and labiate processes,
- the shape and porosity of the girdle bands should also be recorded.

* For calculating the averages and ranges of the measured morphological data at least 10 specimens should be measured. Acid cleaned material allows observation of details but often causes change or collapse of the frustule's three-dimensional ultrastructure, thus the selection of the most suitable cleaning procedure might be a crucial step to observe the diatom ultrastructure properly. It worthy to be noted that the observation of the frustule parts using SEM might be sufficient for genus and species identification without the need of LM; however, some important descriptions for identification are based on the LM appearance of the valve.

6. Observing the ultrastructure using TEM [1.38], if necessary and missing details did not appear under SEM, such as the hymenate pore occlusions of raphid pennate (see, for example, Figure 1.5b).
7. Study the nanotopography of the internal and external surface of the valve using AFM [1.34]. Some features within the valve could be observed through SEM, or even optical microscope if large enough, such as the frustule surface topology, which could be flat or rather undulated, wrinkled, or containing groves; however, other nanotopographical features cannot be revealed without a tool like AFM such as the dome shape of the cribellum in [1.34].
8. To understand the relationships of frustule elements and the inner ultrastructure of a given valve, or girdle bands, FIB-SEM should be considered (see example in Figure 1.6). For more advanced understanding, one of the methods mentioned for 3D reconstruction should be used [1.41] such as the 3D reconstruction of the 2D image series resulting from FIB-SEM. Through these techniques, a clear understanding of, for example, the multilayer nature of the valve could be obtained.

Afterward, all the ultrastructure details from inside and outside become well-known.

1.4 Conclusion

Diatom morphology could be an interesting topic for students; however, many challenges in the field usually keep them away from getting a deeper understanding. This is a major opportunity, since only 15,000–20,000 species have been identified [1.46], out of very large number of extant and fossil species. Recently, the developed tools, especially AFM and FIB-SEM, can make the complex 3D ultrastructure of diatoms more accessible; however, very few attempts have been recorded in literature. For both the beginners and professionals, many unrevealed secrets of diatom morphology and morphogenesis are still kept hidden, which needs more progress in the field. The study of diatom morphology could also inspire material scientists and engineers especially as the functions beyond morphology are revealed. A comprehensive tutorial suitable for different specialists could be helpful to put the beginners on the right track. Here, we only provided a potential introduction for such tutorial.

Acknowledgements

We would like to express our gratitude to the editors for their understanding and help; we are grateful for the meaningful suggestions of reviewers. KMM's work on this project was supported by funding from EPD, State of Georgia, USA. MG is very grateful to Dr. Mary Ann Tiffany for giving him the micrographs in Figure 1.1.

References

[1.1] Bahls, L., Boynton, B., Johnston, B., Atlas of diatoms (Bacillariophyta) from diverse habitats in remote regions of western Canada. *Phytokeys*, 105, 1–186, 2018.

[1.2] Bishop, I.W., Esposito, R.M., Tyree, M., Spaulding, S.A., A diatom voucher flora from selected southeast rivers (USA). *Phytotaxa*, 332, 2, 101–140, 2017.

[1.3] Brotosudarmo, T.H.P., Wibowo, A.A., Heriyanto, Adhiwibawa, M.A.S., Single cells diatom Chaetoceros muelleri investigated by homebuilt confocal fluorescence spectro-microscopy, in: *Third International Seminar on Photonics, Optics, and Its Applications*, A. Nasution and A.M. Hatta (Eds.), p. 11044, 2019.

[1.4] Cohn, S.A., Manoylov, K.M., Gordon, R. (Eds.), *Diatom Gliding Motility[DIGM, Volume in the series: Diatoms: Biology & Applications, series editors: Richard Gordon & Joseph Seckbach, In preparation]*, Wiley-Scrivener, Beverly, MA, USA, 2020.

[1.5] Cox, E.J., *Identification of Freshwater Diatoms from Live Material*, Chapman & Hall, London, 1996.

[1.6] Diatom Flora of Britain and Ireland, Glossary, Wales, UK, 2020, https://naturalhistory.museumwales.ac.uk/diatoms

[1.7] Diatoms of North America, Glossary, USA, 2020, https://diatoms.org/glossary.

[1.8] Falkowski, P.G., Barber, R.T., Smetacek, V., Biogeochemical controls and feedbacks on ocean primary production. *Science*, 281, 5374, 200–206, 1998.

[1.9] Ghobara, M.M., Mazumder, N., Vinayak, V., Reissig, L., Gebeshuber, I.C., Tiffany, M.A., Gordon, R., On light and diatoms: A photonics and photobiology review[Chapter 7], in: *Diatoms: Fundamentals & Applications[DIFA, Volume 1 in the series: Diatoms: Biology & Applications, series editors: Richard Gordon & Joseph Seckbach]*, J. Seckbach and R. Gordon (Eds.), pp. 129–190, Wiley-Scrivener, Beverly, MA, USA, 2019.

[1.10] Gordon, R., Kling, H.J., Sterrenburg, F.A.S., A guide to the diatom literature for diatom nanotechnologists. *J. Nanosci. Nanotechnol.*, 5, 1, 175–178, 2005.

[1.11] Kaczmarska, I., Bates, S.S., Ehrman, J.M., Léger, C., Fine structure of the gamete, auxospore and initial cell in the pennate diatom *Pseudo-nitzschia multiseries* (Bacillariophyta). *Nova Hedwig.*, 71, 3–4, 337–357, 2000.

[1.12] Molino, P.J. and Wetherbee, R., The biology of biofouling diatoms and their role in the development of microbial slimes. *Biofouling*, 24, 5, 365–379, 2008.

[1.13] Romann, J., Valmalette, J.C., Royset, A., Einarsrud, M.A., Optical properties of single diatom frustules revealed by confocal microspectroscopy. *Opt. Lett.*, 40, 5, 740–743, 2015.

[1.14] Sato, S., Nanjappa, D., Dorrell, R.G., Vieira, F.R.J., Kazamia, E., Tirichine, L., Veluchamy, A., Heilig, R., Aury, J.-M., Jaillon, O., Wincker, P., Fussy, Z., Obornik, M., Munoz-Gomez, S.A., Mann, D.G., Bowler, C., Zingone, A., Genome-enabled phylogenetic and functional reconstruction of an araphid pennate diatom Plagiostriata sp. CCMP470, previously assigned as a radial centric diatom, and its bacterial commensal. *Sci. Rep.*, 10, 1, 9449, 2020.

[1.15] Smol, J.P. and Stoermer, E.F., Glossary, acronyms, and abbreviations, in: *The Diatoms: Applications for the Environmental and Earth Sciences*, 2nd, J.P. Smol and E.F. Stoermer (Eds.), pp. 614–654, Cambridge University Press, Cambridge, 2010.

[1.16] Spaulding, S., Esposito, R., Lubinski, D., Horn, S., Cox, M., McKnight, D., Alger, A., Hall, B., Mayernick, M., Whittaker, T., Yang, C., Antarctic Freshwater Diatoms: Glossary, 2009, http://huey.colorado.edu/diatoms/about/glossary.php.

[1.17] Tiffany, M.A. and Nagy, S.S., The beauty of diatoms[Chapter 3], in: *Diatoms: Fundamentals & Applications[DIFA, Volume 1 in the series: Diatoms: Biology & Applications, series editors: Richard Gordon & Joseph Seckbach]*, J. Seckbach and R. Gordon (Eds.), pp. 31–40, Wiley-Scrivener, Beverly, MA, USA, 2019.

[1.18] Witkowski, A., Gomes, A., Mann, D.G., Trobajo, R., Li, C.L., Barka, F., Gusev, E., Dąbek, P., Grzonka, J., Kurzydłowski, K.J., Zglobicka, I., Harrison, M., Boski, T., *Simonsenia aveniformis* sp. nov. (Bacillariophyceae), molecular phylogeny and systematics of the genus, and a new type of canal raphe system. *Sci. Rep.*, 5, #17115, 2015.

[1.19] Tiffany, M.A., Epizoic and Epiphytic Diatoms, in: *The Diatom World. Cellular Origin, Life in Extreme Habitats and Astrobiology*, vol. 19, J. Seckbach and P. Kociolek (Eds.), Springer, Dordrecht, 2011.

[1.20] Gillard, J., Frenkel, J., Devos, V., Sabbe, K., Paul, C., Rempt, M., Inz, D., Pohnert, G., Vuylsteke, M., Vyverman, W., Metabolomics Enables the Structure Elucidation of a Diatom Sex Pheromone. *Angew. Commun.*, 52, 3, 1056–1056, 2013.

[1.21] Bedoshvili, Y.D. and Likhoshway, Y.V., Cellular Mechanisms of Diatom Valve Morphogenesis, in: *Diatoms: Fundamentals and Applications*, Seckbach, J. and Gordon, R. (Eds.), 2020.

[1.22] Anonymous, Two letters from a gentleman in the country, relating to Mr. Leuwenhoeck's letter in Transaction, No. 283. *Philos. Trans. R. Soc. London*, 23, 288, 1494, 1703.

[1.23] Bukhtiyarova, L.N., The genus Eunotia Ehrenb. (Bacillariophyta) in the Cheremsky Nature Reserve, Ukrainian Polissya, and refined terminology relevant to the raphe system morphology. *PhytoKeys*, 128, 1–31, 2019.

[1.24] Sanderson, J.B., The historical use of diatoms as test-objects for Transmission electron microscopy. *Quekett J. Microsc.*, 37, 467–475, 1995.

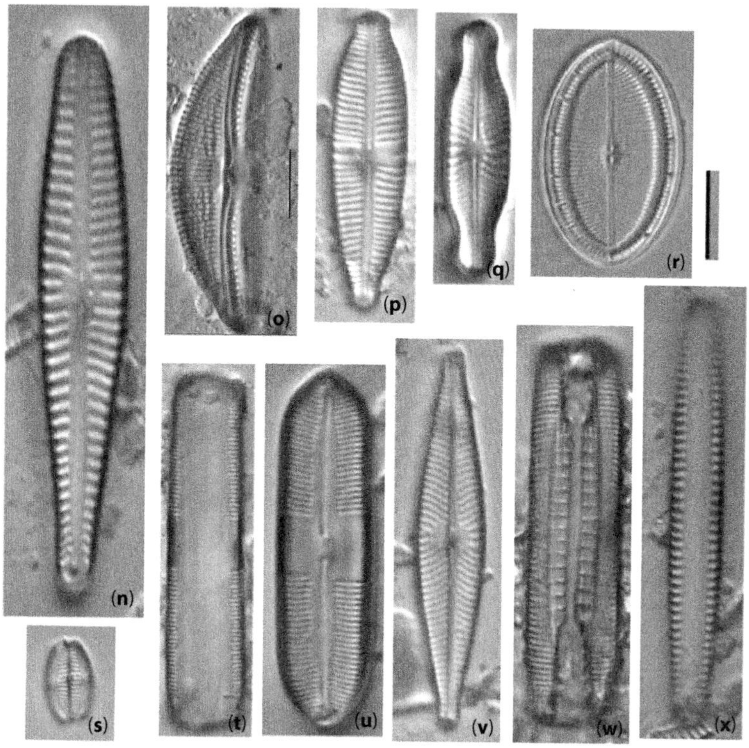

Figure 1.4 (Continued) Cleaned diatoms in valve (g, h–j, m–r, u, v) and girdle views (a–f, k, l, s, t, w, x). (n) *Gomphonema* sp. valve heteropolar wider in the middle, axial area narrow, central area irregular outlined by two shortened striae and opposite to a single striae finishing with an isolated pore, striae parallel toward the headpole, radiate toward the foot pole; (o) *Amphora ovalis*, dorsal fascia visible and dorsal striae interrupted transapically by intercostal ribs; (p) *Gomphonema micropus* Reichardt lanceolate valve with headpole widely drawn out and wider than foot pole, striae radiate, central area unilaterally rectangular with shortened central stria, on the opposite side longer striae finishing with a stigmoid; (q) *Navicula genovefae* Fusey valve linear-lanceolate with rostrate broadly rounded apices, punctate striae radiate and curved, becoming nearly parallel at the apices, less dense around the well-defined central area; (r) *Cocconeis placentula* Ehrenb. Valves elliptic, striae radiate and interrupted by a hyaline ring positioned close to the valve margin, siliceous bridges (imbriae extending from valvocopula) visible; (s) *Amphora pediculus* (Kütz.) Grunow focus from dorsal site of two frustules; (t, u) *Caloneis* sp. on girdle view striae continue on valve mantle, on the linear valve view with rounded apices, axial area is narrow, broadening to a transverse fascia; (v) *Navicula cryptocephala* Kütz. Valve lanceolate with protracted apices and visible large, circular central area; (w) *Mastogloia pseudosmithii* Sylvia S. Lee, E. E. Gaiser, Van de Vijver, Edlund, and S. A. Spaulding, evenly sized partecta (chambers on the valvocopula) on both valves; (x) *Navicula cf. tripunctata* (O.F. Müll.) Bory. Scale bar, 10 μm. These micrographs were obtained and identified by KMM.

In 1992, the first observation of diatoms using an AFM has been done [1.33]. In general, AFM is used as an advanced tool to explore diatom ultrastructure providing information in the Z-direction, with the ability to understand the surface topology of the frustule parts with a nanoresolution. For instance, AFM observations of *Coscinodiscus* sp. clean valves revealed a distinct dome topology for the cribellum, which was not observed before [1.34]. At the beginning of the current century, AFM was used in several works for understanding the nanoscale ultrastructure and topology of frustule surfaces in a 3D manner. Today, AFM is also used to explore the organic envelope, micromechanical properties, and to

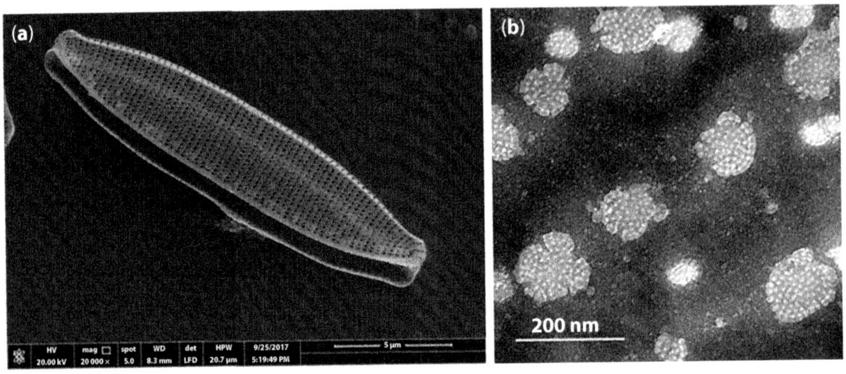

Figure 1.5 (a) SEM of a single cleaned partially open frustule, two overlapping valves, of *Nitzschia palea* (Kützing) W. Smith, and scale bar is 5 μm. The rows of pores (striae) that observed here cannot be observed under LM for this species. (b) TEM of a close-up in *Navicula* sp. valve showing the hymenate pore occlusions that will not be observed under SEM; scale bar is 200 nm. These micrographs were obtained and identified by MG.

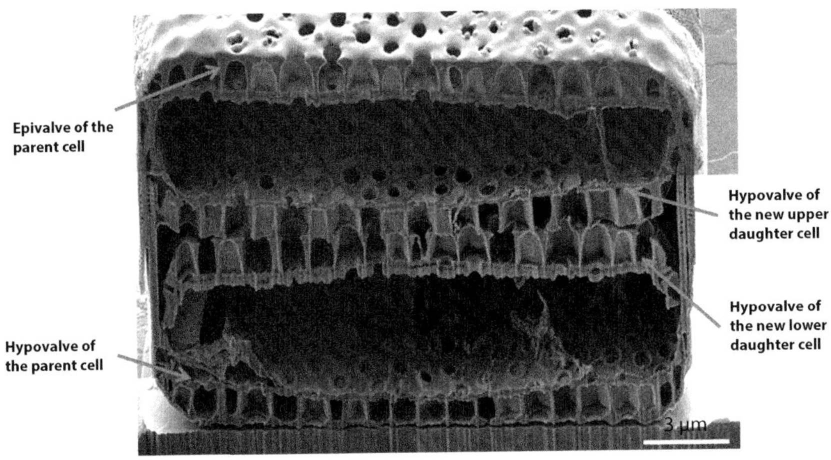

Figure 1.6 A cross-section at the center of *Coscinodiscus* sp. cell collected and treated while binary fission process was in progress, fabricated and captured by FIB-SEM. Reproduced from Xing *et al.* [1.42] under a Creative Commons Attribution 4.0 International license.

understand the biomineralization processes of diatom frustules [1.35]. Luís *et al.* [1.35] can be considered a good review for starting AFM studies on diatom frustules.

Furthermore, diatom valves seem to have a complex inner ultrastructure that cannot be understood completely by observing the internal and external view of a given valve surface using the previously mentioned tools. Although the multilayer, multiscalar porosity can be observed easily using such techniques, the internal anatomy and relations of the siliceous elements of the frustule cannot be understood [1.41]. It was usual to wish that the observation of a broken valve or girdle band at the right site and right angle would help, otherwise, the complex inner structure remained unseen [1.41].

Thus, another advanced method was required for understanding the inner structures and spatial relationships of the siliceous elements of a given diatom frustule. The FIB-SEM was introduced as a solution for such a problem by cutting the diatom frustule parts at nanoresolution to reveal the inner complex ultrastructure of a given valve or frustule (Figure 1.6) [1.41]. Suzuki *et al.* [1.40] was the first work introduced using FIB-SEM for making a cross-section in diatoms. Only a few articles are available using FIB-SEM and the field is still growing. The acquired data using FIB-SEM could be used to reconstruct

Table 1.1 A summary of the major tools used to study diatom frustule morphology and its ultrastructure.

	LM	TEM	SEM	AFM	FIB-SEM
The date of first known observation of diatoms using the tool	Anonymous, 1703 [1.22]	Krause, 1936 [1.27]	Mid of 1960s [1.24]	Linder *et al.*, 1992 [1.33]	Suzuki *et al.*, 2001 [1.40]
Up-to-date resolution	The maximum resolution of the common compound optical microscope can be around 200 nm. Recently, the resolution was enhanced (down to 97 nm) using special kind of lenses [1.37].	Up-to-date, the highest TEM resolution could be down to 50 picometer or even lower [1.29].	The details less than 15 nm was not resolved under most of SEMs. Recently, an outbreak has been achieved, and the resolution of SEM could be below 1 nm [1.39].	Recently, the resolution can be below 1 nm.	Having SEM as the microscope part of the device. Thus, the resolution is dependent on this SEM.
When we should use?	Observation of the presence or absence of diatoms in a sample. Identification of diatoms on the genus level. Enumeration of diatom frustules for different purposes.	Observation of the fine porosity (mesopores) present in some genera, like raphid pennates (Figure 1.5b). Observation of thin cross-sections in a valve or a girdle band. Observation of the cytoplasmic components of thin cross-sections of living cells (living cells anatomy).	Observation of the outer ultrastructure including most porosity. Observation of the overall 3D ultrastructure of the frustule or different parts. Identification at the species and subspecies level.	Observation of the 3D topology of a diatom frustule or its components. Measuring forces related with both living diatoms and its cleaned frustules.	Understanding the inner ultrastructure of diatom frustule or its parts by cutting cross-sections through it. Observation of the siliceous elements structural relations within the frustule. Observation of the whole 3D ultrastructure of the frustule via the 3D reconstruction.

(*Continued*)

Table 1.1 A summary of the major tools used to study diatom frustule morphology and its ultrastructure. (*Continued*)

	LM	TEM	SEM	AFM	FIB-SEM
The disadvantages	The observations for most of the ultrastructure details will be limited. Either the girdle view or the valve view will be available.	Only the tiniest parts of the valve, like pore occlusions, will be observed. The high energy electron beam may damage some sensitive samples, so it should be used wisely.	The samples must be coated with a conductive layer, which in turn could change the nano texture of the frustule silica and probably pore sizes, thus the thickness and smoothness of the conductive layer should be optimized and be thin as possible without getting nanoparticles on the top. The high energy electron beam may also damage some sensitive samples The regular resolution keep the pore occlusions of very fine porosity (below 10 nm) hidden.	The frustules must fix to the substrate before measuring. A very sensitive tool with complicated precautions to follow to get the desired results.	This technique sometimes needs more sophisticated preparation of the samples and more sophisticated work to reconstruct the frustule or its parts, however it worth. Related with the presence of the device, which usually is not available for all research groups.

the overall 3D geometry of diatoms to carry out further computational simulations necessary for diatom nanotechnology applications.

Finally, all the techniques mentioned were summarized in Table 1.1 to help beginners and students choose between different tools on-demand.

1.3 Diatom Frustule 3D Reconstruction

Toward the complete understanding of the 3D structure of a given diatom frustule, a comprehensive 3D model can be created from the data collected from different characterization techniques. This approach, which is designated as the 3D reconstruction of diatom frustules, can be used for different purposes but is mainly for computer modeling. Oncoming tools for the 3D reconstruction of diatom frustules are FIB-SEM [1.32, 1.42] and digital holographic microscopy (DHM) combined with SEM [1.30]. The combination of DHM and SEM or AFM might give the ability to model and visualize microscopic 3D objects with a high resolution in all directions [1.30]. Hildebrand *et al.* [1.32] introduced the ability for the 3D reconstruction of subcellular architecture using FIB-SEM with new insights into the architecture and synthesis process of both the siliceous and organic components inside the

Matching symmetry states to morphogenetic stages may inform the developmental process.

Symmetry is, then, a compilation of perception or observation as well as the arrangement of structural features (internally or externally) defining dimension (3D) and geometry (contours, surfaces, and boundaries) in terms of "balance" and "likeness", and this compilation can be evaluated at any step in the morphogenetic and developmental processes of an organism at the individual, population, or higher taxonomic grouping level.

2.1.1 Recognition and Symmetry

Object recognition is the term applied to our understanding of how to perceive or observe a three-dimensional (3D) object and interpret it from its two-dimensional (2D) projection. Perception studies with Attneave's cat [2.9] or Biederman's cup [2.13] as iconic images to discern the process of recognition utilize the geometric aspect of objects as points connected with straight edges or partial contours, respectively (cf. visual tracking of eye movements [2.4]). These studies showed that recognition is initiated from perception of outlines rather than just points on an object. Information acquisition is accomplished via this perception or observation. One perspective is the preference for the simplest interpretation over others to infer a structured whole, and this forms the basis of structural information theory [2.80]. Another perspective involves preference for a minimum (not necessarily simple) description forming the basis of algorithmic information theory [2.21]. In any case, information theory is at the heart of object recognition. Perceived or observed information may occur as an object that is recognized. Initial recognition may occur of a whole scene as an object. In contrast, individual objects in the scene and their boundary and surface attributes such as shape, transparency, pattern, location, size, or texture may be perceived, and perception violations of individual objects in a scene may be recognized (Figure 2.1, after Mezzanotte's scenes, in [2.14]). Symmetry as a mode of recognition can apply to the scene as an object or the parts of this object in relation to the boundary and surface attributes that comprise the scene. Information about symmetry is acquired with respect to an external or anatomical viewpoint.

Body shape is important in symmetry determination of organisms. Shape deformation relies on a point-based approach with regard to object recognition requiring identifiable equivalences on multiple shapes. The shape decomposition approach represents an implicit reference to contours in recognition as the conduit to assessing symmetry [2.127] (Figures 2.2a–c). Shape decomposition relies on the delamination of surface boundary height via contours with respect to object recognition and implicitly influences shape in symmetry assessment [2.72] (Figures 2.2d–f).

For reflective (also called bilateral, mirror, mirror-image, line, plane, or left/right) symmetry [2.159], shape is considered to be the most ubiquitous among organisms and is especially noticeable in animals and plants. Colored symmetry [2.132] can also occur, as in the left/right asymmetry of otherwise bilaterally symmetric organisms [2.41, 2.144]. However, symmetry is present as more than rotational or reflective in organisms (Figure 2.3). In 2D, dihedral (rotational plus reflective) [Weisstein 2002], translational, and glide (reflective plus translational) [2.159] symmetries are evident. The 2D and 3D quality of whole organisms or their parts lends itself to the related symmetries of knots (including chirality)

Figure 2.1 A drawing composed by JLP that is reminiscent of one of Mezzanotte's scenes [2.15]. Although the scene is the whole perceived object, individual objects diatom (fish, jelly fish, boat, oars, waves, and shore) and their attributes such as shape, transparency, location, size, and texture and perspective of the scene may be viewed as perception violations.

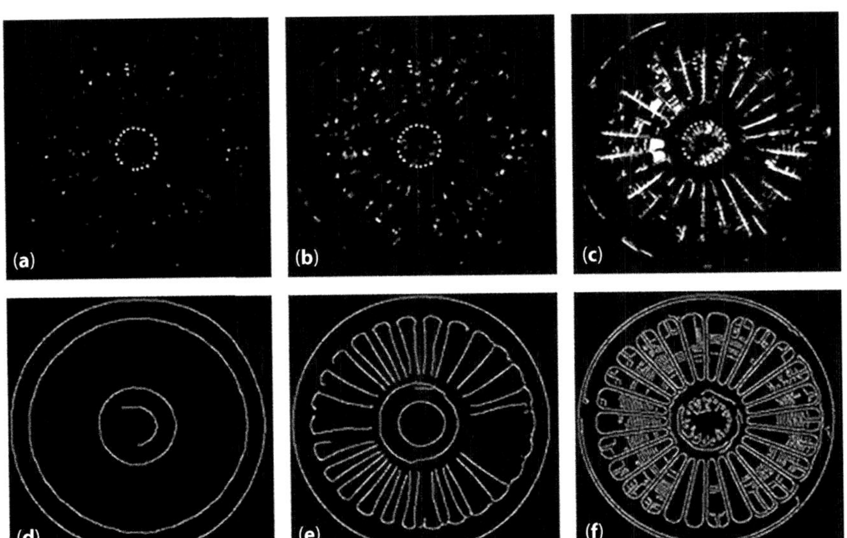

Figure 2.2 Shape perception of *Arachnoidiscus ehrenbergii* from image ProvBay5_12lx450. (a–c) Shape deformation via an increasing number of points defining valve features. (d–f) Shape decomposition via an increasing number of contours defining valve features.

[2.154], helices/spirals (including handedness) [2.159], and compound or multiple spirals (a variant of knot symmetry), [2.159] as well as scale (fractal or dilation) [2.159] and related conformal symmetry [2.159] with associated Möbius transformations [2.159]. Symmetries associated with loop networks [2.36, 2.88, 2.89, 2.159, 2.164] are present at various scales in organisms as well. See Figure 2.3 for examples of all symmetries on the diatom valve surface. Note that 1) rotational and reflective symmetries apply to the whole valve; 2) dihedral, translational, and glide symmetries apply to pores on the valve face; 3) knot, helical, compound spiral, and conformal symmetries apply to valve central areas; 4) scale symmetry applies to the replication of five-fold symmetry within a *Triceratium* valve face; 5) loop networks are seen in the pores on the valve surface of the *Trigonium* and *Triceratium* valves depicted.

Symmetry has implications for evolutionary processes in animals and plants [2.54]. Symmetry may be adaptive so that active organisms or those in rapidly flowing water [2.45] will have reflective symmetry while sedentary organisms in low flow shear regimes will have rotational or dihedral symmetry. Organisms with large surface to volume ratios or repetitive parts tend to have translational symmetry (Figure 2.4). Organisms that exhibit energy efficiency have scale symmetry [2.158]. In general, symmetry is a dynamical property of organisms.

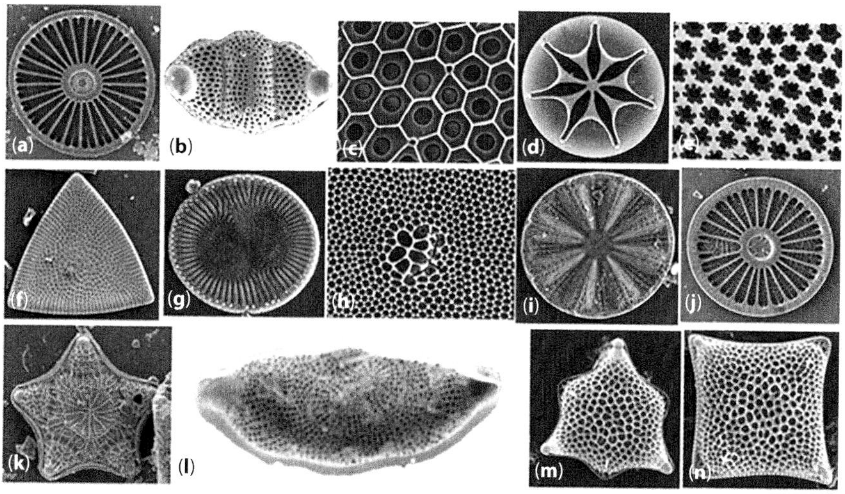

Figure 2.3 Diatom valve face symmetries. (a) *Arachnoidiscus ehrenbergii*-rotational; (b) *Biddulphia* sp.-reflective; (c) *Coscinodiscus* sp. microstructure-translational; (d) *Asterolampra marylandica*-dihedral; (e) *Coscinodiscus* sp. microstructure-glide; (f) *Trigonium americanum*-helical/spiral central area; (g) *abnormal Cyclotella meneghiniana*-multiple helices/spirals central area; (h) *Coscinodiscus* sp.-helical/spiral; (i) *Actinoptychus splendens*-knot (interlooping valve structure); (j) *Arachnoidiscus ornatus*-knot (interlocking rings under valve ribs); (k) *Triceratium pentacrinus* fo. *quadrata*-scale; (l) initial valve of *Biddulphia* sp. 3D valve surface-conformal symmetry or Möbius transformation; (m) *Trigonium dubium*-loop network; (n) *Triceratium* bicorne-loop network. (All SEMs by Mary Ann Tiffany). See "Recognition and symmetry" section in the text.

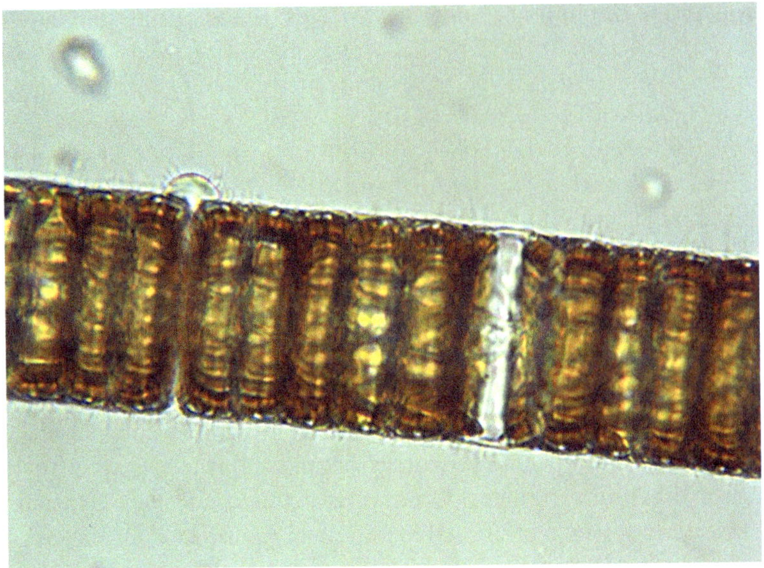

Figure 2.4 Double translational symmetry at two scales exhibited in a *Paralia sulcata* chain colony of repetitive cells. Light micrograph by Mary Ann Tiffany.

2.1.2 Symmetry and Growth

Morphogenesis has been characterized as a system of detached movement of ensembles of cells in animals in contrast to lack of detached movement via ensembles of cells in plants. That is, locomotion is one way to associate a specific kind of symmetry to motile organisms in contrast to sessile organisms which may exhibit another kind of symmetry [2.59]. Modes of growth contribute to the recognition of specific types of symmetry. Branching growth exhibits continuous symmetry decrease and increase in size during growth [2.47]. Stepwise growth and size reduction are modes of discrete growth where size increases or decreases over time. Growth changes are concomitant with symmetry changes as potential indicators of stress associated with development, morphogenesis, cytogenesis, epigenesis [2.154], and embryogenesis [2.43, 2.149], depending on evolutionary or ecological conditions. Levels of environmental or teratological stress are said to influence fitness and adaptability of an organism, although this connection has not been quantitatively or definitively formalized [2.53, 2.54, 2.35, 2.68].

Developmental instability has been associated with fluctuating asymmetry which is generally considered as small random deviations from perfect reflective (bilateral) symmetry [2.7, 2.53, 2.60, 2.87, 2.94, 2.128, 2.152]. Instability is presumed to be associated with either side of an organism and developmental processes, and genetics or environmental conditions may influence the size of the fluctuations in asymmetry [2.53]. No explicit or quantitative relation has been given connecting fluctuating asymmetry with instability as a measure of stress. Assessment using fluctuating asymmetry characterizations have been made on a population (species) level and has not been carried out with individuals or at interspecific or higher taxonomic levels [2.53, 2.54].

2.1.3 Diatom Pattern Formation, Growth, and Symmetry

Diatoms are unicellular microalgae that are known for their exquisitely ornamented hydrated silica frustule (cell wall) with geometric intricacies. At a molecular level, ornamentation on the valve surface emanates from silicic acid precipitation [2.47] and interaction with proteins in the silica deposition vesicle (SDV) and silica deposition through the SDV membrane (known as the silicalemma) [2.50, 2.142]. Valve surface patterning occurs as a diversity of geometric structures [2.98]: the multitude of species-specific forms with various symmetries. Anatomically, a diatom frustule is composed of two close-fitting thecae (valves) subtended by cingulae (girdle bands) [2.119, 2.150, 2.151]. During reproduction, the cell undergoes expansion as two daughter cells are created, with each daughter half-cell acquiring each half of the mother cell and one half (valve) of the frustule [2.119]. Each new valve is constructed within an SDV within each daughter cell. Because one valve fits inside the other (like petri dishes) and each daughter cell builds its new valve within the perimeter of the valve it inherited, one daughter cell is usually smaller in size than the mother cell (but cf. [2.117]), and the size diminution process continues until a developmental limit is reached according to the MacDonald-Pfitzer rule [2.90, 2.111, 2.112], which may correlate with a minimal organelle content [2.97]. At this point, size is restored via sexual reproduction and auxosporulation and from this, an initial cell is produced which in turn produces the vegetative cell [2.119].

Diatom symmetry formation is closely tied to pattern formation. Models of pattern formation have been developed that consider diffusion limited precipitation in non-equilibrium conditions where silica diffuses within the thin SDV depositing silica initially on the midring in centric diatoms or the midrib in pennate diatoms [2.47, 2.109]. This process has been observed to take about 10 minutes and is followed by thickening of the valve surface, which occupies an additional 6 hours [2.47]. Alternatively, pattern formation has been modeled as phase separation of two hypothesized liquids [2.82, 2.138] so that the repetitive honeycomb and pore patterns formed in some diatoms occur because of the pH regime and molecular reaction during silica deposition [2.76]. The third phase, the actual solid precipitated silica, is not included in this model [2.82]. In an alternate computer simulation, albeit confined to pore occlusions, a single fluid is involved (cf. concentrating "mother liquor" [2.47]), with solid silica particles free to "redissolve" after precipitation [2.161, 2.162]. This is similar to a simulation of diatom costae that permitted surface diffusion of silica particles along the precipitate surface, after they adhered to it, to approximate sintering [2.47]. The balance between precipitation and sintering has been simulated, using the language of "diffusion limited" versus "reaction limited" [2.25]. The main difference is whether silica particles continued to move in the medium [2.25] or along the surface of the precipitate [2.47]. Pore occlusion simulations [2.25] parallel simulation of precipitation with a centric diatom's midring [2.47]. Spine formation has been modelled as "a combination of membrane and cytoskeletal activities creating a mold for silica deposition" [2.11], where the silica presumably sinters. A model has been proposed that regards the SDV, in conjunction with the microfilament ring at its perimeter [2.50], as a tensegrity structure that may explain the shape of polygonal centric diatoms [2.48]. All of these models presume that the SDV membrane constrains the silica precipitation, the latter at best somehow enlarging the SDV, by unstated mechanisms. However, the opposite effect, of the precipitated silica altering the shape of the SDV, not just its size, can also be conceived.

Pattern formation diatom studies on cellular organelles during mitotic stages of ontogeny preceding valve formation indicate changes in organelle location and symmetry (cell organization) at the subcellular level as well [2.114, 2.123, 2.155]. For example, during interphase the arrangement of vacuoles on either side of the central area with respect to microtubules and the microtubule organizing center [2.113, 2.122] and spindle formation via the position of the microtubules are indicative of reflective symmetry. As cytokinesis proceeds, at metaphase chromosomes assemble around the spindle during metaphase, while at the end of anaphase complete cleavage of the spindle occurs. The position and cleavage of the spindle exhibit reflective symmetry just prior to the formation of new daughter cells [2.28] (cf. [2.1]).

Each of the modeling systems proposed for diatom valve formation and morphogenesis are incomplete in that they either have not been found to be present during all parts of the process or indirectly attempt to account for morphogenetic processes from a chemical/molecular or cytogenetic viewpoint. As has been emphasized, each **"new model of diatom morphogenesis"**, which, although highly attractive, is a model, awaiting to be substantiated by TEM-micrographs of sufficiently prepared cells, and experiments undertaken with the living cell" [2.124]. This remains the case. Experiments such as manipulation of salinity have morphogenetic effects, which can be interpreted at the level of silica precipitation [2.157], and the presence of (probably) sodium chloride crystals inside the silicalemma [2.46] is consistent with easy access of small ions into the SDV. The same goes for pH [2.58]. Gene regulation of silica transporters may occur at multiple levels [2.120], perhaps corresponding to multiscalar effects on morphogenesis.

In any case, no model so far accurately simulates the mature silica structure at all scales. This is not surprising, since multiscalar diatom silica structures span an extraordinary 8 orders of magnitude [2.37], and it would seem unlikely that the same mechanism prevails at all size scales (although nested hexagons made of silica spheres of decreasing sizes have been postulated, but not computer simulated, to account for three levels of structure in *Coscinodiscus* [2.138, 2.139]. The formation of these nanospheres has also been simulated [2.81]. A seven-state cellular automaton model for pennate diatom morphogenesis sets the striae and pores at specific spacings, by unspecified mechanisms [2.12]. At a lower scale, organic microrings of diameter 6 nm may correspond to fultoportulae [2.75]. By looking at each of the models as potential partial mechanisms for producing different aspects of symmetry during different morphogenetic stages and scales, the models may be implicitly useful in addressing how symmetry becomes evident during morphogenesis and how symmetry may be instrumental in understanding stability changes during morphogenesis. However, we presently have no global models that work at all scales. "Better models are needed, therefore, but for this we need a better formulation of the problems we are trying to solve" [2.93].

None of these models for diatom pattern formation involve the linked chemical reactions of Turing reaction-diffusion processes [2.148] which may lead to instabilities that have been invoked for multicellular morphogenesis [2.91, 2.92, 2.99]. However, Turing models have recently been simulated for multilayer situations [2.153], producing patterns somewhat similar to arrays of diatom pores so that such models are still under consideration [2.43].

In a dynamical system, changes in structural stability may be related to changes in symmetry via pattern formation changes not only over time but also at hierarchical spatial scales [2.166]. Pattern formation as an indicator of symmetry states may be evident as self-similarity as well [2.96]. The role of symmetry in developmental processes is discernable through surface pattern and shape changes over time, and these changes may be used in a morphogenetic dynamical system to explicitly assess the contribution of symmetry to structural stability.

In our work on morphogenesis with the late Antone G. Jacobson, we conceived of and repeatedly traversed an intellectual triangular loop of 1) experiments and observations, 2) computer simulation, and 3) formal mathematics, to converge to an understanding of how nature did it [2.49, 2.61–2.64]. This loop has not been traversed or ever closed even once for diatoms. Curiously, in the research with Jacobson, we invoked the bilateral symmetry of the organism we were simulating, which both reduced computer time and made the simulation match reality much better. A similar two axis symmetry was placed on a pennate diatom simulation [2.12]. However, in a real organism there is no global "force" creating its symmetry. Symmetry is a consequence of morphogenesis, not a cause, and deviations from symmetry or its perfection are hints to the underlying processes.

2.1.4 Diatoms and Uncanny Symmetry

Microorganisms are generally not well studied in terms of quantified symmetry changes over time. Diatoms have distinct amorphous silica frustules that exhibit a variety of geometric shapes and surfaces that lend themselves to analyses of symmetries. Diatoms are pigmented protists that are considered to be a monophyletic phylum. Morphogenesis is a topic of great interest not only to phycologists but also to nanotechnologists [2.42, 2.44, 2.51, 2.85]. In girdle view, diatoms are asymmetrical because of the parent-daughter cell division that occurs within the previously formed cell. In valve view, shape and surface are both open to symmetry considerations.

We have proposed that diatoms possess "uncanny symmetry" [2.134]. By uncanny symmetry is meant that (some) diatoms in valve view (may) exhibit near perfect symmetry. It was shown that by subtracting the rotated image of a given diatom from its original image, an almost completely blacked out image would result, indicating near perfect matching of image pixels, suggesting near perfect symmetry [2.134]. Such results were obtained for *Aulacodiscus oregonus* and *Triceratium formosum* var. *quinquelobata* (Plate 7, Figures 1–4 in [2.134]). "The apparently high degree of perfection of this noncrystalline precipitate deserves quantification, which may prove comparable with the (small) degree of imperfection of crystalline snowflakes [2.83]" [2.134].

Centric diatoms exhibit rotational symmetry in shape and surface, but other symmetries such as dihedral symmetry are present as well. A case in point is *Auliscus* (Plate 5, Figure 1 in [2.134]). From dihedral symmetry, reflective symmetry of this diatom may be characterized as a 180° rotational symmetry, thereby identifying rotational symmetry as a starting point in the measurement of uncanny symmetry. That multiple symmetries occur simultaneously in centric diatom may be recovered by an uncanny symmetry measurement.

As suggested in [2.134], we propose to quantify centric diatom symmetry generally and uncanny symmetry specifically using concepts from information theory and image processing methods. Because symmetry involves the acquisition of information regarding the "balance" of an organism such as a diatom, information theory is used to develop a measure of centric diatom valve uncanny symmetry.

Table 2.1 List of species considered and thumbnail, masked image examples for all of the individual diatom taxa we analyzed, all scaled to the same apparent diameter, along with individual number of rotations. Actual diameters vary with the life cycle. All thumbnails are external valves. All SEMs by Mary Ann Tiffany.

Genus species	Thumbnail images	Rotations
1. *Actinoptychus*		
1.1. *Actinoptychus senarius*		3
1.2. *Actinoptychus splendens*		14
2. *Amphitetras antediluviana*	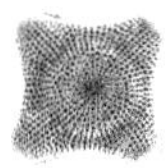	4
3. *Arachnoidiscus*		
3.1. *Arachnoidiscus ehrenbergii*	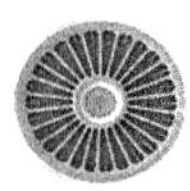  (forming)	14–31
3.2. *Arachnoidiscus ornatus*		17–24
4. *Asterolampra*		
4.1. *Asterolampra grevillei*	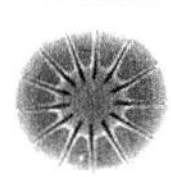	13
4.2. *Asterolampra marylandica*		7
5. *Asteromphalus*		
5.1. *Asteromphalus heptactis*	  (forming)	7
5.2. *Asteromphalus imbricatus*	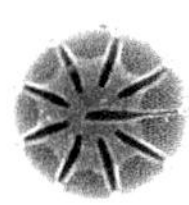	9

(Continued)

Table 2.1 List of species considered and thumbnail, masked image examples for all of the individual diatom taxa we analyzed, all scaled to the same apparent diameter, along with individual number of rotations. Actual diameters vary with the life cycle. All thumbnails are external valves. All SEMs by Mary Ann Tiffany. (*Continued*)

Genus species	Thumbnail images	Rotations
5.3. *Asteromphalus shadboltianus*	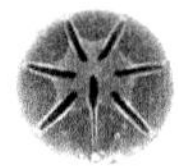	7
5.4. *Asteromphalus vanheurckii*	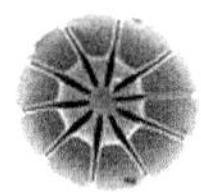	10–11
6. *Aulacodiscus*		
6.1. *Aulacodiscus africanus*	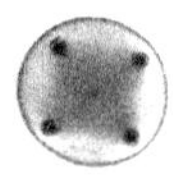	4–5
6.2. *Aulacodiscus kittonii*	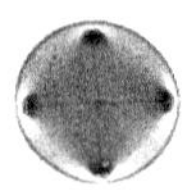	4
6.3. *Aulacodiscus oregonus*	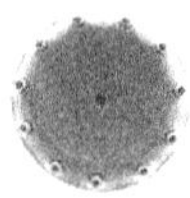	8–13
6.4. *Aulacodiscus petersii*		3
6.5. *Aulacodiscus rogersii*	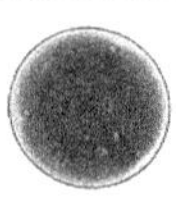	3
6.6. *Aulacodiscus scaber*	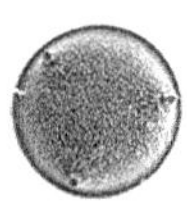	3
7. *Coscinodiscus* sp.		14
8. *Cyclotella meneghiniana*	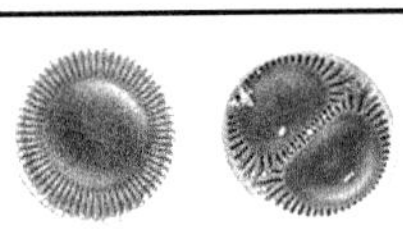 (abnormal, forming)	3–64

(*Continued*)

Table 2.1 List of species considered and thumbnail, masked image examples for all of the individual diatom taxa we analyzed, all scaled to the same apparent diameter, along with individual number of rotations. Actual diameters vary with the life cycle. All thumbnails are external valves. All SEMs by Mary Ann Tiffany. (*Continued*)

Genus species	**Thumbnail images**	**Rotations**
9. *Eupodiscus radiatus*	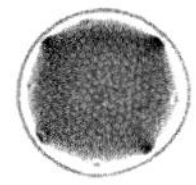	4
10. *Glyphodiscus stellatus*	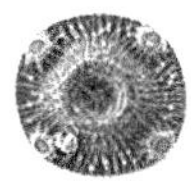	4
11. *Lampriscus shadboltianum*		3
12. *Spatangidium arachne*	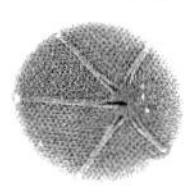	5
13. *Triceratium bicorne*	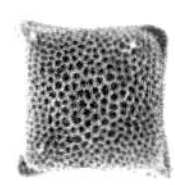	4
14. *Triceratium*		
14.1. *Triceratium castellatum* var. *fractum*		3
14.2. *Triceratium crenulatum*	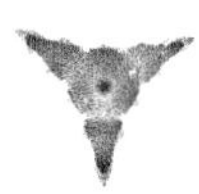 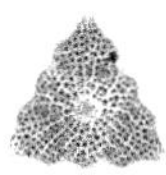	3
14.3. *Triceratium favus*		3
14.4. *Triceratium favus* var. *maxima*	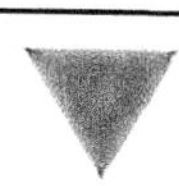	3
14.5. *Triceratium favus* var. *quadrata*	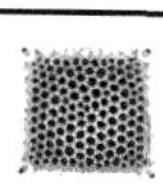	3
14.6. *Triceratium impressum*	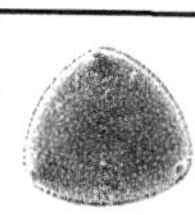	3

(*Continued*)

Table 2.1 List of species considered and thumbnail, masked image examples for all of the individual diatom taxa we analyzed, all scaled to the same apparent diameter, along with individual number of rotations. Actual diameters vary with the life cycle. All thumbnails are external valves. All SEMs by Mary Ann Tiffany. (*Continued*)

Genus species	**Thumbnail images**	**Rotations**
14.7. *Triceratium pentacrinus* fo. *quadrata*		4
14.8. *Triceratium* sp. [fossils]	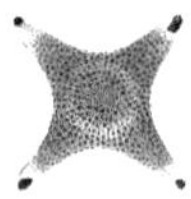 	4–15
15. *Trigonium*		
15.1. *Trigonium alternans*	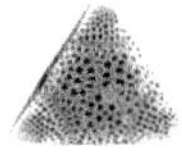	3
15.2. *Trigonium arcticum*		3
15.3. *Trigonium arcticum* var. *kerguelense*		3
15.4. *Trigonium arcticum* var. *quadrata*		4
15.5. *Trigonium dubium*	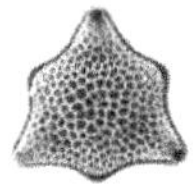	3
15.6. *Trigonium formosum* fo. *quadrata*		4

part of the morphogenetic process. The shape and position of valve features are indicators of symmetry of the valve surface.

Aulacodiscus species can be differentiated by their tube openings, not only in the number on the valve face near the margin but also in terms of the shape and whether the tubes have external flanges or distinct internal structures. Tube openings are positioned on the valve and are indicative of symmetry. Tube opening formation occurs as valve formation proceeds. Externally flanged with tongued and ridged internal tube structures are present in circular tubed *A. scaber* and *A. petersii* in contrast to *A. africanus* and *A. kittonii* with slitted tube openings with no distinct internal structures. The fossil species, *Aulacodiscus rogersii*, has a flanged circular tube opening as does *A. oregonus*. Tube openings are formed near the valve margin, and their number and position on the valve margin are indicators of valve formation [2.144].

Areolae are loculate or poroid in *Aulacodiscus*, and the areas without pores that radiate from the central annulus form the rays of the valve face that extend to the tube openings. Occluded areolae occur as well at the valve margin. Cribra form from internal struts of the areolae, and as bullulae develop in early tube formation, they are covered by silica at the completion of tube formation. Cribra are finished after tube formation, with concentric rings of pores within the cribral surface and siliceous caps when mature [2.144].

Coscinodiscus has radial hexagonal loculate areolae, proximal circular rimmed foramen, and complex domed cribra [2.116, 2.126]. Areolae become higher at the margin with increased silica deposition in contrast to the center [2.116, 2.126]. The cribra are perforated and formed via differentiation of the outer velum covering each areola [2.116], and cribella fill the cribra as a small sieve plate structure [2.126]. *Coscinodiscus* has solid silica ribs outlining areolae radially with variably spotted and striated hyaline rays in the central area [2.126]. A central rosette from which silica strings diverge and branch form a general overall valve pattern [2.116]. There is a rimoportula at the marginal end of every two to three areolae, rimoportulae below the rimmed mantle edge, and other rimoportulae are scattered on the *Coscinodiscus* valve face, including the terminus of hyaline rays [2.126]. The timing of each daughter cell forming within the mother cell may be different for different species [2.116], which may have implications for symmetry during development.

Cyclotella is characterized by its mantle fultoportulae, clear central area with fultoportulae, and distinctive striations as ribs regularly placed at the valve margin covering about half of the valve face [2.56, 2.126]. At the valve margin, rimoportulae are present in varying numbers. The central area may have tubular fultoportulae and associated pores unlike mantle fultoportulae that may have a collar. *Cyclotella* initial cells have an unstructured central area and many valve fultoportulae but are hemispherically shaped, unlike vegetative cells which have an undulated shape. Salinity affects *Cyclotella* and may produce abnormal cells in terms of their internal structure [2.56].

Asterolampra valves form from a raised annulus from which siliceous rays as "spokes on a wheel" are formed, then bifurcated twice so that the spaces between the siliceous "tines" are the sites of areolae formation. Vertically, round pores form prior to the hexagonal honeycomb, and a central area fuses over the rays as a "roof" from the center to the periphery, producing a convex shape. The rays vary in number but are all of similar size and shape [2.146]. The regularity in the valve formation of *Asterolampra* is an indicator of valve symmetry.

Asteromphalus valves form so that a singular ray is slimmer and shorter than the rest of the rays which are uniform in shape and size but vary in number. Hexagonal areolae form between the rays prior to mantle formation, and large columnar structures form on the edges of areolae for all but the singular ray. Cribra are very complex. The central area fuses over the singular ray prior to the roof forming over the rest of the rays, and the overall surface produces an undulating shape. Internal formation of the rimoportulae is larger for the singular ray than the rest of the rays. Valve features point to asymmetry as does the eccentric annulus of *Asteromphalus* in contrast to *Asterolampra* which has a central annulus. Although *Spatangidium arachne* is similar to *Asteromphalus*, this taxon is dissimilar in having a central rimoportulae, a different cribral pattern, and a singular ray that is longer than the remaining four rays. Both genera have asymmetric valve faces unlike *Asterolampra* [2.146].

For species in *Actinoptychus*, there is a hyaline central area on external and internal valves. A narrow hyaline zone is present in *A. splendens* and *A. senarius* that extends from the central area to around halfway to the valve margin. Areolae are irregular in most species, and taxa have a variable number of sectors on the valve face [2.79]. *Actinoptychus* has a sectored valve face and curled rimoportulae. Areolae form a netted or reticulated pattern on the valve surface with a marginal ridge in species such as *A. senarius*, and other species have a loculus with a smooth valve face with no marginal ridge [2.57].

Arachnoidiscus has auxospore attachment to the hypovalve of the mother cell, while auxospore attachment to the mother cell of *Amphitetras* occurs on the epivalve [2.125]. The implication is that valve patterning forms differently depending on the attachment site.

Eupodiscus radiatus has marginal equally spaced ocelli with intercalated rimoportulae extending to the ocelli, a scalloped mantle edge [2.30], and loculate hexagonal areolae and cribral pores arranged in parallel rows [2.30, 2.31]. There are siliceous strips with flanges on the spines and other structures defining the valve mantle as well [2.30]. In contrast, *Amphitetras* has pseudoloculi with siliceous strips on the mantle without other structures present [2.30]. *E. radiatus* and *Amphitetras* symmetry may be based on the number and position of equally spaced ocelli or pseudocelli structures.

Triceratium species have elongated ocelli and poroidal areolae with domed cribra [2.31]. Unlike *Triceratium*, *Lampriscus shadboltianum* has pseudocelli [2.30], and *Glyphodiscus stellatus* has four-part symmetry in which the valve is concentrically undulate. *Triceratium pentacrinus* fo. *quadrata* and *T. bicorne* are uncommon four-part symmetrical species, as most are three-part symmetrical valves [2.31].

Pseudoloculi or loculi are variably present in *Triceratium*, e.g., as in pseudoloculi surrounding areolar clusters on the valve surface of *Triceratium dubium* [2.31]. Rimoportulae openings differentiate species, such that *Triceratium dubium* having an elongated tube opening with apical spinules is different from *T. favus* having a spatulate opening that is hemispherical internally [2.31].

Trigonium have pseudocelli, centrally located rimoportulae with oval openings, and honeycombed loculate areolae [2.143] in a radial arrangement [2.38, 2.143]. In *Trigonium arcticum*, areolae are covered by rota-type vela, and a straight cingular suture is present [2.38]. Toward the end of cribra formation, this structure finishes at valve surface level in *Trigonium arcticum*, and rotae within porelli of the pseudocelli are also being formed [2.143]. For *Trigonium formosum*, rimoportulae are centrally located and cribra are highly domed [2.143].

2.2.2 Centric Diatoms, Morphology, and Valve Formation

Centric diatom valve formation occurs from a central area to the valve margin in three general stages [2.126] and is illustrated in Figure 2.5. First, horizontal silica deposition occurs to form a basal layer. From an annulus in the central area at the site of valve formation initiation, radial rows of silica are deposited as areolae with cross extensions and connections so that a branching pattern of silica strands emerges. Gaps in this overall pattern are filled in with silica, and internal rimoportulae tubes commence [2.126]. Second, vertical silica deposition occurs so that areolae walls increase in height. Round pores are transformed to hexagonal or other shapes, and external rimoportulae are evident [2.126]. Third, areolar

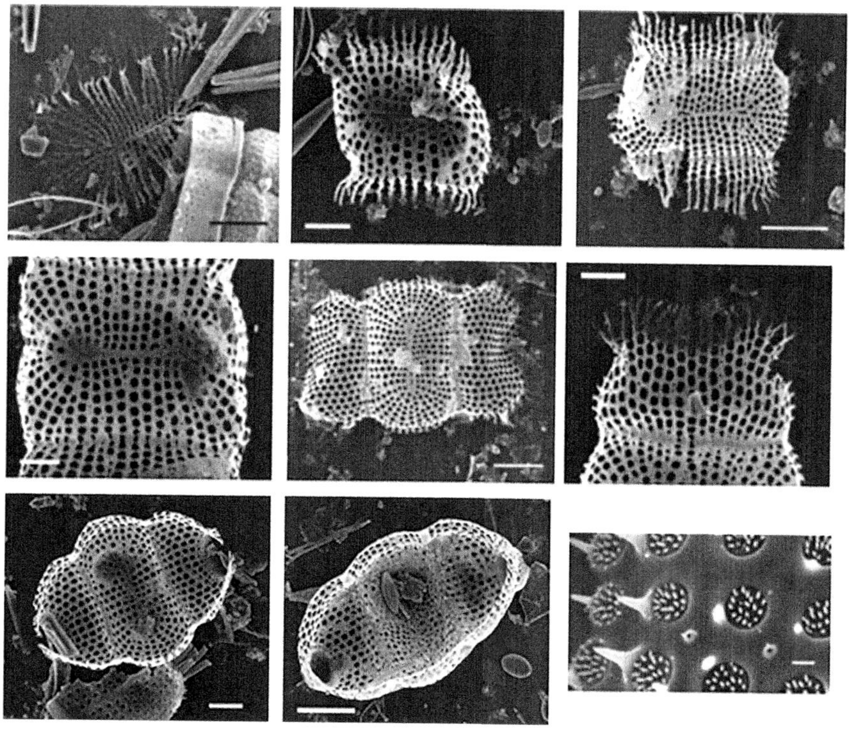

Figure 2.5 Three stages of valve formation (as defined in [2.125]) in *Biddulphia*. Row 1: First stage shows formation of the elongated annulus, branching virgae and vimines producing open holed areolae, and the internal tubes of rimoportulae just beginning. Row 2: Second stage shows the external emergence of rimoportulae, the filling in of areolae, and ridges are more hyaline with increasing silica deposition. Row 3: Third stage shows the beginning of pseudocelli formation, fusing of the valve margin, and the mature cribra. All photos by Mary Ann Tiffany.

associated structures of cribra and cribella are completed during horizontal silica deposition so that the size and spacing of these structures occur as a response to the constraints imposed by the areolae [2.126]. Successive layers of silica are deposited, and the completion of the valve and its structures extends to the margin [2.116].

Prior to valve formation, mitotic and cytokinetic processes occur [2.122]. Silica aggregation occurs within the silicalemma/SDV where new wall formation occurs [2.59, 2.116, 2.119, 2.126]. The general outline of the cell wall is molded via the plasmalemma [2.122].

As silica deposition occurs from center to the periphery, it is evident that various structures are created at various times in a sequence. Major structures common to all species of a given genus, particular valve structures that are important morphologically, and concavity or undulations associated with the valve face (perhaps due to buckling [2.52]) are associated with symmetries of the diatom valve. The species-specific differences in valve structures and their location and timing of formation are indicators of symmetry changes during valve formation.

Sexual reproduction and auxosporulation result in a cell wall that is similar to a diatom spore and takes a few generations to "restore" the vegetative shape [2.69]. The initial valve that is formed, by an unknown mechanism, morphogenetically determines vegetative valve shape via protoplast adhesion and contraction within the spherical auxospore [2.122, 2.156]. Heterovalvy is apparent even within the initial valve [2.125], having implications for symmetry.

2.2.3 Image Entropy and Symmetry Measurement

Symmetry measurement has been accomplished in a variety of ways. For example, rotational symmetry has been measured as a comparison of equal length five-petalled flowers to pentagons [2.33], point-spread function of optical properties within circular shapes exclusively [2.15], landmarks [2.121] or feature points [2.86] used for intraspecies shape comparison, and cylindrical helical diffraction grid patterns via trigonometric polynomials to measure helices [2.136]. All of these methods are inadequate for measuring symmetry in centric diatoms for a number of reasons. Because symmetry has been primarily applied to the understanding of developmental stability, fitness and adaptation [2.54], measurement needs to reflect the connection among all changes during development and should not be merely a matter of geometric convenience. Centric diatoms exhibit multiple symmetries (Figure 2.3) which necessarily requires a flexible measurement tool. From the spherical auxospore to the variably symmetric initial cell to the rotationally and/or reflective symmetric vegetative cell, different symmetries are expressed throughout the diatom life cycle. The sources of symmetry that contribute to pattern formation include perturbations as symmetry breaking instances to preservation of structural elements throughout development [2.40], and these structural elements may have different symmetries. None of the aforementioned methods have utility in comparing multiple symmetries concurrently or have been used in symmetry measurement for multiple genera or species at the same time. Additionally, these methods are used only in intraspecies shape comparisons without surface feature measurement in symmetry assessment.

Alternatively, a more broad-based methodology is necessary to use in symmetry measurement in centric diatoms. Valve shape and surface feature pattern variation as symmetry changes are measurable simultaneously via the amount of information available in an image. Measurement of information is not reliant on specific geometries or the spatial frequency or position of morphological features or scale. Multiple symmetries may be measured based on available information of shape and surface.

Symmetry may be defined as the micro-states of a macro-state dynamical system of an organism which undergoes morphogenetic, embryogenetic, cytogenetic, epigenetic, or developmental changes. When information is known only about the macro-state, probable micro-states are measures of the lack of this information and what is unknown about the actual micro-states [2.100]. Entropy is a measure of the lack of this information [2.66], and the behavior of symmetry changes is measurable as changes in the amount of information available about the arrangement of and distribution of all symmetry micro-states.

Image processing techniques are useful in characterizing information contained in pixels in digital images. The geometry of a diatom is related to pixel intensities in a digital image, and that geometry can be translated into information about diatom shape and surface features. Image entropy is the amount of compression of an image as it relates to the amount of contrast. Low image entropy indicates that there is low contrast with a lot of similar valued pixels in the image, i.e., black dominates the image, and the image is compressible to a small size. High image entropy indicates that there is high contrast, and the image requires a high amount of compression. Entropy is invariant to translation, rotation and scaling [2.5, 2.70].

Image entropy is $H(p)_{Image} = -\sum_{i=1}^{n} p_i \log_2 p_i$, where p is the probability that the difference between adjacent pixels is i and log_2 is the base-2 logarithm, if information is measured in bits [2.129]; $log_2 (p_i)$ is the amount of uncertainty or information associated with a given outcome.

In an image, the probable information in gray-level pixels is equated with the state of the gray-level pixels associated with a given outcome. One such outcome is a symmetry state of an ensemble of gray-level pixels. For an image, information as states is calculated as a Boltzmann entropy [2.16–2.18, 2.84], $S = -k_B \ln w$, where micro-state w is a symmetry state and k_B is the Boltzmann constant [2.16–2.18, 2.84]. For $w = \frac{N!}{\prod_{i=1}^{L} n_i}$ symmetry states with $N = \sum_{i=1}^{L} n_i$ total number of symmetry numbers, log w given by Stirling's approximation is $\log n! = n \log\left(\frac{n}{e}\right) + O(\log n)$ [2.10, 2.67]. From this, $\log w \cong N \log N - \sum_{i=1}^{L} n_i \log n_i$, and for $p_i = \frac{n_i}{N}$, $S = \log w \cong N \sum_{i=1}^{L} p_i \log p_i$ [2.10].

Using the natural logarithm, $H(p)_{Image} = S(w) = -\sum_{i=1}^{w} p_i \ln p_i$, and with $k_B = 1$, $S_i = -\ln w_i$ for a single ith micro-state [2.84]. Symmetry states and their probable presence are connected in this way. Rearranging the Boltzmann entropy equation results in $w = e^S$ [2.66, 2.84].

Rotational symmetry of an image may be dissected iteratively as angular segments representing repeating regular features in a 2D plane whether the shape of the planar figure is circular or an n-polygon [2.160]. The 3D surface contained in a 2D boundary shape has information content that plays into symmetry measurement. For each image, entropy is a measure of information about how well the rotated images are matched after being overlain, i.e., how rotationally symmetrical an image is after subtracting the rotated image from its previously state. The more uniform in grayscale endpoint, i.e., degree of blackness, the better the match, and the entropy is the lowest possible number with the symmetry at its the highest value. A black image will have zero entropy, so that uncanny symmetry is calculated to be the value closest to this perfect symmetry via $S_{Uncanny\ Symmetry} = S_{final\ entropy}$ where final entropy is the entropy value at the last rotation subtracted from the stack of rotated and overlain images so that $w_{Uncanny\ Symmetry} = e^{S_{Uncanny\ Symmetry}}$.

2.2.4 Image Preparation for Measurement

Each original image is masked to eliminate the background. From this, the masked image is used in entropy analysis. The number of rotations per image depends on identifiable regular occurring equally angularly spaced surface features such as rays, ocelli, rimoportulae, areolae, costae, chambers, ribbing, or slits. For example, in *Asterolampra marylandica*, there are large extended chambers on the surface resulting in a star-like shaped central area, and the number of these chambers dictates the number of rotations. Or, for *Aulacodiscus*, the number of rotations is equal to 360° divided by the number of ocelli or rimoportulae on the valve surface. If there are multiple images of the same species with a different number of the same identifiable feature, then the number of rotations is determined for each image, not each species. The number of rotations is based also on shape for non-circular valves. For example, *Triceratium* and *Amphitetras* would receive three and four rotations, respectively.

Number of rotations per image ranged from 3 to 64 as determined by the taxa used in this study (Table 2.1). A rotations test was used to determine if an optimal number of rotations was associated with the lowest possible entropy value for those circular-shaped taxa without distinct identifiable equally spaced surface features. For this test, *Cyclotella meneghiniana* was used.

Entropy is calculated as a cumulative quantity based on successive overlaying of rotated images while matching edges for each original image. After all rotations are completed, the final entropy value is the minimum amount of entropy left after successive subtractions of one image from the previous stack of rotated and overlain images. For entropy measurement, an equivalence between rotational and reflective symmetry also can be made. An equivalence between rotational and reflective symmetries was established as follows: two rotations represent one plane of reflective symmetry; three rotations represent two planes of reflective symmetry; and four rotations represent three planes of reflective symmetry. Rectangular taxa were rotated four times to represent four planes of reflective symmetry, and pentagonal taxa were rotated five times to represent five planes of reflective symmetry.

2.2.5 Image Tilt and Slant Measurement Correction for Entropy Values

Each SEM was taken at 0° stage angle. However, the diatom itself may be tilted or slanted in the stub mountant that may be present between the diatom and stub or sputter-coated medium that covers or infiltrates the diatom or other regions of the stub. Because of this, digital images need to be corrected for tilt and slant. Tilt and slant are measured via the gradient of the pixel gray-levels in each image. Tilt is the amount of change of the normal to a 0° plane or direction of slant [2.135], while slant is the angle of rotation different from the 0° plane [2.20]. The gradient measures the greatest amount of change in *x*- and *y*-directions over the entire image. Often, image changes are most evident at the boundaries or edges between objects or as changes in luminosity or gray scale value. This feature and the image pixel gray scale values contain information so that tilt and slant may be expressed as entropy values. The information contained in the digital image is measurable as the gradient entropy.

For an image with a range of pixel values, r, $\nabla \mathbf{r} = (\nabla_x r, \nabla_y r) = \begin{bmatrix} \frac{\partial r}{\partial x} \\ \frac{\partial r}{\partial y} \end{bmatrix}$. Tilt, the angular direction that changes most within the image plane, is the gradient direction given as $\phi_r = arctan\left({(\nabla_x r)}/{(\nabla_y r)} \right) = arctan\left({(x - S_{x\ direction})}/{(y - S_{y\ direction})} \right)$. Slant, the angular distance (in degrees) change out of the image plane, is the gradient magnitude given as $\theta_r = arctan\sqrt{(\nabla_x r)^2 + (\nabla_y r)^2} = arctan\sqrt{(x - S_{x\ direction}{}^2) + (y - S_{y\ direction}{}^2)}$. The gradient entropies of the image, $S_{x\text{-}direction}$ and $S_{y\text{-}direction}$, are used in the calculation of tilt and slant.

The angular difference between tilt and normal to the *x-y* plane can be corrected by a shear transformation [2.140] with respect to the *y-z* plane. That is, pixel position in the original image is transformed via a tangent function of tilt to new pixel positions. After masking and prior to rotation, each image is corrected for tilt in this way. Angular rotation out of the *x-y* plane is used to correct slant by using *n*-rotations. That is, image rotation in the image preparation process is used to correct for slant.

2.2.6 Symmetry Analysis

The number of rotations to use per image was established using least-squares regression analysis for images without regularly spaced valve features. For each masked image, symmetry state was calculated from and plotted against tilt-corrected final entropy. Average taxon symmetry was measured for all images. Average taxon external valve symmetry was also measured. Comparison was made between symmetric *Arachnoidiscus* and *Aulacodiscus* valves and asymmetric *Asteromphalus* valves. *Cyclotella meneghiniana* was used in comparisons between normal and abnormal valves as well as comparison of initial to vegetative valves. For symmetric/asymmetric and normal/abnormal comparisons, least-squares regression analysis was accomplished. For initial and vegetative valves, a bar graph was used to depict symmetry differences.

Tests were conducted on symmetry changes with respect to valve formation. To model vertical change in morphology and symmetry, external and forming valves were compared. Entropy was measured for external and forming valves of *Arachnoidiscus ehrenbergii*, *Asterolampra marylandica*, *Asteromphalus heptactis*, *Asteromphalus imbricatus*, *Asteromphalus vanheurckii*, *Cyclotella meneghiniana*, *Glyphodiscus stellatus*, and *Triceratium favus*. Comparison among taxa was accomplished with a bar graph of external and forming valve symmetries.

For horizontal change in morphology and symmetry, 24 equally spaced concentric circles divided the valve face in a manner presumed to be similar to the time course of the first stage of valve growth [2.126]. Entropy was measured at each accumulative radius, and the resultant values were analyzed for symmetry of eight different circular-shaped taxon images: *Actinoptychus senarius*, *Actinoptychus splendens*, *Arachnoidiscus ehrenbergii*, *Arachnoidiscus ornatus*, *Asteroplampra marylandica*, *Aulacodiscus oregonus*, *Coscinodiscus* sp., and *Cyclotella meneghiniana*.

Entropy and symmetry were calculated for an elongated centric diatom as well. The image is masked then flipped along a bisecting horizontal axis, and then flipped along a bisecting vertical axis, illustrating the procedure for measurement of dihedral symmetry. The method of rotation and flipping along a bisecting axis enables entropy measurement of any organism with reflective, glide, or dihedral symmetry including pennate diatoms which develop via a sternum or sternum-raphe system [2.119]. *Biddulphia* sp. [2.145] was used for illustration (Figure 2.5).

Measurement error of entropy values for masked image matching after rotation and overlaying was calculated to determine accuracy and validity as $MeasError_{variation} = \frac{2 * Variance_{ImageMatching}}{Number\ of\ Rotations}$. From this, variation bias due to measurement error expressed as a percentage is $\%Bias = \frac{Variance_{ImageMatching}}{RMSE_{ImageMatching}} * 100$, based on the root mean square error, *RMSE* to image matching variance ratio.

For analysis of all images and valve formation simulation, entropy values were characterized by their statistical distributions. A histogram of final entropies was plotted, and a probability density function (PDF) and a cumulative density function (CDF) were determined. Maximum entropy is determined from the PDF [2.65]. For a uniform PDF, maximum entropy indicates that every symmetry state has an equal chance of occurring. For a Gaussian PDF, the least amount of information available means having the lowest probability and is maximum entropy; the highest amount of information available means the lowest entropy and the highest probability of occurring. Maximum entropy is also determined via standardization of the PDF so that a prior probability distribution results [2.65, 2.67].

2.2.7 Entropy, Symmetry, and Stability

During morphogenesis, changes in symmetry may occur. These changes may be thought of as morphogenetic states with respect to the spatial and temporal displacement of cell components and morphological shape and surface structures. These changes occur over time and comprise a dynamical system. Symmetry is important in pattern formation in linear and non-linear dynamical systems [2.130]. The behavior of symmetry changes over time may be characterized in centric diatoms by step-wise valve formation patterns as a growth of distances [2.27] from central area to valve margin and may be construed to be characterizations of a dynamical system. In this system, stability at equilibrium with respect to symmetry state behavior may be assessed in a diatom morphogenetic dynamical system.

To model such a system, valve formation simulation as accretionary growth behavior is used. Centric diatom valve formation on the valve surface commences at an annulus and proceeds roughly as a sequential addition of surface ornamentation to the valve margin. This is consistent with the diffusion limited aggregation model in which most deposition occurs in a narrow zone behind the leading silica growth (Figure 9 in [2.47]). Although the initial state of the system may change slightly, valve formation proceeds to produce the same species-specific valve surface. More generally, changes in this valve formation system exhibit ergodic behavior, and the stability of such behavior in a dynamical system is measurable as Lyapunov exponents.

In a dynamical system, different states occupy different spaces that exist at different times. Changes from state to state can be delineated using a system of differential equations derived from the original relation among the variables of interest. A Lyapunov function of the general form $\dot{x}_i = f_i(x)$ has first derivatives that form the elements of the Jacobian matrix (i.e., Jacobian) $J_{ij}(t) = \left.\frac{df_i(x)}{dx_j}\right|_{x(t)}$, where $x(t)$ is x_0 at equilibrium [2.78]. The Jacobian represents the change in tangent vectors over time t as a linearization of the tangent space system [2.39].

The relation between entropy and symmetry is of the general form of $x_i = f(x_i)$, and from the Boltzmann entropy equation, $x_i = S_i$ and $f(x_i) = -\ln w_i$. A Lyapunov function, $\dot{S}_i = -\ln w$ for each ith state of n-symmetry states at $\dot{\mathbf{S}} = 0$, is a system of homogeneous differential equations representing the change in symmetry states over time as a dynamical morphogenetic system. In matrix form, $\dot{\mathbf{S}} = \boldsymbol{AS}$ is a time evolution equation where $\mathbf{A}$ is the matrix of constant coefficients from $\left\{\begin{array}{l} s_1' = a_{11}s_1 + \cdots + a_{1n}s_n \\ s_2' = a_{21}s_1 + \cdots + a_{2n}s_n \\ \vdots = \vdots \quad + \cdots + \quad \vdots \\ s_n' = a_{n1}s_1 + \cdots + a_{nn}s_n \end{array}\right\}$ consisting of first partial derivatives comprising $\mathbf{A}$, the Jacobian.

To determine the behavior characteristics of symmetry states, solution to the Lyapunov function of the form $\mathbf{A}^{\mathrm{T}}\mathbf{S} + \mathbf{SA} = -\mathbf{Q}$ is evaluated, where $\mathbf{Q}$ is the resultant matrix. If the left side of the equation is determined to be positive definite when the right side of the equation is negative definite, then the symmetry states system is stable; otherwise, the system is unstable [2.23].

Specific characteristics of instability are evaluated via $\mathbf{A}$. At equilibrium, $(\mathbf{A} - \vartheta\mathbf{I}) = 0$ and for $\det(\mathbf{A} - \vartheta\mathbf{I}) = 0$, ϑ are eigenvalues of $\mathbf{A}$, and $tr\ \mathbf{A} = \Sigma\vartheta$ [2.29, 2.55]. For all symmetry states, $\ln w = \ln \prod_{i=1}^{n} e^{\vartheta_i t} = \sum_{i=1}^{n} \vartheta_i t$ [2.23]. Transformation of $\dot{\mathbf{S}} = \mathbf{AS}$ to $\dot{\mathbf{S}} = \Delta\mathbf{S}$ produces $\Delta = \begin{bmatrix} \vartheta_1 & \cdots & 0 \\ \vdots & \ddots & \vdots \\ 0 & \cdots & \vartheta_n \end{bmatrix}$ of $\mathbf{A}$, and diagonalization and a change of basis leads to a decoupled system with the general solution of the time evolution matrix $e^{\Delta t} = \begin{bmatrix} e^{\vartheta_1 t} & \cdots & 0 \\ \vdots & \ddots & \vdots \\ 0 & \cdots & e^{\vartheta_n t} \end{bmatrix}$ [2.22].

Lyapunov exponents [2.88, 2.101] are meaningfully characterized by the real parts of the complex valued eigenvalues of $\mathbf{A}$ [2.29] and are $\lambda_{Lyapunov} = \log \vartheta_i$ where *log* is any base. Lyapunov exponents measure fluctuation as average rates of divergence (expansion and instability) or convergence (contraction and stability) and represent expected values [2.29]. Initial conditions influence the outcome, and changes in initial conditions are measurable as indicators of the behavior of fluctuations. Boltzmann entropy production is necessarily sensitive to exponentially changing initial conditions in tangent space [2.39]. However, slight perturbations in initial conditions will yield measurement of the degree and characterization of symmetry behavior at any point in the dynamical system.

Lyapunov exponents comprise a multidimensional spectrum indicative of the rate of (exponential) divergence of multiple symmetry states with respect to direction of changes from one symmetry state to the next. Lyapunov exponents give bounds on information production of a dynamical system. They measure local behavior of the collective properties of a dynamical system yielding a global characterization of the system [2.8, 2.133]. A Lyapunov function close to equilibrium ensures that entropy tends to move toward a stable maximum in a dynamical system [2.29].

Lyapunov exponents are invariant to coordinate transformations and are ordered from largest to smallest [2.29]. They are a measure of the behavior of vectors in a tangent space. As such, these exponents are a measure of stability or instability at each symmetry state. For Lyapunov exponents that are positive, the dynamical system is chaotically unstable [2.29, 2.118], while negative Lyapunov exponents are indicators of a stable dynamical system [2.29]. A Lyapunov exponent with a value of zero may indicate a dissipative, regular [2.35], weakly chaotic [2.73] or intermittent system [2.74], depending on the particulars defining the initial system.

By quantifying the relation between symmetry states and stability via Lyapunov exponents, instability is characterizable as a specific behavior, either deterministically chaotic or not. However, the picture may be more complicated. Slight changes in initial conditions can precipitate chaotic behavior resulting in instability [2.29] even though the initial conditions may be random. Chaotic systems may appear to be random when considering fewer than all possible states in the dynamical system [2.97]. High-dimensional systems may appear to be ordered despite non-convergent random behavior [2.35]. Different kinds of randomness as "noise" may appear to be chaotic in a given dynamical system [2.118]. Care must be taken to discern the behavior of a dynamical system in terms of randomness or chaoticity.

2.2.8 Randomness and Instability

Lyapunov exponents enable the assessment of randomness with respect to instability in a dynamical system [2.74, 2.118, 2.163]. While local unstable behavior on a state-by-state basis may be chaotic, the macro-state instability of valve formation may be inherently random at a global scale. Alternatively, there may be intermittent chaotic or random unstable states [2.74] during the morphogenetic process. Because of this, a test for randomness [2.34] is necessary.

Thus far, our dynamical system model has been analyzed at equilibrium. However, to detect possible random instability, we need to consider the possible non-equilibrium consequences of such behavior in this system. Non-equilibrium may be a long-term phenomenon in contrast to short-term steady-states present in a dynamical system, for example, as time oscillations create Turing effects of diffusion driven instability [2.6]. Toy models have been used to link time intervals over multiple levels enabling the application of equilibrium dynamics to non-equilibrium behavior [2.96]. They have great potential in modeling complicated biological processes [2.2]. Non-equilibrium dynamics can be assessed in terms of constrained information loss or an increase in entropy [2.65].

As entropy increases over longer and longer sequences for a given dynamical system, entropy blocks or sections that define the structure of these sequences do not characterize the entire system. What remains is randomness that is not taken into consideration and given as the source entropy rate $S_{Randomness} \equiv \lim_{L\to\infty} \frac{S(L)}{L}$, where L is the number of sequence steps [2.26]. One way to test for randomness is to consider the non-deterministic behavior of the valve formation simulated system as a simple renewal process [2.35, 2.74, 2.115]. In a study of morphology and inheritance, *Cyclotella meneghiniana* offspring cell structure was analyzed for partial renewal with regard to the mother cell and complete renewal in terms of genetic or epigenetic inheritance [2.131], so that valve formation was considered to be a renewal process.

Maximizing the rate of variation in the entropy values during valve formation involves converting Boltzmann entropies of symmetry states to Kolmogorov-Sinai (KS) entropies based on probability [2.97]. KS entropies are time-averaged Shannon entropies over joint probability space [2.147] and are used to construct a PDF as a maximum entropy distribution [2.65, 2.137]. Randomness distributed over a probability distribution involves maximum entropy with regard to a PDF.

Evaluating entropy S using the PDF of associated probabilities involves first derivatives of S as entropy rates [2.26, 2.74] corresponding to bandwidth in the histogram used to determine the PDF [2.137]. For the expected value of the probabilities associated to S, $\langle S(w)\rangle = \sum_{i=1}^{n} p_i S(w_i)$, the associated Lyapunov exponents are found via Lagrange multipliers partitioned on the function and given as a maximum entropy probability distribution as $p_i = \exp\{-[\lambda_0 + \lambda_1 S_1(w_i) + \cdots + \lambda_m S_m(w_i)]\}$ [2.65]. With constraint $\sum p_i = 1$, maximum entropy with respect to Lyapunov exponents is $\langle S(w)\rangle = S_{max} = -\frac{\partial}{\partial\lambda}\sum \ln w = \lambda_0 + \lambda_1 S_1(w_i) + \cdots + \lambda_m S_m(w_i)$ [2.65, 2.67]. Over a probability distribution, $S(p_i)_{max} = \sum \lambda_{max}$ [2.84, 2.110], where λ_{max} is a positive Lyapunov exponent. The sum of the positive Lyapunov exponents is a maximization of the possible states of KS entropy, and as such, KS entropy indicates degree of randomness [2.26].

The Lyapunov exponents for KS entropy of a sequence $L(t)$ as a random function over a probability distribution for the αth probability [2.74] are $\lambda_\alpha(w_0) = \frac{1}{t^\alpha}\sum_{i=0}^{t-1} \ln\left|\mathbf{D}_{K-S\,entropy}(w_i)\right|$, where $\mathbf{D}_{KS\,entropy}$ is the density matrix of first partial derivatives from KS entropy where the diagonal elements are probability values and t is the time step of L. Using $S'(w) = \frac{d(\ln w)}{dt} = \frac{1}{w}$ from Boltzmann entropy, Lyapunov exponents from KS entropy become $\lambda_{K-S} = \frac{1}{t}\sum_{i=0}^{t-1}\left|\frac{1}{w_i}\right|$, and from Pesin's identity [2.110], $S_{K-S} = \sum \lambda_{K-S}$. When $S_{K-S} = 0$, the rate of prior information loss is equal to the rate of new information created, and stability is evident [2.35]. As $\lambda_{K-S} \to \infty$ and is no longer a constant when changes in initial conditions approach zero, $S_{K-S} \to \infty$ and randomness occurs [2.35]. For a measure of randomness, $S_{Randomness}(t) = \lim_{t\to\infty}\sum \lambda(t)_{K-S} > 0$, and $S_{Randomness} = \sum \lambda_{Randomness} = \frac{\sum_{i}^{t=\infty} \partial \ln w_i}{\partial\lambda(w_i)}$.

2.3 Results

2.3.1 Symmetry Analysis

For analysis, 151 centric diatom images (Figure 2.6a) were background masked, rotated, and overlain to calculate entropy values using the equation for Boltzmann entropy. From this, 94 external valves and 19 forming valves, along with *Cyclotella meneghiniana* valves: 12 replicates of SClemtbx1800, 13 abnormal valves, 6 forming valves, and 2 initial valves. Entropy filtering applied to the original images enabled detection of those structures that contribute to the measurement of symmetry (Figure 2.6b). Tilt and slant were calculated for each image. Tilt ranged from 0.725 to 0.785, and slant ranged from 1.392 to 1.496° for all images. Tilt correction was applied to each image (Figures 2.6c, d), and slant was corrected via rotation. All initial entropy values were tilt-corrected via tilt entropy to produce final entropy values.

For each image, final entropy (hereafter simply called "entropy") values were plotted against the number of rotations, resulting in $R^2 = 0.0063$, which indicates that number of rotations was not correlated to entropy values (Figure 2.7). Each entropy value was converted to symmetry using the equation for Boltzmann entropy. Entropy vs. symmetry was plotted for each image, and the result was a negative exponential relation between the two variables. An example is plotted for image ProvBay5_12lx450, *Arachnoidiscus ehrenbergii* and depicted in Figure 2.8. Measurement error was calculated as 0.000153 with 0.1068% bias.

The optimum number of rotations that gives the minimum entropy value was tested for those taxa with valve face features that were not easily divisible into equally spaced partitions. Image SClemtbx1800, *Cyclotella meneghiniana*, was tested with 3, 4, 5, 7, 13, 14, 15, 17, 18, 24, 27, 31, and 64 rotations, where 64 represents the actual number of marginal chambered striae in the image. Entropy values were plotted against number of rotations, and from a least-squares regression, no trend was discernable between the minimum entropy value and number of rotations (Figure 2.9). That is, an increasingly larger number of rotations used did not correlate to any improvement in obtaining a minimum entropy value ($R^2 = 0.1097$). Fourteen rotations were used for taxa lacking distinguishing valve face features.

For all image entropies, a histogram and PDF was constructed and depicted a Gaussian distribution (Figure 2.10). The distribution has a slightly long left tail and is therefore slightly positively skewed. A CDF of all entropies was also constructed (Figure 2.11).

Average symmetry was determined and plotted for each taxon (Figure 2.12). A least-squares regression was performed that resulted in a best-fit curve of $w_{taxon} = 0.9051S + 15.654$, and the coefficient of determination was 0.9474. Average taxon symmetry was determined and plotted for all external valves (Figure 2.13). A least-squares regression was performed that resulted in a best-fit curve of $w_{taxon} = 0.9457S + 15.473$, and the coefficient of determination was 0.9451. External and forming valve average symmetries were compared for ten taxa. From the bar graph, *Asteromphalus heptactis*, *Asteromphalus imbricatus*, *Asteromphalus shadboltianus*, *Asteromphalus vanheurckii*, *Asterolampra marylandica*, *Cyclotella meneghiniana*, and *Glyphodiscus stellatus* had higher external valve symmetry than forming valve symmetry (Figure 2.14). *Triceratium favus* and *Triceratium pentacrinus* fo. *quadrata* had higher forming valve symmetry than external valve symmetry. *Arachnoidiscus ehrenbergii* had approximately equal forming and external symmetry (Figure 2.14).

Arachnoidiscus and *Aulacodiscus* with symmetric valve surfaces were compared to *Asteromphalus* with asymmetric valve surfaces using least-squares regression (Figure 2.15). Best-fit curves for species, x, were $w_{taxon} = 1.1878x + 27.27$, $w_{taxon} = 1.276x + 30.67$, and $w_{taxon} = 2.2925x + 23.238$ for *Asteromphalus*, *Arachnoidiscus*, and *Aulacodiscus*, respectively. Coefficients of determination were 0.7622 for *Asteromphalus* and 0.9645 for *Arachnoidiscus*, and *Aulacodiscus*.

A number of analyses were conducted to illustrate symmetry differences in different valves of *Cyclotella meneghiniana*. Normal and abnormal *Cyclotella meneghiniana* valve symmetries were compared using least-squares regression (Figure 2.16). Slope for normal valves was 1.345 compared to a slope of 1.008 for abnormal valves. Coefficients of determination

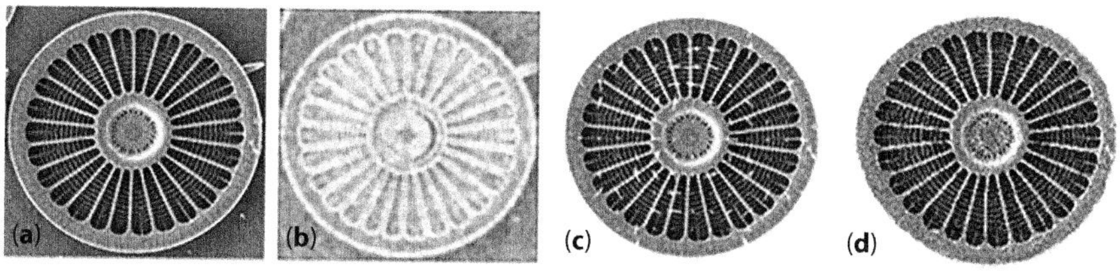

Figure 2.6 ProvBay5_12lx450, *Arachnoidiscus ehrenbergii.* (a) Original scanning electron micrograph image; (b) Entropy filtering applied to image; (c) Vectors showing gradient changes in magnitude and direction; (d) Tilt-corrected image. Original SEM by Mary Ann Tiffany.

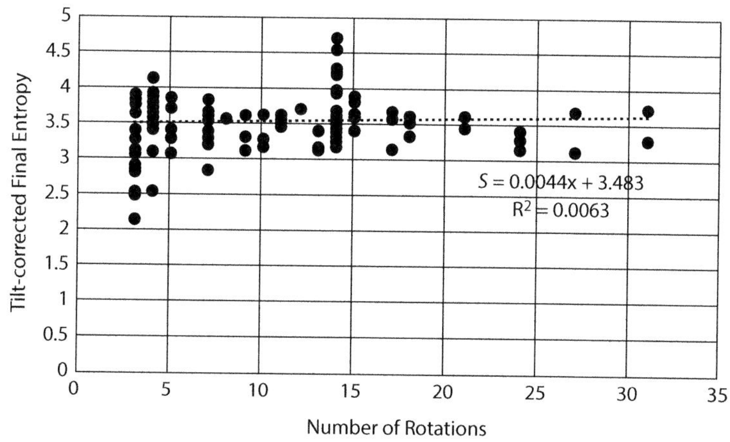

Figure 2.7 Tilt-corrected final entropy vs. number of rotations for all images, omitting 64 rotations used for *Cyclotella meneghiniana.*

were 0.9737 and 0.9284 for normal and abnormal valves, respectively. Masked images of *Cyclotella meneghiniana* initial valves were measured for symmetry and compared to normal vegetative valve symmetry. Average initial valve symmetry had much higher symmetry than normal external and forming vegetative valves (Figure 2.17). Different valve surface morphologies revealed differences in symmetry that were discernable via analysis using entropy.

Measurement error for image matching rotation and overlaying was not a factor in symmetry analysis. Values ranged from 0.0000946 to 0.0184 measurement error with 0.0172 to 1.994% bias. The percentages fall within the range of values that do not affect the measurement of true variation [2.103]. For symmetry analysis, the entropy values measured are not biased by measurement error.

An image of *Biddulphia* sp. was used to illustrate measurement of dihedral symmetry. Entropy and symmetry values were 1.94041 and 6.96162 with a measurement error of 0.0000614 and a 0.00663% bias. *Biddulphia* sp. dihedral symmetry measurement was accomplished using rotation and flipping the masked image about the *x* and *y* axes in the *x-y* plane.

Valve formation was simulated in a simple model by sequentially adding concentric rings of valve structure from an annulus to the valve margin to cumulatively create a valve face surface. Although diatom valve formation is much more complicated

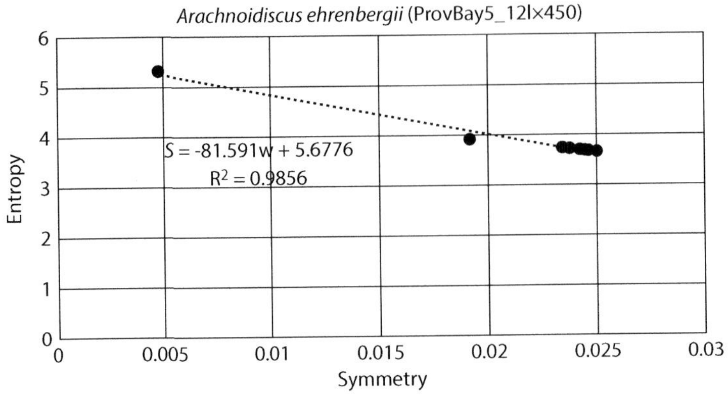

Figure 2.8 Typical entropy vs. symmetry plot. The example is image ProvBay5_12lx450, *Arachnoidiscus ehrenbergii*. The first datum is the original masked tilt-corrected image.

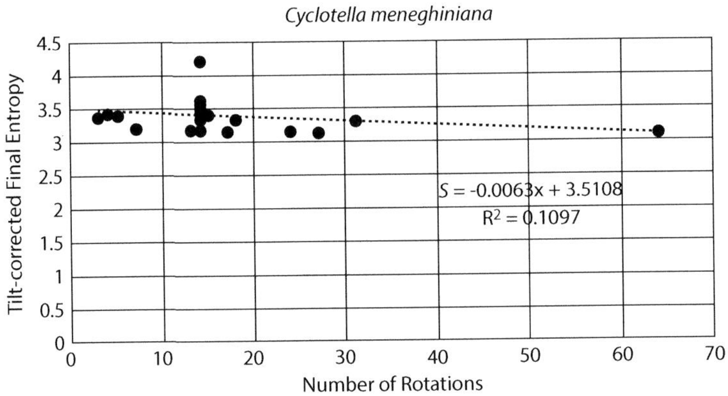

Figure 2.9 Number of rotations vs. tilt-corrected final entropies for Cyclotella meneghiniana external valves using 3 to 64 rotations.

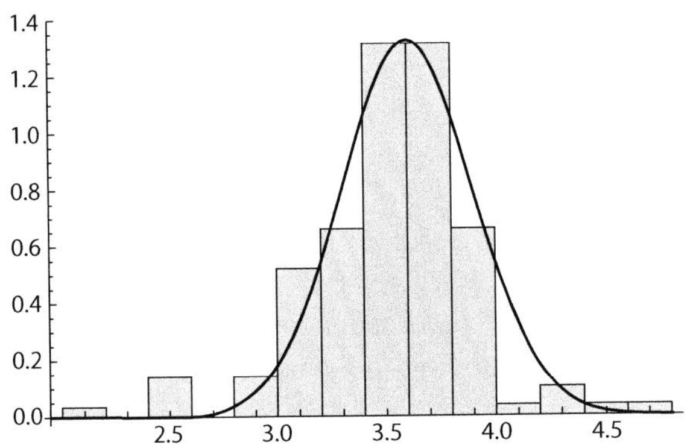

Figure 2.10 Histogram and overlain probability density function for all tilt-corrected final entropies resulting in a Gaussian distribution.

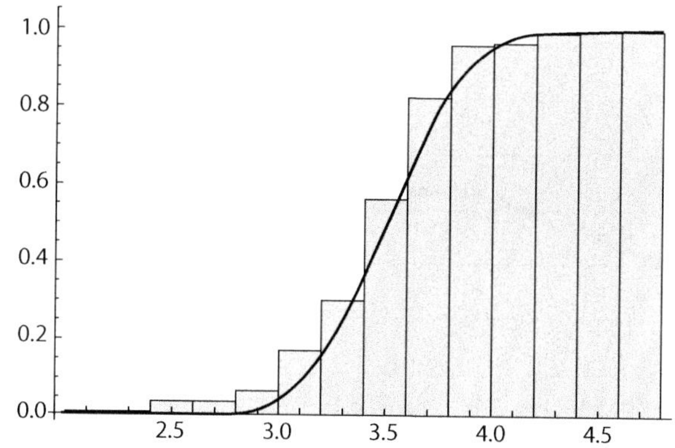

Figure 2.11 Histogram and overlain cumulative density function for all tilt-corrected final entropies.

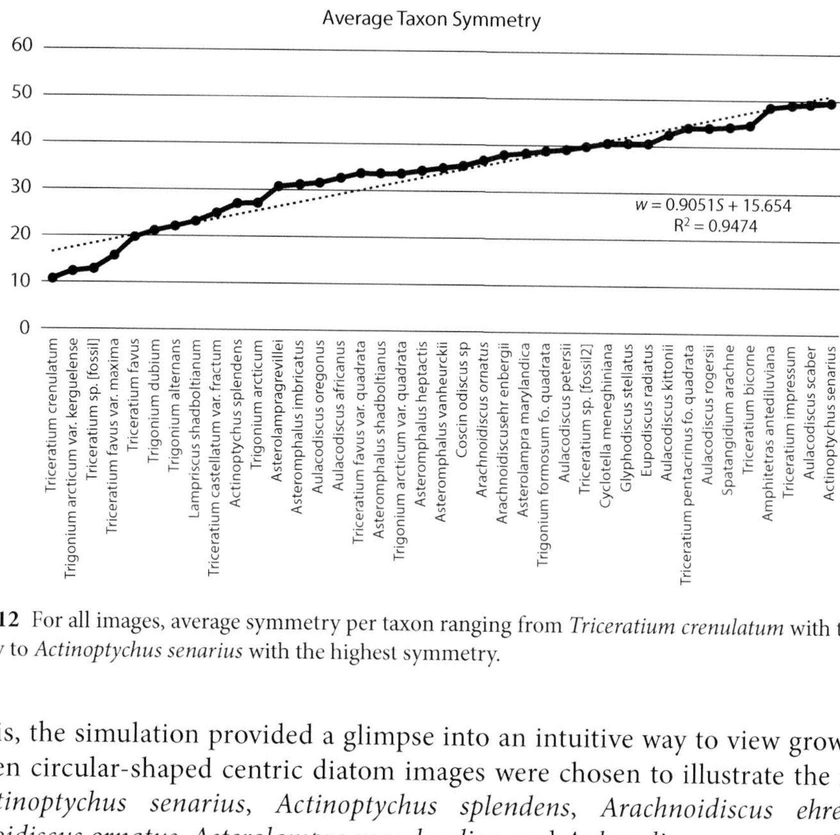

Figure 2.12 For all images, average symmetry per taxon ranging from *Triceratium crenulatum* with the lowest symmetry to *Actinoptychus senarius* with the highest symmetry.

than this, the simulation provided a glimpse into an intuitive way to view growth over time. Ten circular-shaped centric diatom images were chosen to illustrate the method for *Actinoptychus senarius*, *Actinoptychus splendens*, *Arachnoidiscus ehrenbergii*, *Arachnoidiscus ornatus*, *Asterolampra marylandica*, and *Aulacodiscus oregonus*. Twenty-four concentric rings were arbitrarily selected on each image. Each ring was of an arbitrary width and added cumulatively, starting with an annulus at the center and ending with a ring at the valve margin (Figure 2.18).

Valve formation simulation for the eight taxa was plotted against symmetry, and the resultant curves were exponential depicting the change in symmetry from annulus to valve margin (Figure 2.19). The plot includes a second *Arachnoidiscus ehrenbergii* external

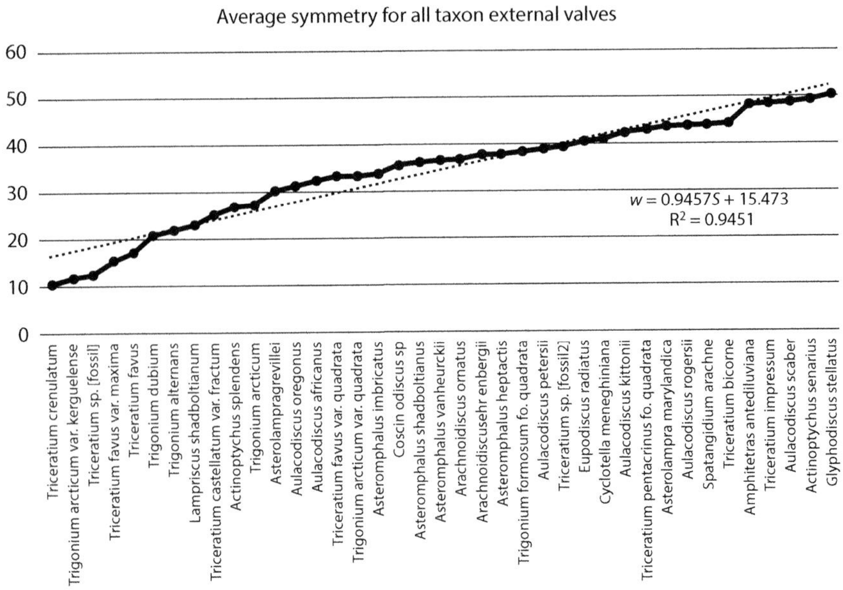

Figure 2.13 Average taxon symmetry for all external centric diatom valves ranging from *Triceratium crenulatum* with the lowest symmetry to *Glyphodiscus stellatus* with the highest symmetry.

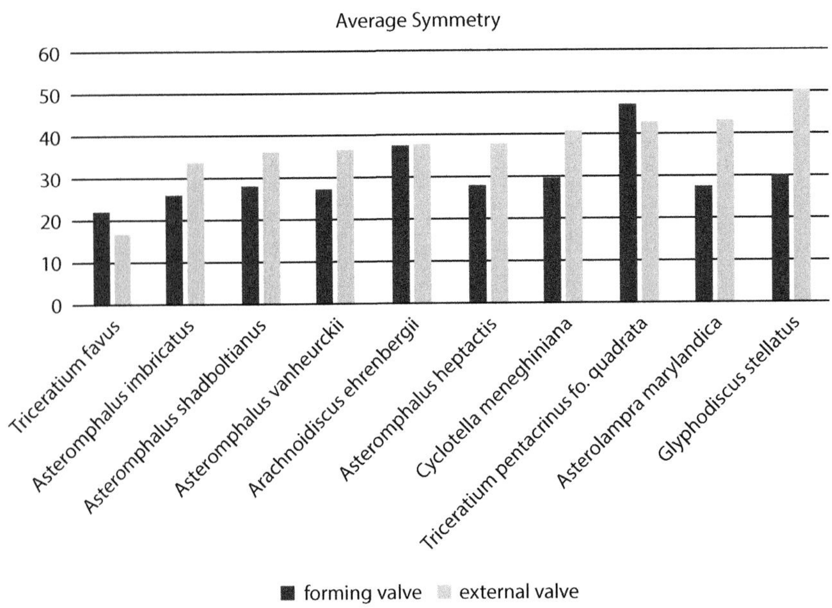

Figure 2.14 Vertical valve formation comparison: average external and forming valve symmetry for ten centric diatoms.

valve and a forming valve of this taxon. For valve formation simulation, the distribution of average entropy values was depicted in a histogram and overlain PDF (Figure 2.20) as a gamma distribution and is a maximum entropy distribution. A histogram and overlain CDF of average entropy values were also constructed (Figure 2.21). Standardizing the PDF

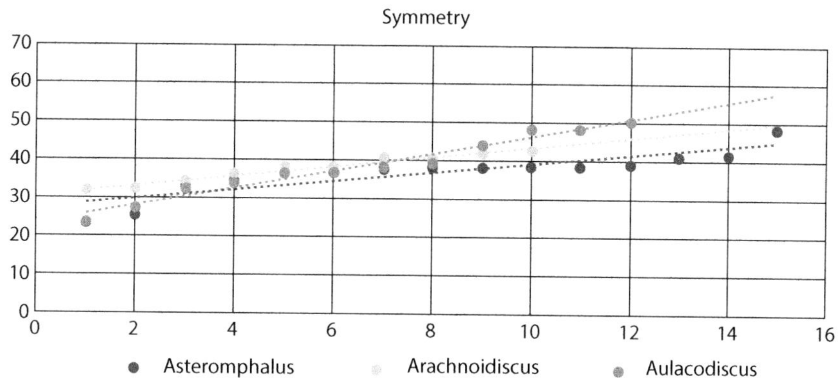

Figure 2.15 Symmetry values for *Asteromphalus*, *Arachnoidiscus*, and *Aulacodiscus* external valves.

produced a skewed-right normal prior probability distribution (Figure 2.22), establishing the relation between available information and its probability of occurrence. The prior distribution is also a maximum entropy distribution as a minimization of prior information given via the PDF.

2.3.2 Valve Formation—Stability and Instability Analyses

Stability analysis was performed using symmetries calculated from the valve formation simulation of cumulative concentric rings from annulus to valve margin for eight taxa. Lyapunov exponents were calculated to determine the behavior of symmetry changes. Initial conditions for valve formation are $v_{Valve\ formation}(0) = v_{annulus}$. Each valve formation simulation produced a series of Lyapunov exponents comprising a Lyapunov spectrum [2.166]. The largest real Lyapunov exponent was positive and characterized the overriding behavior of symmetry changes during valve formation. For all taxa, sum of the positive Lyapunov exponents signified chaotic asymmetric behavior characterizing instability (Table 2.2). The remainder of the real Lyapunov exponents in descending order were very small (large negative values), summed, and signified stability (Table 2.2). For Lyapunov exponents calculated

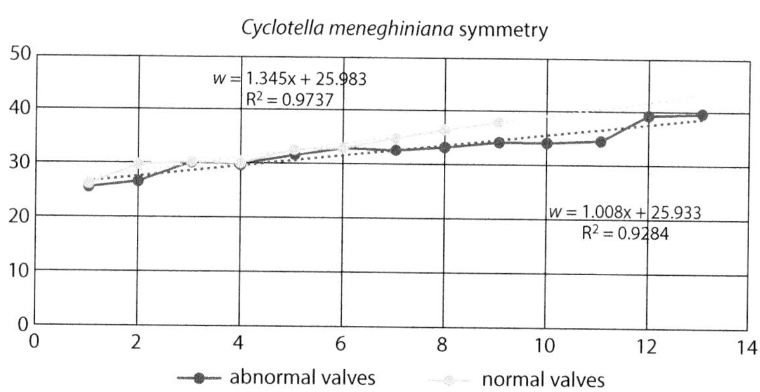

Figure 2.16 *Cyclotella meneghiniana* symmetry for normal and abnormal valves.

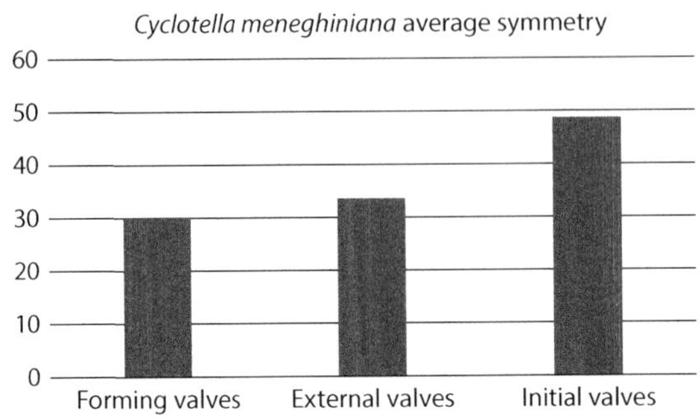

Figure 2.17 *Cyclotella meneghiniana* symmetry values for normal forming and external vegetative and initial valves.

from KS entropy, some random instability was present during valve simulation as well (Table 2.2).

The change in symmetry was compared with the change in chaotic instability for taxa used in the valve formation simulation. Symmetry increased monotonically in the order of least symmetric to most symmetric: *Actinoptychus splendens*, *Cyclotella meneghiniana*, *Arachnoidiscus ornatus*, *Aulacodiscus oregonus*, *Coscinodiscus* sp., *Arachnoidiscus ehrenbergii* (1), *Arachnoidiscus ehrenbergii* forming valve, *Asterolampra marylandica*, *Arachnoidiscus ehrenbergii* (2), and *Actinoptychus senarius* (Table 2.2). Rescaling symmetry and chaotic instability values, a comparative bar graph showed that chaotic instability was more evident in less symmetric taxa *Actinoptychus splendens*, *Cyclotella meneghiniana*,

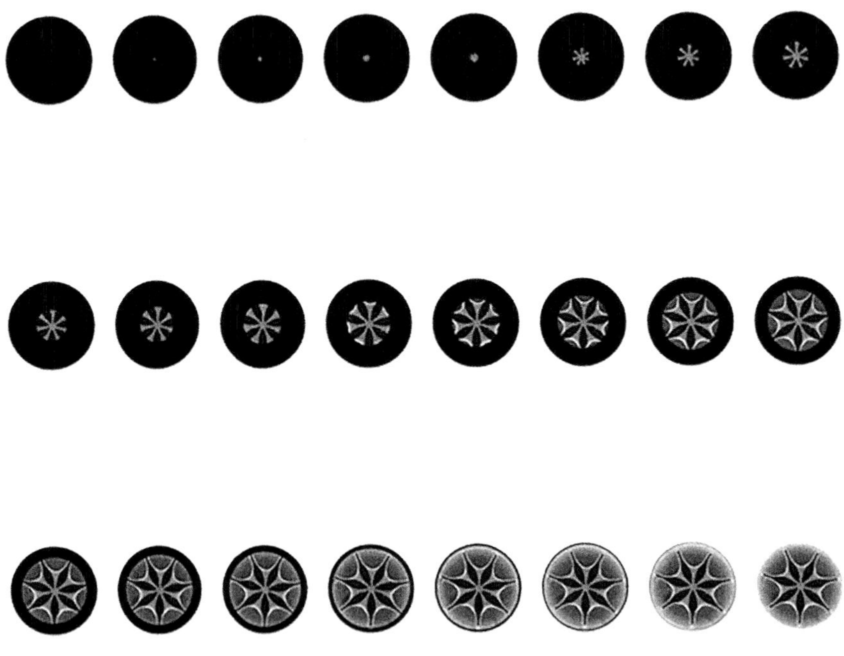

Figure 2.18 *Asterolampra marylandica* (SanDBay02_03dx1100) as a sequence of 24 cumulative concentric rings simulating valve formation from annulus to valve margin.

Arachnoidiscus ornatus, and *Aulacodiscus oregonus* than higher symmetric taxa with the exception of *Arachnoidiscus ehrenbergii* (2) (Figure 2.23). Rescaled random instability was more evident in less symmetric taxa *Actinoptychus splendens*, *Cyclotella meneghiniana*, and *Aulacodiscus oregonus* as well, in contrast to higher symmetric taxa, with the exception of the *Arachnoidiscus ehrenbergii* forming valve (Figure 2.24). Overall, less symmetry is associated with higher chaotic and random instability (Figures 2.23 and 2.24).

2.4 Discussion

From average symmetry per taxon for all valves, triangular and non-regular structured surface circular taxa exhibited less symmetry than highly regularly structured circular and four-edged or greater polygonal taxa (Figure 2.12). By inspection, the result is evident, and the method of using entropy to determine symmetry states has explicit utility. Additionally, although we measured rotational symmetry, other symmetries are present as well (Figure 2.3). Separate tests for each type of symmetry could be made, and quantification of each

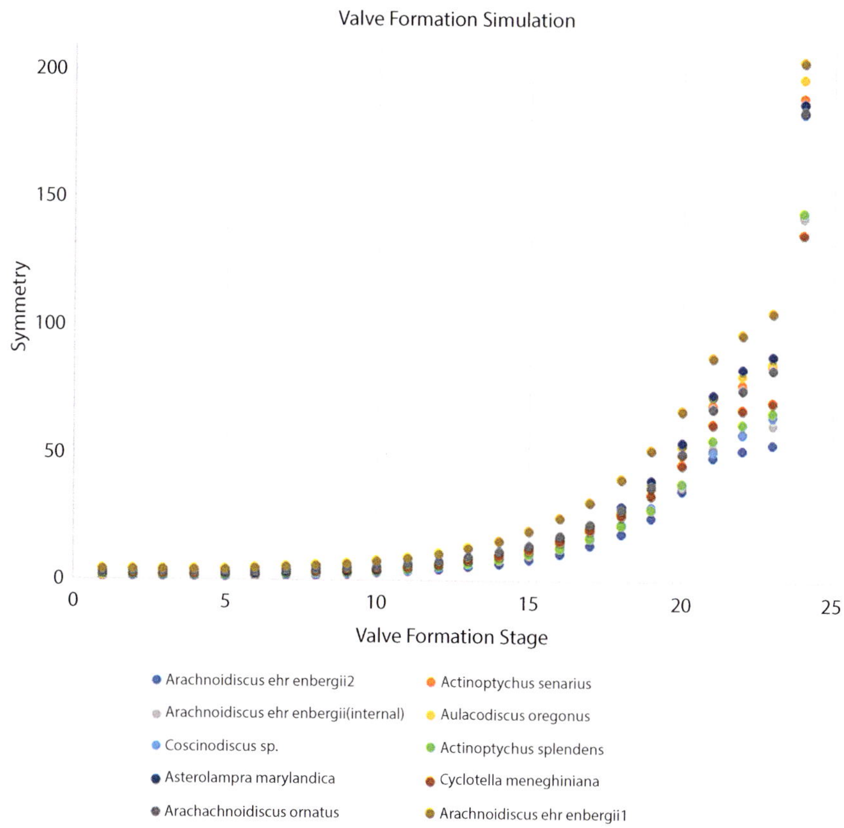

Figure 2.19 Plots of symmetry vs. 24-valve formation simulation steps from annulus to valve margin for eight centric diatom taxa. Order from least to highest symmetry curve per taxon is read left to right in legend. Symmetry increases exponentially from the annulus to the completed valve.

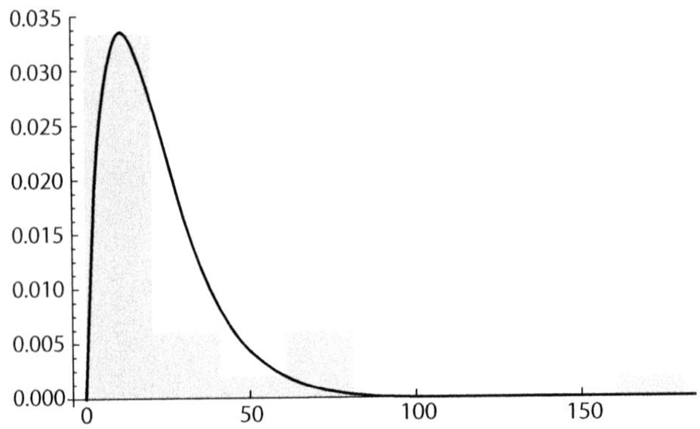

Figure 2.20 Valve formation steps for eight taxa as a probability density function of entropy values resulting in a gamma distribution.

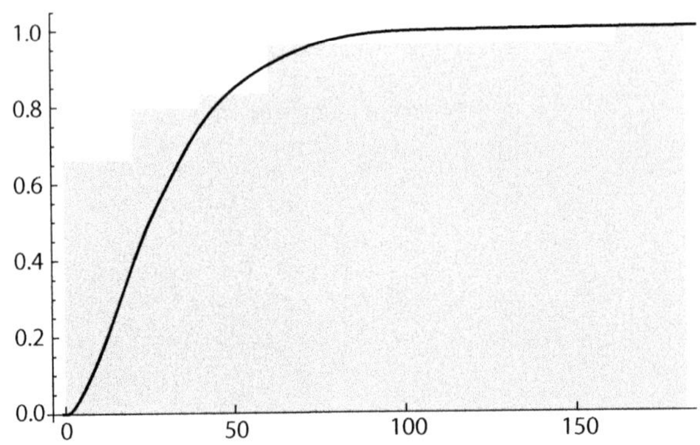

Figure 2.21 Valve formation steps for eight taxa as a cumulative density function of entropy values.

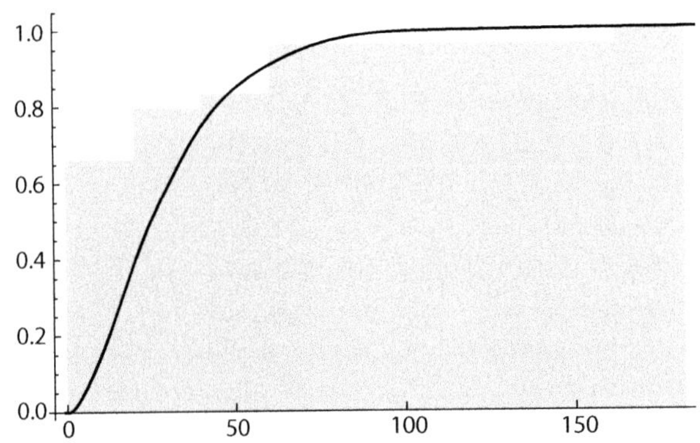

Figure 2.22 Valve formation steps for eight taxa as a skew-right normal prior probability density function of entropy values.

Table 2.2 Valve formation simulation for eight centric diatom taxa and characterization of instability and stability from Lyapunov exponents. Positive sums indicate chaotic instability, negative sums indicate stability, and Lyapunov exponents from Kolmogorov-Sinai (KS) entropy indicate random instability.

Taxon*	File name	Chaotic Instability: Sum of + Lyapunov exponents	Stability: Sum of – Lyapunov exponents	Random Instability: Lyapunov exponent from K-S entropy
Actinoptychus senarius	motewx1500_2	5.0775	−429.424	4399.39
Actinoptychus splendens	MisBy5_1hx2000	4.987	−388.348	3079.95
Arachnoidiscus ehrenbergii (forming valve)	halfmax500	4.96527	−390.747	3186.1
Arachnoidiscus ehrenbergii (1)	provbay5_12ix400	5.36916	−379.328	2359.42
Arachnoidiscus ehrenbergii (2)	ProvBay5_12lx450	5.04099	−398.893	4656.9
Arachnoidiscus ornatus	PoiDumox700	5.18264	−454.589	2669.33
Asterolampra marylandica	SanDBay02_03dx1100	5.15695	−383.425	3354.2
Aulacodiscus oregonus	PoiDumjx600www	5.1158	−424.086	4033.01
Coscinodiscus sp.	MisBayhx500	5.00421	−465.124	3277.96
Cyclotella meneghiniana†	SClemtbx1800	5.09775	−422.176	2448.34

* All taxa are external valves unless otherwise noted.
† Symmetry value is the average of all rotations test results.

type of symmetry would need to be instituted to determine the contribution of a particular symmetry throughout centric diatom development.

The results for external valves was somewhat different from those for all valves combined. Average symmetry per taxon revealed that most of the triangular taxa or those with three-part valve features were the least symmetric in contrast to the most symmetric taxa, including species from *Aulacodiscus*, taxa with four-part valve features or shape, *Glyphodiscus stellatus*, *Eupodiscus radiatus*, *Amphitetras antediluviana*, and *Triceratium*

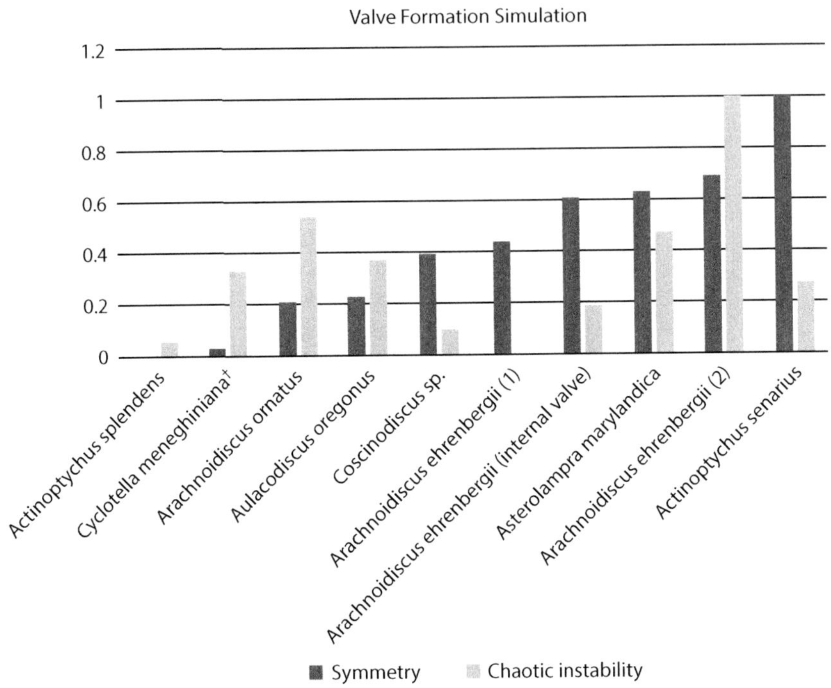

Figure 2.23 Valve formation simulation comparison of symmetry and chaotic instability. See Table 2.2 for numerical values.

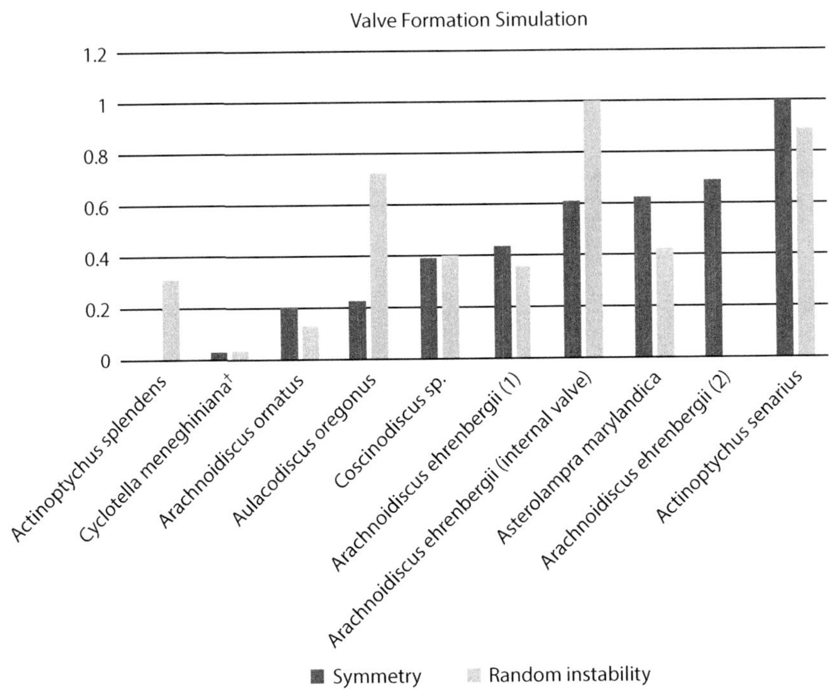

Figure 2.24 Valve formation simulation comparison of symmetry and random instability. See Table 2.2 for numerical values.

pentacrinus fo. *quadrata*). *Actinoptychus senarius*, *Spatangidium arachne*, *Asterolampra marylandica*, *Arachnoidiscus ehrenbergii*, and *Arachnoidiscus ornatus* were notable exceptions to the four-part rule in the highly symmetric group (Figure 2.13). As expected, species of *Asteromphalus* were nearer in symmetry to *Asterolampra grevillei* and less symmetric than *Asterolampra marylandica*. Although, fewer taxa were analyzed with respect to forming valves, a similar result emerged when compared to external valves. Four-part valve or greater featured or shaped taxa were the most symmetric (Figure 2.14). Taxa with higher forming than external valve symmetry were *Triceratium favus* and *Triceratium pentacrinus* fo. *quadrata* (Figure 2.14). Different species of *Asteromphalus* forming valves were found to be similar in symmetry to those of *Asterolampra marylandica*, contrary to the situation with external valves.

A comparison of symmetric and asymmetric external valve surfaces resulted in higher symmetry for *Arachnoidiscus* and *Aulacodiscus* over *Asteromphalus* (Figure 2.15). Steeper slopes for the symmetric genera of 1.276 and 2.2925 also indicate more symmetry change among species than *Asteromphalus* species with a slope of 1.1878. Generally, rotationally asymmetric diatom surfaces exhibit less symmetry than morphologically similar taxa such as *Asterolampra marylandica* (Figure 2.13). However, asymmetric surfaced *Spatangidium arachne* has higher symmetry than *Asteromphalus* species (Figure 2.13). The five rays of *Spatangidium arachne* are thin, straight and regularly spaced and sized, which may be recovered as more dihedrally symmetric than *Asteromphalus* species which are more variable in these surface features [2.146].

Normal and abnormal valves for *Cyclotella meneghiniana* were compared in terms of symmetry (Figure 2.16). Normal valves had a steeper slope than abnormal valves, indicating a greater change in symmetry. That is, abnormal valves have less symmetry or greater asymmetry than normal valves of *Cyclotella meneghiniana*.

Initial valves were compared to normal vegetative valves of *Cyclotella meneghiniana*. The average symmetry of the initial valves was much higher than average symmetry for external and forming vegetative valves (Figure 2.17). The difference in symmetry of the central area was the key to the difference between initial and vegetative valves. The undulated central area of the vegetative valve has less symmetry than the symmetry of the more uniform convex shape in initial valves.

In the valve formation simulation, least symmetric taxa ranged from *Actinoptychus splendens* to the highest symmetric taxon, *Actinoptychus senarius* (Figures 2.23 and 2.24). The more distinct and regular the valve surfaces features were present, the more symmetric was the taxon. Less symmetric taxa had the highest chaotic and random instability (Figures 2.23 and 2.24). This relation between asymmetry and instability is supported for other organisms [2.54]. The main difference in our analysis is that the sources of instability are different for different degrees of symmetry.

2.4.1 Symmetry and Scale in Diatoms

Symmetry in diatoms is a characteristic of morphology that is evident at various scales. Discrete or continuous symmetry [2.167] as well as local or global symmetry are issues of scale. In our study, we used discrete measures at a local scale. Symmetry may be scale dependent over the 3D morphology of the organism. For different structural parts, e.g., honeycomb, repeating hexagonal pores in layers [2.138], of the diatom valve, there may be a hierarchical cumulative symmetry of morphological characters at the nano to micro to macro scale, across the valve face as a result of epigenesis with respect to morphogenesis, differentiation, and growth [2.32]. Hierarchical cumulative symmetry may also change from organelle to the whole diatom cell as a result of cytokinesis and mitosis with respect to morphogenesis. At the 3D surface and boundary shape of the cell, diatoms have structural symmetry changes at each developmental stage as each subsequent whole daughter cell undergoes size diminution, auxosporulation, and production of an initial cell. Our focus on centric diatom vegetative cells and valve formation as well as vegetative and initial cell symmetry was used in assessing symmetry during diatom morphogenesis.

Nano-, micro-, or macro-scale dependent symmetry may be evident as individual or groups of structural features that may not have the same symmetry at the same time (Figure 2.3). Concomitantly, these different kinds of symmetry are measurable implicitly at multiple scales. By isolating sections of the valve face in terms of micro or nanostructure, each kind of symmetry can be measured. Scale dependency occurs if a valve formation sequence is not monotonic or linear. For example, centric diatom symmetry in *Cyclotella meneghiniana* may be an example of scale dependency (Figure 2.19). In some cases, scale symmetry as scale invariance may be present (Figure 2.3k). Pattern repetition at multiple scales and fractal scaling may be determined by testing for the Hausdorff dimension [2.29].

2.4.2 Valve Formation and Stability

We modeled our diatom morphogenetic valve formation system as dynamical and in equilibrium in order to test for stability behavior as it is associated to symmetry. Uncanny symmetry as a measure was useful in determining the degree of stability in valve formation. We determined that valve formation simulation proceeded primarily as deterministic chaotic instability mixed with lesser periods of stability. While instability may look like randomness, we found that for our simulation of valve morphogenesis, accretionary behavior is essentially chaotic. Valve formation PDF depicted a gamma distribution with a stretched exponential tail (Figure 2.20), potentially indicating intermittency in this dynamical system [2.115], provided that as time approaches infinity, the Lyapunov exponent is equal to zero [2.74].

Stability analysis in the form of Lyapunov systems provided a way to assess the sources of instability with regard to valve formation in diatom morphogenesis. Finding a deterministic chaotic component is noteworthy because the assumption is that only random behavior dictates instability in developmental systems [2.54], or that if chaos is present, it cannot be quantified separately from randomness [2.53]. Our results indicate that the behavior of instability varies chaotically and randomly throughout the valve formation process; however, less symmetric forms have the highest instabilities. In spite of the presence of chaotic and random instabilities, valve formation overall is a regular, stable process when considering the end product. The chaotic component of instability may be indicative of multi-scalability of symmetry during valve formation. Fluctuations of chaotic and random instability may be embedded in the valve formation system so that at times, scale symmetry as well as scale-dependent symmetries may be present.

Diatom valve formation in morphogenesis may be viewed as exhibiting different dynamical system qualities. Some elements of valve formation may be dissipative at equilibrium. Other elements may exhibit a reactive-diffusive, non-equilibrium state. Self-assembly at the molecular level may induce periodic micro- and nanoscale pore assemblies that can be modified via nanoparticles in a prescribed way in diatoms [2.165]. Silica deposition may exhibit phase separation at the chemical level of valve formation [2.138] leading to pore formation [2.154]. Elastic instability may induce buckling [2.95] as a phenomenon in diatoms during morphogenesis [2.52] in contrast to Turing instabilities via reaction-diffusion [2.91, 2.99]. Buckling may be characterized by Lyapunov exponents to determine degree of chaotically unstable behavior [2.141].

Valve formation as a dynamical system has aspects of regularity as well. Cross-costae formation exhibits a regularly spaced homogeneous growth pattern, with initiation of the valve formation process starting with a central organizing structure [2.154]. The dynamic growth of a diatom valve is spatially controlled, and sequential deposition contributing to the height of silica on the valve surface contributes to species-specific pattern formation [2.154]. Given constraints in the process, at the valve micro-scale surface, chaotic or random instability may be present, while at the nano-scale level, hexagonal or round pores with varying patterns of cribra within areolae pores may be in a regular periodic stabilizing pattern. Morphometric noise of multiple traits has weak control over pore shape and size. Development of a morphological character such as a pore is affected by inhibitors in contrast to the spatial arrangement of pores [2.154]. With further study, parsing the relation between unstable and stable states at multiple scales may be useful in understanding symmetry breaking [2.41, 2.102] during valve morphogenesis.

2.4.3 Symmetry, Stability and Diatom Morphogenesis

Diatoms and valve morphogenesis are a combination of static structure formation at specific time junctures and a dynamic progression to the formation of the vegetative cell. As static structures, diatoms have high symmetry, termed uncanny symmetry. As self-organizing siliceous structures, they are some of the most stable static structures. Entropy as an ensemble measure [2.84] is instrumental in measuring of silica-deposited dynamics during valve formation that lead to the various regularly created geometric patterns defining species-specific valve surfaces.

The dynamical process from a more to less chaotic unstable state may be a characterization of self-organization. In the valve formation process, self-organization is the recovery from a chaotic unstable state, even though the chaotic state could be construed to be "normal." Symmetry is a reflection of these states. In contrast, abnormally developed valves could be construed to be in a higher "abnormal" chaotic state, and stress during valve formation is evidenced by the resultant vegetative cell and its increasingly asymmetry state. Self-organization as the appearance of new structures may be a transition from a more chaotic, unordered to less chaotic, more ordered symmetry state, or vice versa in the case of a final abnormal symmetry state. In this way, equilibrium states during non-linear dissipative fluctuations in a dynamical system may end up being the most chaotic states [2.71].

The picture that symmetry and its relation to stability exhibits during morphogenesis is not straight-forward. In fact, no quantitative generalized overall framework for evolutionary dynamics, including stability of structural inheritance or structural integrity during reproduction for changing environmental conditions, exists with respect to diatom

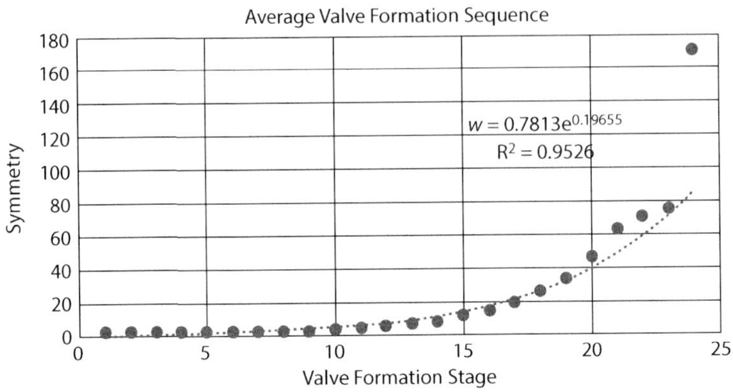

Figure 2.25 Average sequence of 24 simulated valve formation steps of symmetry for eight centric diatom taxa. Symmetry increases exponentially from the annulus to the completed valve.

morphogenesis [2.131]. From our study, we give a speculative understanding and interpretation of our results on symmetry and stability as they pertain to the three major steps of centric diatom morphogenesis [2.126].

Overall, symmetry increased over the valve formation simulation with highest symmetry at the finished vegetative valve (Figure 2.19). The basal layer that forms from silica deposition horizontally during the first major step of diatom valve morphogenesis, starting with the annulus, is roughly covered by the first 8 of 24 "stages" of simulation (Figure 2.18). Symmetry with the lowest values occurring at approximately the same value (Figure 2.25). For roughly "stages" 9–16 representing the second major step, vertical silica deposition is represented by the changing luminosity of pixels in the central area of the images as the indication of the formation of rays and areolae emerges (Figure 2.18). Symmetry changes become more evident, going from low to higher symmetry, yet still within a small range (Figure 2.25). For the final major step, horizontal silica deposition is represented by even more changes in pixel luminosity over the valve face in "stages" 17 to 24 where indications of the formation of cribra going toward the valve margin is present (Figure 2.18). Symmetry changes are highest in these "stages", representing completion of silica deposition of the valve (Figure 2.25). The valve forms present during the three major valve formation steps may be viewed as symmetries functioning as silica depositional ensembles changing in direction and magnitude (horizontally to vertically to horizontally again) on the valve face. This dynamical system operates at equilibrium with exponentially changing symmetry occurring via chaotic instability during valve formation.

2.4.4 Future Research—Symmetry, Stability and Directionality in Diatom Morphogenesis

We measured and characterized instability, but what are the characterizations of stability—i.e., is stability periodic, monotonic, or something else? Characterization of stability may be informative concerning the end products of epigenetic, morphogenetic, or other processes, or at times during the morphogenetic process when apparent stasis occurs as no discernable change in morphology, symmetry, or other structural state. Characterization of the behavior of stability during morphogenesis may enable inferences about directionality and irreversibility at an evolutionary level. Research along this line has implications for studies of diatom valve formation and morphological evolution.

Improvements to assessing such systems for the factors enabling directionality with respect to stability or instability are in the offing by using Lyapunov co-vectors [2.39]. These co-vectors are associated with stable and unstable manifolds via Lyapunov exponents along trajectories in a dynamical system [2.39, 2.101], and they are also associated with entropies with respect to Lyapunov exponents [2.39, 2.110]. There are positive Lyapunov exponents associated with forward Lyapunov co-vectors, and negative Lyapunov exponents associated with backward Lyapunov co-vectors [2.77]. These Lyapunov co-vectors do not directly indicate time; however, they can indicate directionality, and inferences about directionality with respect to time may be determined. Directionality with respect to valve formation changes at species-level or higher taxonomic category may be informative about diatom evolution.

To determine directionality, irreversibility and degree of indistinguishability among symmetry states need to be assessed within a dynamical morphogenetic system. Changes in instability behavior, at which points, stability occurs, are potentially a multidimensional and multiscale problem. Chaotic instability may be a part of directionality for shorter time spans in contrast to long term random instability potentially for all dimensions and scales during the morphogenetic process. More detailed work is necessary to determine the contributions of stability and instability as well as symmetry at each morphogenetic stage. A more exacting representation is needed of diatom valve formation in morphogenesis concerning irreversibility of the process and indistinguishability of symmetry states and their associated valve formation stages.

If successive states of instability via symmetry states are found during diatom valve formation or morphogenesis more broadly, how does this impact diatom morphological complexity [2.107] over time? Additionally, what happens to morphological complexity over long time periods at stationarity? The expectation is that dynamical complexity is related to chaotic instability [2.22] and that increasing complexity occurs over time. Algorithmic information theory may be used to tie symmetry to complexity and to determine the role of instability over time, and in turn, gain an understanding of another facet of diatom morphogenesis.

References

[2.1] Albrecht-Buehler, G., Daughter 3T3 cells. Are they mirror images of each other? *J. Cell Biol.*, 72, 3, 595–603, 1977.

[2.2] Alicea, B. and Gordon, R., Toy models for macroevolutionary patterns and trends. *BioSystems*, 122, Special Issue: Patterns of Evolution, 25–37, 2014.

[2.3] Alicea, B. and Gordon, R., Cell differentiation processes as spatial networks: Identifying four-dimensional structure in embryogenesis. *BioSystems*, 173, 235–246, 2018.

[2.4] Alvare, G. and Gordon, R., CT Brush and CancerZap!: Two video games for computed tomography dose minimization. *Theor. Biol. Med. Modell.*, 12, 1, 7, 2015.

[2.5] Amr, I.I., Amin, M., El-Kafrawy, P., Sauber, A.M., Using Statistical Moment Invariants and Entropy in Image Retrieval. *Int. J. Comput. Sci. Inf. Secur.*, 7, 1, 160–164, 2010.

[2.6] Aragón, J., Barrio, R., Woolley, T., Baker, R., Maini, P., Nonlinear effects on Turing patterns: Time oscillations and chaos. *Phys. Rev. E*, 86, 2, 026201, 2012.
[2.7] Astauroff, B.L., Analyse der erblichen Störungsfälle der bilateralen Symmetrie [Analysis of hereditary disorders of bilateral symmetry] [German]. *Z. Indukt. Abstamm. Vererbungsl.*, 55, 1, 183–262, 1930.
[2.8] Atay, F.M., Jalan, S., Jost, J., Randomness, Chaos, and Structure. *Complexity*, 15, 1, 29–35, 2009.
[2.9] Attneave, F., Some informational aspects of visual perception. *Psychol. Rev.*, 61, 3, 183–193, 1954.
[2.10] Benci, V. and Menconi, G., Some remarks on the definition of Boltzmann, Shannon and Kolmogorov entropy. *Milan J. Math.*, 73, 1, 187–209, 2005.
[2.11] Bentley, K., Clack, C., Cox, E.J., Diatom colony formation: A computational study predicts a single mechanism can produce both linkage and separation valves due to an environmental switch. *J. Phycol.*, 48, 3, 716–728, 2012.
[2.12] Bentley, K., Cox, E.J., Bentley, P.J., Nature's batik: A computer evolution model of diatom valve morphogenesis. *J. Nanosci. Nanotechnol.*, 5, 1, 25–34, 2005.
[2.13] Biederman, I., Recognition-by-components: A theory of human image understanding. *Psychol. Rev.*, 94, 2, 115–147, 1987.
[2.14] Biederman, I., Mezzanotte, R.J., Rabinowitz, J.C., Scene perception: Detecting and judging objects undergoing relational violations. *Cogn. Psychol.*, 14, 2, 143–177, 1982.
[2.15] Bissantz, N., Holzmann, H., Pawlak, M., Testing for image symmetries-with application to confocal microscopy. *IEEE Trans. Inf. Theory*, 55, 4, 1841–1855, 2009.
[2.16] Boltzmann, L., Über die Mechanische Bedeutung des Zweiten Hauptsatzes der Wärmetheorie [On the Mechanical Meaning of the Second Law of Heat Theory] [German]. *Wien. Ber.*, 53, 195–220, 1866.
[2.17] Boltzmann, L., *Vorlesungen über Gastheorie, vol. I [Lectures on Gas Theory] [German]*, vol. 1, J. A. Barth, Leipzig, Germany, 1896.
[2.18] Boltzmann, L., *Vorlesungen über Gastheorie, [Lectures on Gas Theory] [German]*, vol. II, J. A. Barth, Leipzig, Germany, 1898.
[2.19] Bredov, D. and Volodyaev, I., Increasing complexity: Mechanical guidance and feedback loops as a basis for self-organization in morphogenesis. *BioSystems*, 173, 133–156, 2018.
[2.20] Burge, J., McCann, B.C., Geisler, W.S., Estimating 3D tilt from local image cues in natural scenes. *J. Vision*, 16, 13, 2, 1–25, 2016.
[2.21] Chaitin, G.J., Algorithmic information theory. *Ibm J. Res. Dev.*, 21, 4, 350–359, 1977.
[2.22] Chakrabarti, C.G. and Ghosh, K., Biological evolution: Entropy, complexity and stability. *J. Mod. Phys.*, 2, 6, 621–626, 2011.
[2.23] Chen, C.T., *Linear System Theory and Design*, 4th ed, Oxford University Press, New York, NY, USA, 2013.
[2.24] Chen, Q., Shi, J.C., Tao, Y., Zernicka-Goetz, M., Tracing the origin of heterogeneity and symmetry breaking in the early mammalian embryo. *Nat. Commun.*, 9, 1819, 2018.
[2.25] Cox, E.J., Willis, L., Bentley, K., Integrated simulation with experimentation is a powerful tool for understanding diatom valve morphogenesis. *BioSystems*, 109, 3, Special Issue on Biological Morphogenesis, 450–459, 2012.
[2.26] Crutchfield, J.P. and Feldman, D.P., Regularities unseen, randomness observed: Levels of entropy convergence. *Chaos*, 13, 1, 25–54, 2003.
[2.27] Dämmig, M. and Mitschke, F., Estimation of Lyapunov exponents from time series: The stochastic case. *Phys. Lett. A*, 178, 5–6, 385–394, 1993.
[2.28] De Martino, A., Amato, A., Bowler, C., Mitosis in diatoms: Rediscovering an old model for cell division. *Bioessays*, 31, 8, 874–884, 2009.

[2.29] Eckmann, J.P. and Ruelle, D., Ergodic theory of chaos and strange attractors. *Rev. Mod. Phys.*, 57, 3, 617–656, 1985.

[2.30] Fernandes, L.F., New observations on frustule morphology of *Eupodiscus radiatus* Bailey and *Fryxelliella floridana* Prasad. *Braz. J. Biol.*, 63, 3, 411–421, 2003.

[2.31] Fernandes, L.F. and de Souza-Mosimann, R.M., *Triceratium moreirae* sp. nov. and *Triceratium dubium* (Triceratiaceae-Bacillariophyta) from estuarine environments of southern Brazil, with comments on the genus *Triceratium* C. G. Ehrenberg. *Braz. J. Biol.*, 61, 1, 159–170, 2001.

[2.32] Ferrell Jr., J.E., Bistability, bifurcations, and Waddington's epigenetic landscape. *Curr. Biol.*, 22, 11, R458–R466, 2012.

[2.33] Frey, F.M., Robertson, A., Bukoski, M., A method for quantifying rotational symmetry. *New Phytol.*, 175, 4, 785–791, 2007.

[2.34] Gan, C.C. and Learmonth, G., Comparing entropy with tests for randomness as a measure of complexity in time series. *arXiv preprint arXiv:1512.00725*, 2015.

[2.35] Gao, J., Hu, J., Tun, W-w., Blasch, E., Multiscale analysis of biological data by scale-dependent Lyapunov exponent. *Front. Physiol.*, 2, 110, 2012.

[2.36] Garrido, A., Symmetry in complex networks. *Symmetry-Basel*, 3, 1, 1–15, 2011.

[2.37] Ghobara, M.M., Mazumder, N., Vinayak, V., Reissig, L., Gebeshuber, I.C., Tiffany, M.A., Gordon, R., On light and diatoms: A photonics and photobiology review [Chapter 7], in: *Diatoms: Fundamentals & Applications [DIFA, Volume 1 in the series: Diatoms: Biology & Applications, series editors: Richard Gordon & Joseph Seckbach]*, J. Seckbach and R. Gordon (Eds.), pp. 129–190, Wiley-Scrivener, Beverly, MA, USA, 2019.

[2.38] Gibson, R.A. and Mahoney, R.K., Comparative valve and cingular structure in *Biddulphia titiana* (Grunow) Grunow in Van Heurck and *Trigonium arcticum* (Brightwell) Cleve (Bacillariophyceae). *Proc. Acad. Nat. Sci. Philadelphia*, 136, 200–217, 1984.

[2.39] Ginelli, F., Chaté, H., Livi, R., Politi, A., Covariant Lyapunov vectors. *J. Phys. a-Math. Theor.*, 46, 25, 254005, 2013.

[2.40] Golubitsky, M. and Stewart, I., Recent advances in symmetric and network dynamics. *Chaos*, 25, 9, 097612, 2015.

[2.41] Gordon, R., *The Hierarchical Genome and Differentiation Waves: Novel Unification of Development, Genetics and Evolution*, World Scientific & Imperial College Press, Singapore & London, 1999.

[2.42] Gordon, R., Diatoms and nanotechnology: Early history and imagined future as seen through patents, in: *The Diatoms: Applications for the EnVironmental and Earth Sciences*, 2nd, J.P. Smol and E.F. Stoermer (Eds.), pp. 585–602, Cambridge University Press, Cambridge, 2010.

[2.43] Gordon, R., Part Three: The reverse engineering road to computing life. Chapter 10: Walking the tightrope: the dilemmas of hierarchical instabilities in Turing's morphogenesis [invited], in: *The Once and Future Turing: Computing the World*, S.B. Cooper and A. Hodges (Eds.), pp. 144–159, Cambridge University Press, Cambridge, 2016.

[2.44] Gordon, R. and Aguda, B.D., Diatom morphogenesis: Natural fractal fabrication of a complex microstructure, in: *Proceedings of the Annual International Conference of the IEEE Engineering in Medicine and Biology Society, Part 1/4: Cardiology and Imaging*, New Orleans, LA, USA, 4-7 Nov. 1988, Institute of Electrical and Electronics Engineers, New York, pp. 273–274, 1988.

[2.45] Gordon, R., Björklund, N.K., Robinson, G.G.C., Kling, H.J., Sheared drops and pennate diatoms. *Nova Hedwig.*, 112, Festschrift for Prof. T.V. Desikachary, 287–297, 1996.

[2.46] Gordon, R. and Brodland, G.W., On square holes in pennate diatoms. *Diatom Res.*, 5, 2, 409–413, 1990.

[2.47] Gordon, R. and Drum, R.W., The chemical basis of diatom morphogenesis. *Int. Rev. Cytol.*, 150, 243–372421–422, 1994.
[2.48] Gordon, R., Hanczyc, M.M., Denkov, N.D., Tiffany, M.A., Smoukov, S.K., Chapter 18: Emergence of polygonal shapes in oil droplets and living cells: The potential role of tensegrity in the origin of life, in: *Habitability of the Universe Before Earth [Volume 1 in series: Astrobiology: Exploring Life on Earth and Beyond, eds. Pabulo Henrique Rampelotto, Joseph Seckbach & Richard Gordon]*, R. Gordon and A.A. Sharov (Eds.), pp. 427–490, Elsevier B.V., Amsterdam, 2017.
[2.49] Gordon, R. and Jacobson, A.G., The shaping of tissues in embryos. *Sci. Am.*, 238, 6, 106–113, 160, 1978.
[2.50] Gordon, R., Losic, D., Tiffany, M.A., Nagy, S.S., Sterrenburg, F.A.S., The Glass Menagerie: Diatoms for novel applications in nanotechnology. *Trends Biotechnol.*, 27, 2, 116–127, 2009.
[2.51] Gordon, R., Sterrenburg, F.A.S., Sandhage, K., A Special Issue on Diatom Nanotechnology. *J. Nanosci. Nanotechnol.*, 5, 1, 1–4, 2005.
[2.52] Gordon, R. and Tiffany, M.A., Possible buckling phenomena in diatom morphogenesis, in: *The Diatom World*, J. Seckbach and J.P. Kociolek (Eds.), pp. 245–272, Springer, Dordrecht, The Netherlands, 2011.
[2.53] Graham, J.H., Freeman, D.C., Emlen, J.M., Antisymmetry, directional asymmetry, and dynamic morphogenesis. *Genetica*, 89, 1–3, 121–137, 1993.
[2.54] Graham, J.H., Raz, S., Hel-Or, H., Nevo, E., Fluctuating asymmetry: Methods, theory, and applications. *Symmetry-Basel*, 2, 2, 466–540, 2010.
[2.55] Guan, K., Important notes on Lyapunov exponents. *arXiv preprint arXiv:1401.3315.*, 2014.
[2.56] Håkansson, H. and Chepurnov, V., A study of variation in valve morphology of the diatom *Cyclotella meneghiniana* in monoclonal cultures: effect of auxospore formation and different salinity conditions. *Diatom Res.*, 14, 2, 251–272, 1999.
[2.57] Hasle, G.R. and Sims, P.A., The diatom genera *Stellarima* and *Symbolophora* with comments on the genus *Actinoptychus*. *Br. Phycol. J.*, 21, 97–114, 1986.
[2.58] Herve, V., Derr, J., Douady, S., Quinet, M., Moisan, L., Lopez, P.J., Multiparametric analyses reveal the pH-dependence of silicon biomineralization in diatoms. *PLoS One*, 7, 10, e46722, 2012.
[2.59] Hildebrand, M., Lerch, S.J.L., Shrestha, R.P., Understanding diatom cell wall silicification—moving forward. *Front. Mar. Sci.*, 5, 125, 2018.
[2.60] Holló, G. and Novák, M., The manoeuvrability hypothesis to explain the maintenance of bilateral symmetry in animal evolution. *Biol. Direct*, 7, 7, 22, 2012.
[2.61] Jacobson, A.G., Some forces that shape the nervous system. *Zoon*, 6, 13–21, 1978.
[2.62] Jacobson, A.G., Computer modeling of morphogenesis. *Am. Zool.*, 20, 4, 669–677, 1980.
[2.63] Jacobson, A.G. and Gordon, R., Changes in the shape of the developing vertebrate nervous system analyzed experimentally, mathematically and by computer simulation. *J. Exp. Zool.*, 197, 2, 191–246, 1976.
[2.64] Jacobson, A.G. and Gordon, R., Nature and origin of patterns of changes in cell shape in embryos. *J. Supramol. Struct.*, 5, 4, 371–380, 1976.
[2.65] Jaynes, E.T., Information theory and statistical mechanics. *Phys. Rev.*, 106, 620–630, 1957.
[2.66] Jaynes, E.T., Gibbs vs Boltzmann entropies. *Am. J. Phys.*, 33, 5, 391–398, 1965.
[2.67] Jaynes, E.T., Prior probabilities. *IEEE Trans. Syst. Sci. Cybern. SSC*, 4, 3, 227–241, 1968.
[2.68] Jones, J.S., Evolution: An asymmetrical view of fitness. *Nature*, 325, 6102, 298–299, 1987.
[2.69] Kaczmarska, I. and Ehrman, J.M., Auxosporulation in *Paralia guyana* MacGillivary (Bacillariophyta) and possible new insights into the habit of the earliest diatoms. *PLoS One*, 10, 10, e0141150, 2015.

[2.70] Kadir, T., Boukerroui, D., Brady, M., *An analysis of the scale saliency algorithm [Technical report TR-2264-03]*, Robotics Research Laboratory, Department of Engineering Science, University of Oxford, Oxford, UK, 2003.

[2.71] Klimontovich, Y.L., Entropy, information and ordering criteria in open systems, in: *Nonlinear Dynamics in the Life and Social Sciences*, vol. 320, W. Sulis and I. Trofimova (Eds.), pp. 13–32, IOS Press, Amsterdam, 2001.

[2.72] Koenderink, J.J., vanDoorn, A.J., Kappers, A.M.L., Todd, J.T., The visual contour in depth. *Percept. Psychophys.*, 59, 6, 828–838, 1997.

[2.73] Korabel, N. and Barkai, E., Pesin-type identity for intermittent dynamics with a zero Lyaponov exponent. *Phys. Rev. Lett.*, 102, 5, 050601, 2009.

[2.74] Korabel, N. and Barkai, E., Separation of trajectories and its relation to entropy for intermittent systems with a zero Lyapunov exponent. *Phys. Rev. E*, 82, 016209, 2010.

[2.75] Kotzsch, A., Pawolski, D., Milentyev, A., Shevchenko, A., Scheffel, A., Poulsen, N., Shevchenko, A., Kröger, N., Biochemical composition and assembly of biosilica-associated insoluble organic matrices from the diatom *Thalassiosira pseudonana*. *J. Biol. Chem.*, 291, 10, 4982–4997, 2016.

[2.76] Kröger, N., Prescribing diatom morphology: toward genetic engineering of biological nanomaterials. *Curr. Opin. Chem. Biol.*, 11, 6, 662–669, 2007.

[2.77] Kuptsov, P.V. and Parlitz, U., Theory and computation of covariant Lyapunov vectors. *J. Nonlinear Sci.*, 22, 5, 727–762, 2012.

[2.78] Laffargue, T., Tailleur, J., van Wijland, F., Lyapunov exponents of stochastic systems-from micro to macro. *J. Stat. Mech.-Theory Exp.*, 2016, 034001, 2016.

[2.79] Lee, J.-H. and Chang, M., Morphological variations of the marine diatom genus *Actinoptychus* in the coastal waters of Korea. *Algae*, 11, 4, 365–374, 1996.

[2.80] Leeuwenberg, E., The perception of assimilation and brightness contrast as derived from code theory. *Percept. Psychophys.*, 32, 4, 345–352, 1982.

[2.81] Lenoci, L. and Camp, P.J., Self-assembly of peptide scaffolds in biosilica formation: computer simulations of a coarse-grained model. *J. Am. Chem. Soc.*, 128, 31, 10111–10117, 2006.

[2.82] Lenoci, L. and Camp, P.J., Diatom structures templated by phase-separated fluids. *Langmuir*, 24, 1, 217–223, 2008.

[2.83] Libbrecht, K. and Rasmussen, P., *The Snowflake, Winter's Secret Beauty*, Voyageur Press, Stillwater, Minnesota, USA, 2003.

[2.84] Lin, S.K., Correlation of entropy with similarity and symmetry. *J. Chem. Inf. Comput. Sci.*, 36, 3, 367–376, 1996.

[2.85] Losic, D. (Ed.), *Diatom Nanotechnology: Progress and Emerging Applications*, Royal Society of Chemistry, London, 2018.

[2.86] Loy, G. and Eklundh, J.O., Detecting symmetry and symmetric constellations of features. *Lect. Notes Comput. Sci.*, 3952, 508–521, 2006.

[2.87] Ludwig, W., *Das Recht-Links-Problem im Tierreich und beim Menschen: Mit einem Anhang: Rechts-Links-Merkmale der Pflanzen [The Right-Left Problem in the Animal Kingdom and in Humans: With an Appendix: Right-Left Characteristics of Plants] [German]*, Verlag von Julius Springer, Berlin, Germany, 1932.

[2.88] Lyapunov, A.M., The general problem of the stability of motion [Reprinted translation of: Probleme generale de la stabilite du mouvement. *Commun. Soc. Math. Kharkov*, 2, 1892, 265–272, 1992.

[2.89] MacArthur, B.D., Sánchez-García, R.J., Anderson, J.W., Symmetry in complex networks. *Discrete Appl. Math.*, 156, 18, 3525–3531, 2008.

[2.90] MacDonald, J.D., On the structure of the diatomaceous frustule, and its genetic cycle. *Ann. Mag. Nat. Hist. Ser.*, 4, 3, 1–8, 1869.

[2.91] Maini, P.K., Baker, R.E., Chuong, C.M., Developmental biology. The Turing model comes of molecular age. *Science*, 314, 5804, 1397–1398, 2006.

[2.92] Maini, P.K., Woolley, T.E., Baker, R.E., Gaffney, E.A., Lee, S.S., Turing's model for biological pattern formation and the robustness problem. *Interface Focus*, 2, 4, 487–496, 2012.

[2.93] Mann, D.G., Specifying a morphogenetic model for diatoms: an analysis of pattern faults in the Voigt zone. *Nova Hedwig.*, Suppl., 130, 97–117, 2006.

[2.94] Mather, K., Genetical control of stability in development. *Heredity*, 7, 3, 297–336, 1953.

[2.95] Matsumoto, E.A. and Kamien, R.D., Elastic-instability triggered pattern formation. *Phys. Rev. E*, 80, 2, 15, #021604, 2009.

[2.96] Meinhardt, H. and Gierer, A., Pattern formation by local self-activation and lateral inhibition. *Bioessays*, 22, 8, 753–760, 2000.

[2.97] Mihelich, M., Dubrulle, B., Paillard, D., Herbert, C., Maximum entropy production vs. Kolmogorov-Sinai entropy in constrained ASEP model. *Entropy*, 16, 2, 1037–1046, 2014.

[2.98] Mitchell, J.G., Whence is the diversity of diatom frustules derived?, in: *Diatom Nanotechnology: Progress and Emerging Applications*, D. Losic (Ed.), pp. 1–13, Royal Society of Chemistry, Cambridge, 2018.

[2.99] Nesterenko, A.M., Kuznetsov, M.B., Korotkova, D.D., Zaraisky, A.G., Morphogene adsorption as a Turing instability regulator: Theoretical analysis and possible applications in multicellular embryonic systems. *PLoS One*, 12, 2, e0171212, 2017.

[2.100] Ornstein, D.S., Ergodic theory, randomness, and "chaos". *Science*, 243, 4888, 182–187, 1989.

[2.101] Oseledec, V.I., A multiplicative ergodic theorem. Lyapunov characteristic numbers for dynamical systems. *Trans. Moscow Math. Soc.*, 19, 197–231, 1968.

[2.102] Palmer, A.R., Symmetry breaking and the evolution of development. *Science*, 306, 5697, 828–833, 2004.

[2.103] Palmer, A.R. and Strobeck, C., *Fluctuating asymmetry analyses revisited*, Oxford University Press, Oxford, UK, 2003.

[2.104] Pappas, J.L., Geometry and topology of diatom shape and surface morphogenesis for use in applications of nanotechnology. *J. Nanosci. Nanotechnol.*, 5, 1, 120–130, 2005.

[2.105] Pappas, J.L., Theoretical morphospace and its relation to freshwater gomphonemoid-cymbelloid diatom (Bacillariophyta) lineages. *J. Biol. Syst.*, 13, 4, 385–398, 2005.

[2.106] Pappas, J.L., More on theoretical morphospace and its relation to freshwater gomphonemoid-cymbelloid diatom (Bacillariophyta) lineages. *J. Biol. Syst.*, 16, 1, 119–137, 2008.

[2.107] Pappas, J.L., Multivariate complexity analysis of 3D surface form and function of centric diatoms at the Eocene-Oligocene transition. *Mar. Micropaleontol.*, 122, 67–86, 2016.

[2.108] Pappas, J.L. and Miller, D.J., A generalized approach to the modeling and analysis of 3D surface morphology in organisms. *PLoS One*, 8, 10, e77551, 2013.

[2.109] Parkinson, J., Brechet, Y., Gordon, R., Centric diatom morphogenesis: A model based on a DLA algorithm investigating the potential role of microtubules. *Biochim. Biophys. Acta – Mol. Cell Res.*, 1452, 1, 89–102, 1999.

[2.110] Pesin, J.B., Families of invariant manifolds corresponding to nonzero characteristic exponents. *Math. USSR-Izv.*, 10, 1261–1305, 1976.

[2.111] Pfitzer, E., Untersuchungen über Bau und Entwicklung der Bacillariaceen (Diatomaceen) [Studies on construction and development of Bacillariaceae (Diatomaceae)] [German], in: *Botanische Abhandlungen aus dem Gebiete der Morphologie und Physiologie [Botanical Treatises in the Field of Morphology and Physiology]*, vol. [i]-vi, J.L.E.R. von Hanstein (Ed.), pp. 1–189, 186 pls, Adolph Marcus, Bonn, Germany, 1871.

[2.112] Pfitzer, E., Die Bacillariaceen (Diatomaceae) [Bacillariaceae (Diatomaceae)] [German], in: *Encyklopaedie der Naturwissenschaften. I. Abteilung. I. Thiel: Handbuch der Botanik*, vol. 2, A. Schenk (Ed.), pp. 403–445, Verlag von Eduard Trewendt, Breslau, 1882.
[2.113] Pickett-Heaps, J.D., The evolution of the mitotic apparatus: An attempt at comparative ultrastructural cytology in dividing plant cells. *Cytobios*, 1, 3, 257–280, 1969.
[2.114] Pickett-Heaps, J.D., Schmid, A.M.M., Edgar, L.A., The cell biology of diatom valve formation. *Prog. Phycol. Res.*, 7, 1–168, 1990.
[2.115] Prasad, A. and Ramaswamy, R., Characteristic distributions of finite-time Lyapunov exponents. *Phys. Rev. E*, 60, 3, 2761–2766, 1999.
[2.116] Rogerson, A., Defreitas, A.S.W., McInnes, A.G., Observations on wall morphogenesis in *Coscinodiscus asteromphalus* (Bacillariophyceae). *Trans. Am. Microsc. Soc.*, 105, 1, 59–67, 1986.
[2.117] Rose, D.T. and Cox, E.J., Some diatom species do not show a gradual decrease in cell size as they reproduce. *Fundam. Appl. Limnol.*, 182, 2, 117–122, 2013.
[2.118] Rosenstein, M.T., Collins, J.J., De Luca, C.J., A practical method for calculating largest Lyapunov exponents from small data sets. *Physica D-Nonlinear Phenomena*, 65, 1–2, 117–134, 1993.
[2.119] Round, F.E., Crawford, R.M., Mann, D.G., *The Diatoms, Biology & Morphology of the Genera*, Cambridge University Press, Cambridge, 1990.
[2.120] Sapriel, G., Quinet, M., Heijde, M., Jourdren, L., Tanty, V., Luo, G.Z., Le Crom, S., Lopez, P.J., Genome-wide transcriptome analyses of silicon metabolism in *Phaeodactylum tricornutum* reveal the multilevel regulation of silicic acid transporters. *PLoS One*, 4, 10, e7458, 2009.
[2.121] Savriama, Y. and Klingenberg, C.P., Beyond bilateral symmetry: Geometric morphometric methods for any type of symmetry. *BMC Evol. Biol.*, 11, 280, 2011.
[2.122] Schmid, A.-M.M., Morphogenetic forces in diatom cell wall formation, in: *Cytomechanics: The Mechanical Basis of Cell Form and Structure*, J. Bereiter-Hahn, O.R. Anderson, W.E. Reif (Eds.), pp. 183–199, Springer-Verlag, Berlin, 1987.
[2.123] Schmid, A.-M.M., Aspects of morphogenesis and function of diatom cell walls with implications for taxonomy. *Protoplasma*, 181, 1–4, 43–60, 1994.
[2.124] Schmid, A.-M.M., The wall & membrane systems in diatoms: Comment in reply to Medlin (2004). *Diatom Res.*, 20, 1, 211–216, 2005.
[2.125] Schmid, A.-M.M. and Crawford, R.M., *Ellerbeckia arenaria* (Bacillariophyceae): Formation of auxospores and initial cells. *Eur. J. Phycol.*, 36, 4, 307–320, 2001.
[2.126] Schmid, A.-M.M. and Volcani, B.E., Wall morphogenesis in *Coscinodiscus wailesii* Gran and Angst. I. Valve morphology and development of its architecture. *J. Phycol.*, 19, 4, 387–402, 1983.
[2.127] Schmidt, F. and Fleming, R.W., Visual perception of complex shape-transforming processes. *Cogn. Psychol.*, 90, 48–70, 2016.
[2.128] Shadrina, E.G. and Vol'pert, Y.L., Experience of applying plant and animal fluctuating asymmetry in assessment of environmental quality in terrestrial ecosystems: Results of 20-year studies of wildlife and anthropogenically transformed territories. *Russ. J. Dev. Biol.*, 49, 1, 23–35, 2018.
[2.129] Shannon, C.E., A mathematical theory of communication [corrected version]. *Bell Syst. Tech. J.*, 27, 379–423, 623–656, 1948.
[2.130] Shimada, Y., Tagagi, E., Ikeguchi, T., Symmetry of Lyapunov exponents in bifurcation structures of one-dimensional maps. *Chaos*, 26, 123119, 2016.

[2.131] Shirokawa, Y. and Shimada, M., Cytoplasmic inheritance of parent - offspring cell structure in the clonal diatom *Cyclotella meneghiniana*. *Proc. R. Soc. B-Biol. Sci.*, 283, 1842, 20161632, 2016.

[2.132] Shubnikov, A.V. and Belov, N.V., *Colored Symmetry*, Pergamon Press, New York, 1964.

[2.133] Steinsaltz, D., Random logistic maps and Lyapunov exponents. *Indagationes Mathematicae*, 12, 4, 557–584, 2001.

[2.134] Sterrenburg, F.A.S., Gordon, R., Tiffany, M.A., Nagy, S.S., Diatoms: Living in a constructal environment, in: *Algae and Cyanobacteria in Extreme Environments. Series: Cellular Origin, Life in Extreme Habitats and Astrobiology*, vol. 11, J. Seckbach (Ed.), pp. 141–172, Springer, Dordrecht, The Netherlands, 2007.

[2.135] Stevens, K.A., Surface tilt (the direction of slant): A neglected psychophysical variable. *Percept. Psychophys.*, 33, 3, 241–250, 1983.

[2.136] Stewart, M., Introduction to the computer image processing of electron micrographs of two-dimensionally ordered biological structures. *J. Electron Microsc. Tech.*, 9, 4, 301–324, 1988.

[2.137] Sui, N., Li, M., He, P., Statistical computation of Boltzmann entropy and estimation of the optimal probability density function from statistical sample. *Mon. Not. R. Astron. Soc.*, 445, 4, 4211–4218, 2014.

[2.138] Sumper, M., A phase separation model for the nanopatterning of diatom biosilica. *Science*, 295, 5564, 2430–2433, 2002.

[2.139] Sumper, M. and Brunner, E., Learning from diatoms: Nature's tools for the production of nanostructured silica. *Adv. Funct. Mater.*, 16, 1, 17–26, 2006.

[2.140] Sun, C.M. and Si, D.Y., Skew and slant correction for document images using gradient direction, in: *Proceedings of the Fourth International Conference on Document Analysis and Recognition*, vol. 1 and 2, IEEE, Computer Society Press, Los Alamitos, California, USA, pp. 142–146, 1997.

[2.141] Taylor, R.L.V., Attractors: nonstrange to chaotic. SIAM Undergraduate Research Online (SIURO). *Soc. Ind. Appl. Math.*, 21, 6, 72–80, 2011.

[2.142] Tesson, B., Lerch, S.J.L., Hildebrand, M., Characterization of a new protein family associated with the silica deposition vesicle membrane enables genetic manipulation of diatom silica. *Sci. Rep.*, 7, 13457, 2017.

[2.143] Tiffany, M.A., Development of valves in the marine diatom genus *Trigonium*. *J. Phycol.*, 38, s1, 35, 2002.

[2.144] Tiffany, M.A., Valve development in *Aulacodiscus*. *Diatom Res.*, 23, 1, 185–212, 2008.

[2.145] Tiffany, M.A., Valve and girdle band morphogenesis in the pseudocellate diatom species *Biddulphia biddulphiana* J.E. Smith (Boyer) and *Isthmia nervosa* Kütz. *Nova Hedwig. Beih.*, 144, 61–95, 2015.

[2.146] Tiffany, M.A. and Hernández-Becerril, D.U., Valve development in the diatom family Asterolampraceae H. L Smith 1872. *Micropaleontology*, 51, 3, 217–258, 2005.

[2.147] Tonelli, R., Mezzorani, G., Meloni, F., Lissia, M., Coraddu, M., Entropy production and Pesin-like identity at the onset of chaos. *Prog. Theor. Phys.*, 115, 1, 23–29, 2006.

[2.148] Turing, A.M., The chemical basis of morphogenesis. *Philos. Trans. R. Soc. Lond. Ser. B, Biol. Sci.*, B237, 641, 37–72, 1952.

[2.149] Tuszynski, J.A. and Gordon, R., A mean field Ising model for cortical rotation in amphibian one-cell stage embryos. *BioSystems*, 109, 3, Special Issue on Biological Morphogenesis, 381–389, 2012.

[2.150] Van De Meene, A.M.L. and Pickett-Heaps, J.D., Valve morphogenesis in the centric dilatom *Proboscia alata* Sundstrom. *J. Phycol.*, 38, 2, 351–363, 2002.

[2.151] Van De Meene, A.M.L. and Pickett-Heaps, J.D., Valve morphogenesis in the centric diatom *Rhizosolenia setigera* (Bacillariophyceae, Centrales) and its taxonomic implications. *Eur. J. Phycol.*, 39, 1, 93–104, 2004.
[2.152] Van Valen, L.M., A study of fluctuating asymmetry. *Evolution*, 16, 125–142, 1962.
[2.153] Vanag, V.K. and Epstein, I.R., Pattern formation mechanisms in reaction-diffusion systems. *Int. J. Dev. Biol.*, 53, 5–6, 673–681, 2009.
[2.154] Vartanian, M., Desclés, J., Quinet, M., Douady, S., Lopez, P.J., Plasticity and robustness of pattern formation in the model diatom *Phaeodactylum tricornutum*. *New Phytol.*, 182, 2, 429–442, 2009.
[2.155] Volcani, B.E., Cell wall formation in diatoms: Morphogenesis and biochemistry, in: *Silicon and Siliceous Structures in Biological Systems*, T.L. Simpson and B.E. Volcani (Eds.), pp. 157–200, Springer-Verlag, New York, 1981.
[2.156] von Stosch, H.A., On auxospore envelopes in diatoms. *Bacillaria*, 5, 127–156, 1982.
[2.157] Vrieling, E.G., Sun, Q.Y., Tian, M., Kooyman, P.J., Gieskes, W.W.C., van Santen, R.A., Sommerdijk, N., Salinity-dependent diatom biosilicification implies an important role of external ionic strength. *Proc. Natl. Acad. Sci. U. S. A.*, 104, 25, 10441–10446, 2007.
[2.158] Weibel, E.R., Fractal geometry: a design principle for living organisms. *Am. J. Physiol.*, 261, 6.1, L361–L369, 1991.
[2.159] Weisstein, E.W., *CRC Concise Encyclopedia of Mathematics*, 2nd ed., Chapman & Hall/CRC, London, United Kingdom, 2002.
[2.160] Weyl, H., *Symmetry*, Princeton University Press, Princeton, New Jersey, USA, 1952.
[2.161] Willis, L., Cox, E.J., Duke, T., A simple probabilistic model of submicroscopic diatom morphogenesis. *J. R. Soc. Interface*, 10, 83, 20130067, 2013.
[2.162] Willis, L., Page, K.M., Broomhead, D.S., Cox, E.J., Discrete free-boundary reaction-diffusion model of diatom pore occlusions. *Plant Ecol. Evol.*, 143, 3, 297–306, 2010.
[2.163] Wolf, A., Swift, J.B., Swinney, H.L., Vastano, J.A., Determining Lyapunov exponents from a time series. *Physica D-Nonlinear Phenomena*, 16, 3, 285–317, 1985.
[2.164] Xiao, Y.H., Xiong, M.M., Wang, W., Wang, H., Emergence of symmetry in complex networks. *Phys. Rev. E*, 77, 6, 066108, 2008.
[2.165] Yang, W., Lopez, P.J., Rosengarten, G., Diatoms: Self assembled silica nanostructures, and templates for bio/chemical sensors and biomimetic membranes. *Analyst*, 136, 1, 42–53, 2011.
[2.166] Yang, X.-Y., Chen, L.-H., Li, Y., Rooke, J.C., Sanchez, C., Su, B.-L., Hierarchically porous materials: Synthesis strategies and structure design. *Chem. Soc. Rev.*, 46, 2, 481–558, 2017.
[2.167] Zabrodsky, H., Peleg, S., Avnir, D., Symmetry as a continuous feature. *IEEE Trans. Pattern Anal. Mach. Intell.*, 17, 12, 1154–1166, 1995.

3

On the Size Sequence of Diatoms in Clonal Chains

Thomas Harbich

Independent Researcher, Am Brüdenrain, Weissach im Tal, Germany

Abstract

The formation of clonal chain-shaped diatom colonies is a widespread phenomenon. It has already been shown that with synchronous division of the diatoms, the size sequence of the diatoms in such a colony can be described by a Lindenmayer system, as it is particularly known from plant growth modeling. Against this background, this paper addresses the question of whether the theoretically derived sequence can be confirmed by observed fragments of diatom chains. For this purpose, an alternative algorithm for calculating the size sequence is derived, from which essential properties of this sequence result. It is shown that an identified coincidence of the size sequence of a fragment with the theoretical sequence of a certain generation is also preserved in its subsequent generations. In addition, the number of matches may also increase from generation to generation. As it is typically not feasible to assign an absolute size index to the diatoms of a found fragment, the sequence of differences in the size indices of neighboring diatoms is investigated. This sequence can be visualized as a Heighway dragon curve which exhibits exact self-similarity and fractal structure. An unknown absolute size increases the number of possible localizations of the fragment within the theoretical size sequence. If no synchronization of the cell divisions is achieved by external constraints, then the occurrence of the loss of synchronicity of cell divisions in a considered segment of the chain is characterized by random processes. The number of generations within which the diatoms in a section of the diatom chain divide synchronously can vary considerably. Assuming a normal distribution for the generation times, a rule of thumb for the beginning of the loss of synchronicity is given. The observability of the size sequence by means of fragments of a diatom chain using light microscopy depends on the species and is limited by disturbing effects. Exemplarily the evaluation of a fragment consisting of 25 diatoms of the genus Eunotia is demonstrated. Another example shows the beginning loss of synchronicity of cell divisions.

***Keywords*:** Clonal chain, size sequence, Lindenmayer system, synchronous cell division, Heighway dragon curve

Email: thomas.harbich@diatoms.de
Thomas Harbich: https://www.researchgate.net/profile/Thomas_Harbich, https://diatoms.de/en/

Vadim Annenkov, Joseph Seckback and Richard Gordon (eds.) Diatom Morphogenesis, (69–92) © 2022 Scrivener Publishing LLC

A string (or "sequence") of symbols represents the sizes of a complete chain-shaped colony or a section thereof. Under the assumption of synchronous divisions, the string itself, which is formed after n iterations, is subsequently referred to as the "nth generation" for the sake of simplicity. The replacement rules P specify how this string changes from generation to generation. The initial sequence determines with which string the calculation starts. The following notation for the symbols is adopted from Ussing *et al.* [3.32]. For a better understanding, imagine the colony and its string positioned horizontally. A diatom whose left valve is larger than its right valve and whose larger valve is characterized by the size index k is referred to as L^k. Accordingly, a diatom whose right valve is larger than its left valve and whose larger valve is given by the size index k is referred to as R^k. The set of symbols consists of the union of the sets $\{L^k \mid k = 0, 1, 2, \ldots, k_{max}\}$ and $\{R^k \mid k = 0, 1, 2, \ldots, k_{max}\}$. Sizes and diatom orientations of a diatom chain are characterized by a string of symbols $G = g_1\, g_2\, g_3 \cdots g_j$. If a diatom divides whose orientation and size is given by L^k, then two diatoms are formed for $k < k_{max}$ according to the following replacement rule (production):

$$L^k \mapsto L^k\, R^{k+1}. \tag{3.2}$$

When the larger valve of the diatom is on its right side, it is merely oriented differently to the observer and one gets for $k < k_{max}$:

$$R^k \mapsto L^{k+1}\, R^k. \tag{3.3}$$

With these rules, the Lindenmayer system is deterministic and context-free. The term "context-free" means that the replacement rules only refer to an individual symbol and not to its neighbors. The initial sequence corresponds to a single diatom of maximum size

$$\omega = L^0 \tag{3.4}$$

where the orientation L is chosen. Starting with the initial sequence, the productions are iteratively applied in parallel to all elements of the string. Thus, a sequence of strings G_j is created, which describes the colony after the j-th iteration, corresponding to the j-th generation. Using the initial sequence according to (3.4) one obtains

$$\begin{aligned} G_0 &= \omega = L^0 \\ G_1 &= L^0\, R^1 \\ G_2 &= L^0\, R^1\, L^2\, R^1 \\ G_3 &= L^0\, R^1\, L^2\, R^1\, L^2\, R^3\, L^2\, R^1 \end{aligned} \tag{3.5}$$

etc.

The generated strings correspond to snapshots between cell divisions.

3.2 Mathematical Analysis of the Size Sequence

3.2.1 Alternative Method for Calculating the Size Sequence

In this part, the size sequence of diatoms in a clonal chain under the condition of synchronous divisions shall be examined mathematically in more detail. We build on the analysis of Ussing *et al.* [3.32] to derive an alternative presentation that allows a better understanding of their characteristics. It is advantageous to write the production rules (3.2) and (3.3) using an operator P in the following form:

$$\mathrm{P}(\mathrm{L}^k) = \mathrm{L}^k\,\mathrm{R}^{k+1} \tag{3.6}$$

$$\mathrm{P}(\mathrm{R}^k) = \mathrm{L}^{k+1}\,\mathrm{R}^k. \tag{3.7}$$

In the case of synchronous divisions, the next generation size sequence is calculated by applying the production rules to all diatoms:

$$\mathrm{P}(\mathrm{G}) = \mathrm{P}(g_1\, g_2\, g_3 \,\ldots.\, g_j) = \mathrm{P}(g_1)\,\mathrm{P}(g_2)\,\mathrm{P}(g_3)\,\ldots.\,\mathrm{P}(g_j) \tag{3.8}$$

To derive a representation different from the L-system, two operators are introduced. The operator S mirrors a string with elements of the set of symbols. It needs to be noted that the alignment of the diatoms must also be mirrored:

$$\mathrm{S}(g_1\, g_2\, g_3 \,\ldots.\, g_j) = \mathrm{S}(g_j)\,\mathrm{S}(g_{j-1})\,\ldots.\,\mathrm{S}(g_3)\,\mathrm{S}(g_2)\,\mathrm{S}(g_1) \tag{3.9}$$

$$\mathrm{S}(\mathrm{R}^k) = \mathrm{L}^k \tag{3.10}$$

$$\mathrm{S}(\mathrm{L}^k) = \mathrm{R}^k. \tag{3.11}$$

In Figure 3.3, the mirroring is shown by the example of generation G_2.

By using Equations (3.6) to (3.11), one shows that

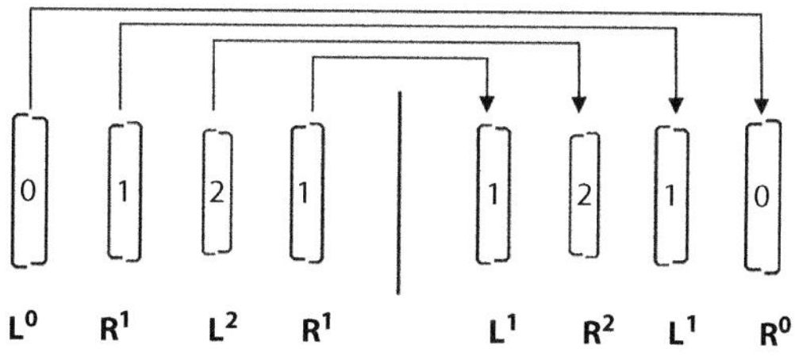

Figure 3.3 In the left half of the figure, diatoms of the chain $\mathrm{L}^0\,\mathrm{R}^1\,\mathrm{L}^2\,\mathrm{R}^1$ are shown. A mirror image with respect to the line in the center of the figure can be seen in the right half of the picture. The mirrored chain is obtained by reversing the sequence as indicated by the arrows. In addition, L must be exchanged with R and R with L.

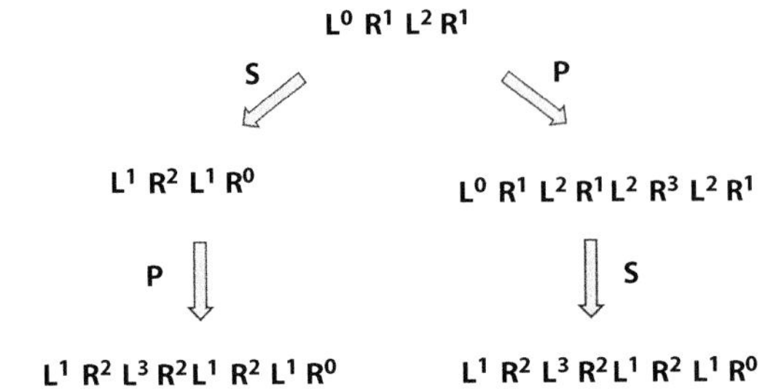

Figure 3.4 The string L^0 R^1 L^2 R^1 is in the left branch first mirrored and then the production rules are applied. In the right branch, the operations are performed in reverse order. The result is identical.

$$\mathrm{S\,P} = \mathrm{P\,S}. \tag{3.12}$$

Like in the usual multiplication of numbers, the order of the operands is irrelevant. As is to be expected, it makes no difference whether one mirrors a diatom chain and then calculates the next generation or whether one first determines the next generation and mirrors the result afterwards. This is illustrated by an example in Figure 3.4.

Furthermore, the operator I is defined, which increments all size indices of a string:

$$\mathrm{I}(g_1\, g_2\, g_3 \, \ldots .\, g_j) = \mathrm{I}(g_1)\, \mathrm{I}(g_2)\, \mathrm{I}(g_3) \ldots \mathrm{I}(g_j) \tag{3.13}$$

$$\mathrm{I}(\mathrm{R}^k) = \mathrm{R}^{k+1} \tag{3.14}$$

$$\mathrm{I}(\mathrm{L}^k) = \mathrm{L}^{k+1}. \tag{3.15}$$

As with the productions, $k < k_{max}$ must be assumed. Apparently, these commutation relations hold:

$$\mathrm{I\,P} = \mathrm{P\,I} \tag{3.16}$$

$$\mathrm{I\,S} = \mathrm{S\,I}. \tag{3.17}$$

In order to clarify which string an operator applies to, the concatenation of two strings is symbolized by the sign "$\frown$". With the initial sequence $G_0 = \omega = L^0$ one obtains according to the above equations for the $(n + 1)$th generation

$$\mathrm{G}_{n+1} = \mathrm{P}^{n+1}(\omega) = \mathrm{P}^n\, \mathrm{P}(\mathrm{L}^0) = \mathrm{P}^n\, (\mathrm{L}^0\, \mathrm{R}^1) = \mathrm{P}^n(\mathrm{L}^0) \frown \mathrm{P}^n(\mathrm{R}^1) \tag{3.18}$$

$$= \mathrm{P}^n(\mathrm{L}^0) \frown \mathrm{P}^n(\mathrm{S\,I}(\mathrm{L}^0\,)) = \mathrm{P}^n(\mathrm{L}^0) \frown \mathrm{I\,S}\,(\mathrm{P}^n(\mathrm{L}^0\,))$$

where $R^1 = S\,I\,(L^0)$ has been used. Thus,

$$\mathrm{G}_{n+1} = \mathrm{G}_n\, \mathrm{I\,S\,G}_n. \tag{3.19}$$

As long as the maximum size index is not exceeded in one generation, i.e., as long as $n < k_{max}$, Equation (3.19) is valid. This is assumed in the following. The relationship can be executed iteratively or recursively. To proceed from one generation to the next, one mirrors and increments its string and attaches it to the existing one. It is therefore sufficient to consider a single sequence in which the string of the nth generation is obtained by cutting off after 2^n symbols.

The initial sequence L^0 was explicitly used in the derivation of (3.19), but this choice does not represent a significant limitation. If you choose R^0 as the initial value, then the result would be the mirrored strings due to (3.12). However, L^0 proves to be a more clever choice if you are used to writing from left to right, because then you can advance to the next generation by extending the existing string at its right end. If you start the calculation with a size index $j > 0$, then the indices of the whole chain are increased by j because of (3.16). Besides the aspect of a simple calculation of the generations, the representation (3.19) offers the advantage that essential relationships can be derived easily. In the examples (3.5), it is noticeable that the alignments of diatoms alternate from the first generation onwards. This can be seen from the replacement rules, but it also follows from (3.19) very simply by induction. Thereby, the statement for G_1 is the base case. If one keeps the alignment of the initial value in mind, then one can restrict oneself to the sequence of the size indices. Equation (3.19) can be used without change to calculate the size indices. When mirroring, just the order of the indices must be reversed. Denoting the size indices of the n-th generation with a_i^n, where i takes the values 1, ..., 2^n, it follows from (3.19)

$$a_i^{n+1} = \begin{cases} a_i^n & for\ i = 1, 2, \ldots, 2^n \\ a_{2^{n+1}-i+1}^n + 1 & for\ i = 2^n + 1, \ldots, 2^{n+1} \end{cases} \tag{3.20}$$

where the initial value corresponding to the chosen initial sequence is $a_1^0 = 0$.

The differences of the size indices of adjacent diatoms can only take the values +1 or −1, which can as well be proven from (3.19) or (3.20) by induction, starting again with the proof for G_1.

The maximum size index of the n-th generation equals n. It appears one time and is in the upper half of each sequence because of incrementing. In the limit $n \rightarrow \infty$ the position of the maximum index converges to 2/3 of the length of the string. At this location, the applicability of the model ends after the k_{max}-th generation.

3.2.2 Self-Similarity and Fractal Structure

A remarkable property of the generations is the self-similarity which is increasing from generation to generation. At each division, one diatom is replaced by two adjacent ones, one with the same size index and the other with the incremented size index. A generation shown in half width will differ only slightly in relative size from its predecessor if the number of generations is large enough. As the first half of a sequence is identical to the previous sequence, it follows that with increasing generation, it becomes similar to the entire generation. Accordingly, the similarity of the upper half of a sequence to that sequence can be

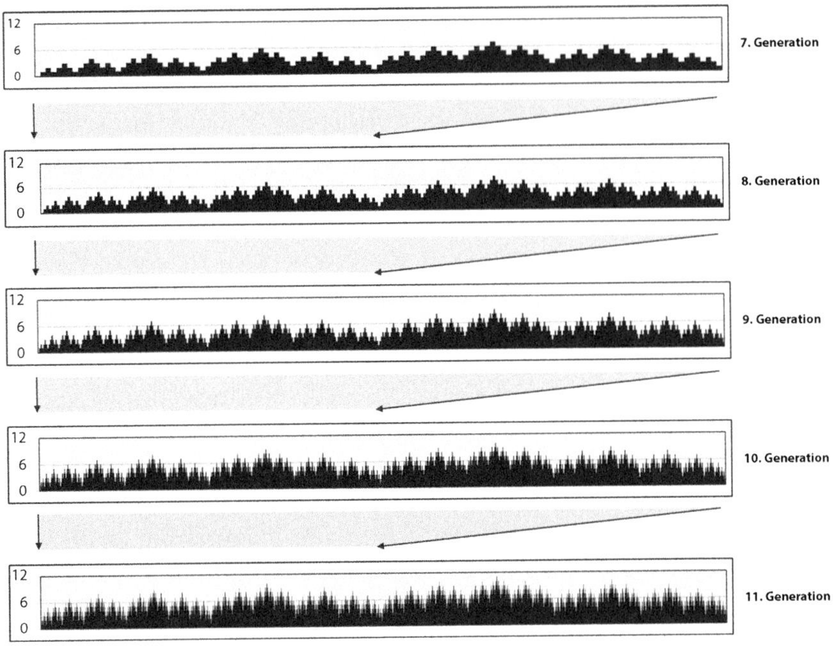

Figure 3.5 Bar chart of the size indices from the 7th generation (128 diatoms) to the 11th generation (2,048 diatoms), whereby the diagrams have the same widths. The first half of each sequence is identical to the previous generation. In view of the resolution of the picture, a difference from the 11th to the following generation would be difficult to see.

proved. Figure 3.5 shows the size of diatoms from the 7th to the 11th generation as a bar chart.

A fractal structure can further be seen here, which is addressed again in the context of the sequence of differences between neighboring size indices. Self-similarity and fractal structure are typical for L-systems (e.g., Prusinkiewicz and Lindenmayer [3.26]).

3.2.3 Matching Fragments to a Generation Based on Known Size Indices of the Fragment

The measurement of the cell sizes of a complete colony would make it possible to test the L-system theory. Since in practice only fragments of complete colonies are found, this will only be possible in sections, at least for higher generations. In addition, the question arises as to whether one can draw conclusions from the fragment regarding a certain generation of development and an unambiguous location within the theoretical sequence. It is assumed that the string of the size indices of such a fragment was determined by measurement. Now, it can be checked whether it can be found in G_0, G_1, G_3, … etc. If no match is found, the string must be mirrored and matching reattempted. If the diatom follows the MacDonald-Pfitzer rule and the cell divisions take place synchronously, then the string M should be found in a certain sequence G_j for the first time. With longer strings, one can assume that they are not mirror symmetric and that one finds a match only in one orientation (left/right reading direction). According to the calculation scheme (3.19), for all

$$d_i^n = a_{i+1}^n - a_i^n \tag{3.23}$$

where i takes the values 1, 2, …, $2^n - 1$. It follows from Equation (3.20) that

$$d_i^{n+1} = \begin{cases} d_i^n & \text{for } i = 1, 2, \ldots, 2^n - 1 \\ 1 & \text{for } i = 2^n \\ -d_{2^{n+1}-i}^n & \text{for } i = 2^n + 1, \ldots, 2^{n+1} - 1. \end{cases} \tag{3.24}$$

The starting value results from the first generation of size indices G_1 in accordance with the chosen initial sequence to $d_1^1 = 1$. From the sequence according to Equations (3.23) and (3.24), the original size sequence can be obtained by calculating the partial sums starting with the initial value $a_1^0 = 0$ As the size indices of neighboring diatoms always differ by only one size value, the elements of the difference sequence consist only of the values +1 and −1. The sequence of differences can also be created by hand easily. The next generation is obtained by first appending a 1 to the existing generation and then the mirrored inverted generation.

The sequence of the differences has a striking geometric representation. A curve consisting of straight segments of equal length is created by drawing one segment and continuously appending a segment for each element of the sequence (3.24). Thereby, one performs in the sense of a turtle graphics a 90° clockwise rotation at the value +1 and a 90° counterclockwise rotation at the value −1. In this way, the so-called dragon curve (Figure 3.8) is obtained (see Tabachnikov [3.31]).

Since there are different variants of dragon curves, one calls this dragon curve after the persons who described it first, more exactly as the Heighway dragon or Harter-Heighway dragon. In Sloan's On-Line Encyclopedia of Integer Sequences (OEIS), you can find the sequence of the Heighway dragon curve under A014577 (Sloane [3.29]). Instead of "−1", the character "0" is used in A014577. Obviously, the size sequence G_n can also be directly interpreted as a dragon curve. One passes through the sequence, takes a step forward for each element, and performs a 90° clockwise rotation when the next element is larger than the previous one and a counterclockwise rotation when it is smaller.

The theory of the dragon curve was developed in particular by Davis Chandler and Donald J. Knuth (Chandler and Knuth [3.2]; reprinted with revisions in Knuth [3.12]). The occurrence of the mathematical structure of the dragon curve shows that it plays a role in biological systems, albeit not in their geometric representation. Similar to Pascal's triangle, which serves as a visualization of the number of diatoms of a certain size depending on the generations, the dragon curve illustrates the size differences of adjacent diatoms in a chain-shaped diatom colony. The interpretation of (3.24) as a dragon curve makes it possible to use known knowledge about the dragon curve. For more information on its often astonishing properties, see Tabachnikov [30]. The dragon curve is an exact self-similar geometric fractal. Therefore, fractal structure and self-similarity of the size sequence are related to these properties of the dragon curve.

the theoretical sequence according to (3.24). As a result of the exact self-similarity of the dragon curve, this pattern and its mirrored pattern (the signs are reversed by the mirroring) can be found in the next generation. Then, the fragment has two positions with a match. With each generation, the number of matches doubles. The uncertainty of the absolute assignment of length to a size index has its price in the form of additional positions of matching with the theoretical size sequence.

3.2.6 Synchronicity of Cell Divisions

If the cell divisions occur synchronously and no cells die, then the described size sequence is maintained until the smallest possible size is reached and cell enlargement becomes necessary. As this is an idealization, the question arises how to choose the conditions favorably. When inoculating a culture with a single diatom, the cells will develop in parallel over a period of time, i.e., they will have a similar physiological stage of development. Under natural conditions, the changing of daylight and darkness may favor synchronous cell division. With a suitable process, such synchronicity could be maintained over a longer period of time, at least for a large proportion of the diatoms of a culture. For this purpose, an accumulation of physiologically young cells is achieved by means of a long period in darkness. Subsequent cultivation in a light-dark cycle at a photoperiod and intensity favorable to the species may result in synchronization of the vegetative cell divisions (Pirson and Lorenzen [3.25]; Paasche [3.21]; Paasche [3.22]). In cultures which are kept under silica deprivation, synchronization can be achieved by targeted silica addition (Lewin *et al.* [3.13]; Busby and Lewin [3.1]; Darley and Volcani [3.3]; Davis, Harrison and Dugdale [3.4]). However, such forced synchronicity does not include all cells of a culture, which limits the lengths of diatom chains according to theory. In a natural environment and certainly in most cultures, there will be a loss of synchronicity after some time due to fluctuations in generation time, assuming no long-range communication in a colony. Subsequently, a simple heuristic consideration will be presented, which relates the size of these fluctuations to the number of synchronous divisions. Strictly speaking, it only applies to lighting of constant intensity. If the cell division comes to a standstill in the dark phase, then one can limit oneself in a first approximation to the observation of the light phases, i.e., mathematically remove the dark phases from the time axis. It is assumed that the periods t_1, t_2, t_3, ... required for successive cell divisions are statistically independent random variables. It has to be noted that this is not generally true, because in different types of cells and various organisms there exist correlations between generation times (see, e.g., Hejblum *et al.* [3.11]). A dependence of the generation time on the size of the cell will be excluded, an assumption that is not fulfilled in every case (Laney, Olson, and Sosik [3.16]). Since the generation time depends on many external parameters such as irradiation intensity and nutrient concentrations, these are assumed to be constant over time. Furthermore, a normal distribution is anticipated. With the expectation value of the generation time μ and the standard deviation σ, the density function for the duration t of a complete vegetative cycle is

$$\mathrm{f}_1(t)=\frac{1}{\sqrt{2\pi\sigma^2}}e^{-\frac{(t-\mu)^2}{2\sigma^2}}. \tag{3.25}$$

The density function $f_2(t)$ for the period $t = t_1 + t_2$, which is required for two consecutive divisions is also normally distributed with the expectation value 2μ and the standard deviation $\sqrt{2}\sigma$. In the general case of k consecutive cell divisions, the density function for the duration $t = \sum_{i=1}^{k} t_i$ of k completed cell divisions is calculated accordingly:

$$f_k(t) = \frac{1}{\sqrt{2\pi k\sigma^2}} e^{-\frac{(t-k\mu)^2}{2k\sigma^2}}. \tag{3.26}$$

Thereby, the expectation value is $k\mu$ and the standard deviation $\sqrt{k}\sigma$. Consider the daughter cells of a diatom after several divisions. As long as μ is large compared to $\sqrt{k}\sigma$, there is no significant overlap between adjacent density functions $f_k(t)$ and $f_{k\pm1}(t)$. Daughter cells of the k-th generation are completely developed with only minor deviations at the time $t = (k + 1/2)\mu$. The $(k + 1)$-th cell division, on the other hand, is almost never completed. However, if μ is small compared to $\sqrt{k}\sigma$, there is a strong overlap of $f_k(t)$ with the neighboring density functions $f_{k\pm1}(t)$ and the cells belong with non-negligible probability to different generations, which means loss of synchronicity. There is a wide time range in which the loss of synchronicity can arise. It can be characterized by noting that $\mu/\left(\sqrt{k}\sigma\right)$ lies between these extremes, approximately in the range from 2 to 4. It is remarkable that synchronicity loss would happen with a generation number k four times as large if the ratio σ/μ could be halved. In Appendix 3B, the probability is derived that at a given time a diatom is found after the k-th completed division and before the $k + 1$-th completed division. Therewith, the given rule of thumb can be substantiated.

Starting with a single diatom within a chain, a section with 2^k diatoms is formed after k generations of synchronous cell divisions. The common ancestors of neighboring diatoms date back 1 to k generations in this section (binary tree). The greater k, the higher is the probability of asynchronous divisions within the segment. On the other hand, if the $\mu/\left(\sqrt{k}\sigma\right)$ is sufficiently large, then it is probable that the considered section fulfills the synchronicity condition overall. In growing chain-shaped diatom colonies, therefore, synchronous sections should exist whose typical length grows with μ/σ. An opportunity for further study is a Monte Carlo simulation of the temporal evolution of a chain. It could answer the question of the statistical distribution of these synchronous sections.

3.3 Observations

3.3.1 Challenges in Verifying the Sequence of Sizes

As a basic prerequisite for the size sequence according to Equation (3.19), the synchronicity of the cell divisions was addressed. In the practical verification of the match of an observed size sequence based on a fragment of a colony, the problem arises in a different form. The presence of different division stages shows the loss of synchronicity, on the other hand, almost identical division stages do not prove that the fragment developed exclusively by synchronous divisions. Further properties of the theoretical sequence can be utilized as indicators for synchronous divisions. One expects an alternation of the orientations of the diatoms and size changes of neighboring diatoms around the minimum length difference. Both criteria provide necessary but not sufficient prerequisites for synchronous divisions. As one cannot assume the synchronicity of an entire clonal colony, it is to be expected that the theoretical sequence of sizes can only be found within certain sections, not in all fragments found.

The assignment of size indices to the measured size was discussed as a major difficulty in proving the size sequence, which led to the analysis of differences in size. Besides this, various other challenges may arise:

- Small differences in length in certain species that make light microscopic examination difficult.
- Occurrence of such short fragments that a match of the size sequence with the theory is not conclusive. For only a few diatoms, coincidence with the theoretical sequence can also be based on probability.
- Difficulties in measuring sizes due to non-planar contact of the fragment to the substrate.
- Dynamic changes of colonies in motile diatoms. These include in particular the migration of diatoms.

The choice of the species studied and the boundary conditions of cultivation must be chosen accordingly. In *Fragilaria* sp. and *Euntia* sp. it was not difficult to measure the lengths and size differences of neighboring diatoms, whereas the size differences in *Melosira varians* turned out to be too small. In the examined *Eunotia* sp. the absolute differences of the distance from apex to apex of neighboring diatoms were about 1 μm and thus within a light microscopically well measurable range. Fan-shaped *Gomphonema* colonies examined by the author were ruled out as a study object as they contained only a small number of diatoms that did not overlap in the image. The specific characteristics of the investigated *Eunotia* sp. will be addressed later.

For a measurement of the size sequence of a fragment, samples from waters or cultures can be used. In cultures, there are more favorable conditions for the formation of long chains. In particular, the breaking of chains by natural water flow can be prevented by cultivation. In addition, many fragments of a clone of a species are available in culture.

3.3.2 Materials and Methods

Diatoms of the genus *Eunotia* were isolated from a pond in Stuttgart (Hohenheim), Germany (48°42'34.0"N 9°12'29.1"E). It is to be assumed that this is *Eunotia glacialis* Meister 1912. Since no SEM images were available, a residual doubt remains. However, the documented ranges (Mattern *et al.* [3.19]) support this identification. The diatoms were cultivated clonally in Petri dishes made of polystyrene. Every other week a new culture was inoculated with diatoms from the respective parent culture. The diatoms were grown in sterilized fresh water enriched with F/2 according to Guillard [3.8]. The cultures were kept at light intensities ranging from 350 lx to 450 lx, with the light phase lasting 11 hours and the dark phase 13 hours.

The images of fragments of the diatom chains were taken directly in culture using a 40× objective (planar; NA = 0.55) of an inverted microscope (Zeiss Axio Vert.A1) in bright-field. This avoided the risk of the chains breaking when transferred to a microscope slide. Methods of image processing, in particular focus stacking and photo stitching were used. The software to increase the depth of focus was CombineZP (Hedley [3.10]). The images were combined using the Microsoft Image Composite Editor (Microsoft [3.20]). This second image processing is not necessary for the numerical evaluations but enables the overview images (Figures 3.9 and 3.12).

3.3.3 Investigation of the Size Sequence of a *Eunotia* sp.

The examined *Eunotia* sp. shows large absolute differences in the length of neighboring diatoms, which makes it a suitable object of investigation. In culture, the length of the band-shaped colonies is limited by various effects. *Eunotia* is motile and therefore there is a migration especially at the ends of a chain. It has been observed (Harbich [3.9]) that occasionally diatoms within the chain detach and move away, causing a break in the chain. As a result of the lateral expansion during cell division, the chain can close again visually, whereby an originally alternating alignment of the diatoms is lost (Harbich [3.9]). The number of diatoms that leave the colony depends significantly on the light intensity. The activity of movement increases with increasing light intensity and comes to a complete standstill in the dark (Harbich [3.9]). Time-lapse photographs show that under the described culture conditions the vegetative divisions of *Eunotia* sp. continue in the dark phase. I would like to mention that it depends on the diatom species and possibly experimental parameters whether cell divisions occur during the light or dark phase (Paasche [3.21]). With the light intensity used, a small net increase in the length of the chain was observed in the bright phase. Above a critical intensity, the increase in length as a result of cell division can exceed the shortening of the chains due to migration on a daily average. The photoperiod and light intensity used represent a compromise that enables sufficiently long chains.

The strong adhesion of the investigated *Eunotia* sp. in petri dishes made of polystyrene often prevents the expansion of long chains parallel to the substrate. When they expand after cell division, they do not slide over the substrate, but detach locally and form a bulge which is high in comparison to the cell size. After that, the fragment usually breaks, causing the fragments of the original chain to lie on top of each other. Sometimes, long chains also tip to the side and are no longer observable in girdle band view. In some densely populated cultures, there were chains of diatoms in spiral structures. In this situation, there are only a few contact points to the substrate, which probably explains the often long chains. In any case, migration seems to limit the length of the colonies. In *Melosira varians* cultures, on the other hand, chains with several thousand diatoms could be observed when water movement was avoided. An example of a *Eunotia* chain of 25 diatoms is shown in Figure 3.9.

Several focal planes were combined and two such superimposed images were composed by means of photo stitching. Since reference values for the allocation of lengths to size index are missing, only the length differences are examined. One recognizes that the diatoms have the required alternating orientation. Different transapical sizes indicate slightly different developmental stages of the diatoms, which indicate an incipient loss of synchronicity. The measurement of the lengths of the diatoms, i.e., the apical lengths of the larger valves and

Figure 3.9 *Eunotia* chain at the bottom of a Petri dish of a culture. Looking at the contours, one can recognize that the alignment of the diatoms alternates.

subsequent calculation of the length differences of neighboring diatoms leads to 24 values, which are illustrated in Figure 3.10 as a bar chart.

The average of the amounts of these differences is 1.01 µm, and the standard deviation is 0.17 µm. In evaluating the fluctuations in the length of the bars, it should be remembered that they were obtained by calculating differences. The causes for the fluctuations are natural deviations and measurement errors. The low optical resolution as well as a perspective shortening due to slightly different inclinations of the diatoms to the substrate can be taken into consideration. The differences in length of neighboring diatoms are about 1.3% of the length of the diatoms. A foreshortening in this magnitude can also be obtained by an inclination angle of approx. 9 degrees. The values of the sequence of differences which are taken from Figure 3.10 by rounding to the absolute value 1 are as follows:

1 1 −1 1 1 −1 −1 1 1 1 −1 −1 1 −1 −1 1 1 1 −1 1 1 −1 −1 −1

This string coincides with the theoretical sequence of differences according to (3.24), where the fragment begins at the first element of the sequence. By supplementing this string up to the complete fifth generation, one obtains

1 1 −1 1 1 −1 −1 1 1 1 −1 −1 1 −1 −1 1 1 1 −1 1 1 −1 −1 −1 ***1 1 −1 −1 1 −1 −1***

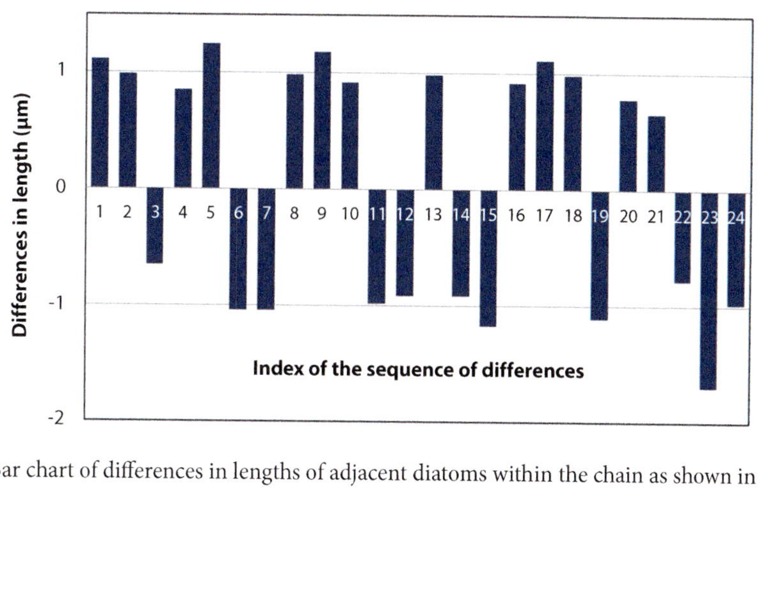

Figure 3.10 Bar chart of differences in lengths of adjacent diatoms within the chain as shown in Figure 3.9.

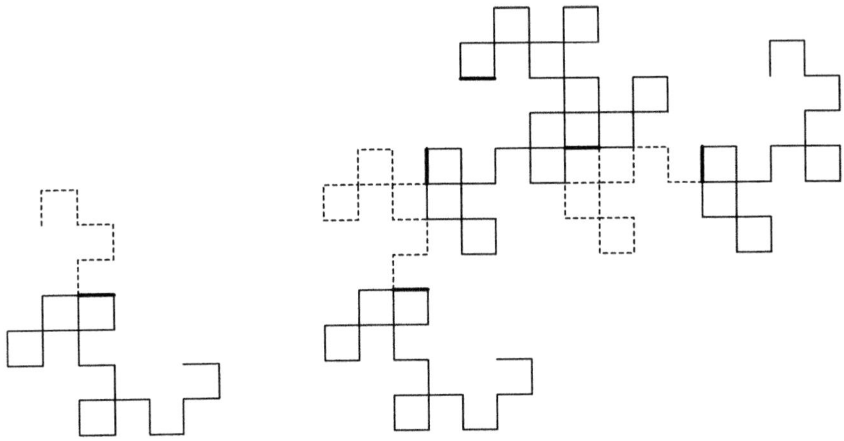

Figure 3.11 Interpretation of the observed sequence of differences (Figure 3.10) as a dragon curve. Since neighboring size indices differ by only one step, the values are rounded to +1 and −1 for this purpose. On the left, the dragon curve is shown, in which the fragment appears for the first time (fifth generation), whereby the dotted part of the curve corresponds to the added values. To the right is the seventh generation, which consists of four such dragon curves of the fifth generation.

The added values are typed in bold italics. For illustration, the fifth generation is visualized as a dragon curve on the left side of Figure 3.11.

The dotted part of the curve corresponds to the supplemented values and the bold connecting piece links the two sections. To the right is the example of the seventh generation dragon curve, in which the string of the fragment appears four times.

3.3.4 Synchronicity

A synchronization of cell divisions by the light-dark cycle could not be observed in *Eunotia* sp. cultures. This could be due to the low light intensities used to achieve long chains. With single diatoms or chains of a few individuals, synchronicity has been maintained over several generations. In contrast, in the case of longer chains in time-lapse shots, one can usually see sections of different lengths whose cells divide almost simultaneously and then expand in a transapical direction. The length of sections with uniform division stages can be very different. The investigated chain containing 25 diatoms had similar division stages and was apparently formed by synchronous divisions, but chains of double or even triple length can also show a comparable pattern. Due to non-uniform alignment to the objective and strong differences in distance, however, it is very difficult to prove the theoretically expected size sequence for such long fragments. On the other hand, even short fragments often exhibit different cell stages. An example is shown in Figure 3.12.

In the left and right image parts are diatoms after a recent division (Figure 3.12). The assignment to the parent diatom is indicated by brackets. The diatoms in between are either fully developed or have already started a new cell division. It is not possible to discern from the appearance whether they lag behind the development of the outer diatoms or whether they are already more advanced. A consistent size sequence with respect to the theory is

Figure 3.12 *Eunotia* chain at the bottom of a Petri dish of a culture. A chain was chosen that exhibited different cell division stages.

achieved under the assumption that the young pairs marked with brackets originated from parental cells belonging to the same generation as the diatoms in between. The unmarked cells in the middle region then lag behind the outer cells. Using the outer thecae at the ends of the brackets to determine the size differences, this string of 18 length differences is obtained:

1 1 −1 −1 1 1 1 −1 −1 1 −1 −1 1 1 1 −1 1 1

which fits correctly into the theoretical sequence of the fifth generation:

1 1 −1 1 1 −1 −1 1 1 1 −1 −1 1 −1 −1 1 1 1 −1 1 1 ***−1 −1 −1 1 1 −1 −1 1 −1 −1 1 1 1 −1***.

The added differences are again marked in bold italics. As this chain was created by five divisions from a single cell, synchronicity was obviously lost in less than five generations. In the case of the investigated chain of 25 diatoms, 7 diatoms were missing in the fragment to prove a synchronous division over five generations. These exemplary studies are not sufficient to estimate μ/σ.

3.4 Conclusions

By a suitable algorithm for the calculation of the size sequence of the diatoms in a clonal chain, which is derived from the L-system, different characteristics of this sequence can be derived in a simple way. In particular, one can answer the question of multiple matches of the size sequence of a given fragment of a diatom chain with the theoretical size sequence. If the assignment of size index to measured length from apex to apex is not known, then it is recommended to consider the differences between adjacent size indices. With this method, it is possible in practice to assign a fragment of a diatom colony to the theoretical sequence. The occurrence of the mathematical structure of the dragon curve in the sequence of the differences of neighboring size indices underlines the importance of fractal structures in general and the dragon curve in biological systems in particular.

The loss of synchronicity limits the occurrence of size sequences that correspond to the theory, but it is to be expected that synchronous conditions prevail in certain sections of a diatom chain. *Eunotia* sp. proves to be well suited for size sequence investigations, but due to its motility, shortenings at the ends of chains and breaks within chains often occur. In order to produce sufficiently long fragments, the culture conditions must be chosen appropriately.

The experimental verification of the theoretical size sequence also represents a proof of the MacDonald-Pfitzer rule, whereby the expected size reduction for all daughter cells can be shown over several generations.

Acknowledgements

I would like to thank Hans Fischer for his support in the probabilistic considerations of synchronization and the confirmation of the results presented. Furthermore, I would like to thank Diethard Michaelis for his hints on the generation of dragon curves with the help of L-systems and on the literature regarding these curves.

Appendix 3A L-System for the Generation of the Sequence of Differences in Size Indices of Adjacent Diatoms

The sequence of differences in size indices of neighboring diatoms according to (3.24) was derived from an L-system for size indices. However, this sequence can alternatively be generated directly by various L-systems. To derive such an L-system from the replacement rules, we consider the set of symbols, which comprises the characters R, L, +, and –. In relation to a chain of diatoms, the symbol L shall again designate a diatom whose left valve is larger than its right and R a diatom whose right valve is larger than its left. The sign + refers to the sizes of neighboring diatoms and means that the diatom on the right has a higher size index than the diatom on the left. Correspondingly, the symbol – stands for the opposite case. We insert the signs + and – between the characters for the diatoms, so that R + L stands for two diatoms in which the right diatom (with orientation L) is larger by one size index than the left diatom (with orientation R). In analogy the meaning of R – L, L + R, and L – R is defined. The replacement rule (3.2) states that a diatom with the orientation L produces two diatoms in which the right one has a size index one step higher than the left one. Taking into account the orientations, the replacement rule reads

$$\mathrm{L} \mapsto \mathrm{L} + \mathrm{R}. \tag{A3.1}$$

The information of the absolute size of the diatoms is omitted. Analogously, we derive from (3.3)

$$\mathrm{R} \mapsto \mathrm{L} - \mathrm{R}. \tag{A3.2}$$

Based on the explicit consideration of the possibilities according to (3.2) and (3.3), one can easily convince oneself that existing + and – signs remain valid when applying the replacement rules, so that

$$+ \mapsto + \tag{A3.3}$$

$$- \mapsto - \tag{A3.4}$$

Using ω = L as initial sequence, these sequences result

$$\begin{gathered}
L \\
L + R \\
L + R + L - R \\
L + R + L - R + L + R - L - R \\
L + R + L - R + L + R - L - R + L + R + L - R - L + R - L - R
\end{gathered} \tag{A5}$$

etc.

Since the alternating symbols R and L are not relevant for the dragon curve, one can drop them. With the symbols + and − instead of +1 and −1, the sequence that results is according to (3.24). This L-system can also be found in OEIS (Sloane [3.30]). Prusinkiewicz and Lindenmayer [3.26] and other authors, on the other hand, describe an L-system for the dragon curve according to the rules

$$L \mapsto L + R + \tag{A3.6}$$

$$R \mapsto -L - R. \tag{A3.7}$$

(A3.3) and (A3.4) shall continue to hold. This produces a different sequence than the development according to (A3.1) and (A3.2). In the geometric interpretation and in the application to diatom chains, both L-systems are equivalent, as the successive + and − that occur when (A3.6) and (A3.7) are applied cancel each other out.

Appendix 3B Probability Consideration for Loss of Synchronicity

The following analysis assumes random and uncorrelated generation times for the species under consideration, so that synchronization effects are explicitly excluded. A diatom is continuously dividing with the generation times t_1, t_2, t_3 ... which are positive random variables. We first determine the probability $P_k(t)$ that the k-th division has occurred at a certain time t, but the k + 1-th division has not yet been completed. This is a so-called "renewal process" (see, e.g., Feller [3.7]). We define an event $A_k(t)$ by

$$A_k(t) = \left\{ \sum_{j=1}^{k} t_j \le t \right\}. \tag{B3.1}$$

The complementary event to this is $\overline{A_k(t)} = \left\{ \sum_{j=1}^{k} t_j > t \right\}$. So, we are looking for the probability of the event

$$\left\{\sum_{j=1}^{k} t_j \le t\right\} \cap \left\{\sum_{j=1}^{k+1} t_j > t\right\}$$
$$= A_k(t) \cap \overline{A_{k+1}(t)} = A_k(t) \backslash A_{k+1}(t) \tag{B3.2}$$

where \ denotes the set difference. Therefore, the following applies for the required probability:

$$P_k(t) = P(A_k(t) \backslash A_{k+1}(t)) = P(A_k(t)) - P(A_{k+1}(t)). \tag{B3.3}$$

In the case of a normal distribution, is $P(A_k(t)) = P\left\{\sum_{j=1}^{k} t_j > t\right\}$ the distribution function of the probability density according to (3.26). By transforming the arguments of Equation (3.26), one obtains

$$P_k(t) = \Phi\left(\frac{t - k\mu}{\sqrt{k}\sigma}\right) - \Phi\left(\frac{t - (k+1)\mu}{\sqrt{k+1}\sigma}\right) \tag{B3.4}$$

where Φ is the cumulative standard normal distribution function.

When t is in the middle between $k\mu$ and $(k + 1)\mu$ and $\sqrt{k}\sigma/\mu \ll 1$, $P_k(t)$ is close to 1, because the k-th cell division is with high probability completed and the following cell division is about half accomplished. This changes with an increasing number of generations, i.e., higher values of k. Within the scope of the estimate, we set $k + 1 \approx k$. The probability at the time t, we are looking for thus follows from Equation (B3.4)

$$P_k(t_k) = \Phi(q/2) - \Phi(-q/2) \tag{B3.5}$$

where the variable

$$q = \mu/(\sqrt{k}\sigma) \tag{B3.6}$$

was introduced. In this approximation, $P_k(t_k)$ is plotted against q in Figure 3.13.

If, for example, $q = 2$, then the diatom is with a probability of 0.7 in the interval between k completed generations and before completion of the $k + 1$-th generation. Considering two diatoms, whose common ancestor dates back k generations, these two diatoms are both in this interval with the squared probability (because of the assumed statistical independence of the random variables). This is below ½, so that a loss of synchronicity is quite likely. The value $q = 4$ gives a probability of about 0.95 for one diatom in the mentioned interval and 0.91 for two diatoms simultaneously in this interval. A loss of synchronicity is at least not

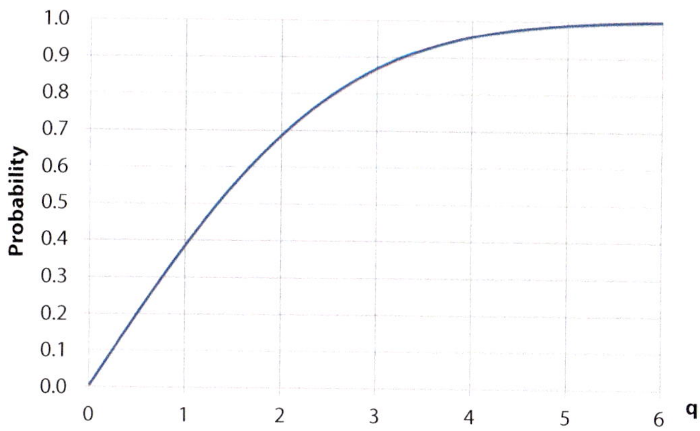

Figure 3.13 Probability $P_k(t)$ that the k-th division has occurred at the time $t_k = (k + ½)\mu$, but the k + 1-th division has not yet been completed plotted against q.

unlikely. It should be recalled that only a rough estimate was given and that a closer look must include all diatoms that have emerged from a first cell.

References

[3.1] Busby, W.F. and Lewin, J., Silicate uptake and silica shell formation by synchronously dividing cells of the diatom *Navicula pelticutosa* (Breb.) Hilse. *J. Phycol.*, 3, 127–131, 1967.

[3.2] Chandler, D. and Knuth, D.J., Number representations and dragon curves. *J. Recreat. Math.*, 3, 66-81 (Part 1), 133–149 (Part 2), 1970.

[3.3] Darley, W.M. and Volcani, B.E., Role of silicon in diatom metabolism. A silicon requirement for desoxyribonucleic acid synthesis in the diatom *Cylindrotheca fusiformis* Reimann and Lewin. *Exp. Cell Res.*, 58, 334–343, 1969.

[3.4] Davis, C.O., Harrison, P.J., Dugdale, R.C., Continuous culture of marine diatoms under silicate limitation I. Synchronized life cycle of *Skeletonema costatum*. *J. Phycol.*, 9, 175–80, 1973.

[3.5] De Koster, C.G. and Lindenmayer, A., Discrete and continuous models for heterocyst differentiation in growing filaments of blue-green bacteria. *Acta Biotheor.*, 36, 249–273, 1987.

[3.6] Edlund, M.B. and Stoermer, E.F., Ecological, evolutionary, and systematic significance of diatom life histories. *J. Phycol.*, 33, 897–918, 1997.

[3.7] Feller, W., *An introduction to probability theory and its applications*, 2nd edition, vol. 2, Wiley and Sons, New York, 1971.

[3.8] Guillard, R.R.L., Culture of phytoplankton for feeding marine invertebrates, in: *Culture of Marine Invertebrate Animals*, W.L. Smith and M.H. Chanley (Eds.), p. 26–60, Plenum Press, New York, 1975.

[3.9] Harbich, T., Size of the diatoms in a chain-like colony, https://diatoms.de/en/chain-like-colonies, 2020.

[3.10] Hedley, A., CombineZP Version 1.0, https://alan-hadley.software.informer.com, 2010.

[3.11] Hejblum, G., Costagliola, D., Valleron, A.J., Mary, J.Y., Cell-Cycle Models and Mother Daughter Correlation. *J. Theor. Biol.*, 131, 2, 255–262, 1988.

[3.12] Knuth, D., *Selected papers on fun & games*, pp. 571–614, CSLI Publications, Stanford, CA, 2011.

[3.13] Lewin, J.C., Reimann, B.E., Busby, W.F., Volcani, B.E., Silica shell formation in synchronously dividing diatoms, in: *Cell Synchrony*, I.L. Cameron and G.M. Podilla (Eds.), p. 169–188, Academic Press, New York, 1966.
[3.14] Lindenmayer, A., Mathematical models for cellular interactions in development: I. Filaments with one-sided inputs. *J. Theor. Biol.*, 18, 290–299, 1968.
[3.15] Lindenmayer, A., Mathematical models for cellular interactions in development: II. Simple and branching filaments with two-sided inputs. *J. Theor. Biol.*, 18, 300–315, 1968.
[3.16] Laney, S.R., Olson, R.J., Sosik, H.M., Diatoms favor their younger daughters. *Limnol. Oceanogr.*, 57, 5, 1572–1578, 2012.
[3.17] Locker, F., Beiträge zur Kenntnis des Formwechsels von Diatomeen an Hand von Kulturversuchen. *Österr. Bot. Z.*, 97, 322–2, 1950.
[3.18] Macdonald, J.D., On the structure of the diatomaceous frustule, and its genetic cycle. *Ann. Mag. Nat. Hist.*, 4, 1–8, 1869.
[3.19] Mattern, H., Stutz, S., van de Vijver, B., King, L., Klasse Bacillariophyceae, in: *Beiträge zu den Algen Baden - Württembergs. Band 2: Spezieller Teil: Euglenozoa und Heterkontobionta*, S. Stutz and H. Mattern (Eds.), p. 94–371, Verlag Manfred Hennecke, Remshalden, 2019.
[3.20] Microsoft, Image Composite Editor (ICE) Version 2.0.3.0, https://www.microsoft.com/en-us/research/product/computational-photography-applications/image-composite-editor, 2015.
[3.21] Paasche, E., Marine Plankton Algae Grown with Light-Dark Cycles. I. *Coccolithus huxleyi. Physiol. Plant*, 20, 946–956, 1967.
[3.22] Paasche, E., Marine plankton algae grown with light-dark cycles. 2. *Ditylum brightwellii* and *Nitzschia turgidula. Physiol. Plant*, 21, 66–77, 1968.
[3.23] Pfitzer, E., Über den Bau und die Zellteilung der Diatomeen. *Bot. Ztg.*, 27, 774–776, 1869.
[3.24] Pfitzer, E., Untersuchungen über Bau und Entwicklung der Bacillariaceen (Diatomeen). *Bot. Abh.*, 2, 1–189, 1871.
[3.25] Pirson, A. and Lorenzen, H., Synchronized dividing algae. *Annu. Rev. Plant Physiol.*, 17, 439–58, 1966.
[3.26] Prusinkiewicz, P. and Lindenmayer, A., *The Algorithmic Beauty of Plants*, Springer, New York, 1990.
[3.27] Richter, O., Zur Physiologie der Diatomeen (II. Mitteilung). Die Biologie der *Nitzschia putrida* Benecke. *Denkschr. Math.-Naturwiss. Kl. Akad. Wiss.*, 84, 660–772, 1909.
[3.28] Round, F.E., Crawford, R.M., Mann, D.G., *The diatoms: biology and morphology of the genera*, Cambridge University Press, Cambridge, 1990.
[3.29] Sloane, N.J.A., Online encyclopedia of integer sequences (OEIS), https://oeis.org/A014577, 1964.
[3.30] Sloane, N.J.A., Online encyclopedia of integer sequences (OEIS) https://oeis.org/A005811, 1964.
[3.31] Tabachnikov, S., Dragon curves revisited. *Math. Intell.*, 1, 36, 13–17, 2014.
[3.32] Ussing, A.P., Gordon, R., Ector, L., Buczko, K., Desnitskiy, A.G., Vanlandingham, S.L., *The colonial diatom "Bacillaria paradoxa": chaotic gliding motility, Lindenmeyer Model of colonial morphogenesis, and bibliography, with translation of O.F. Müller (1783), 'About a peculiar being in the beach-water'. Diatom Monographs*, vol. 5, Koeltz Scientific Books, Koenigstein, 2005.
[3.33] Wiedling, S., Die Gültigkeit der MacDonald-Pfitzerischen Regel bei der Diatomeengattung Nitzschia. *Naturwissenschaften*, 31, 114–115, 1943.

Figure 3.8 Dragon curve of the 13-th generation. It contains $2^{13} - 1 = 8{,}192$ angles. Dragon does not cross itself and no segment is run twice.

The dragon curve can be generated in different ways. The most common method for its production is paper folding (e.g., Knuth [3.12]). There are a number of variants of generation by L-systems. Appendix 3A shows how such an L-system can be derived directly from the replacement rules (3.2) and (3.3). Thus, it would also have been possible to begin with the sequence of the differences in the size indices and to derive the sequence of the size indices in the form (3.20) from it. In fact, the sequence of size indices is known in dragon curve theory. It results when one successively calculates the excess of the turns to the left over the turns to the right, which corresponds to partial sums. It is registered as A005811 in OEIS (Sloane [3.30]).

3.2.5 Matching Fragments to a Generation Based on Unknown Size Indices of the Fragment

We assume a correctly determined sequence of size differences of neighboring diatoms of a fragment, which were assigned the values +1 or −1. Assuming synchronicity of the cell divisions, this or its mirror image should occur for the first time in a certain generation in

4

Valve Morphogenesis in *Amphitetras antediluviana* Ehrenberg

Mary A. Tiffany[1]* and Bonnie L. Hurwitz[2,3]

[1]*Department of Biology and Center for Inland Waters, San Diego State University, San Diego, CA, United States*
[2]*Department of Biosystems Engineering, University of Arizona, Tucson, AZ, United States*
[3]*BIO5 Institute, University of Arizona, Tucson, AZ, United States*

Abstract

Amphitetras antediluviana is a marine chain-forming epiphytic diatom. Here we provide the first study of its valve and girdle band development. Each valve has four ocelli at the corners with gently curving sides and a circular elevation in the center. In early valves, we observed a circular to oval annulus with radiating costae (virgae) that occasionally bifurcate before forming the ocellus. The costae interconnect by crossbars (vimines) to form areolae. Virgae at the periphery of the valve bend in the direction of the ocelli, eventually fusing to produce their rims. Once the rim is fully enclosed, small pores start to form centripetally until they meet at the center. Two types of cribra develop within the areolae. The most common form consists of small poroids supported by tiny struts on the edge of the areola. The struts repeatedly bifurcate to produce a mesh of poroids. The other type of cribrum begins with sturdier struts that broaden to become spathulate, adding flattened structures as they expand, leaving slits instead of pores. Finally a marginal flange is added. Future research is needed to determine how this species forms two very different types of cribra, possibly using transmission electron microscopy of dividing cells.

Keywords: Southern California, *Amphitetras antediluviana*, valve development in diatoms, ocellus, cribrum development, alternate cribra, girdle bands

4.1 Introduction

Amphitetras antediluviana is a marine epiphytic centric diatom. The processes in genus *Amphitetras* are true ocelli (possessing a rim of silica around a mesh of tiny pores) [4.22]. Cells attach to their sisters at the ocelli to form zigzag colonies. The last detailed study of *Amphitetras* was in 1928 [4.15]. Mayama and Fujita [4.16] documented the presence of two types of cribra by scanning electron microscopy (S.E.M.).

Currently, only a handful of studies on valve morphogenesis exist for the many centric diatom genera (e.g., [4.14, 4.22, 4.23, 4.27–4.29, 4.30, 4.32]). In general, studying

**Corresponding author*: mtiffany24@yahoo.com

Vadim Annenkov, Joseph Seckback and Richard Gordon (eds.) Diatom Morphogenesis, (93–104) © 2022 Scrivener Publishing LLC

morphogenesis in cultured species is preferred to ensure its identity [4.22]. However, few of the heavily silicified diatoms can be successfully maintained in culture. The forming valves of some centric genera species, because of their unique shape and structures, can be recognized even in field samples. For example, *Asterolampra* [4.28], *Asteromphalus* [4.28], *Aulacodiscus* [4.29], *Aulacoseira* [4.2], *Isthmia* [4.30], and *Biddulphia* [4.30] are easily recognized based on their structure and the sequence of silica deposition.

This chapter aims to document the morphology of this interesting and beautiful species and to demonstrate the sequence in the morphogenesis of its valves and girdle bands. The terminology is based on Anonymous [4.1], Ross and Sims [4.19], Ross and Sims [4.20], Ross *et al.* [4.17], and Cox and Ross [4.4]. A glossary of terms is appended to this chapter to aid the reader.

In addition to the intrinsic interest in diatom valve structures and their morphogenesis, nanotechnologists and material scientists are interested in applying information on the micro-and nanostructures of mature as well as nascent diatom architectures [4.31]. Hopefully, this study will provide a useful addition for these specialists.

4.2 Material and Methods

Approximately, 10 grams of *Plocamium coccineum* var. *pacificum* (fresh seaweed drift) with visibly attached diatoms were collected from the beach at Point Dume State Preserve, Southern California, on 10 June 2001. It is unknown at what depth this seaweed was originally growing. To remove organic material, the sample was acid cleaned according to the von Stosch method [4.9]. For S.E.M., cover slips of dried material were sputter-coated with Au/Pd and examined with a Hitachi S-2700 scanning electron microscope at an accelerating voltage of 10 kV. In addition to *Amphitetras antediluviana*, a number of other epiphytic diatoms were also imaged including: *Arachnoidiscus ornatus*, *Trigonium arcticum*, *Biddulphia biddulphiana*, and *Climacosphenia moniligera*. Also in the sample were some "surf" diatoms: *Aulacodiscus oregonus*, *Aulacodiscus scaber*, and *Aulacodiscus kittonii*. These are usually solitary diatoms found in the surf zone. It is likely they were swept up as the floating seaweed drifted toward shore. A number of the valves of *A. antediluviana* were immature, forming valves and these were documented at various magnifications.

4.3 Observations

This living California taxon conforms to the type species of the genus *Amphitetras antediluviana* described by Ehrenberg [4.8] with slightly concave sides and four ocelli. However, the sides are more concave in valves depicted by Jahn and Kusber [4.13] and described as *Amphitetras antediluviana*. The valves measured 54–114 μm in width.

4.3.1 *Amphitetras antediluviana* Mature Valves

A complete frustule with the epivalve on the left is shown in Figure 4.1; the two bands to the left belong to the epivalve, and the single band seen to the right is the valvocopula of the hypovalve. All specimens found in the sample had four gently concave sides with raised ocelli at each corner. Girdle bands are complete (Figure 4.2). A solid rim

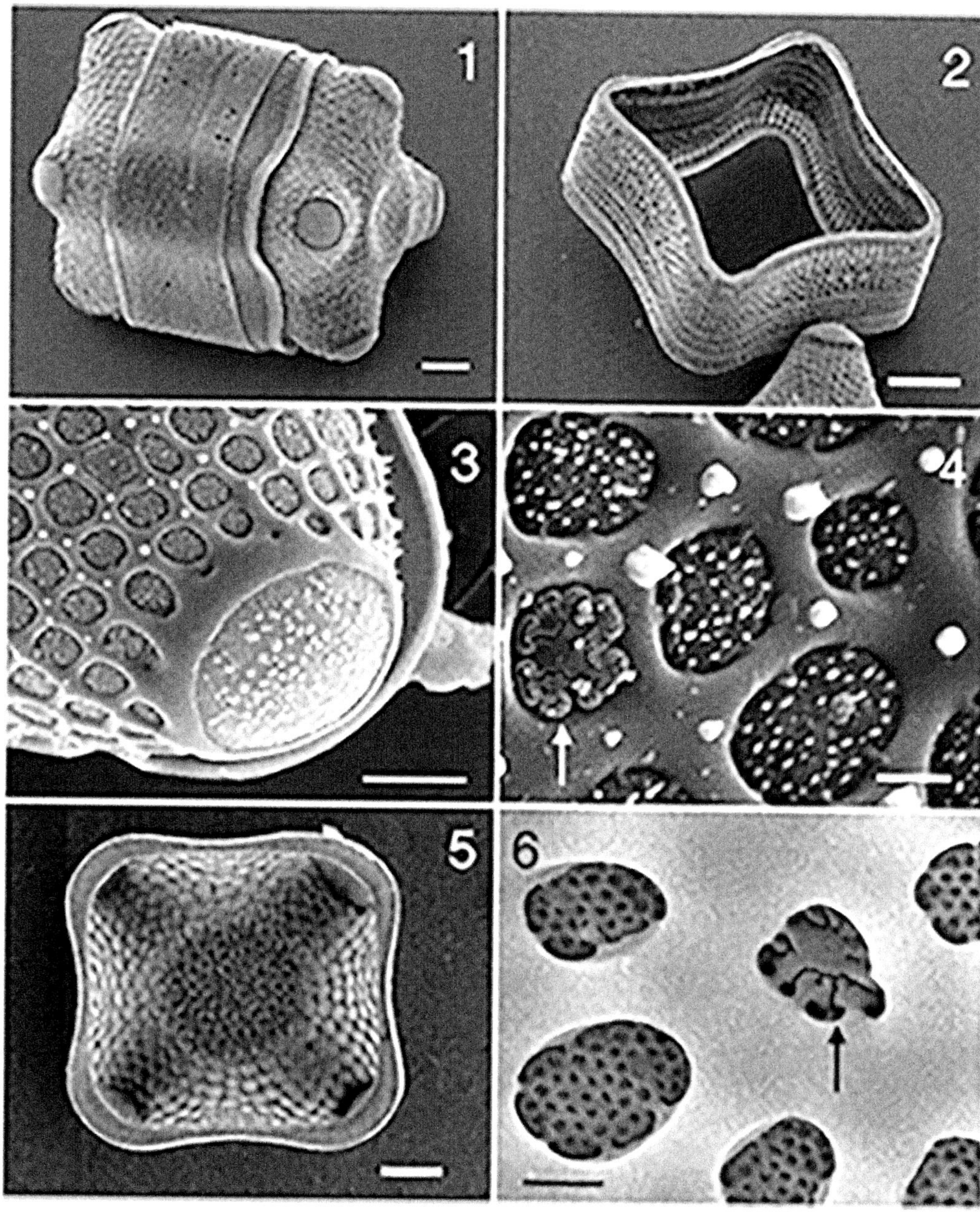

Figures 4.1–4.6 *Amphitetras antediluviana*, mature valves, S.E.M. **Figure 4.1** Complete frustule, epivalve is to the left, tilted 45°. **Figure 4.2** Set of girdle bands, tilted 45°. **Figure 4.3** Close-up of ocellus at corner of valve. Note large granules occluding some porelli. The mantle is surrounded by a flange. **Figure 4.4** Detail of valve surface, external view. "Usual" cribra with small granules. Alternate cribra with "T" shaped slits (arrow). **Figure 4.5** Valve, internal view. **Figure 4.6** Detail of valve surface, internal view. "Usual" cribra have small poroids supported by struts. Alternate cribrum lacking pores (arrow). Scale bars = 1 μm (Figures 4.4 and 4.6); 5 μm (Figure 4.3); 10 μm (Figure 4.5); 20 μm (Figures 4.1–4.2).

of silica delineates the four ocelli, and a flange surrounds the periphery of the mantle (Figure 4.3). Areolae are approximately 1 ½ to 3 µm in diameter and are about 4.0–4.3 µm apart. There are two types of cribra on the surface of the valve (Figure 4.4). The "usual" cribra have a sprinkling of small granules or spines outside (Figure 4.4). The inside view of a mature valve is shown in Figure 4.5. When viewed from the inside, the "usual" cribra consists of small pores (poroids) supported by struts (Figure 4.6). In the mature cribrum, even tinier pores can be seen on the interior view, further subdividing the cribrum lacework (Figure 4.6). Alternate cribra have a coarser structure with slits rather than pores (Figures 4.4 and 4.6, arrows). Granules of varying sizes are sprinkled on the smooth exterior surface of the valve, often on the interstices of the areolae (Figure 4.4).

The small pores (porelli) within the ocellus are in close-packing arrangement with tiny granules on most of the interstices and larger globular ones randomly on the surface of the ocellus (see Figure 4.26).

4.3.2 *Amphitetras antediluviana* Forming Valves

An early forming valve of *A. antediluviana* reveals an oval pattern center (annulus) with radiating costae that occasionally bifurcate (Figure 4.7). Near the center, crossbars connect the costae forming the earliest areolae. The ocelli have not yet developed nor their rims. Still, indications of their future locations are seen between the four corners of the nascent valve, where the costae gently curve inwardly (indicated by arrows in Figure 4.7). The annulus has a bar across it in this specimen (Figure 4.8). As the valve matures, circular openings form at each of the four corners where the ocelli will develop, but the rim is not enclosed (Figures 4.9 and 4.13). At the point where the rim has been completed (Figures 4.10 and 4.14), the valve's three-dimensional structure is already evident (Figure 4.10). Once the rim of the ocelli has fused, the valve's overall three-dimensional structure is evident, and the basic framework is complete (Figure 4.10). The next structure to develop is the ocellus that forms centrifugally (Figures 4.11 and 4.15 to 4.17). When the porelli of the ocelli meet at the center (Figures 4.12 and 4.18), the cribra begin to form on the valve's main surface. Small granules do not yet adorn the ocellus' porelli, but larger granules are randomly sprinkled on its surface (Figure 4.25). In the mature valve, the larger granules on the ocellus have expanded in size and number (Figure 4.26). Small granules are seen at most of the corners of the porelli (Figure 4.26).

Cribrum development occurs in two manners (Figures 4.19 to 4.24). In early valves, few areolae show hints of future cribrum structures (Figure 4.19). At later stages, tiny struts appear on most, but not all areolae. The areolae without struts are destined to house the alternative form of cribrum (arrow, Figure 4.20). A wavy design appears at the periphery of the alternate cribra, perhaps the site of incipient struts. At this stage, small granules start to appear on the external surface of the costae (Figure 4.20). In the next inferred stages, the struts have bifurcated once or twice (Figures 4.21 to 4.23). The cribra begin to meet in the center (Figure 4.24). The alternate cribra form slightly later, starting with thicker struts and becoming spathulate (Figures 4.21 to 4.23, arrows). As the development of the alternative cribra proceeds, flat structures expand and form a network of "T"-shaped slits (arrow, Figure 4.24).

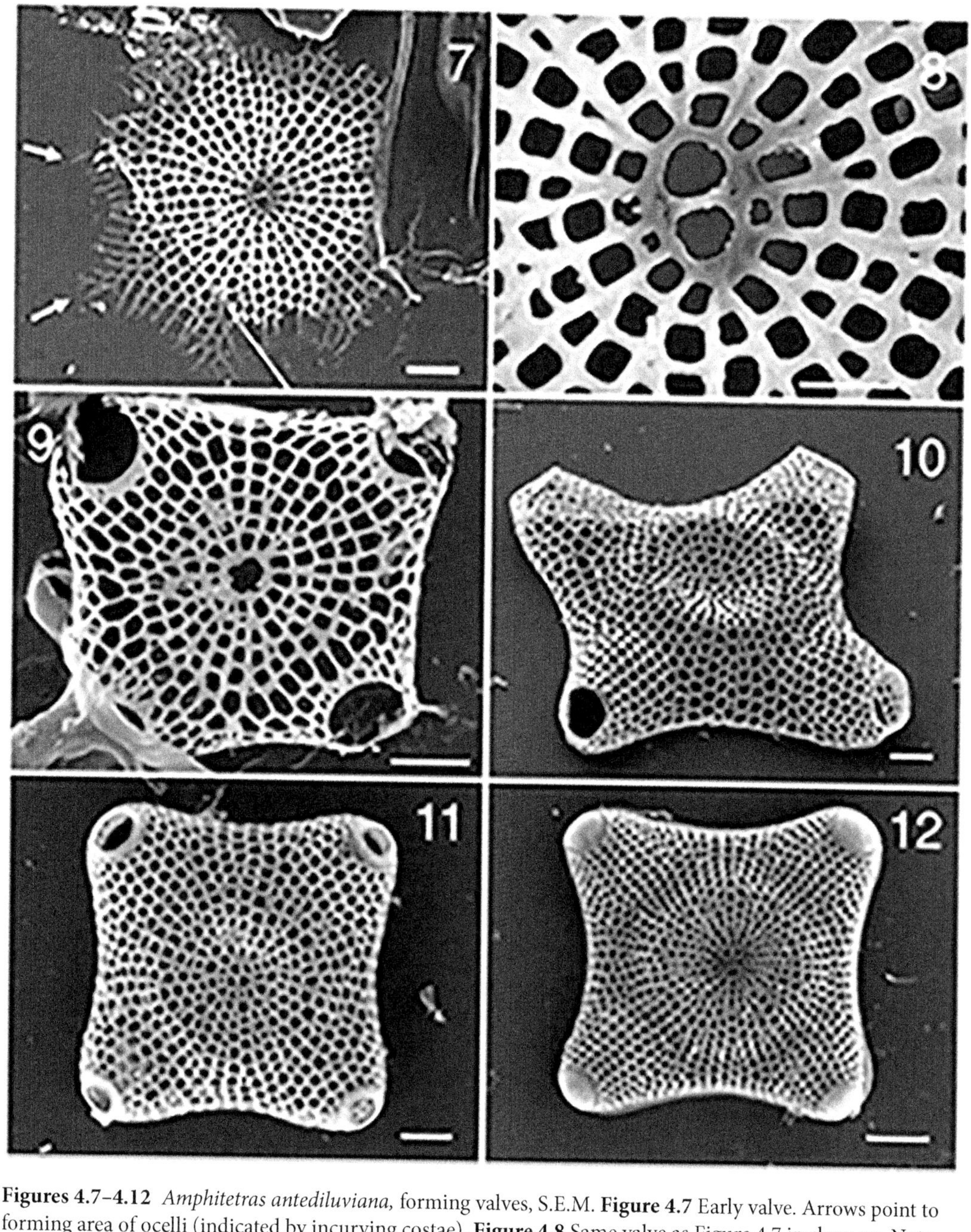

Figures 4.7–4.12 *Amphitetras antediluviana,* forming valves, S.E.M. **Figure 4.7** Early valve. Arrows point to forming area of ocelli (indicated by incurving costae). **Figure 4.8** Same valve as Figure 4.7 in close-up. Note annulus in center with supporting bar subtending it. **Figure 4.9** Valve at later stage of development. Margin of edge of valve is fused, but the rims of the ocelli are not fused (arrow lower right hand corner). **Figure 4.10** Three-dimensional structure of valve is evident, but cribra and ocelli porelli have not yet formed, tilted 45°. **Figure 4.11** Valve at slightly later stage of development. Porelli of ocelli are beginning to form. **Figure 4.12** Porelli of ocelli have completed formation, but cribra yet to begin to form. Scale bars = 5 μm (Figure 4.8); 20 μm (Figures 4.7, 4.9–4.12).

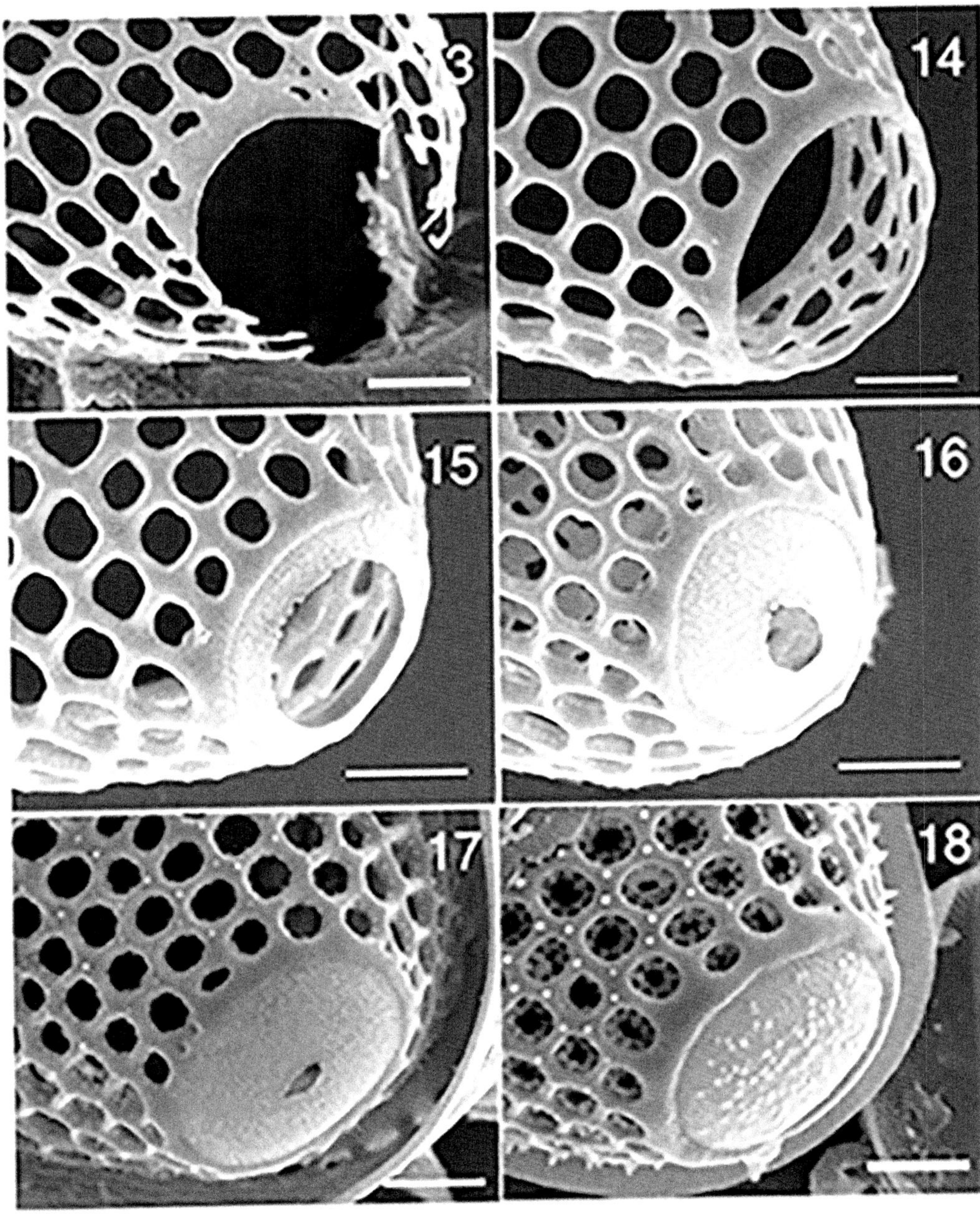

Figures 4.13–4.18 *Amphitetras antediluviana,* forming ocelli, S.E.M. **Figure 4.13** Forming ocellus, rim not yet fused. Same valve as Figure 4.9. **Figure 4.14** Early valve. Porelli of ocellus not yet begun, but rim is fused. Same valve as Figure 4.10. **Figure 4.15** Porelli of ocellus forming centrifugally. Same valve as Figure 4.11. **Figure 4.16** Porelli of ocellus continuing to form centrifugally. Hints of struts of cribra. **Figure 4.17** Porelli nearly at center of ocellus, cribra just beginning to form. Structure surrounding valve is a girdle band. **Figure 4.18** Nearly mature ocellus with globular granules, cribra partially formed. Flange at edge of mantle is forming. Same valve as Figure 4.12. For mature ocellus see Figure 4.3. Scale bars = 5 μm.

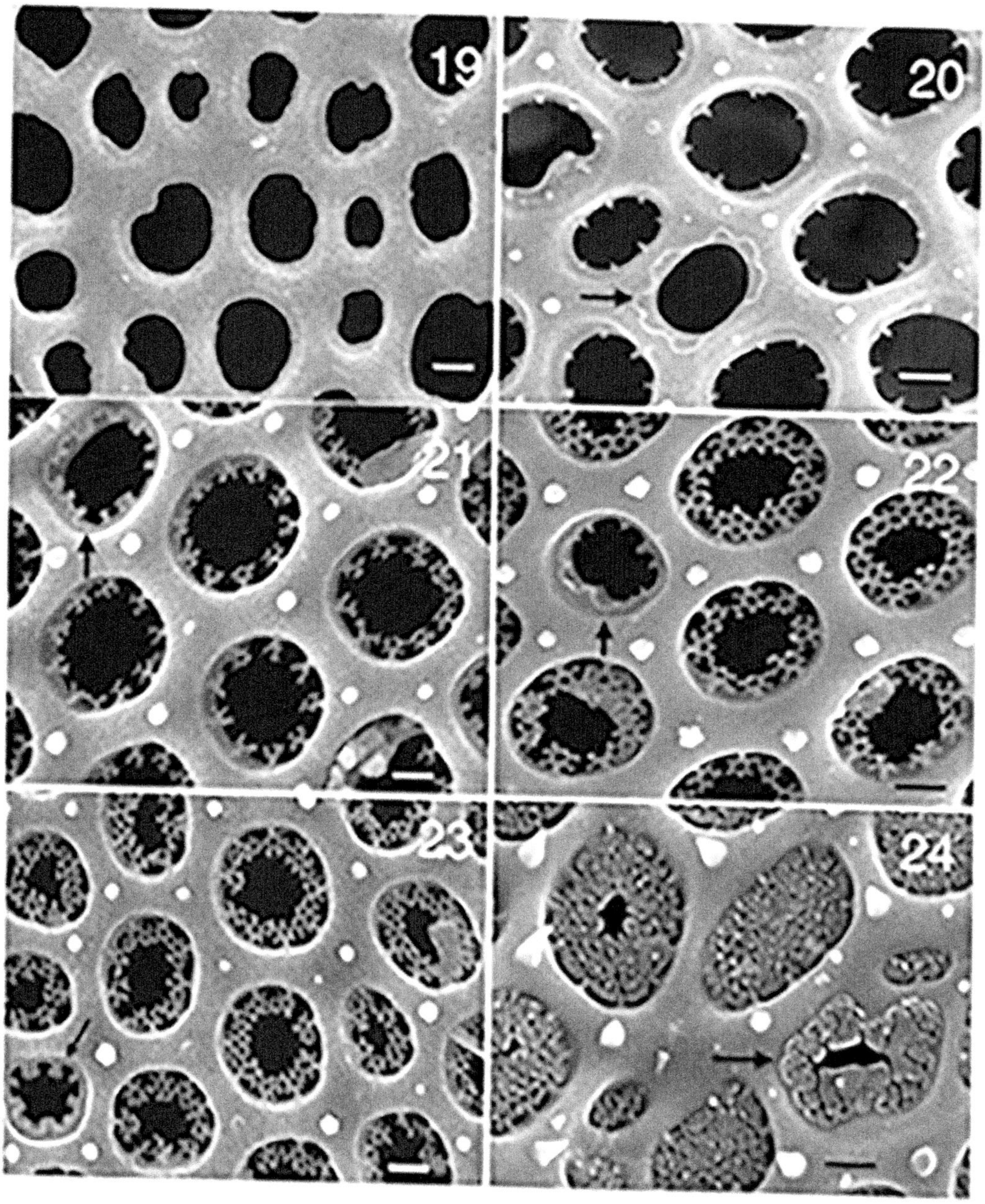

Figures 4.19–4.24 *Amphitetras antediluviana,* forming cribra, S.E.M. **Figure 4.19** Early valve. Tiny struts just barely visible, alternate cribra not evident yet. **Figure 4.20** Struts increasing in length. Alternate cribra have no struts (arrow). Granules beginning to form between areolae. **Figure 4.21** Bifurcations forming first row of poroids. Alternate cribrum has formed struts (thicker than those of usual cribra, arrow). **Figure 4.22** Continued formation of poroids. Alternate cribrum starting to form struts (arrow). **Figure 4.23** Rows of poroids nearly reaching center of cribra. Struts of alternate cribrum becoming spathulate (arrow). **Figure 4.24** Nearly mature cribra. Poroids of usual cribra becoming occluded, less obvious. Alternate cribrum growing flat plate-like structures with slits between them (arrow). See Figures 4.4 and 4.6 for images of mature cribra. Scale bars = 1 μm.

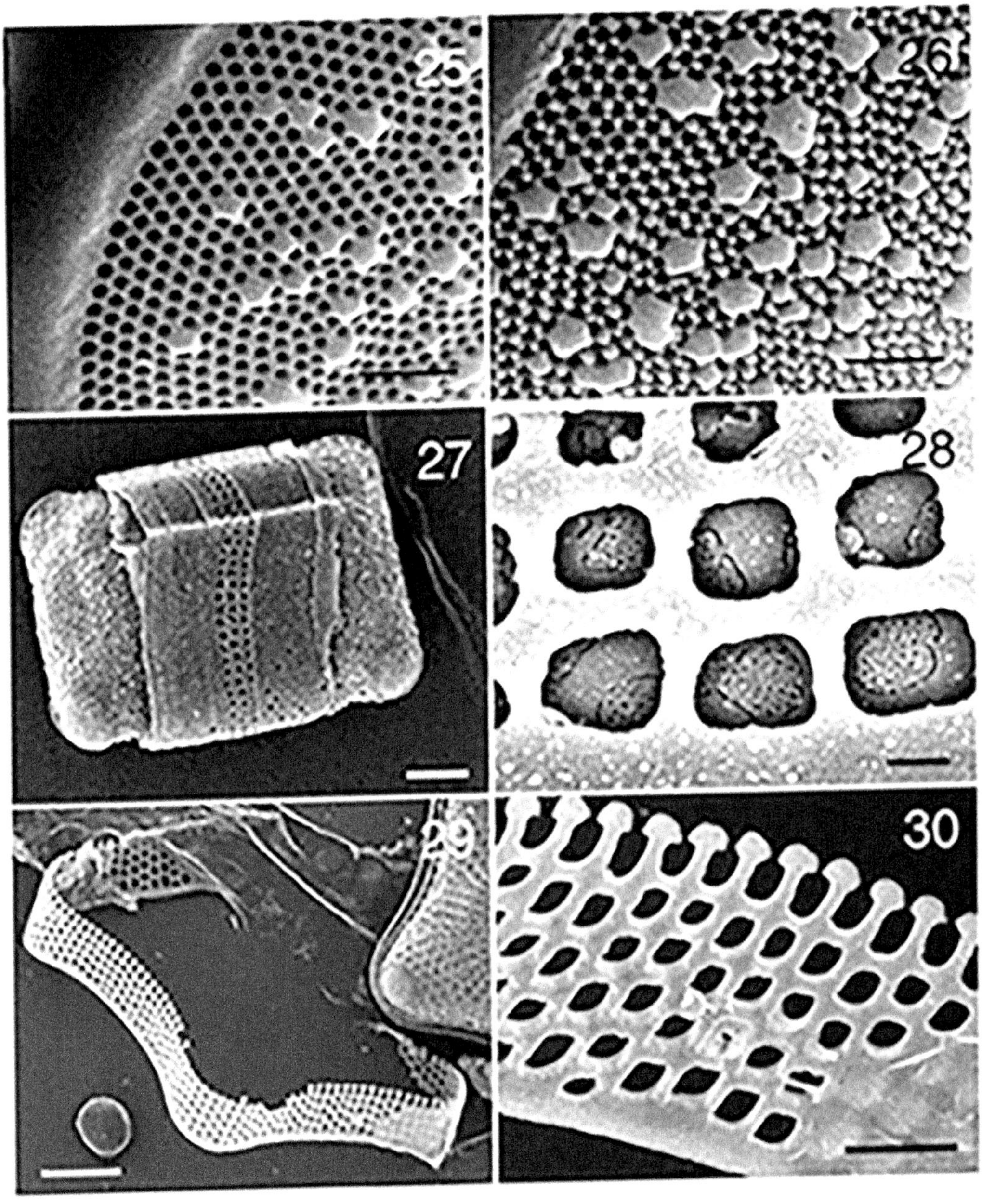

Figures 4.25–4.30 *Amphitetras antediluviana*, S.E.M. **Figure 4.25** Detail of nearly mature ocellus with growing large granules occluding some porelli. **Figure 4.26** Detail of mature ocellus. Note larger granules more-or-less randomly spread on surface. Small granules on interstices of most porelli. **Figure 4.27** Frustule with incomplete girdle band in the hypotheca. **Figure 4.28** Detail of Figure 1.27. Girdle band with incomplete cribra overlying band with complete cribra, all of the usual variety. **Figure 4.29** Forming girdle band (copula). **Figure 4.30** Detail of forming girdle band. Scale bars = 1 μm (Figures 4.25–4.26, 4.28); 5 μm (Figure 4.30); 10 μm (Figure 4.27); 20 μm (Figure 4.29).

4.3.3 *Amphitetras antediluviana* Girdle Band Formation

We encountered two forming girdle bands during this study, one of which was still within its frustule (Figures 4.27 and 4.28). The forming band in Figures 4.27 and 4.28, evidently the newest in the hypotheca, has no cribrum development; this is likely its next stage of development. The loose band in Figures 4.29 and 4.30 has areolae arrayed in a rectangular fashion emanating from a solid siliceous strip. Four rows of pores are complete, and the fifth at the distal (upper) side is beginning to grow crossbars.

4.4 Conclusion

Diatoms are ubiquitous primary producers in the ocean [4.12]. Therefore, documenting the steps in valve and girdle band morphogenesis in various species of diatoms is invaluable in determining how algae deposit silica as they fashion their ornate structures [4.5].

Centric diatoms begin their formation with a pattern center called an annulus (a ring of silica) [4.2, 4.3, 4.21, 4.25, 4.27]. In *Amphitetras*, this ring is oval (Figure 4.10) or almost circular (e.g., Figure 4.11). Further siliceous development within the annulus can occur. For example, in Figure 4.24 of Round *et al.* [4.21], a forming valve of *Melosira* has pores within its annulus. In the case of *Amphitetras antediluviana*, one or more bars usually form across the annulus. These bars can be seen in the mature valves via an inside view, as externally the spaces they form become occluded with cribra.

Cribrum development in many species is similar to that of the ordinary "usual" cribra in *Amphitetras* ([4.27], Figure 12), including *Asteromphalus heptactis* ([4.29], plate 20, Figures 1–6), and *Coscinodiscus wailesii* ([4.23], Figures 26–30; [4.26], Figure 5e–h). These genera do not, however, have two types of cribra.

A few other species have alternate cribra resembling those in *Amphitetras*, including *Amphipentas boryana* and *Diommatetras grossa* [4.24]. How these cells form quite different cribra in neighboring areolae is also unknown. Cox [4.6] suggested that, "morphologically distinct external pore occlusions may be parts of the same morphogenetic pathway." However, this might not be the case in *Amphitetras*, as there is no difference in the size or shape of areolae destined to have alternate cribra. The placement of alternate cribra appears to be random. The function of the alternate cribra in *A. antediluviana* and the two genera named above is also not clear. One possibility is that the slits allow for greater access to nutrients from outside the cell. How the cell controls the production of two very different, but neighboring cribra requires further investigation.

The sequence of formation of the ocelli is of taxonomic interest, as genera possessing these may not use the same cellular mechanism to produce them, as do genera with pseudocelli [4.11, 4.18]. In the ocellate *Amphitetras*, the rim fuses first, completing the basic framework of the valve. The pores then fill in the opening within the rim proceeding centripetally. The pseudocellus, lacking a rim, forms centrifugally as the areolae at a corner of the nascent valve become smaller while the valve increases in girth (e.g., *Isthmia* and *Biddulphia* [4.30]).

To date, few studies have investigated the formation of girdle bands [4.7, 4.10, 4.22]. These bands are essential in the cell cycle. As bands are added to the hypotheca, the cell volume increases in preparation for division. Whether the bands form in an advalvar or abvalvar direction has not been investigated. However, in the case of *Isthmia nervosa*, the valvocopula and the copula form in opposite directions. The valvocopula forms in an advalvar and the copula in an abvalvar direction [4.30].

Acknowledgments

We thank Steve Barlow for the generous use of the Electron Microscope Facility at San Diego State University.

References

[4.1] Anonymous, Proposals for a standardization of diatom terminology and diagnoses. *Beih. Nova Hedwig.*, 53, 232–354, 1975.

[4.2] Bedoshvili, Y.D., Kaluzhnaya, O., Likhoshway, Y.V., The frustule morphogenesis of *Aulacoseira baicalensis* in the natural population. *J. Adv. Microsc. Res.*, 7, 218–224, 2012.

[4.3] Bedoshvili, Y.D. and Likhoshway, Y.V., Cellular mechanisms of diatom morphogenesis, in: *Diatoms: Fundamentals and Applications*, pp. 99–114, Scrivener Publishing LLC, Hoboken, New Jersey, 2019.

[4.4] Cox, E.J. and Ross, R., The striae of pennate diatoms, in: *Proceedings of the 6th Diatom Symposium on Recent and Fossil Diatoms*, Koeltz Scientific Books, Koenigstein, pp. 267–278, 1981.

[4.5] Cox, E.J., Variations in patterns of valve morphogenesis between representatives of six biraphid diatom genera (Bacillariophyceae). *J. Phycol.*, 35, 1297–1312, 1999.

[4.6] Cox, E.J., Morphogenetic information and the selection of taxonomic characters for raphid diatom systematics. *Plant Ecol. Evol.*, 143, 271–277, 2010.

[4.7] Cox, E.J., Ontogeny, homology, and terminology—wall morphogenesis as an aid to character recognition and character state definition for pennate diatom systematics. *J. Phycol.*, 48, 1–31, 2012.

[4.8] Ehrenberg, C.G., *Mikrogeologie. Das Erden und Felsen schaffende Wirken des unsichtbar kleinen selbstständigen Lebens auf der Erde*, pp. 1–40, Atlas. T, Leipzig, 1854, (In German).

[4.9] Hasle, G.R. and Syvertsen, E.E., Chapter 2. Marine Diatoms, in: *Identifying Marine Phytoplankton*, C. Tomas (Ed.), pp. 5–385, Academic Press, San Diego, 1997.

[4.10] Hildebrand, M., Kim, S., Shi, D., Scott, K., Subramaniam, S., 3D Imaging of Diatoms with Ion-abrasion Scanning Electron Microscopy. *J. Struct. Biol.*, 166, 316–328, 2009.

[4.11] Hoban, M.A., Biddulphioid Diatoms. II: The Morphology and Systematics of the Pseudocellate Species, *Biddulphia biddulphiana* (Smith) Boyer, *B. alternans* (Bailey) Van Heurck, and *Trigonium arcticum* (Brightwell) Cleve. *Bot. Mar.*, 26, 271–284, 1983.

[4.12] Hustedt, F., Die Kieselalgen Deutschlands, Österreichs und der Schweiz mit Berücksichtigung der übrigen Länder Europas sowie der angrenzenden Meeresgebeite, in: *Dr. L. Rabenhorst's Kryptogamen-Flora von Deutschland, Österreich und der Schweiz*, Band VII, Leipzig, 1959, (In German).

[4.13] Jahn, R. and Kusber, W.-H., Algae of the Ehrenberg collection-3. Typification of seven diatom taxa described by C G. Ehrenberg predominantly from Greece. *Wildenowia*, 6, 527–535, 2006.

[4.14] Li, C.-W. and Volcani, B.E., Studies on the biochemistry and fine structure of silica shell formation in diatoms. VIII. Morphogenesis of cell wall in a centric diatom, *Ditylum brightwellii*. *Protoplasma*, 124, 10–29, 1985.

[4.15] Liebisch, W., *Amphitetras antediluviana* Ehrbg., sowie einige Beiträge zum Bau und zur Entwicklung der Diatomeenzelle. *Z. Bot.*, 20, 225–271, 1928, (In German).

[4.16] Mayama, S. and Fujita, D., Two types of areola occlusion in the valve of *Amphitetras antediluviana* Ehrenberg. *Diatom*, 11, 25–29, 1995, (In Japanese).

[4.17] Ross, R., Cox, E.J., Karayeva, N.I., Mann, D.G., Paddock, T.B.B., Simonsen, R., Sims, P.A., An amended terminology for the siliceous components of the diatom cell. *Beih. Nova Hedwig.*, 64, 513–530, 1979.

[4.18] Ross, R. and Sims, P.A., Generic limits in the Biddulphiaceae as indicated by the scanning electron microscope, in: *Scanning Electron Microscopy: Systematic and Evolutionary Applications*, V.H. Heywood (Ed.), pp. 155–177, Academic Press, London, 1971.

[4.19] Ross, R. and Sims, P.A., The fine structure of the frustule in centric diatoms: a suggested terminology. *Br. Phycol. J.*, 7, 139–163, 1972.

[4.20] Ross, R., Cox, E.J., Karayeva, N.I., Mann, D.G., Paddock, T.B.B., Simonsen, R., Sims, P.A., A suggested terminology for the siliceous components of the diatom cell. *Beih. Nova Hedwig.*, 64, 513–533, 1979.

[4.21] Round, F.E., Crawford, R.M., Mann, D.G., *The diatoms. Biology and morphology of the genera*, 747 pp, Cambridge University Press, Cambridge, 1990.

[4.22] Sato, S., Valve and girdle band morphogenesis in a bipolar centric diatom *Plagiogrammopsis vanheurckii* (Cymatosiraceae, Bacillariophyta). *Eur. J. Phycol.*, 45, 167–176, 2010.

[4.23] Schmid, A.-M. and Volcani, B., Wall morphogenesis in *Coscinodiscus wailesii* Gran and Angst. I. Valve morphology and development of its architecture. *J. Phycol.*, 19, 387–402, 1983.

[4.24] Sims, P.A., *Ceratangula* gen. nov. is proposed together with a discussion on the genera *Cerataulus*, *Eupodiscus*, *Amphitetras*, *Amphipentas* and *Diommatetras*. *Diatom Res.*, 23, 435–444, 2008.

[4.25] von Stosch, H.A., Observations on *Bellerochea* and *Streptotheca*, including descriptions of three new planktonic species. *Beih. Nova Hedwig.*, 54, 113–166, 1977.

[4.26] Sumper, M. and Brunner, E., Learning from diatoms: nature's tools for production of nanostructured silica. *Adv. Struct. Mater.*, 16, 17–6, 2006.

[4.27] Tesson, B. and Hildebrand, M., Dynamics of silica cell wall morphogenesis in the diatom *Cyclotella cryptica*: Substructure formation and the role of microfilaments. *J. Struct. Biol.*, 169, 1, 62–74, 2010.

[4.28] Tiffany, M.A., Valve development in the diatom family Asterolampraceae H.L. Smith. *Micropaleontology*, 51, 3, 2005.

[4.29] Tiffany, M.A., Valve development in *Aulacodiscus*. *Diatom Res.*, 23, 85–212, 2008.

[4.30] Tiffany, M.A., Valve and girdle band morphogenesis in the pseudocellate diatom species *Biddulphia biddulphiana* J.E. Smith (Boyer) and *Isthmia nervosa* Kütz .*Nova Hedwig. Beih.*, 144, 61–95, 2015.

[4.31] Tiffany, M.A., Diatom auxospore scales and early stages in diatom frustule morphogenesis: their potential for use in nanotechnology. *J. Nanosci. Nanotechnol.*, 5, 1, 131–139, 2005.

[4.32] Tiffany, M.A. and Hernandez-Becerril, D.U., Valve morphogenesis in the diatom family Asterolampraceae H. L. Smith. *Micropaleontology*, 51, 217–258, 2005.

Glossary

Abvalvar growth away from the valve
Advalvar growth toward the valve
Annulus a silica structure, often ring-shaped, the first part of a centric diatom to form [4.21]
Centric diatoms a group of diatoms with radial or polygonal symmetry
Costa, pl. costae secondarily thickened rib-like structures [4.7]
Cribrum, pl. cribra usually only visible in electron microscopy, a sieve-like covering of a pore [4.21]
Epiphytic dwelling upon a plant (e.g. seaweed)
Epivalve the outer, slightly larger valve of a diatom frustule
Frustule the complete siliceous cell wall of a diatom, including the two valves and any girdle bands
Hypovalve the inner, slightly smaller valve of a diatom frustule
Ocellate resembling an small eye
Ocellus, pl. ocelli a porefield with a silica rim at the outer part of some centric diatom valves [4.21]
Pseudocellus, pl. pseudocelli a porefield without a silica rim at the outer part of some centric diatom valves [4.21]
Spathulate having a broad rounded end and a narrower base
Valve one of two silica parts of a diatom frustule (often petri dish-like with one, the epivalve, overlapping the hypovalve)
Valvocopula, pl. valvocopulae the girdle band (copula) in direct contact with the valve, the first girdle band formed during morphogenesis [4.21]
Virga, pl. virgae early structures emanating from the annulus [4.5]

Part 2
SIMULATION

5

Geometric Models of Concentric and Spiral Areola Patterns of Centric Diatoms

Anton M. Lyakh

A.O. Kovalevsky Institute of Biology of the Southern Seas of RAS, Sevastopol, Russia

Abstract

In this chapter, geometric models that simulate concentric and spiral areolae distributions on valves of some centric diatoms are described. The first model simulates a concentric pattern of areolae. The formation of the pattern includes the repeating process of sequential divergence of adjacent rows. In every local region of the pattern, there are two adjacent rows that gradually diverge. They mechanically press on neighboring rows and shift them toward sector borders. When the distance between divergent rows exceeds the size of an areola, the divergence stops and the new row starts to form. On the next transect, another two randomly selected rows start to diverge and the process repeats. The second model simulates a spiral pattern of areolae. The spiral patterns consist only of spiral rows of areolae and do not contain any radial rows. Two kinds of spiral patterns are distinguished. In unidirectional patterns, rows of areolae are curved either in a counterclockwise or in a clockwise direction; in bidirectional patterns, both types of spirals are visible. All helices are simulated with the help of Archimedean spirals. These principles of construction of geometric models can be extended to a diversity of diatom patterns.

***Keywords*:** Centric diatoms, valve morphogenesis, valve pattern simulation, geometric model, areolae-based model, frame-based model, perfect pattern

5.1 Introduction

The regular pattern of areolae raises the interest of many researchers. They try to understand how diatoms reproduce beautiful forms and textures. Mathematical models, which to a different degree of success imitate valve morphogenesis, help to cognize this process.

The morphogenesis of a diatom valve finally leads to the formation of a valve pattern. It is the most visually apparent result of valve formation. The pattern consists of areolae arranged in a manner specific for each species. Therefore, one of the goals of the modeling of valve morphogenesis is to simulate the pattern constructed. Geometric models are suitable for solving this task.

Email: me@antonlyakh.ru

Vadim Annenkov, Joseph Seckback and Richard Gordon (eds.) Diatom Morphogenesis, (107–116) © 2022 Scrivener Publishing LLC

Hotton *et al.* [5.6] defined the essence of geometric models. The authors talked about phyllotaxis, but their statement is suitable for any geometric model used in biology: "… the most important contribution of geometrical models is *to reduce [a]complex* […] *pattern to a set of local rules* that a repeated recursively. *These local rules*, rather than the overall patterns, *are ultimately what needs to be explained* at the mechanistic level".

I distinguish two classes of the models that simulate areolae patterns. Their main difference is in the object modeled. The first class of models constructs a siliceous frame around space occupied by a chloroplast. They are named *frame-based models*. The second class mimics the distribution of areolae on the valve surface. They are named *areolae-based models*.

The simulation of a siliceous frame is more commonly used in the modeling of diatom patterns. Gordon and Drum [5.5] described a model, where fine branching patterns of silica, observed on early stages of valve morphogenesis, are constructed because of random diffuse precipitation of siliceous particles (Diffusion-Limited Aggregation, DLA-model). Parkinson *et al.* [5.14] include in a DLA-model additional characteristics, which allow taking into account physical conditions of an environment where pattern is formed. Their model helped "to explain the formation of radial costae, but the complex […] arrays of pores are […] unexplained" [5.8]. Sumper [5.18] and Sumper and Kröger [5.19] presented a model of valve pattern formation where droplets of long-chain polyamides recursively separate in the silica deposition vesicle into smaller droplets. Silica precipitates between droplets and forms a hierarchial siliceous pattern observed in many locular areolae. As Sumper said "the model can explain the complex patterns produced during formation of the cribrum and cribellum", but for other aspects of valve morphogenesis, we need to introduce additional mechanisms [5.18]. Lenoci and Camp [5.8] made a quantitative model where condensing silica is deposited in a quasi-static mold created during phase-separation processes. This leads to the generation of verisimilar areola cribrae in a diverse range of diatom species. Willis *et al.* [5.21] presented a model that very plausible reproduces most morphological features of diatom areola occlusions. It is a nonlinear stochastic model based on continuous-time Markov chains. Lyakh [5.11] used an L-system to simulate the radial-parallel pattern of areolae.

The areolae-based models are rarely used for simulating diatom patterns. The major example of such a model is a conformal mapping presented by Longuet-Higgins [5.9]. It reproduces the polygonal areolae pattern [5.12] of some diatom valves. The second model is presented by Lyakh [5.10]. It is a pure geometric model that constructs radial rows of areolae testing areola intersections. The model replicates the mix of radial and radial-parallel patterns of areolae which do not exist in nature.

The frame-based models more naturally reproduce the process of valve pattern development, while the areolae-based models use an artificial approach, which can produce areolae patterns that look like natural or are very similar to them.

In the paper, I present areolae-based geometric models that simulate two types of areolae distribution on the valves of some centric diatoms. I consider only those patterns where the most of the areolae have equal sizes. For easy reading, the phrase "pattern of areolae on the valves of centric diatoms" is shortened to "areolae pattern" or "pattern".

5.2 Set of Common Rules Used in the Models

I use geometric models that are based on the set of local rules. The rules determine how to construct each row of areolae in every specific pattern. The similar rule-based approach that Mann [5.13] used to analysis of pattern faults in raphid diatoms.

The main rule of simulation is that a diatom tends to place areolae in an optimal way. It strives to leave as little space free of areolae as possible under the action of given geometrical constraints. I designate this rule as *R.0.*

The other four rules are also in common for the simulated patterns.

R.1. Any pattern consists of a set of areolae rows[1].

R.2. Any row is formed from the valve center toward the valve margin.

R.3. All areolae are circular and have equal size.

R.4. Distances between neighboring areolae in any row are equal.

These rules determine two common parameters of geometric models: areolae diameter and the distance between centers or borders of areolae. The other two parameters are the shape and size of a valve. In this chapter, I consider only models with circular valves.

5.3 Concentric Pattern of Areolae

The pattern consists of closely fitting radial rows of areolae. All areolae have almost equal sizes. They are located on predefined discrete levels equidistant from the valve center. That leads to the formation of concentric circles consisting of transversely curved rows of areolae (Figure 5.1a).

The presence of rings of areolae is the most distinctive feature of a perfect concentric pattern. However, diatoms can also produce an imperfect concentric pattern, where, instead of right circles, closed wavy lines occur (Figure 5.1b).

Another visible feature of the concentric pattern is the grouping of rows into sectors. Each sector is bounded by hyaline areas prolonged from the valve center to the valve margin. The widths of hyaline areas are bigger than the distance between rows for areolae, but slightly smaller than the areola diameter (Figure 5.1).

The perfect concentric pattern is simulated with the help of the following rules:

Con.1. Pattern is divided into sectors and areola rows are formed inside each sector independently.

Con.2. Sectors are bordered by radial primary rows. Between borders (primary rows) of neighboring sectors there are hyaline areas.

Con.3. Primary rows are steady, and their direction stays radial during simulation.

Con.4. Every row inside a sector also tends to be radial, if it is not influenced by any forces.

Con.5. The first areola (the initial position) of each row can take only a predetermined distance from the valve center.

---

[1]I consider only patterns which consist of the rows of areolae. Other types of patterns also exist where areolae do not form rows and are ordered in another manner.

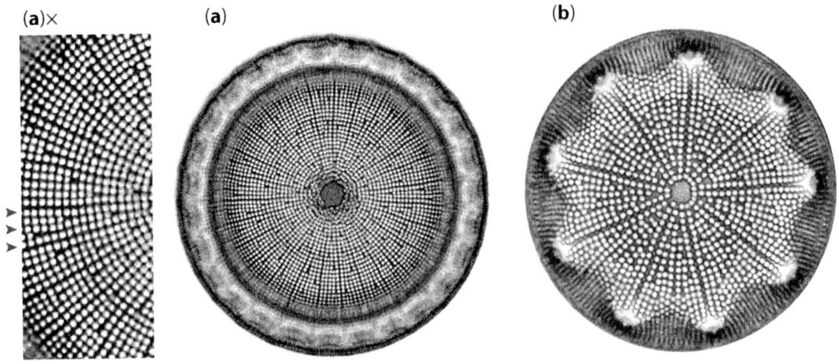

Figure 5.1 (a and b) Concentric pattern of areolae on the valve of *Aulacodiscus orientalis*. (a) The perfect concentric pattern [5.2]. (a×) Zoomed fragment of the pattern (a). The arrows indicate divergent rows of areolae inside one sector. Black elongated triangles visible on the pattern are hyaline areas that appear between the divergent rows before a new row starts. Many first areolae of rows are slightly shifted from a perfect position or have smaller size. (b) An example of the deformed concentric pattern where, instead of right circles, we see wavy curved lines [5.20].

These rules produce the pattern shown on Figure 5.2a. Due to preset starting positions of rows (Con.5) and a constant distance between areola centers (R.4), areolae are located on discrete layers and form concentric circles. However, this pattern looks unnatural. There is a large amount of hyaline areas between rows (Figure 5.2a×), while on a real valve, areolae are packed more densely (Figure 5.1a).

Detailed investigation of the concentric pattern shows that the rows of areolae are influenced by each other. In every local region of the pattern, there are two adjacent rows that gradually diverge (Figure 5.1a×). These rows press on neighbor rows and shift them toward sector borders. Because of that, a triangular hyaline area is formed between divergent rows, while other areolae in a transect become arranged tightly to one another. When the distance between divergent rows exceeds the size of an areola, the divergence stops and a new row starts to form. On the next transect another two randomly selected rows start to diverge and the process repeats (Figure 5.3). Thus, the formation of a concentric pattern includes the repeating process of sequential divergence of adjacent rows.

The described algorithm produces the pattern on Figure 5.2b, which is very similar to the perfect concentric pattern of diatom valves (Figure 5.1a). The difference between them is that on the natural pattern the first areola of a new row can have a smaller size or it can be formed below a predefined level (Figure 5.1a×).

5.4 Spiral Patterns of Areolae

This pattern consists of spiral rows of areolae. In contrast to the other described pattern, the original spiral pattern does not contain radial rows (Figures 5.4b and 5.5b). Only spirals of areolae are visible in it.

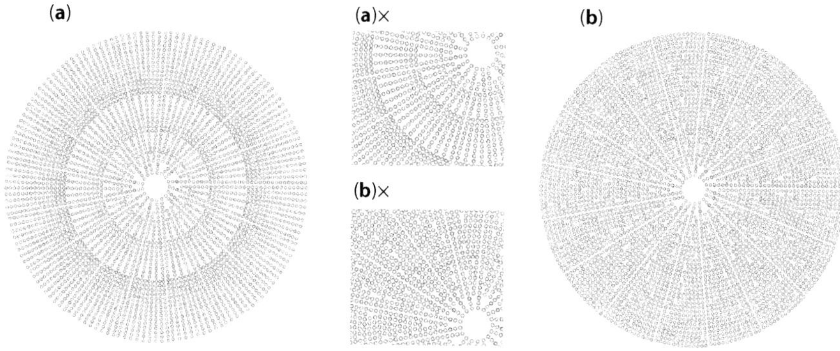

Figure 5.2 (a) An artificially generated concentric pattern where all rows do not interact with each other and stay straight. (a×) Zoomed fragment of pattern (a). There are many free spaces between rows thence the pattern looks rarefied. (b) An artificially generated concentric pattern where the process of sequential divergences of two adjacent rows took place. (b×) Zoomed fragment of pattern (b). Here, we see the elongated triangular hyaline areas, as in a real concentric pattern, and all areolae are densely packed.

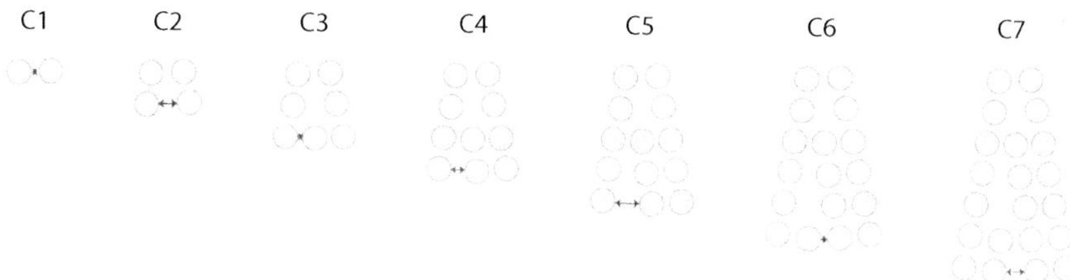

Figure 5.3 The repeating stages of the formation of the rows of a concentric pattern. When areolae optimally fill a transverse row (the bottom row in C1, C3, and C6), a pair of adjoining areolae is randomly selected (it is marked by a small black bar). The transverse forces act between this pair and push areolae away toward sector borders (C2, C4, C5, and C7). When there is enough of space between pushed areolae, the new areola is formed (C3 and C6), and the process repeats.

The original spiral patterns are rare. A very small number of diatom species have a spiral pattern on their valves. Spirals are sometimes visible on diatom valves with radial rows, as in *Coscinodiscus* species. But that pattern is constructed from radial rows, and spirals appear as a secondary structure. Therefore, I named this pattern as *pseudospiral* and do not consider it in this chapter.

A spiral pattern can be differentiated by the number of spiral groups that have different directions. Two kinds of spiral patterns are possible under such conditions: *unidirectional* (Figure 5.4b) and *bidirectional* (Figure 5.5b).

5.4.1 Unidirectional Spiral Pattern

In a unidirectional spiral pattern, all rows are curved in one direction. This pattern appears in species of the genus *Brightwellia* Ralf, especially well visible on the valve of *B. spiralis* Gleser (Figure 5.4b).

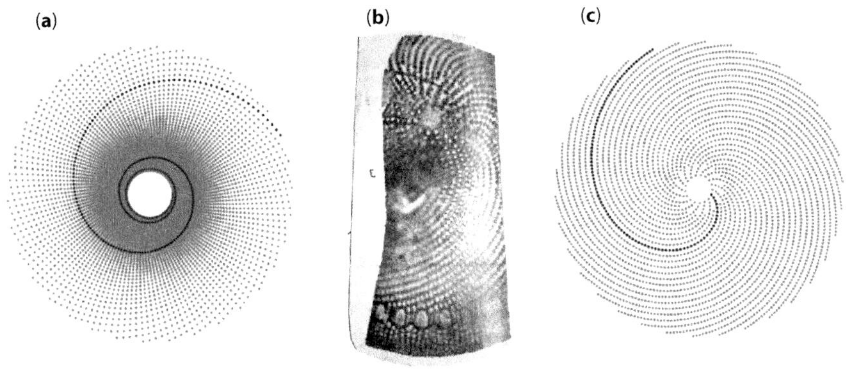

Figure 5.4 Natural (b) and artificial (a, c) unidirectional spiral patterns. (a, c) Geometric model of one-sided spiral pattern consisted of logarithmic (A) and Archimedean spirals (c). On each pattern, one spiral is highlighted to illustrate its shape. Logarithmic spirals are denser near the center, while Archimedean spirals more uniformly cover the surface. (b) Part of the valve of a fossil specimen of *Brightwellia spiralis* ([5.16]: pl. II, fig. 3) with a one-sided spiral pattern of areolae.

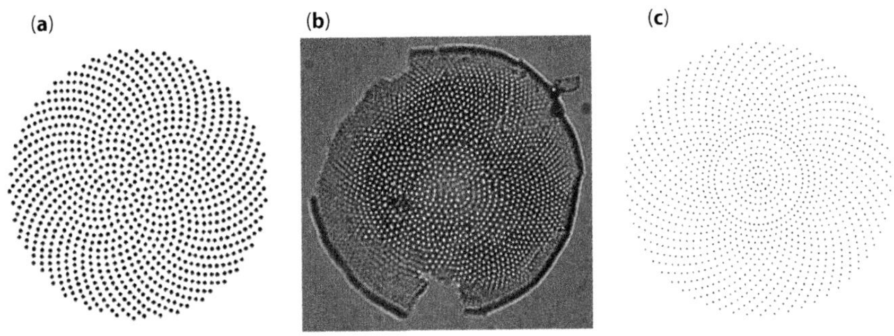

Figure 5.5 Natural (b) and artificial (a, c) bidirectional spiral pattern. (a) The model of spiral phyllotaxis. It contains some spiral families in which the curvature of visible spirals changes. (b) Specimen of *Roperia marginata* with a bidirectional spiral pattern of areolae [5.1]. (c) The bidirectional spiral pattern generated by a geometric model. The model produces only clockwise-oriented spirals, which are nicely visible in the pattern center. However, due to the carefully selected parameters, a group of counterclockwise-oriented spirals also appears.

The pattern includes primary and secondary spiral rows. Primary spirals start from the annulus (valve center) and continues toward the valve margin. Secondary spirals are located between them.

I have two hypotheses concerning the type of spirals. They can be either logarithmic or Archimedean [5.3]. Both are common in nature. To test which spiral is present on the diatom valves, I have constructed two different patterns and visually compared them with natural one (Figure 5.4).

When spiral rows are logarithmic spirals (Figure 5.4a), the distance between the turns is very small near the valve center and significantly increases toward the valve margin, which is not observed on a real valve. Moreover, a logarithmic spiral makes more than two turns around the center, which is also not true for the observed pattern (Figure 5.4b).

When the spirals are Archimedean, they uniformly cover the valve surface and make less than one turn around the valve center. So the generated pattern (Figure 5.4c) more closely corresponds to the observed one (Figure 5.4b).

5.4.2 Bidirectional Spiral Pattern

A bidirectional spiral pattern contains helices directed both in clockwise and counterclockwise directions (Figure 5.5b). This pattern looks similar to spiral phyllotaxis [5.4, 5.7]. It is a perfect pattern convenient for describing spiral arrangement of organs in higher plants [5.7, 5.15].

In previous work, I suggested that valves of diatom *Roperia marginata* contain a real spiral phyllotactic pattern ([5.11]: Figure 5.3). But this is not true. A detailed visual inspection shows that the pattern on the valve is different from a spiral phyllotaxis (compare Figures 5.5a and b). On the phyllotactic pattern, the spirals form well separated families where the curvature of visible spirals changes (Figure 5.5a), whereas *R. marginata* contains only one family of oppositely directed spirals (Figure 5.5b).

The observed natural spirals do not make a full rotation around the pattern center. Only 0.1–0.2 part of full coils are found on the valve. This leads to three consequences:

- Spiral rows form a small angle with the radial direction and are practically straight.
- New spiral rows are formed close to the valve margin, because only there is there enough free space.
- The concentric rows of areolae can be distinguished. Their appearance is caused by almost equal distances between areolae in a row and a nearly straight direction of spiral rows. But these concentric rows are not right circles.

To simulate a bidirectional spiral pattern, I suggest that diatoms do not form oppositely directed groups of spirals. They create only unidirectional spirals and make distances between spiral rows bigger than the distances between areolae in rows. Because of that, the second group of oppositely directed spirals appears automatically.

This explanation is used as an algorithm for the creation of an artificial bidirectional spiral pattern. For the simulation, I use single-screw Archimedean spirals. When the distance between spiral rows is about two times larger than the interareolae distance, the model produces bidirectional spiral patterns (Figure 5.5c) that look natural.

5.4.3 Common Genesis of Unidirectional and Bidirectional Spiral Patterns

The model used shows a unified mechanism of spiral pattern morphogenesis. Diatoms, changing the distance between spiral rows and areolae in rows, achieve construction of visually different but structurally identical spiral patterns.

5.5 Conversion of an Areolae-Based Model Into a Frame-Based Model

Schmid and Volcani [5.17] described the stages of the development of valves of some centric diatoms. On one of the stages, the valve pattern is formed from "spaghetti-like siliceous strings" that start to grow from the valve center and each of which "radiates to the margin". Each string continuously bifurcates allow them to cover all of valve surface by a tree-like structure. At uniform intervals between each pair of adjacent strings, the transverse ribs are branched, which creates the areola walls.

Thus, every row of areolae starts to form when a siliceous string bifurcates. The two branches of the bifurcated string limit the row. Transverse ribs between these parts form areolae.

To convert an areolae-based model into a frame-based model, it is necessary:

- to stretch siliceous strings from the pattern center (annulus) to the starting areola of every row;
- to bifurcate every siliceous string into a pair of strings prior to every first areola of a row, and to replace each row by a pair of siliceous strings;
- to replace each areola by transverse ribs between this pair.

The resulting construction can be reproduced using any suitable model, for example, with the help of L-system [5.11].

5.6 Conclusion

The principles of construction of areolae-based models described here can be adopted for the simulation of a diversity of diatom patterns.

The areolae-based models can be translated into frame-based models. The last ones are naturalistic and simulate the process of the formation of centric diatom valves from bifurcated siliceous strings.

The model presented simulates perfect patterns. However, in nature, most centric diatoms patterns are imperfect. They contain different defects: areolae shift from a perfect position or imperfectly change their size. But when an object is perfect, it does not change. The presence of defects in patterns allows them to change over time, that is, an integral part of diatom evolution.

Acknowledgements

I would like to thank Richard Gordon for his help in English proof reading of the chapter. This work was supported by the scientific research theme of the A.O. Kovalevsky Institute of Biology of the Southern Seas of RAS (No. 121030100028-0).

References

[5.1] California Academy of Sciences, Geology: *Roperia marginata*. https://flickr.com/photos/casgeology/4526334861/in/album-72157626507832405/, 2019.

[5.2] de Ville, F. and Voisin, D., *Aulacodiscus orientalis* Grev., http://forum.mikroscopia.com/topic/4243-aulacodiscus-orientalis-grev/, 2006.

[5.3] Diedrichs, D.R., Archimedean, Logarithmic and Euler spirals - intriguing and ubiquitous patterns in nature. *Math. Gaz.*, 103, 556, 52–64, 2019.

[5.4] Dixon, R., Spiral phyllotaxis. *Comput. Math. Appl.*, 17, 4–6, 535–538, 1989.

[5.5] Gordon, R. and Drum, R.W., The chemical basis of diatom morphogenesis. *Int. Rev. Cytol.*, 150, 243–372, 421–422, 1994.

[5.6] Hotton, S., Johnson, V., Wilbarger, J., Zwieniecki, K., Atela, P., Golé, C., Dumais, J., The possible and the actual in phyllotaxis: Bridging the gap between empirical observations and iterative models. *J. Plant Growth Regul.*, 25, 4, 313–323, 2006.

[5.7] Jean, R.V., *Phyllotaxis: A Systemic Study in Plant Morphogenesis*, Cambridge University Press, New York, NY, USA, 1994.

[5.8] Lenoci, L. and Camp, P.J., Diatom structures templated by phase-separated fluids. *Langmuir*, 24, 1, 217–223, 2008.

[5.9] Longuet-Higgins, M.S., Geometrical constraints on the development of a diatom. *J. Theor. Biol.*, 210, 1, 101–105, 2001.

[5.10] Lyakh, A.M., The model of areolae distribution and surface area of pores located on the valves of centric diatoms *Coscinodiscus* Ehr. and *Thalassiosira* Cl. *Mod. Phytomorphol.*, 3, 213–218, 2013.

[5.11] Lyakh, A.M., Morphogenesis of areolae pattern of Centric diatoms valve as L-system. I. Radial-parallel and spiral patterns. In: Modern trends in biological physics and chemistry, in: *BPPC-2016: proceedings of XI International Science-Technica Confernce. Sevastopol*, 25-29 April, 2016, vol. 1, Sevastopol State University, Sevastopol, pp. 56–59, 2016.

[5.12] Lyakh, A.M. and Bedoshvili, Y.D., Tiny diatoms *Minidiscus vodyanitskiyi* sp. nov. (Bacillariophyta) from the Sea of Azov and consideration of polygonal areolae pattern. *Phytotaxa*, 375, 2, 171–181, 2018.

[5.13] Mann, D.G., Specifying a morphogenetic model for diatoms: an analysis of pattern faults in the Voigt zone. *Nova Hedwig.*, 130, 97–117, 2006.

[5.14] Parkinson, J., Brechet, Y., Gordon, R., Centric diatom morphogenesis: A model based on a DLA algorithm investigating the potential role of microtubules. *Biochim. Biophys. Acta-Mol. Cell Res.*, 1452, 1, 89–102, 1999.

[5.15] Pennybacker, M.F., Shipman, P.D., Newell, A.C., Phyllotaxis: Some progress, but a story far from over. *Physica D-Nonlinear Phenomena*, 306, 48–81, 2015.

[5.16] Scheschukova-Poretzkaja, V.S. and Gleser, S.I., Novie vidy morskikh paleogenovikh diatomovikh vodoroslei U.S.S.R. (Diatomeae Marinae novae e Paleogeno Ucrainiae). *Novitates Systematicae Plantarum Non Vascularium (Academia Scientiarum URSS Institutum Botanicum Nomine V.L. Komarovii) (Nov. Sist. Nizsch. Rast.)*, 1, 78–92, 1964, 74 pls.

[5.17] Schmid, A.M.M. and Volcani, B.E., Wall morphogenesis in *Coscinodiscus wailesii* Gran and Angst. I. Valve morphology and development of its architecture. *J. Phycol.*, 19, 4, 387–402, 1983.

[5.18] Sumper, M., A phase separation model for the nanopatterning of diatom biosilica. *Science*, 295, 5564, 2430–2433, 2002.

[5.19] Sumper, M. and Kröger, N., Silica formation in diatoms: The function of long-chain polyamines and silaffins. *J. Mater. Chem.*, 14, 14, 2059–2065, 2004.

[5.20] Walker, D., John Albert Long, 1863–1945, http://microscopist.net/LongJA.html, 2012.
[5.21] Willis, L., Cox, E.J., Duke, T., A simple probabilistic model of submicroscopic diatom morphogenesis. *J. R. Soc. Interface*, 10, 83, 20130067, 2013.

6

Diatom Pore Arrays' Periodicities and Symmetries in the Euclidean Plane: Nature Between Perfection and Imperfection

Mohamed M. Ghobara[1*], Mary Ann Tiffany[2†], Richard Gordon[3,4‡] and Louisa Reissig[1§]

[1]*Department of Physics, Freie Universität Berlin, Berlin, Germany*
[2]*Department of Biology, San Diego State University, San Diego, CA, United States*
[3]*Gulf Specimen Marine Laboratory, Panacea, FL, United States*
[4]*C.S. Mott Center for Human Growth & Development, Department of Obstetrics & Gynecology, Wayne State University, Detroit, MI, United States*

Abstract

Diatom frustules feature species-specific, multilayer, multiscalar, hierarchical porosity, making them a gift for (nano)technological applications. Their precision challenges the present nanofabrication techniques of similar structures, e.g., photonic crystals or diffraction gratings. Understanding the geometry of the pore arrays present in diatom frustules might seem an easy task, however, imperfections and the overlay of various symmetric features complicate the picture. Nevertheless, evaluating the periodic regularities of such pore arrays along with its degree of perfection in a reliable, comparable, and easy method is important to critically assess different frustule designs for certain applications, especially in photonics and optoelectronics. In this chapter, we provide an in-depth description of the geometry of the 2D hierarchical porosity patterns, which appear in and are representative for diatom frustules. Using image processing and analysis techniques, much information can be extracted from scanning and transmission electron micrographs, atomic force micrographs, or even high-resolution optical micrographs. We present solutions to the development of a standardized and easy technique to observe and measure (i) possible planar periodicities, (ii) possible presence of different 2D lattices, and (iii) to assess the degree of perfection. Within this chapter, we also present a wide range of geometrical phenomena found in diatom frustules.

Keywords: Periodic porous silica, diatom frustule, 2D FFT analysis, 2D ACF analysis, natural two-dimensional lattice, planar symmetry groups, Bravais lattice

**Corresponding author:* mohamedghobara@rocketmail.com
†*Corresponding author:* mtiffany24@yahoo.com
‡*Corresponding author:* DickGordonCan@xplornet.ca
§*Corresponding author:* louisa_reissig@gmx.de

Vadim Annenkov, Joseph Seckback and Richard Gordon (eds.) Diatom Morphogenesis, (117–158) © 2022 Scrivener Publishing LLC

6.1 Introduction

Periodicity can be defined as a certain value that occurs at regular intervals in space and/or time. It is a common phenomenon in our universe found everywhere around us with a countless number of applications [6.2, 6.47]. For instance, in solid-state physics, atoms and molecules are frequently found tightly arranged with highly ordered two- or three-dimensional periodicities that can give a material unique physical and chemical properties depending on the various geometrical parameters associated with these periodicities. These regular two- or three-dimensional periodic arrangements of atoms or molecules are called "crystals" with characteristic translational symmetry between the positions of their atoms. In order to understand crystals and their types, the following concepts should be considered.

A crystal structure is usually described with the help of a lattice, a basis and a unit cell [6.4, 6.20]. The lattice is a purely mathematical concept used in geometry, which can be defined as a two or three-dimensional infinite periodic arrangement (array) of points, connected by lattice vectors, resulting in a structure similar to a mesh or net. In the case of a real crystal, the lattice points need to represent one or more atoms. How each lattice point is "decorated" with atoms or even entire molecules is given by the basis. Given that the crystal is characterized by translational symmetry, also a lattice representing a crystal must have translational symmetry, and therefore there must be a smallest entity or "unit" composed of lattice points, which when repeated along the lattice vectors recreates the entire crystal. This repeat unit is called the unit cell of the crystal. The exact choice of lattice vectors, basis and unit cells is conventionally defined by the desire to represent the crystal's symmetry and minimize the number of atoms and space in it.

When considering a lattice in 2D, as done in most parts of this chapter, there are exactly 5 possible types of lattices featuring the required translational symmetry (see Figure 6.1), the so-called "Bravais lattices". If the structure of the real 2D crystal can however only be represented by a lattice, when each lattice point represents more than 1 atom or different atom types, i.e., the basis is containing more than 1 atom or changing type depending on lattice position, it is conventionally considered to be non-Bravais in nature. Such lattices, such as the honeycomb-like lattice in graphene [6.3, 6.24] and Penrose tilings [6.54], are then often displayed with non-equivalent lattice points and lattice vectors, which require not only translational but also rotational operations or a change in "type" of lattice point to recreate the entire crystal structure. One should however note that when choosing a broader basis, leading to equivalent lattice points, such a non-bravais lattice can always be represented as a bravais lattice.

The concepts of periodicity, lattices, and crystal structures are not limited to the atomic or molecular levels but have been extended to larger structures, for instance, metamaterials or photonic crystals that strongly rely on their periodicities and their level of perfection for their intended applications [6.21, 6.28]. When expanding to biological structures, one can find various examples for 1D periodicities, as in diatom colonies (see Table 2 and related references inside [6.11]), but also 2D or 3D periodicities starting from small macromolecules within the cells of a living organism [6.9] up to large structures, such as the naturally formed biological photonic crystals [6.19, 6.47], pollen [6.38], plant stem cross-sections [6.15, 6.41], and diatom frustules, which are the focus of this study. However, in all those cases, the lattice points do not require to be equivalent on an atomic level, but equivalent in the existence of larger structural characteristics, depending on the system and application of interest.

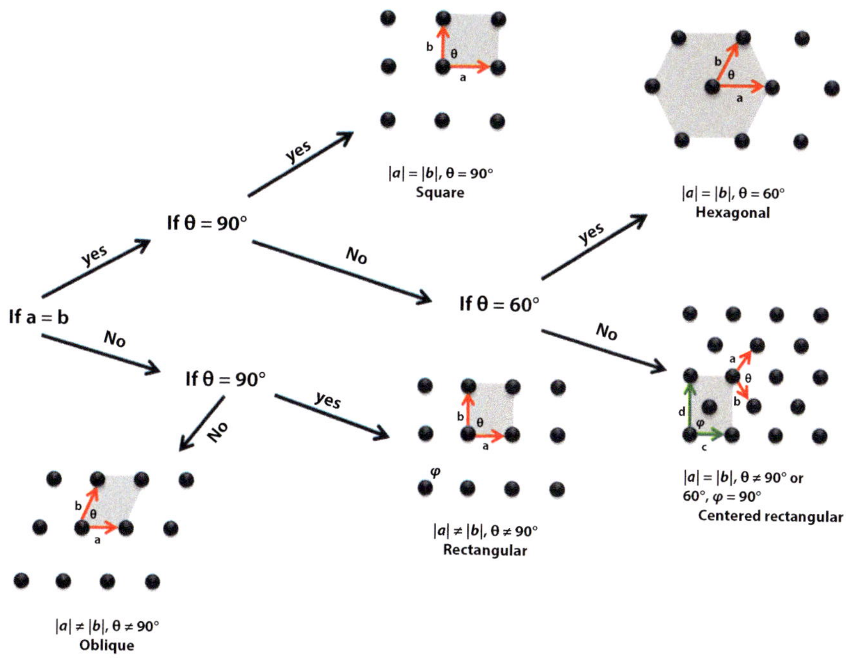

Figure 6.1 A flow chart illustrating the methodology for assigning the most likely 2D Bravais lattice to a measured 2D lattice during our study. All the 2D lattices in this figure are composed of periodic arrays of small simple spherical points. The imaginary unit cells of the different lattices are highlighted in gray, while a, b, and θ are the lattice parameters or constants, while c, d, and φ are given as additional related lattice parameters for the centered rectangular lattice. The red arrows represent the measured lattice vector lengths. It should be noted that the nomenclature used here follows that of the International Union of Crystallography (IUCr) [6.8, 6.20]. It should however be noted for completion that the hexagonal lattice is often also referred to as the "equilateral triangular lattice", the centered rectangular lattice as the "rhombic" lattice and the oblique lattice as the "parallelogrammic" lattice [6.59]. In planar wallpaper groups, those are further classified according to additional geometrical criteria [6.53, 6.59].

In this chapter, we are focusing, as already mentioned, mainly on periodicities in the 2D (the Euclidean) plane, describing the periodicity of pores on the 2D surface of diatom frustules. It should be noted that the intrinsic surface curvature of the diatom frustule needs to be also considered, when discussing periodicities and when interpreting the projected images obtained by ordinary imaging techniques. In general in this study, the lattice points are considered to be equivalent, when they represent pores of similar shape and size. In this case, all periodic structures should be able to be described with the five Bravais lattices (Figure 6.1). However, for comparison with other 2D materials we will also look at characteristic non-Bravais lattices, such as the honeycomb-like lattice.

Diatoms are photosynthetic aquatic microorganisms having a siliceous frustule surrounding their protoplast of design similar to a petri-dish, consisting of two overlapping thecae (epitheca and hypotheca) [6.37]. Each theca is made up of a valve and at least one girdle band (Figure 6.2). The two, or more, girdle bands together compose what is known as the girdle. That the frustule is porous in order to guarantee the exchange of the cell with the outer medium for food and signaling is reasonable, but interestingly, these pore arrays are multiscalar and often occur in many layers, and the pore arrays are of a species-specific arrangement [6.37]. In the case of valves, this porosity often indicates the presence of a buried 3D structure

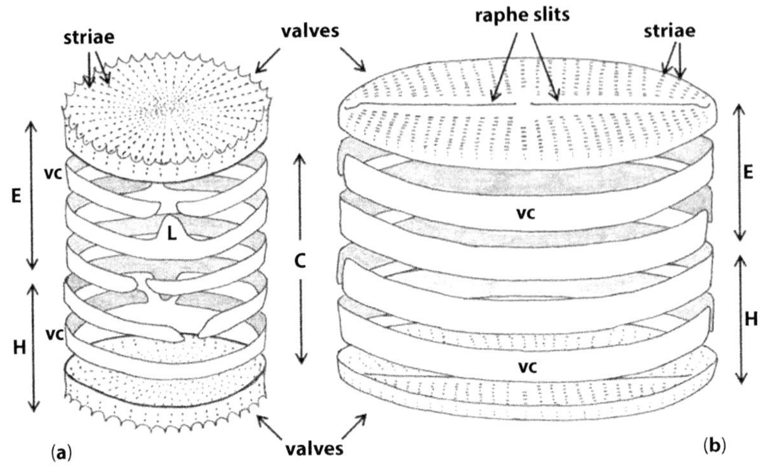

Figure 6.2 Diagram showing the frustule structure of (a) a centric frustule and (b) a raphid pennate frustule. E means an older epitheca; H means younger hypotheca; C means girdle bands; L means tongue-like extension; and VC means valvocopula. Reproduced from [6.6] under the Creative Commons Attribution 4.0 License.

called areolae. The openings of the areolae are often covered with internal and/or external porous layers, similar to 2D thin slabs of silica. The pore arrays which appear at the external or internal face of a valve are usually arranged in so-called striae (Figure 6.2). According to the symmetry of the distribution of these striae and giving space for imperfections, diatoms are usually divided into two major groups: centrics and pennates. Centrics have a radial symmetry [6.33, 6.44], while pennates usually have bilateral symmetry.

Diatomists consider the geometrical distribution of such striae as a taxonomic criterion but are not much concerned about the geometrical relations between its discrete pores and potential 2D planar symmetry groups present between these pores. Instead of paying attention to the periodicity of the pore arrays, they focus efforts on the description of the overall frustule geometry, symmetry of the valves, symmetry of striae, number of striae per micron, and on measuring the area of porosity [6.29, 6.37].

With the advent of diatom nanotechnology, many material scientists, physicists, and chemical engineers became interested in the unique properties of diatom frustules, which are associated with their architecture, the size and shape of pores (including nanopores), and their periodicities and symmetries [6.7, 6.16, 6.18, 6.26, 6.27]. The interaction with light, i.e., the frustule photonics, is expected to be affected significantly, especially in a medium with high refractive index contrast such as air, by many structural factors including the multilayer nature of the valve or the girdle band with focus on the thickness, pore diameters, possible 2D lattice type, and its associated parameters, including pore spacing, present in each layer [6.5, 6.11, 6.13, 6.14, 6.32]. As with any 2D photonic crystal or 2D diffraction grating, these factors would affect their photonic properties [6.22, 6.58]. That is one of the reasons why the 2D pore arrays' periodicities and symmetries demand increasing attention.

To date, only a few published studies focused on the diatom's 2D pores' periodicities, for instance [6.5] and [6.25]. In general, the 2D hexagonal Bravais lattice was the most common 2D lattice structure associated in literature with the 2D periodic pore arrays within various diatom valves [6.10, 6.12, 6.25, 6.27, 6.32]. For example, *Coscinodiscus granii* and *C. walesii* are two well-known centric diatoms acknowledged to have 2D hexagonal lattices [6.7, 6.10].

Interestingly, in [6.10], it has been mentioned that the girdle bands of *Coscinodiscus granii* have another 2D lattice structure (square lattice). Similar results were obtained in [6.14] in greater detail. They also used this square lattice type, but with different lattice parameters, to describe the 2D pore arrays in the girdle bands of other centrics, including *Thalassiosira pseudonana*, *Coscinodiscus radiatus*, and *Coscinodiscus wailesii* [6.14]. Moreover, in [6.12], the 2D square and quasi-square lattices were observed in pennate diatoms: *Gyrosigma fasciola* and *Gyrosigma balticum*, respectively. Thus, examples are reported for two 2D Bravais lattices, so could all 2D lattice types, the five Bravais lattices as well as some non-Bravais lattices, be present in diatoms?

Before answering this question, it should be noted that in most studies, little has been mentioned regarding the methods to assign the lattice and its parameters, and those were often based on bare hand measurements. In [6.27], Fourier analysis has been used to more accurately define the hexagonal lattice of the pore arrays in the atomic force micrographs of *Coscinodiscus* sp. and *Thalassiosira eccentrica*. This method was also used in [6.14] for the same purpose. It is worth mentioning that Fourier analysis of optical micrographs of diatom valves was also carried out in [6.28] for a different goal: the automatic identification of diatom genera and species based on their characteristic patterns in the frequency domain. Interestingly, this analysis could not only be done through image analysis of micrographs, but in [6.1] also experimentally using phase contrast diatom micrographs and an optical diffractometer with a laser source to get its Fourier transform pattern. Another experimental approach, based on using the diffraction patterns of single diatom valves instead of using micrographs, showed the hexagonal pattern of *Coscinodiscus* valves associated with the hexagonal pore arrangement [6.60-6.62]. However, the patterns obtained from both experimental approaches were based on the entire input valve and did not contain selected information of specific pore arrays on the surface or the exact representation of its basis. Alternatively, Cohoon *et al.* [6.5] used Matlab software to carry out the Delaunay triangulation of segmented SEMs for statistically estimating pore spacing (from center to center), providing average spacing and its deviation. However, they did not provide information regarding vector angles and thus could not give a clear image about the 2D lattice within the pore arrays and its perfection.

Although, when studying the lattice type and its associated parameters at the atomic scale, X-ray diffraction techniques are usually used to find out the reciprocal lattice. In larger structures such as in diatoms, scanning and transmission electron micrographs (SEMs and TEMs), and atomic force micrographs (AFMs) are generally of sufficient resolution to study and understand the 2D periodicities of pore arrays. It should be noted that optical micrographs with high contrast and enough resolution could be suitable to study large pore arrays; however, it is not sufficient for most of diatoms. Using bare hands combined with a basic image analysis software (such as ImageJ [6.36]) could be sufficient, if structures are clear and lattices easy to spot. However, it could also be illusive and time consuming to get statistically representative values. Therefore, it is better to use more advanced techniques for image analysis and data acquisition such as the 2D fast Fourier analysis to reveal the periodicity accurately and determine its type and its degree of perfection.

The 2D Fast Fourier Transform (2D FFT) is a useful technique for image analysis to produce the Fourier or frequency domain (sine and cosine components of periodic signals) for a given spatial domain (the original image). One of the important applications for 2D FFT is its pattern or texture recognition ability [6.28]. Thus, it could provide us with desired information about the periodicity, lattice parameters, and symmetry within 2D images.

Besides Fourier analysis, the cross-correlation function is a standard tool in (periodic) signal processing to study the coherence between two data sets depending on a temporal delay or spatial shift [6.42]. In case the coherence within a single data set is studied, to, for example, study periodicities within a structure, the autocorrelation function (ACF) is used. Applications of this methodology include analysis of spatial data distribution [6.46], Fluorescence Correlation Spectroscopy [6.30], and pattern recognition in 2D and 3D micrographs [6.31]. In the case of image analysis, the ACF methodology is based on the correlation of a signal with a shifted copy of itself as a function of extend of shift in different directions of the image. It can be used to reveal and compare characteristic length scales between repeating patterns otherwise obscured by noise and depending on the correlation direction. Thus, it is potentially a suitable tool for the study of periodicities and their coherence within a given micrograph and thus to extract the periodic pattern parameters if present. 2D ACF calculations can be derived from 2D FFT equations according to the convolution theorem [6.35]. 2D ACF image analysis results are generally visualized as a 2D map of ACF (or coherence) values. Any observable periodicity in this image reveals information about occurring periodicities in the original image, with the spatial distribution corresponding to the noise-free (averaged) lattice parameters in the original image. The strength of the coherence signal and its decay provides information about the spatial variability of the periodic structure in the original image. However, this information is among other things also dependent on the finite size of the input signal.

It should also be noted that, although it is widely accepted that diatom frustules have a regular to quasi-regular porosity, it is not well-studied how close this periodicity is to perfection when compared to the calculated lattice of averaged or optimal parameters of the assigned lattice type. Furthermore, little is known about how the periodicity (and thus coherence with the assigned lattice) is dependent on position (e.g., center of valve compared to the edge). Additionally, there is a lack of discussion on the natural defects that occur even in the highly regular periodic pores of some diatom frustules. Longuet-Higgins [6.25] studied the central defects present in some centric models. He mentioned that the pore array arrangement within *Thalassiosira oestrupii* is hexagonal with the exception of pores near the center of the valve, very similar to what is known as the central lattice defect. A number of articles including [6.5, 6.11, 6.13] have since then discussed and introduced the presence of lattice defects in diatom pore arrays.

In this chapter, we only focus on the 2D periodicities of pore arrays within a single silica layer of the projected 2D surface of the diatom frustule and do not aim to explore the multilayer nature of porosity of the valves. By doing so, we seek for the standardization and establishment of a clear methodology to reveal the periodicity of planar pore arrays in diatoms. Further, we try to answer a number of questions including: the possible presence of different 2D lattice types in diatoms, how perfect diatoms could be in the fabrication of such 2D lattice structures compared to perfect 2D Bravais lattices, and how perfect diatoms could be within the fabrication of different parts of a single valve. Finally, we discuss the planar symmetry groups present in diatoms through considering the additional symmetrical relations between pores within different diatom frustule parts.

6.2 Materials and Methods

About 53 SEMs belonging to nine diatom species were carefully selected from Mary Ann Tiffany's catalog of SEMs. Additionally, five SEMs belonging to five different diatom species were chosen from the literature. It should be noted that the results, presented in this chapter, are just examples and may not be representative for the general morphometric criteria of these species.

Using the raw micrographs produced from SEM (or TEM and AFM) directly without any modifications for the analysis of the targeted pore arrays is expected to be unclear and thus unsuitable for reliable analysis. Therefore, in order to verify this hypothesis, analysis was carried out for both raw SEMs and for micrographs resulting from image segmentation, which was carried out to extract the targeted pore patterns.

6.2.1 Micrograph Segmentation

All the segmentation processes were carried out using ImageJ software [6.52]. Histogram equalization was carried out for the raw micrographs to enhance the contrast of the image through adjusting the image histogram's intensity distribution without loss of details. Afterward, if necessary, background subtraction was carried out to lower the noise. Then, the micrographs were converted into binary images to maximize contrast of the structures. After that, segmentation was carried out for the binary images through thresholding. The threshold value was variable and chosen to optimize the output image. Finally, the targeted porosity pattern was selected with the "clear surrounding" option. This method produced white pores on a black valve and black background, which we found gave optimal results, similar to the case of black pores on white valves and white background. It should also be noted that, in some cases, any still present noise or undesired details (such as impurities or structures of the layers below the pores) were manually covered with black color.

6.2.2 Two-Dimensional Fast Fourier Analysis and Autocorrelation Function Analysis

The original or segmented micrographs were individually imported into Gwyddion software (version 2.51) [6.49], and the correct image dimensions specified in pixels or micrometers. The corresponding 2D FFT pattern was calculated automatically with a Hann window selection to obtain the frequency domain, containing information about possible periodicity and symmetry groups within the micrograph. For comparison, the 2D ACF was also calculated using suitable refinements to study the corresponding 2D ACF pattern for the possible periodicity and symmetry groups within a given pattern. Comparisons between these methods have been made to optimize the results and interpretations and to understand the limitations of these two methods.

6.2.3 Lattice Measurements and Recognition

Using the "measure lattice" function in the Gwyddion software, the unit cell's parameters can be measured utilizing either 2D ACF or 2D FFT patterns (specifically 2D PSDF, i.e., the normalized version of the 2D FFT spectra). Nevertheless, in our study, we preferentially used the 2D ACF pattern for lattice measurements for the reasons discussed in Section 6.3.1.2. In case of a lattice structure observed in a given 2D ACF graph, the structure around the central point of the 2D ACF pattern usually can be considered as a representative of the unit cell of a given 2D lattice (see Figure 6.3). It should be noted that the program, through

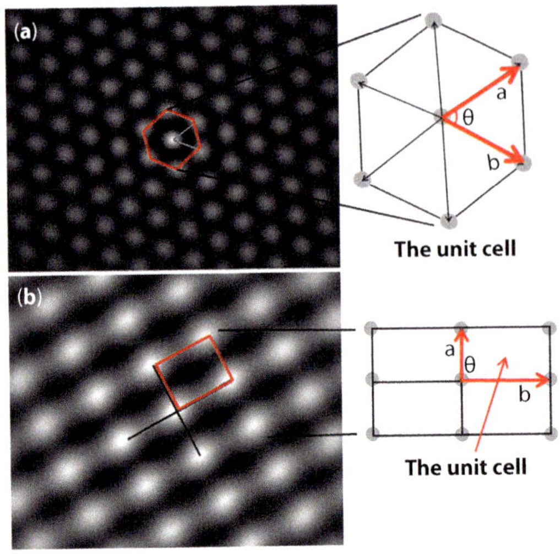

Figure 6.3 Diagram illustrating two types of 2D lattices, the hexagonal and rectangular, appearing in the corresponding 2D ACF patterns of thresholded micrographs of the internal view of a single valve of *Roperia tesselata* (a) and the internal view of a single valve of *Haslea* sp. (b), respectively, both with 4× magnification. The red shapes can be seen representative for the unit cell of the 2D lattice. To assign the most similar 2D Bravais lattice, vector lengths a and b, as well as the angle θ were extracted and compared with the flow cart in Figure 6.1.

its measure lattice window, enables only the measurements starting from the central point of the unit cell at the center of the pattern. To define the lattice type, three parameters were measured the vector lengths a and b, and the angle θ between these two vectors (Figure 6.3). The measured values can be considered averaged values between all the lattice points in the respective direction. As already mentioned, the height of the ACF at the central peak, as well as the decay length of the height of the consecutive peaks, can be related to the coherence of the autocorrelation of the pattern. If the lattice parameters are with high precision replicated in the valve, then the height of the central peak will approach unity, and the larger the extent of the autocorrelation length, the shallower the decay will be. If the central peak is low, then the autocorrelation coefficient and thus the coherence of the data is low. In contrast to the decay length and the height of the signal, the general width of the autocorrelation pattern is associated with about 2× the width of the pattern of the input image.

Following parameter extraction, the results of these measurements were compared to the flow cart in Figure 6.1 and the type of 2D lattice was assigned based on best fit to a standard lattice for different degrees of lattice imperfection. In all cases, the values of vector length a did not have the same numerical values as the vector length b, and the angles were neither 60° nor 90°, leading to an assignment of oblique pattern, if no deviation from the standard values was accepted. However, it was noted that often the numerical values could within a given precision window (5%, 10%, or 20%) be related to the standard lattice parameters for one or several 2D Bravais lattices of corresponding dimensions. To judge the values within an accepted deviation level, the numerical difference between vector lengths a and b was calculated and normalized to the vector length a, to obtain the percentage of the vector length deviation. In analogy, the difference between the measured angle θ and the closest angle (either 60° or 90°) was calculated and normalized to its standard value, to obtain the percentage angle deviation. The results obtained were compared to the flow chart in Figure 6.1, with accepted deviation level lower than 5%, 10%, and 20%, respectively.

$$\textit{deviation of vector length}\,(b)\textit{ from vector length}\,(a)\textit{ in percentage}\,(\%) = \frac{|b-a|}{a}\times 100 \tag{6.1}$$

$$\textit{deviation of angle}\,(\theta)\textit{ from the closest angle}\,(60^\circ\textit{ or }90^\circ)\textit{ in percentage}\,(\%) = \frac{|\theta - 60^\circ\ (or\ 90^\circ)|}{60^\circ\ (or\ 90^\circ)}\times 100 \tag{6.2}$$

6.2.4 Accuracy of 2D ACF-Based Calculations

Accuracy calculations of the 2D lattice parameters obtained using 2D ACF analysis were carried out using both Gwyddion and ImageJ software. A comparison between the bare-hand measurements of pore spacing (from center to center) in a specific direction as well as the angles between them using ImageJ and the measurement of the averaged value of the corresponding lattice parameters obtained by the 2D ACF analysis for the segmented micrographs using Gwyddion software was carried out for three selected examples, including; (i) the internal view of a single valve of *Roperia tesselata*, (ii) the internal view of a half of a single valve of *Haslea* sp., and (iii) the internal view of *Haslea feriarum*. The 2D ACF measurements were carried out three times separately for each original micrograph/example, in order to obtain a magnitude for the standard deviation associated with image processing and 2D ACF analysis. With this deviation, we can estimate how reproducible the 2D ACF results are, and it was found that the error was considerably lower than the error of bare-hand measurement. It should be noted, however, that the origin of the standard deviation in the bare-hand measurements lies mostly in the variation of lattice constants across the chosen area of the valve and is therefore rather correlated with the coherence of the autocorrelation pattern, than the precision of the detected period and thus lattice constants. It is expected that the standard deviation associated with the coherence of the pattern should lie within the order of magnitude of the standard deviation of the bare-hand measurements. Future work will be carried out in order to confirm this expectation and provide an "easy" method for extracting the values directly from 2D ACF analysis.

The comparison between the two measurements has been carried out via calculating the probability of the measured 2D ACF value to fall within the range of the measured value obtained by bare-hand measurements. To do this, first, the deviation between the values was calculated for each parameter separately: $|x_{hands} - x_{ACF}|$. Then, the width of the probability window t_σ of the normal distribution with standard deviation σ, obtained from the bare hand measurements was calculated;

$$t_\sigma = \frac{|x_{hands} - x_{ACF}|}{\sigma} \tag{6.3}$$

Then, the probability of the 2D ACF value to lie outside the normal distribution of the bare hand measurements was extracted from Table A (Appendix A) in [6.45] and the probability of a statistical match between the 2D ACF value and the bare hand measured value calculated.

$$Prob(inside\ t_{\sigma}) = 100\% - Prob\ (outside\ t_{\sigma}) \tag{6.4}$$

If the probability is large, then the accuracy of the 2D ACF method as compared to the bare hand measurements was seen satisfactory. It should be noted that this probability decreased (for the same deviation) if the bare hand measurements are more precise, corresponding to a smaller variation in lattice parameters across the sample.

6.2.5 The Perfection of the Unit Cell Parameters Between Different Parts (Groups of Pore Arrays) of the Same Valve and the Same Micrograph

In order to compare the variation of the unit cell parameters between different parts (groups of pore arrays) of the same diatom valve, two examples were considered as illustrated in Table 6.5. The same methodology mentioned in Section 6.2.3 was followed to extract, measure, and recognize of the 2D lattice for each selected group of pore arrays separately. Such analysis should provide information about the diatom's ability to reproduce a pattern with similar degree of perfection within different places on the same valve.

In future, this analysis should be extended to study the overall variation with position of a lattice as compared to a calculated lattice of parameters according to the unit cell obtained. This could be done by taking the difference between a generated image of the same dimensions with the original image.

6.3 Results and Discussion

6.3.1 Toward Standardization of the Methodology for the Recognition of 2D Periodicities of Pore Arrays in Diatom Micrographs

6.3.1.1 Using Two-Dimensional Fast Fourier Transform Analysis

The Fourier domain, obtained from 2D FFT analysis, is a micrograph of equal size to the original one, which illustrates potentially present 2D periodicities. The frequency domain is a reciprocal of the real 2D space with internal lengths measured as (1/original length scale) units, as shown in the scale bar of Figure 6.4. That means the smaller the period of a repetitive pattern is in the spatial domain (the original micrograph), the farther the observed positions are located from the center in the frequency domain (2D FFT domain) and *vice versa*. The number of rings in the 2D FFT pattern is on the one hand associated with the coherence of the pattern (as in the 2D ACF analysis, see Section 6.3.1.2). However also here, the limited sampling size artificially further reduces the decay length, and thus, comparison can only be made easily on the same micrographs. During our study, some important points have been observed regarding this method discussed in the following paragraphs to show its limitations and best procedures.

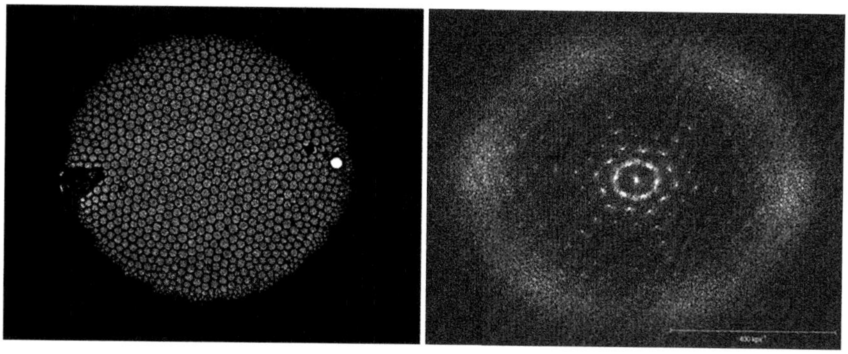

Figure 6.4 A segmented micrograph of the external view of a single valve of *Roperia tesselata* (1,024 × 800 pixels) with its corresponding 2D FFT (not magnified), illustrating the scale bar (400 Kpx^{-1}).

If the 2D FFT analysis was carried out on the original SEMs, i.e., without prior image processing (Figure 6.5), only for a few examples the 2D pore periodicities were relatively clear and did not require further processing. However, in most cases, when analyzing raw SEMs without segmentation or enhancement, confusing or unclear results were obtained. This could be associated with several sources of noise including: the resolution of the scanning electron microscope, its magnification, the stability of the applied voltage to the device,

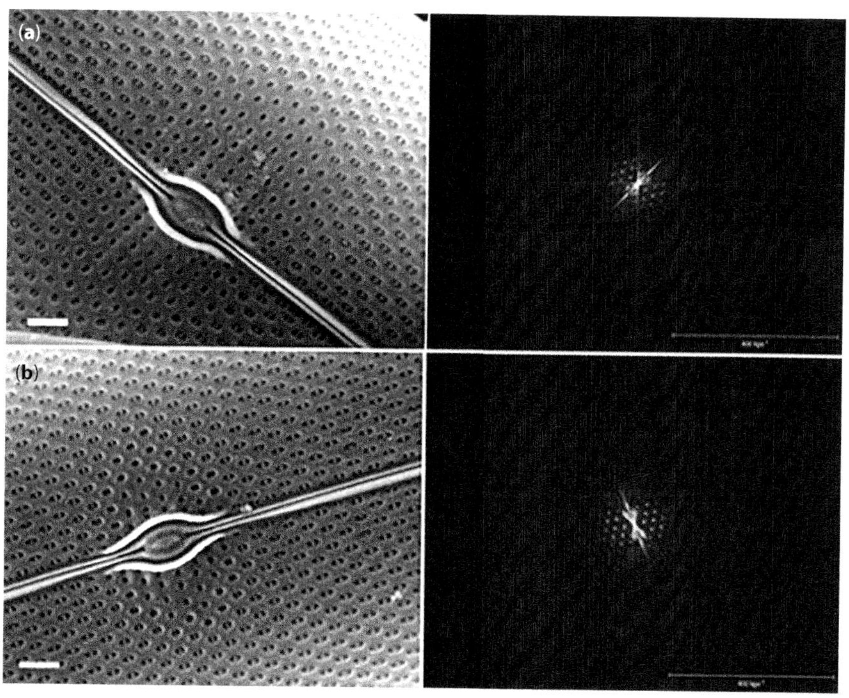

Figure 6.5 Two SEMs of the internal view of a single valve of *Pleurosigma* sp. (a and b) (both have the same dimensions, 1,024 × 800 pixels) and their corresponding 2D FFT domain. The major difference between the two micrographs is the rotational angle of the valve, which appears clearly from the raphe position and the highly bright axis at the center of the corresponding 2D FFT patterns. Scale bars of SEMs are 1 μm.

the image background, the presence of debris on the diatom frustules, and specular reflectance (Pappas, 2020, personal communication). For instance, the 2D FFT images of the original micrograph of the external view of a *Roperia tesselata* valve did not show a distinctive pattern, however a clear pattern appeared in the processed micrograph (Figures 6.6a vs. b). Also, the 2D periodic patterns, within the internal view of a single valve of *Asteromphalus hookeri*, became clearer as seen in the increased number of rings that appeared in the 2D FFT output image after image processing (Figures 6.7a vs. b). That is why segmentation through, for example, thresholding is recommended before calculating the 2D FFT and using its output images in periodicity recognition studies.

Moreover, a difference in the 2D FFT pattern was noticed when comparing processed micrographs with different brightness/darkness levels of the features in the background or the porosity (Figure 6.11). In general, the clearest 2D FFT patterns were obtained after maximizing the contrast between the pores and a homogeneous background, from which all remaining information (such as the presence of a valve edge) was removed. This can be either achieved via white pores on a black valve and black background, or via black pores on a white valve and white background, as it removes all the interference from the background and the valve with the periodic signals from pores (Figure 6.11). It should, however, be noticed that for Gwyddion an image of white pores on black background corresponds to "particles on a substrate" with height given when importing the image, and black pores on white background corresponds to "pores/holes on a substrate" with given depth.

Furthermore, it should be noted that the orientation of the 2D FFT pattern depends on the orientation of the valve in the micrograph (Figures 6.5a and b). Thus, when comparing

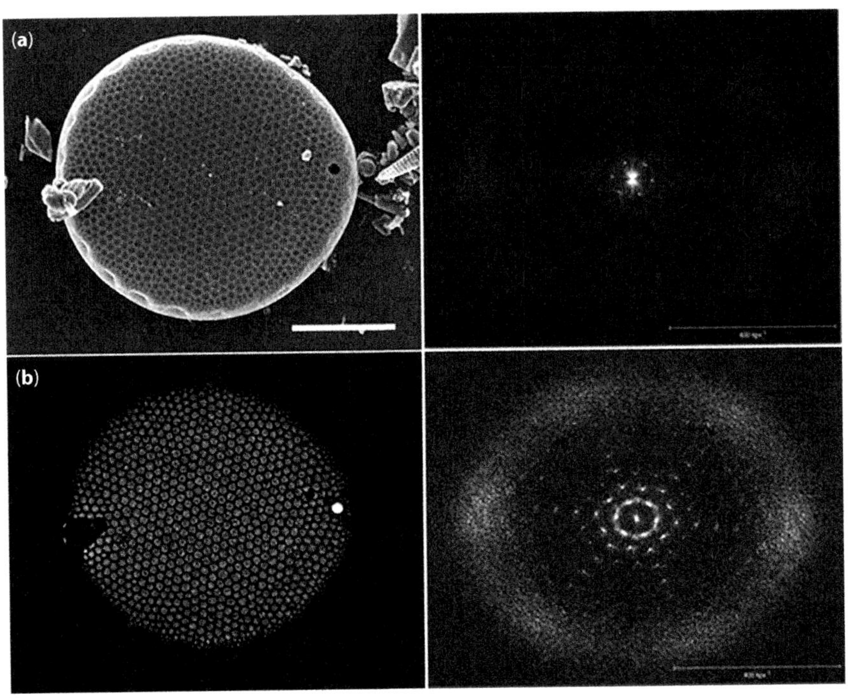

Figure 6.6 SEM of the external view of a single valve of *Roperia tesselata* (1,024 × 800 pixels) (a), its threshold (b) and their corresponding 2D FFT pattern. Scale bar of the SEM = 20 μm.

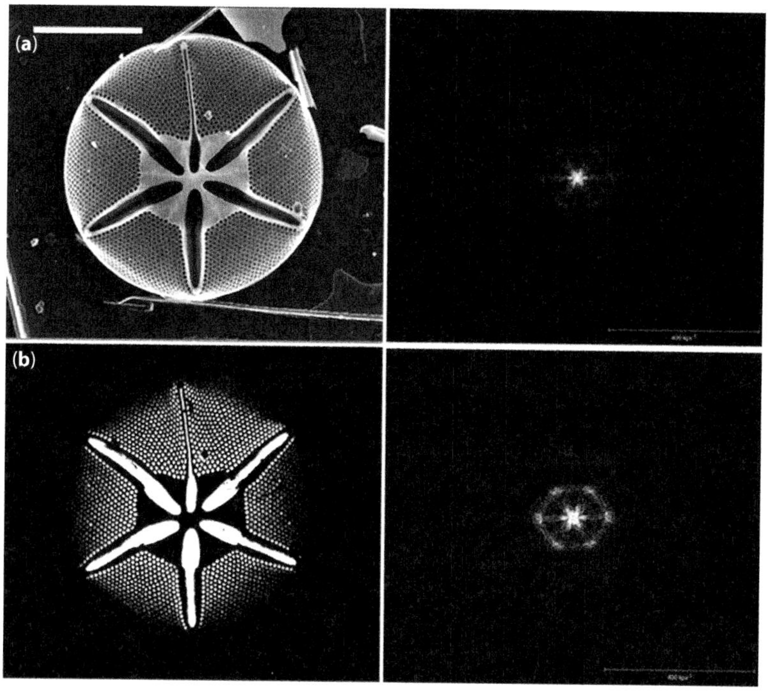

Figure 6.7 SEM of the internal view of a single valve of *Asteromphalus hookeri* (1,021 × 885 pixels) (a), its threshold (b) and their corresponding 2D FFT pattern. Scale bar of the SEM = 30 μm.

images, the origin of orientation should be standardized. Additionally, a tilt of the valve in a micrograph could lead to a distortion of its Fourier domain pattern (as seen in Figure 6.8), and thus, a potentially wrong pattern assignment. Thus, using tilted valves is not recommended for the study of 2D pore arrays; otherwise, a correction for the tilt should be carried out [6.33].

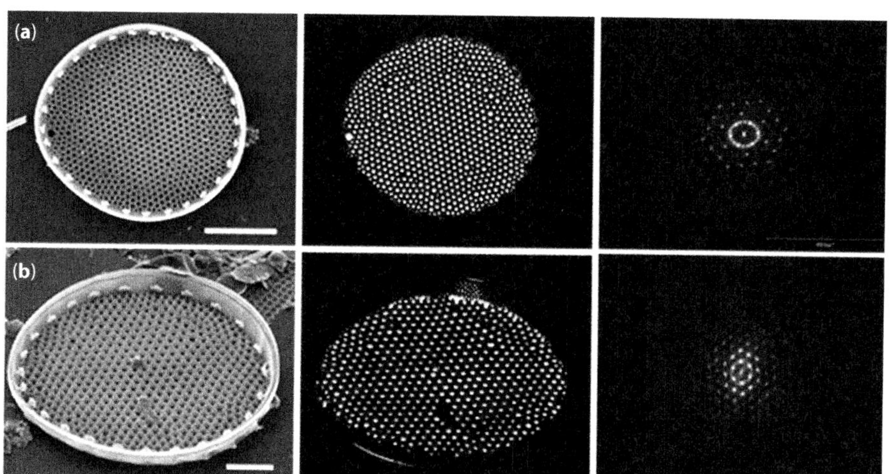

Figure 6.8 SEM of an internal view of a single valve of *Roperia tesselata* in upright position (a), scale bar = 20 μm, in a tilted position (b), scale bar = 10 μm (both have the same dimensions, 1,024 × 800 pixels) and their corresponding segmented micrographs and 2D FFT domains.

Apart from the tilt during acquisition, the natural curvature of the surface of interest can also hamper the analysis when using the 2D projection obtained in the micrographs. In such a case, it would be necessary to reduce the area to only contain the pores of a almost flat region which is not affected by the curvature too much.

In general, diatom valves often contain not only one but several scales of porosity and each can have its associated level of periodicity and symmetry (see various genera in [6.37]). In such cases, using the whole valve for the 2D FFT analysis may not reveal the true nature of the periodicity of the 2D pore arrays within such a diatom valve (as shown in Figures 6.9 and 6.10). In this case, the target group of pore arrays should be selected for study after segmentation. Luo *et al.* [6.28] followed a similar strategy in the segmentation for diatom valves, which helped them to obtain a clearer pattern in the Fourier domain (see Figure 4 in [6.28]). In other words, rather than using the whole valve for the 2D FFT analysis, it is better to isolate different parts that have distinct symmetry, e.g., through cutting the image or using the "clear surrounding" option in ImageJ and performing a separate 2D FFT analysis for each part to determine the actual 2D pore periodicity associated with each group. For example, in Figure 6.10, the periodicity pattern of the pores became clearer after isolation of different parts and carrying out the 2D FFT separately for each one. Nevertheless, a change in the degree of perfection of the lattice and its vectors' directions has been noticed in the Fourier domain, of the various selected (rotated) input regions (see further Section 6.3.3.3).

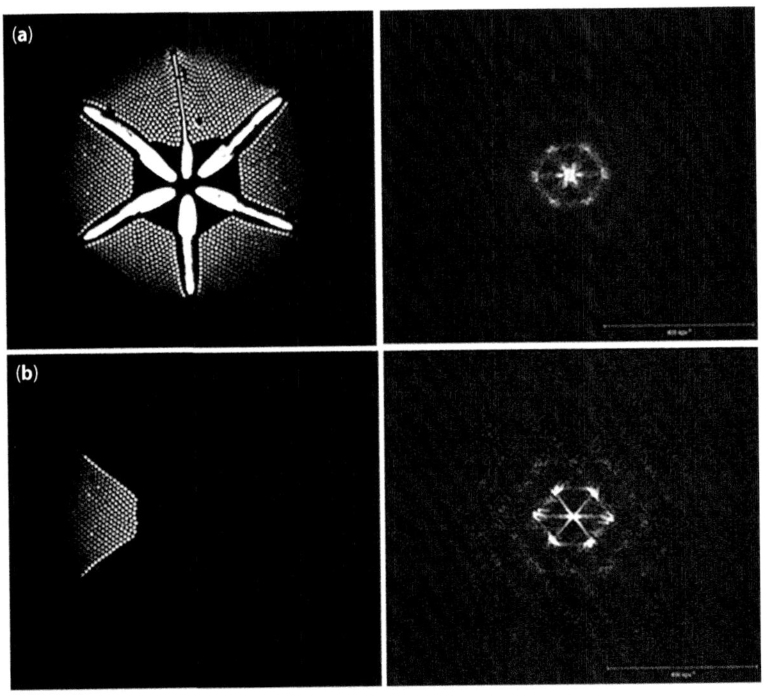

Figure 6.9 (a) A segmented SEM of the internal view of a single valve of *Asteromphalus hookeri* with six separate parts of 2D pore arrays that have rotational relations with each other (1,021 × 885 pixels), (b) a selected separated part, and their corresponding 2D FFT domain. The fact that the 2D FFT of the complete valve is not clear may be due to the interference emerging from the differences (and rotational transformation) of the periodic pattern between the six parts (see Figure 6.10). The 2D FFT pattern becomes very clear with distinct bright dots and more rings, when the 2D FFT analysis was carried out for the separated part.

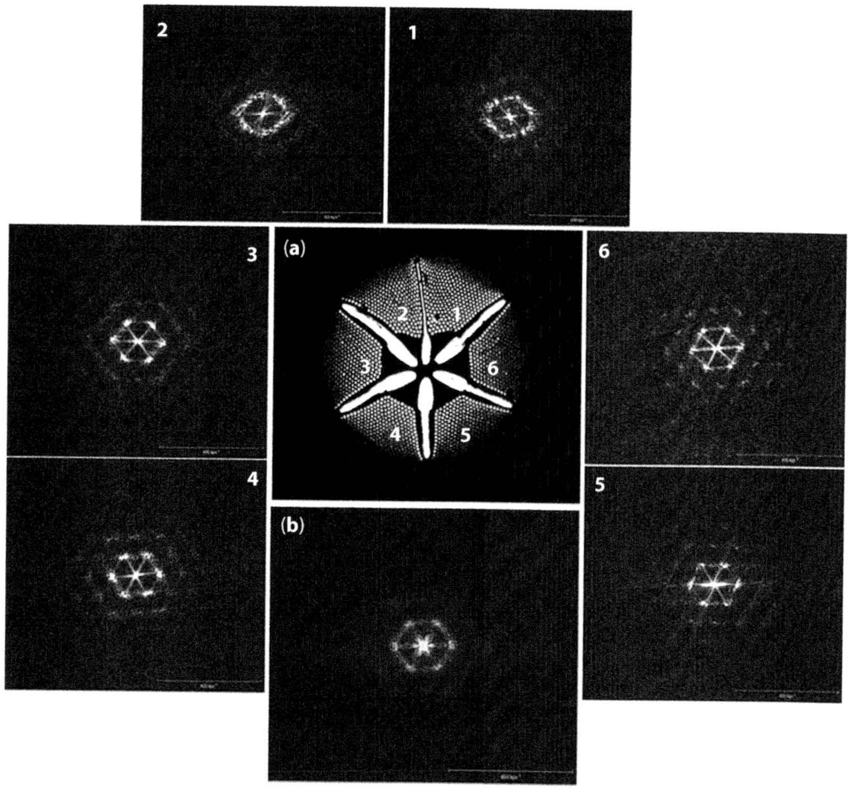

Figure 6.10 A segmented SEM of the internal view of a single valve of *Asteromphalus hookeri* at the center (1,021 × 885 pixels) (a), the complete valve 2D FFT pattern (b), and the separated different parts of the valve (marked with white numbers). These are in 1,...,6. This procedure avoids interference and ensures that the 2D FFT pattern originates from the 2D pore array pattern.

6.3.1.2 *Using Two-Dimensional Autocorrelation Function*

The 2D lattice parameters, i.e., lattice spacing and vector angle, can be retrieved from the Fourier domain (2D FFT) or its normalized form (2D PSDF); however, it was often found of comparably low resolution especially for large periods, when the reciprocal spacing becomes very close, which hampers its use for lattice recognition and measurements especially by non-mathematicians [6.23]. Additionally, the direction of the lattice vectors in the Fourier domain correspond to the inverse lattice (Figures 6.12a vs. b), which complicates assignment. In contrast, 2D ACF analysis offers not only better resolution but also a direct relation of the plane vector directions with those in the original micrograph (Figure 6.11; Figures 6.12a vs. c; Figures 6.13a vs. d). Also, for this method, some important points have been observed during our study, discussed in the following paragraphs to show its limitations and best procedures.

The 2D ACF pattern might include additional details about the lattice, for instance, the appearance of reflectional symmetry within the basis at each single lattice point as shown in Figure 6.12c, which can be associated with an additional planar symmetry group and is further discussed in Section 6.3.4. Nevertheless, it should also be noted that the 2D ACF will not demonstrate any lattice defects as this method oversees through "averaging" the

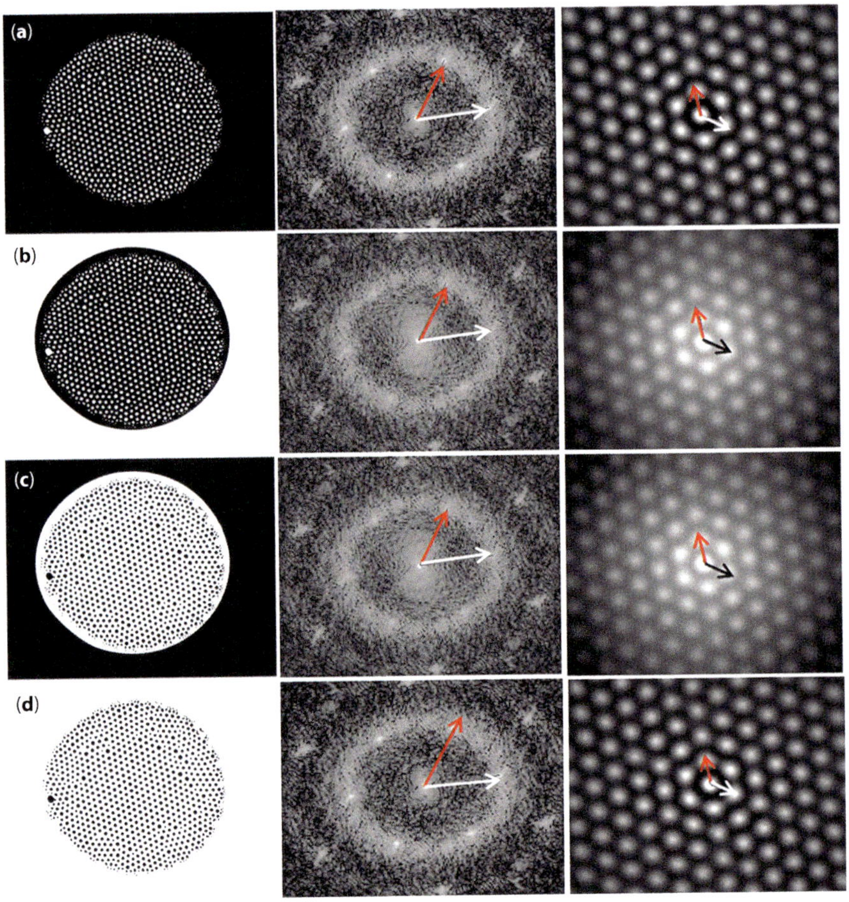

Figure 6.11 A threshold of the same micrograph of the internal view of a single valve of *Roperia tesselata* with different brightness/darkness of background/porosity (1,024 × 800 pixels). (a) A black background and valve, with white pores. (b) A white background and pores, with black valve. (c) A white valve, with black background and pores. (d) A white background and valve, with black pores. Each micrograph has its corresponding normalized Fourier domain (2D PSDF) and 2D ACF shown to its right. Both 2D PSDF and 2D ACF are with 4× magnification. The white and red arrows refer to the lattice vectors in the real and reciprocal space.

missing values in the lattice. Thus, this method will not be suitable for finding defects, i.e., missing points (pores) within the lattice, in the periodic pore arrays. Additionally, the close but distinct relation between 2D ACF and the double 2D FFT is illustrated in (Figures 6.12c vs. d; Figures 6.13c vs. d).

It should be noted that also 2D ACF analysis is very sensitive to the contrast between the pores and features in the background of the micrograph, sometimes even more than 2D FFT. Also, here white pores on a homogeneous black background (or black pores on a homogenously white background) gave the clearest results (Figure 6.11). There was no difference in the 2D ACF extracted profiles in both cases; white pores on black background or black pores on white background.

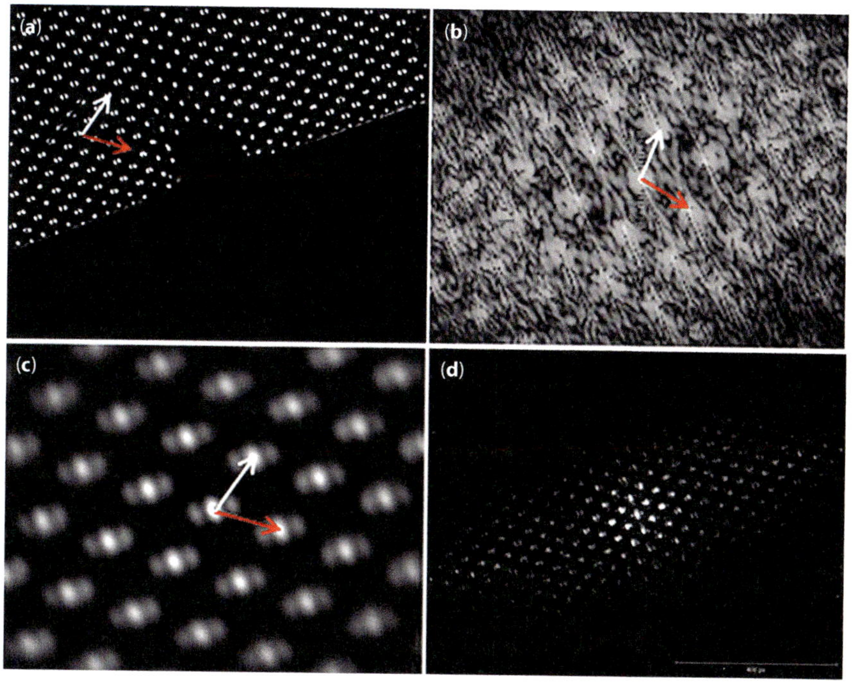

Figure 6.12 (a) A segmented selected part of the SEM of an internal view of a single valve of *Pleurosigma* sp. (1,024 × 800 pixels). (b) The corresponding normalized Fourier domain (2D PSDF), 4× magnified. (c) The corresponding 2D ACF, 4× magnified. (d) The double 2D FFT. The two arrows represent the lattice vectors. While it is a reciprocal value with different directions in the 2D FFT domain, in comparison to the original micrograph, it is the same direction in the 2D ACF.

Furthermore, using the complete valve in 2D ACF is usually not suitable due to possible interference effects between distinct regions that could lead to wrong conclusions as shown in Figure 6.14. The presence of radial or bilaterally additional symmetry between separated 2D pore array groups within the valve may lead to incorrect conclusions from the 2D ACF analysis. Therefore, the best approach is also here to select the group of distinct target pore arrays and remove any symmetrically similar patterns before the study (Figure 6.14).

In future work, the following points should be considered when applying 2D ACF analysis to micrographs of diatom frustules in order to get the best results:

1. The targeted part of the frustule should be clean, free of organic matter and other debris as much as possible before any analysis.
2. The detector should be perpendicular to the valve's face for correct determination of pore size and lattice parameters, or a tilt correction should be applied [6.33].
3. The contrast and brightness level within the micrograph, especially for the targeted pore array, should be adjusted to minimize noise during processing that leads to unsharp results.
4. The image target area should be placed on a homogenous background of at least 4× the size of the target area.

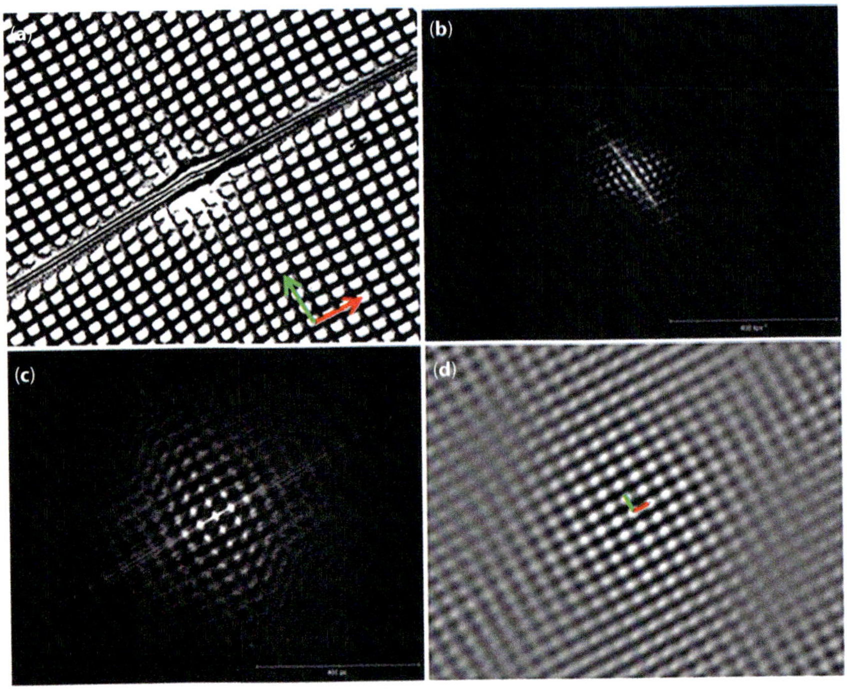

Figure 6.13 A segmented micrograph of a zoom in the internal view of a single valve of *Haslea* sp. (1,024 × 800 pixels) showing 2D pore arrays groups with a bilateral symmetry around the raphe axis (a), its corresponding 2D FFT (b), the double 2D FFT (c), and the corresponding 2D ACF (d).

6.3.1.3 *The Accuracy of Lattice Parameters' Measurements Using the Proposed 2D ACF Analysis*

The precision of the proposed methodology, from the segmentation to parameter measurements, was in the range of four digits in both the case of lengths and in the case of angles. This means that, if we repeat the whole process for a given pattern, then it will repeat the results with these levels of uncertainty. Thus, our results are highly reproducible for the same micrograph and the same pattern.

In order to explore the accuracy of the 2D ACF analysis for measuring lattice parameters using Gwyddion software, compared to the bare hand measurements of the 2D lattice parameters, using ImageJ, the probability of a match between the accurate 2D ACF value and the normal distribution of the bare hand measurements was calculated for three distinct examples: (i) 2D pore arrays at a selected circle from the internal view of a single valve of *Roperia tesselata* (Figure 6.15), (ii) the internal view of a half of a single valve of *Haslea* sp. (Figure 6.22b), and (iii) the linear spacing between the successive striae in the internal view of a single valve of *Haslea feriarum* (Figure 6.14b). The results were summarized in Table 6.1, Table 6.2, and Table 6.3, respectively. Although, the probability was variable, it was deemed acceptable for all studied parameters. This implies that the match between the bare hand measurements and the 2D ACF extracted values is perfectly reasonable, and the agreement between the two methods is

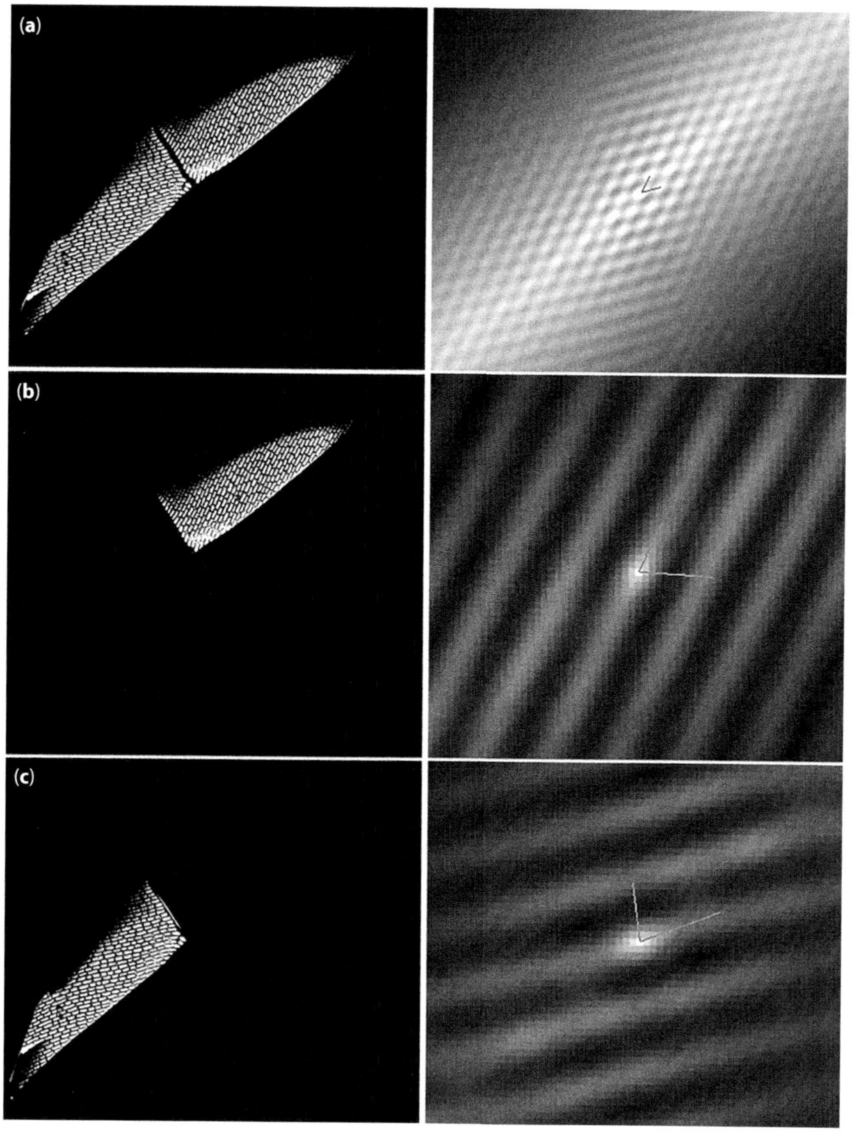

Figure 6.14 (a) A thresholded micrograph of the internal view of a single valve of *Haslea feriarum* with clear bilateral symmetry (1,019 × 861 pixels), (b) the separated upper part, (c) the separated lower part, and their corresponding 2D ACF analysis showing an intereference in (a) between the two patterns emerging from both (b) and (c) patterns, resulting in a seemingly 2D lattice. Although, a 1D periodicity can be seen in the arrangement of the successive striae, no further 2D lattices can be obtained for the isolated two parts of the valve. Scale bar is 20 μm.

satisfactory. Thus, we can depend on the 2D ACF method in the lattice recognition and measurements with a high accuracy. However, it should be noted that the information obtained easily with the bare hand measurements of the lattice variation within the valve is less accessible in the 2D ACF measurements, as the coherence is also influenced by the sampling size.

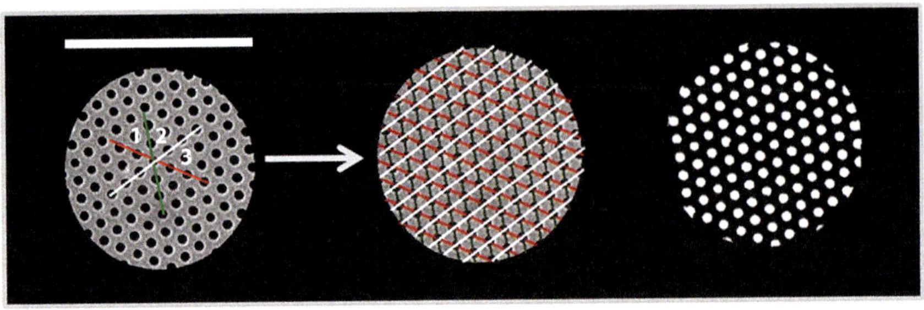

Figure 6.15 A selected circle of an SEM of the internal view of a single valve of *Roperia tesselata* showing the 2D pore arrays at the center, and the corresponding thresholded micrograph. The red, green, and white lines represent the different periodic vectors $\mathbf{a}_1$, $\mathbf{a}_2$, and $\mathbf{a}_3$ respectively. The angles between the different lines were marked as 1, 2, and 3, respectively. The accuracy measurements in Table 6.1 is dependent on these figures. Scale bar is 20 μm.

Table 6.1 The comparison of the bare hand measurements of 2D lattice parameters, and the 2D ACF analysis of the thresholded micrograph of a selected circle from the SEM of the internal view of a single valve of *Roperia tesselata* illustrated in Figure 6.15.

Method of measurement	**Lattice Spacing a_1 (Red Line) (μm)**	**Lattice Spacing a_2 (Green Line) (μm)**	**Lattice Spacing a_3 (White Line) (μm)**	**Angle 1 (degree)**	**Angle 2 (degree)**	**Angle 3 (degree)**
Measurements via bare hand for the original micrograph (ImageJ software) (center to center)	1.95 ± 0.05	1.83 ± 0.07	1.85 ± 0.06	60 ± 3	63 ± 3	58 ± 2
Measurements via 2D ACF analysis for the thresholded micrograph (Gwyddion software)	1.914 ± 0.002	1.816 ± 0.001	1.845 ± 0.001	59.17	63.09 ± 0.04	57.77 ± 0.03
Prob(inside $t\sigma$)	51.85%	84.15%	93.62%	77.95%	97.61%	90.45%

Table 6.2 The comparison of the bare hand measurements of 2D lattice parameters, and the 2D ACF analysis of thresholded micrographs of the SEM of the internal view of a single valve of *Haslea* sp. illustrated in Figure 6.22b.

Method of measurement	Lattice spacing a (px)	Lattice spacing b (px)	Angle 1	Angle 2
Measurements via bare hand for the original micrograph (ImageJ software) (center to center)	38 ± 1	46 ± 2	92 ± 3	90 ± 3
Measurements via 2D ACF analysis for the thresholded micrograph (Gwyddion software)	37.305 ± 0.006	46.076 ± 0.008	90.37 ± 0.01	89.63 ± 0.01
Prob(inside $t\sigma$)	48.39%	96.81%	58.92%	90.45%

Table 6.3 The comparison of the bare hand measurements of 2D lattice parameters, and the 2D ACF analysis of the thresholded micrograph of the SEM of the internal view of *Haslea feriarum* illustrated in Figures 6.14b and 6.29b.

Method of measurement	Linear spacing (μm)
Measurements via bare hand for the original micrograph (ImageJ software) (center to center)	0.46 ± 0.04
Measurements via 2D ACF analysis for the thresholded micrograph with black background (Gwyddion software)	0.454 ± 0.004
Prob(inside $t\sigma$)	88.08%

6.3.2 Exploring the Periodicity in Our Studied Micrographs and the Possible Presence of Different Types of 2D Lattices in Diatoms

6.3.2.1 *Irregular Pore Scattering (Non-Periodic Pores)*

Most diatoms, when cultivated in optimal conditions, synthesize a well-engineered frustule with periodic to quasi-periodic porosity usually arranged in the so-called striae. However,

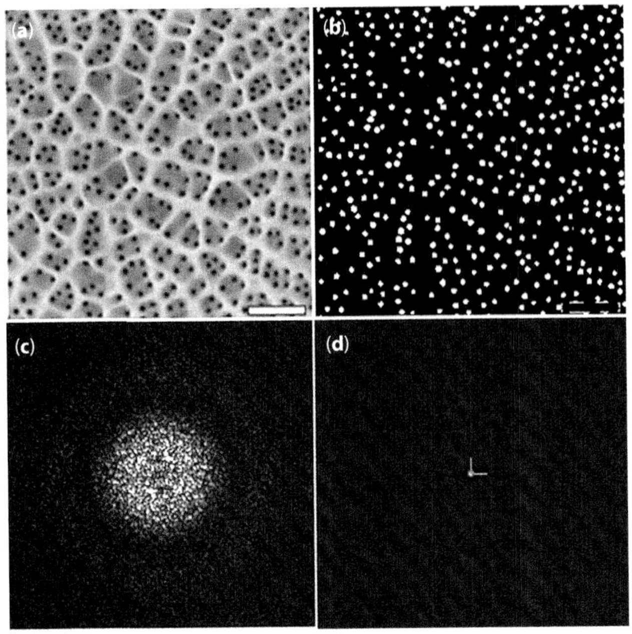

Figure 6.16 (a) SEM of the external view at the center of a single valve of *Thalassiosira pseudonana* (scale bar is 0.2 μm) with its corresponding threshold micrograph (b), 2D PSDF (c), and 2D ACF (d), showing no regular periodicity pattern in the 2D pore distribution. Reproduced from [6.17] with the kind permission of Cambridge University Press.

as became obvious throughout our study, a few species exhibit irregular pore arrangements such as in the case of the pores scattered at the external view of a single valve of *Thalassiosira pseudonana* (Figure 6.16). As can be seen, no regular periodic relations are observed between its pores.

6.3.2.2 *Linear Periodicity of Pores in Striae (1D Periodicity)*

Most diatom species, centrics and pennates, show regular to quasi-regular linear striae, extending from the annulus (in case of centrics) or from the sternum (i.e., a longitudinal area free of pores at the middle of the valve in case of pennates, which sometimes contains a raphe) reaching the outline of the valve (as shown in Figure 6.2). These striae may be considered to be the first level of pore periodicity (1D periodicity) that is present in diatom silica layers. Figure 6.17 is an example of linear periodicity within a single stria in pennates.

Moreover, according to the distance and angle relations between the pores of neighboring striae, the presence or absence of extra periodicities can be determined. An additional circular periodicity (Figure 6.18) (branching striae [6.17, 6.34]), additional linear periodicity (Figure 6.14), or even 2D lattice structures (see Section 6.3.2.3) usually appear through 2D FFT and/or 2D ACF analysis of some centric and pennate valves. Analysis of branching striae is left for future work.

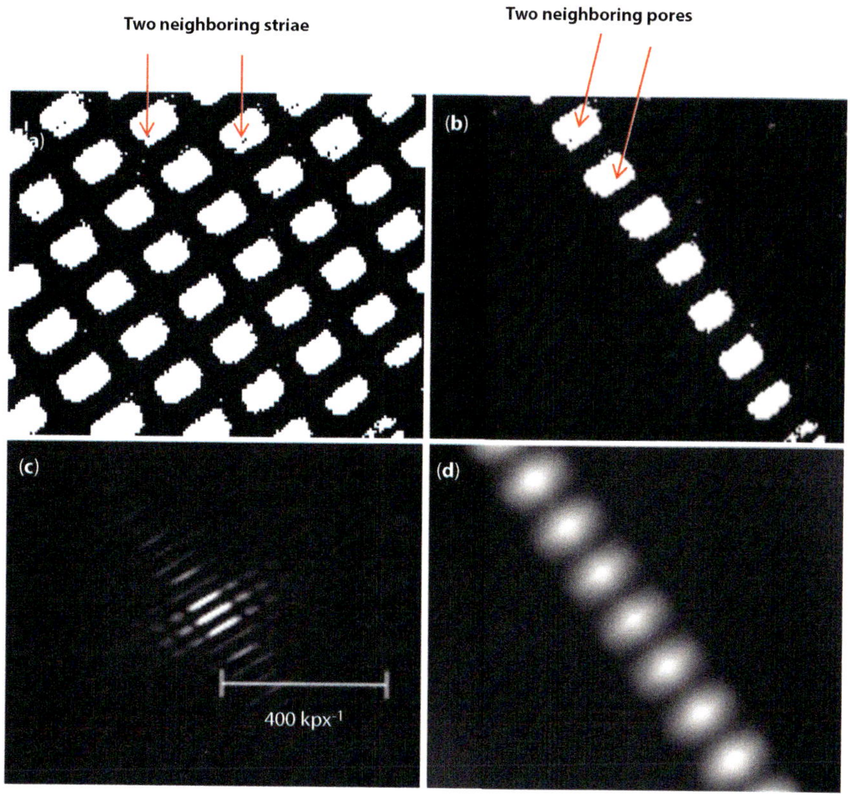

Figure 6.17 (a) A thresholded SEM of a zoom-in of the internal view of a single valve *of Haslea* sp. (170 × 137 pixels) (b) an isolated single stria showing linear periodicity between its neighboring pores. This linear periodicity appears clearly in the corresponding 2D FFT (c) and 2D ACF (d) patterns.

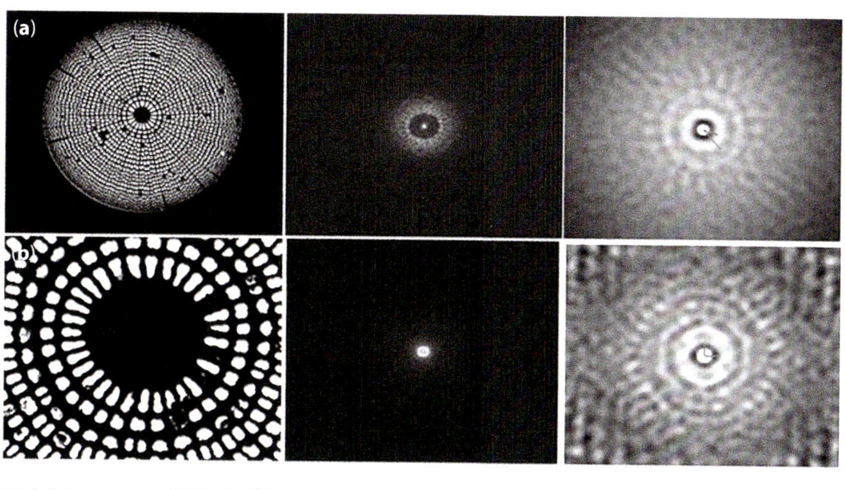

Figure 6.18 (a) A segmented SEM of the external view of a single immature valve of *Arachnoidiscus ehrenbergii* (1,024 × 800 pixels) with its corresponding 2D FFT and 2D ACF analysis (magnified 4×) respectively, showing a ring-shaped pattern reflecting the circular relations between the 2D pore arrays. While the 1D linear periodicity of the quasi-regular branched striae also appears. (b) The zoom-in at the center (1,024 × 800 pixels) with the circular periodicity is more dominant on the 2D FFT and 2D ACF pattern compared to the linear periodicity, which leads to some interference in both 2D FFT and 2D ACF analysis.

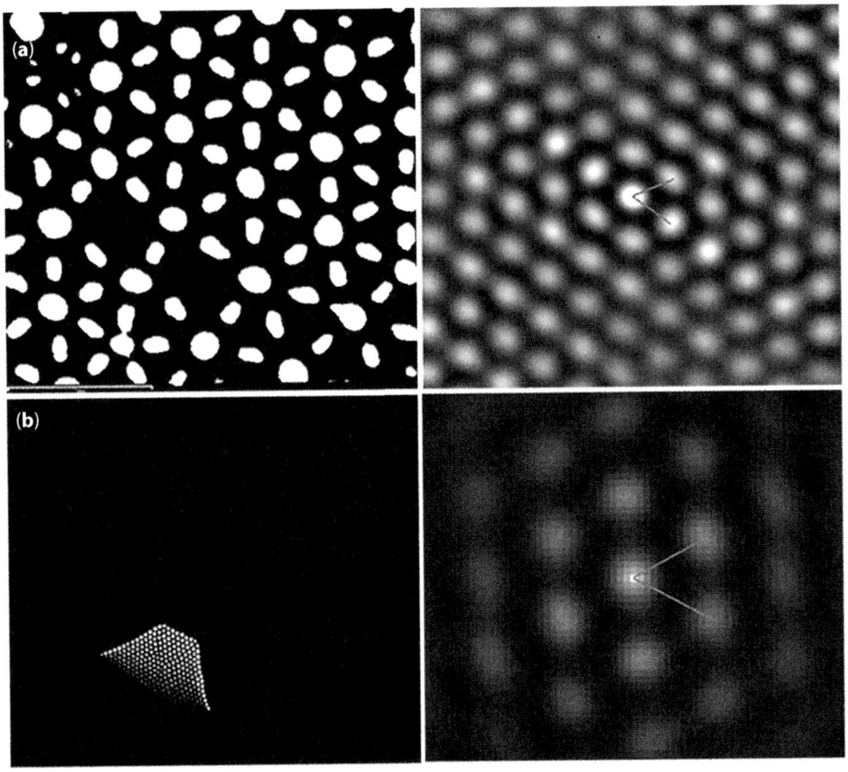

Figure 6.19 (a) A segmented SEM micrograph of a zoom-in external view of *Asterolampra marylandica* (512 × 459 pixels) and its corresponding 2D ACF analysis. (b) A segmented separated part of an SEM of the internal view of *Asteromphalus hookeri* (1,021 × 785 pixels) and its corresponding 2D ACF analysis (16× magnified).

6.3.2.3 *The Different 2D Lattices in Diatom Pore Arrays*

Beside linear and circular periodicities present in diatom frustules, other 2D periodicities can be found, evolving from, as mentioned, the distance and angle relationships between pores of neighboring striae. Through our study of 2D periodicities of pore arrays, different examples of 2D lattices were observed in different diatom valves with different levels of imperfections. Using bare hands to manually measure the lattice parameters and finding the respective vector orientations, or spot a lattice type can in theory be done. However, it can be illusive and time consuming. Various articles use this technique (see Figure 2 in [6.12]; Figure 3b in [6.39]). In contrast, 2D FFT analysis has been carried out in other articles [6.12, 6.14, 6.27], which might be a suitable approach to reveal the periodicity and lattice type in selected segmented micrographs, as discussed in Section 6.3.1.1. Alternatively, we suggested in our study the use of the 2D ACF method on diatom patterns to measure the average values of the length of the lattice vectors as well as the angles between them according to the method discussed in Section 6.2.3 and the results explained using the logic of the flow chart in Figure 6.1 and then illustrated in Table 6.4. Using this simple method, better and clearer results could be obtained concerning the periodic arrangements as discussed in Section 6.3.1.2. It accurately measures the average lattice parameters as discussed in Section 6.3.1.3. Based on these measurements, one can detect the presence of possible 2D lattice types of a given 2D regular pore array and compare them to the five 2D Bravais lattices (Figure 6.1).

Table 6.4 The lattice parameters and lattice types found in some of the studied examples using 2D ACF method. Length is in pixels, while angle is in degrees. The most relevant 2D Bravais lattices are given here according to the accepted deviation value.

Species	Figure	Vector length a (px)	Vector length b (px)	Angle between the two vectors (degree)	Deviation of vector length b compared to vector length a (%)	Deviation of angle θ compared to closest perfect angle (%)	Lattice type with a deviation up to 5%	Lattice type with a deviation up to 10%	Lattice type with a deviation up to 20%
Asterolampra marylandica (external view)	6.19a	52.72	55.72	57.7°	5.7	3.8	Oblique	Hexagonal	Hexagonal
Asteromphalus hookeri (internal view)	6.10a, part 4	12.09	13.26	56.1°	9.7	6.5	Oblique	Hexagonal	Hexagonal
Roperia tesselata (external view)	6.20a	23.18	23.65	57.9°	2	3.5	Hexagonal	Hexagonal	Hexagonal
Roperia tesselata (internal view)	6.20b	23.18	24.01	57.6°	3.6	4	Hexagonal	Hexagonal	Hexagonal
Cocconeis sp.	6.21a	32.78	38.03	54.5°	16	9.2	Oblique	Oblique	Hexagonal
Haslea sp.	6.22b	37.3	46.08	90.4°	23.5	0.4	Rectangular	Rectangular	Rectangular

(Continued)

Table 6.4 The lattice parameters and lattice types found in some of the studied examples using 2D ACF method. Length is in pixels, while angle is in degrees. The most relevant 2D Bravais lattices are given here according to the accepted deviation value. (*Continued*)

Species	Figure	Vector length a (px)	Vector length b (px)	Angle between the two vectors (degree)	Deviation of vector length b compared to vector length a (%)	Deviation of angle θ compared to closest perfect angle (%)	Lattice type with a deviation up to 5%	Lattice type with a deviation up to 10%	Lattice type with a deviation up to 20%
Pleurosigma sp.	6.12a	39.55	41.99	69.3°	6.2	15.5	Oblique	Centered rectangular	Hexagonal
Triceratium favus	6.21b	25.06	25.74	61.1°	2.7	1.8	Hexagonal	Hexagonal	Hexagonal
Gyrosigma balticum	6.24a	34.88	29.58	96.5°	15.2	7.2	Oblique	Rectangular	Square
Gyrosigma fasciola	6.23	39.52	42.45	87.3°	7.4	3	Rectangular	Square	Square
Pleurosigma angulatum	6.24b	25.67	24.89	63.6°	3	6	Centered rectangular	Hexagonal	Hexagonal

As can be seen, in all cases, the numerical values of the lattice vector lengths a and b are not equal, and the angle θ is equal to neither 60° nor 90°. However, in most cases the results were close to a certain 2D Bravais lattice type. Thus, additionally the best fitting 2D Bravais lattices were assigned for a given 2D periodicity at different levels of deviation of 5%, 10%, and 20% from the perfect 2D Bravais lattices (Table 6.4).

It should be noted that in all cases the trends follow as expected an increase in associated symmetry of a given lattice when allowing for greater imperfections. Either from oblique (no symmetry) over centered rectangular (lateral symmetry) to hexagonal (lateral and rotational symmetry) or from oblique (no symmetry) over rectangular (lateral symmetry) to square (lateral and rotational symmetry). However, when speaking about this symmetry, only the symmetry of the lattice is considered. When using a complex basis (like when considering the pattern of a pore as seen in Figure 6.12) certain symmetric features might not exist in the real pattern.

6.3.2.3.1 Examples of 2D Hexagonal Lattice

According to the measurements that were carried out through 2D ACF analysis (Table 6.4), two centric diatoms, among the species studied here, have pore arrays with a deviation up to 5% from the perfect 2D hexagonal lattice. This includes: the pore arrays at the external and internal views of a valve of *Roperia tesselata* (Table 6.4; Figure 6.20), and the pore arrays at the

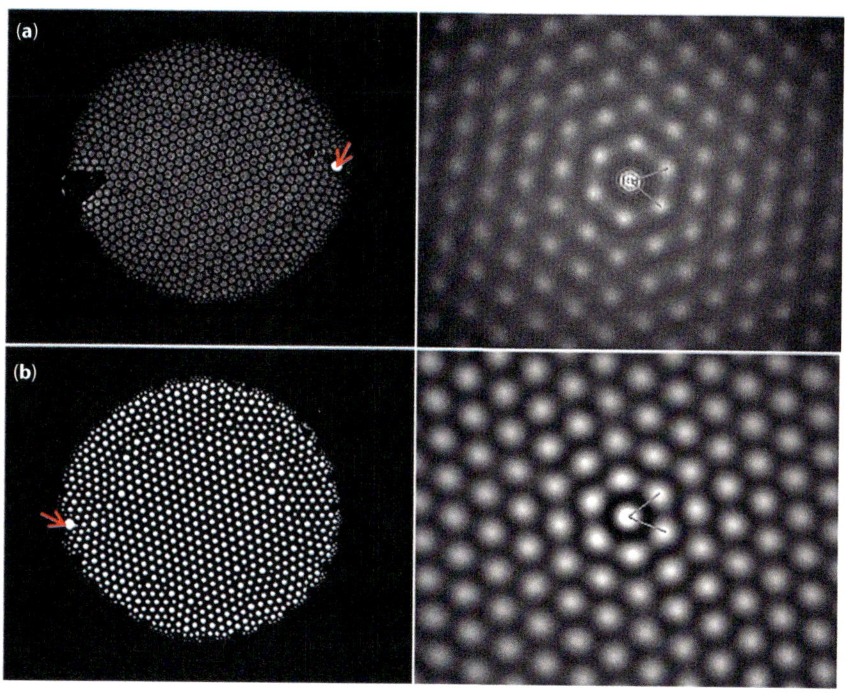

Figure 6.20 (a) A segmented SEM micrograph of the external view of *Roperia tesselata* (1,024 × 800 pixels) and its corresponding 2D ACF analysis (4× magnified). (b) A segmented SEM micrograph of the internal view of *Roperia tesselata* (1,024 × 800 pixels) and its corresponding 2D ACF analysis (4× magnified). The 2D ACF analysis of the external view has an intereference due to the remaining presence of the pore occlusion in the thresholded micrograph; however, the larger pores are still present. It should also be noted that the two valves belonged to two different frustules, however, showed very similar pattern with a clear orientation dependent on the place of the pseudonodulus (red arrows).

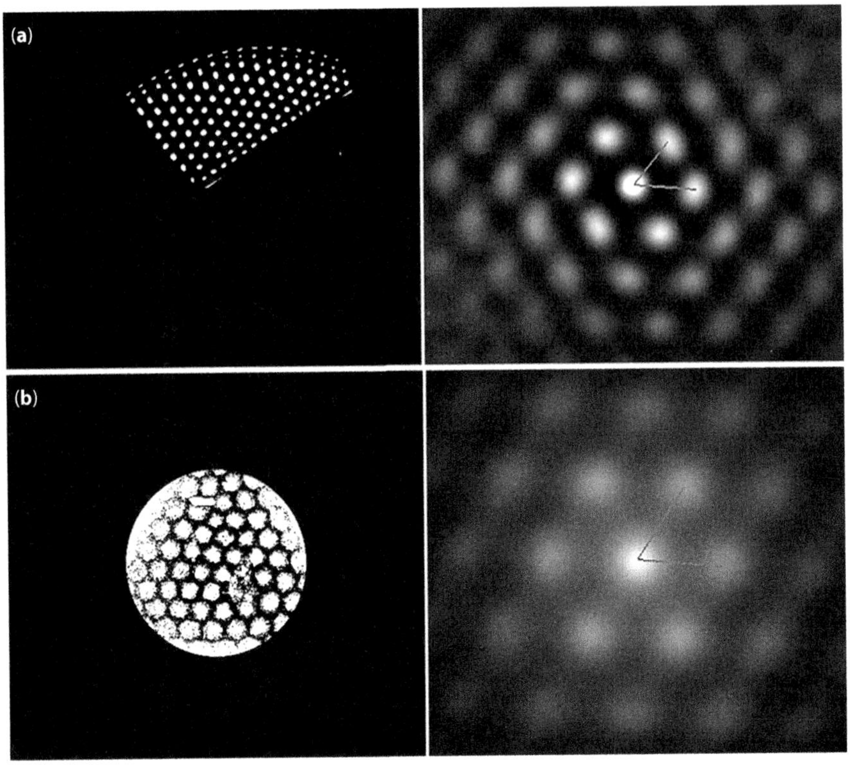

Figure 6.21 (a) A segmented SEM micrograph of a selected 2D pore group at the external view of *Cocconeis* sp. (1,021 × 847 pixels) and its corresponding 2D ACF analysis (4× magnified). (b) a segmented SEM micrograph of the internal view of *Triceratium favus* (512 × 459 pixels), with its sharp edges removed, and its corresponding 2D ACF analysis (4× magnified).

internal view of a valve of *Triceratium favus* (Table 6.4; Figure 6.21b). Other examples, found to have 2D hexagonal lattices, if we can accept a deviation reaching 10% or 20%, as illustrated in Table 6.4. Thus, from the literature and our results, the 2D hexagonal lattice might be very common in centrics possibly due to the rotational symmetry of the lattice, but also in some pennates such as *Pleurosigma* spp. with different degrees of imperfection.

6.3.2.3.2 An Example of a 2D Rectangular Lattice

The 2D rectangular lattice might be very common in pennates, where the neighboring striae tend to be parallel to each other and perpendicular to the sternum (the raphe area in case of raphid pennates) (Figure 6.2), thus leaving an angle of almost 90° between the lattice vectors, while the spacing between the two neighboring striae is usually larger than the spacing between the neighboring pores in the same striae. This means, in this type, the lattice vector length between the two parallel striae is higher than that between two neighboring pores in the same stria, which is characteristic of a 2D rectangular lattice. A clear example of this lattice was observed during the study of the 2D pore arrays appearing at the internal view of a valve of *Haslea* sp. (Table 6.4; Figure 6.22).

6.3.2.3.3 Examples of 2D Square Lattice

In literature, this type of 2D lattice was reported for both centrics and pennates, as it on the one hand also has rotational symmetry, but also has the 90° angle characteristic for

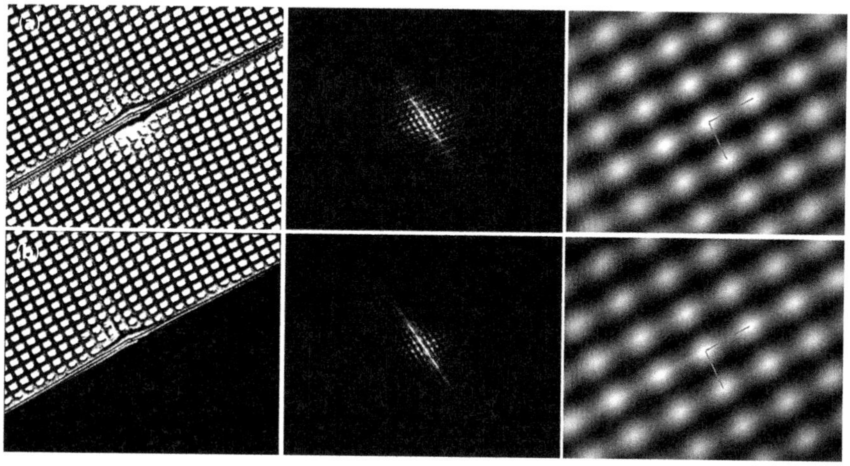

Figure 6.22 (a) A segmented SEM micrograph of the internal view of *Haslea* sp. (1,024 × 800 pixels) and its corresponding 2D FFT and 2D ACF analysis (4× magnified), respectively. (b) a cut from A, with its corresponding 2D FFT and 2D ACF analysis (4× magnified), respectively.

the striae orientation with respect to the sternum in pennates. However, it seems not be very common, even in the case of pennates. The pennate diatom *Gyrosigma fasciola* from [6.12] has been reported here as an example for the presence of pore arrays with 2D square lattice arrangements in pennates, if we accept the deviation in lattice parameters up to 10% compared to the perfect 2D square Bravais lattice (Table 6.4; Figure 6.23). Additionally, *Gyrosigma balticum* from [6.12] showed also a square 2D lattice if we accept deviation up to 20%. In centrics, the 2D square lattice seems to only occur within the pore arrays of girdle bands. For instance, Fruhmann *et al.* [6.10] recorded the 2D square lattice in the pore arrays of girdle bands of the centric diatom *Coscinodiscus granii*, while the valves have a 2D hexagonal lattice. This was further confirmed by [6.14] as they revealed the presence of 2D square lattices in the pore arrays of girdle bands of *Thalassiosira pseudonana, Coscinodiscus granii, C. radiates*, and *C. wailesii*. Thus, it might be that the 2D square lattice is frequent in the pore arrays of girdle bands of centrics.

6.3.2.3.4 Examples of 2D Centered Rectangular Lattice

Interestingly, one of the studied examples, *Pleurosigma angulatum* from [6.12], Figure 6.24b has a 2D centered rectangular lattice, if we accept a deviation of up to 5%. Additionally, the *Pleurosigma* sp. showed also a 2D centered rectangular lattice, if we accept a deviation up to 10%.

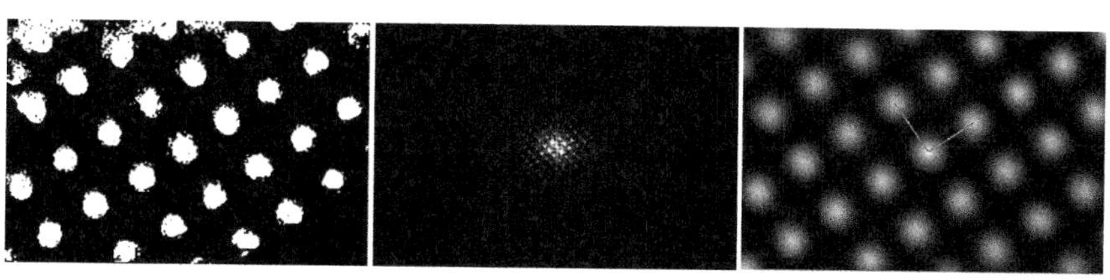

Figure 6.23 A segmented SEM of a zoom-in of a *Gyrosigma fasciola* valve (276 × 179 pixels), adapted from [6.12] with the kind permission of Inter-Research, with its corresponding 2D FFT (left) and 2D ACF analysis (right).

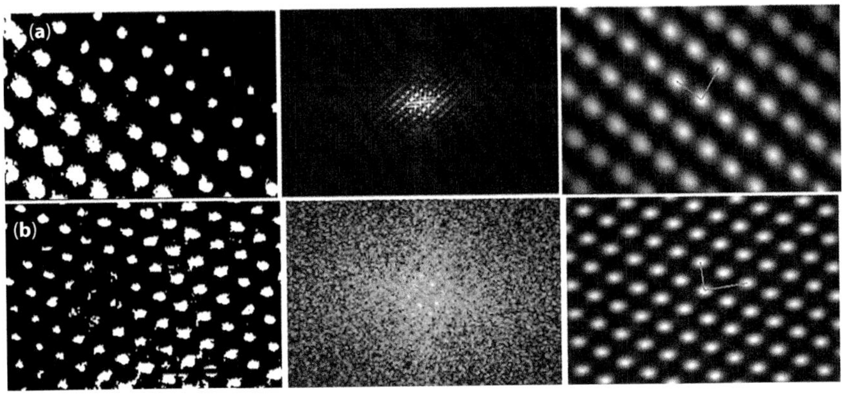

Figure 6.24 (a) A segmented SEM of a zoom-in of a valve of *Gyrosigma balticum*, adapted from [6.12] with the kind permission of Inter-Research (276 × 182 pixels), with its corresponding 2D FFT (left) and 2D ACF (right) analysis. (b) A segmented SEM of a zoom-in of a valve of *Pleurosigma angulatum*, adapted from [6.12] with the kind permission of Inter-Research (276 × 180 pixels), with its corresponding 2D FFT (left) and 2D ACF (right) analysis.

6.3.2.3.5 Examples of 2D Oblique Lattice

All the numerical values of lattice parameters were to some degree imperfect compared to the above mentioned 2D Bravais lattices, which means that all mentioned examples would be assigned to the 2D oblique lattice, if no imperfection levels were tolerated, as this lattice has different values of vector lengths a and b, and the angle θ that is not equal to 90°. However, in most cases, allowing a small degree of imperfection, the lattice constants were very close to other types of 2D lattices. As an example of a 2D oblique lattice, after the acceptation of deviation up to 10%, the 2D pore arrays appearing at the internal valve view of *Cocconeis* sp. (Table 6.4; Figure 6.21a) can be given. If we considered the deviation only up to 5%, then the 2D pore arrays of the internal view of *Asteromphalus hookeri*, the external view of *Asterolampra marylandica*, *Pleurosigma* sp., and *Gyrosigma balticum* from [6.12] would be other examples for 2D oblique lattices.

6.3.2.3.6 Examples of 2D Non-Bravais Lattice

Figure 6.25 provides an example of a 2D-honeycomb structure, if we do not consider the arrangements of the pores, but of the walls of the loculate areolae in the valve of *Asterolampra marylandica* (which became visible in a damaged valve). The 2D FFT and ACF patterns are comparably complex and indeed similar to the patterns of an artificially generated perfect 2D honeycomb-like structure (Figure 6.26). It is worthy to mention that a degree of distortion in the diatom honeycomb structure became, however, clear. Thus, diatoms can build structures (when not limiting ourselves to pore periodicities), which can fit also the more complex 2D non-Bravais lattices.

6.3.3 How Perfectly Can Diatoms Build Their 2D Pore Arrays?

6.3.3.1 Variation of the 2D Lattice Within the Connected Pore Array of the Valve

The 2D lattice parameters of a given 2D pore periodicity within a diatom valve, for a continuous pore array group, are not exactly the same within the whole area of the valve, i.e.,

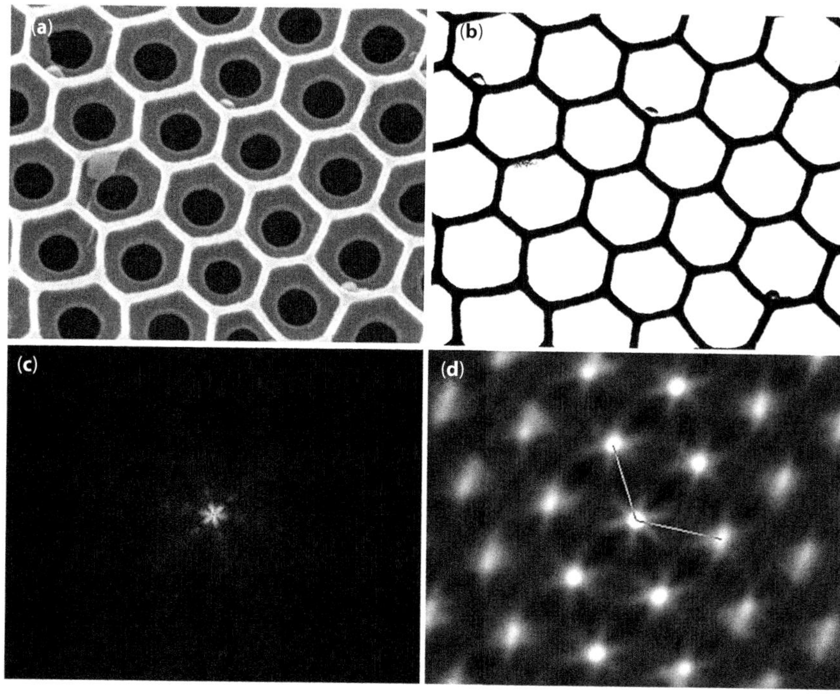

Figure 6.25 (a) The areolae chambers appearing from a crack in the valve of *Asterolampra marylandica*, (b) its thresholded micrograph, (c) 2D FFT pattern, and (d) 2D ACF pattern.

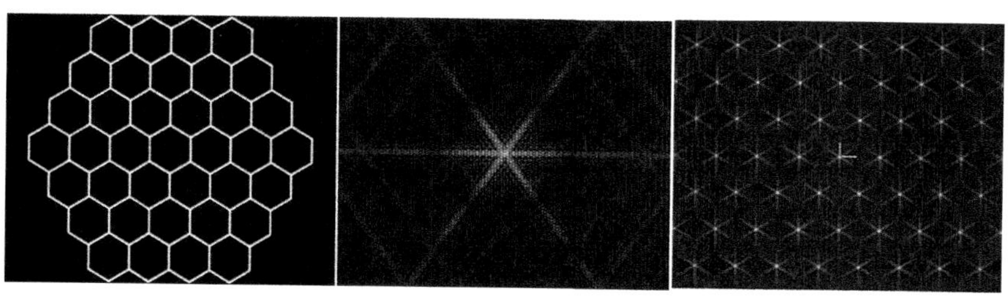

Figure 6.26 An artificial honeycomb-like structure and its correpsonding 2D FFT and 2D ACF patterns.

the lattice parameters might vary, for example, from the valve center compared to the edges, a phenomenon that was clear through changing the sampling size of pores in the 2D ACF analysis, which will be analyzed in more detail in our future work. It should be noted that a variation in the pore spacing was observed in the different parts of *Coscinodiscus wailesii* valve and was experimentally correlated with a variation in the associated laser transmission power spectrum, which was consistent with the diffraction created by different pore spacing [6.62]. Such analysis might be done by correlating the autocorrelation coherence peak height and decay with the lattice standard deviation and establishing a reliable methodology for this independent of sampling size, or via calculating the difference image or the cross-correlation between the original image (thus the unaveraged 2D lattice of the diatom part) and an artificially built perfect 2D lattice. Using a number of refinement steps, slightly changing the lattice parameters, a method of finding the lattice with the best fit (not purely based on averaging) could be developed (as is standardly done in crystallography). It should also be noted that, for such more complex analysis, the pattern should be centered on a homogenous background with at least 4× the size of the pattern itself from all directions. Providing such a methodology will be a key to future work.

6.3.3.2 *Comparison of 2D Lattice Parameters and Degree of Perfection of Distinct Pore Array Groups in the Same Micrograph and Valve but With Different Rotational or Reflection Symmetry*

Two examples (from Figures 6.10 and 6.29c) were studied in order to see if there is a similar trend in the reproducibility of the average lattice parameters within pore array groups within the same valve, but distinct parts as illustrated in Table 6.5. In the case of the centric diatom *Asteromphalus hookeri*, the deviation values for the lattice parameters between the three studied parts (which appeared in Figure 6.10) did not follow the same trend or degree of perfection. This leads to the conclusion that this species might not reproduce the lattice structure perfectly, when synthesizing the distinct parts of the valve, however to quantify this claim more studies need to be carried out including the extraction of the standard deviation of each parameter and comparing the measured difference with the standard deviation using equations, similar to Equations (6.3) and (6.4), to conclude the matching probability window. In contrast, the deviation of the lattice parameters in the two halves of the single valve of the pennate *Haslea* sp. followed the same trend. This implies that this species might be able to reproduce its lattice and the degree of perfection. Thus, diatoms might not be as perfect as we thought, with a variable ability to reproduce its degree of perfection. This study should be further extended to investigate the degree of perfection and similarity between diatoms of the same culture, as also here the common assumption is that the reproduction of the structure is of high reproducibility. Comparisons of patterns between valves in heterovalvy or both sides of a nonbilaterally symmetric pennate valve could also be interesting.

6.3.3.3 *The Perfection of 2D Lattices of Diatom Pore Arrays Compared to Perfect (Non-Oblique) 2D Bravais Lattices*

Through the study, it became clear from Table 6.4 that some diatom examples approach perfection in lattice parameters (others than oblique) with a deviation level below 5%, while other examples can be considered imperfect 2D lattices (other than oblique). This variation between species can be expected when surfing a book like Round *et al.* [6.37]. Although, our suggested simple method could provide a judgment tool for the study of the diatom lattice perfection that can be used by anyone, it is important to mention that in order to judge the perfection of a given lattice in a more precise way, a modification of this method should be considered in the future to include the standard deviation of the lattice, i.e., the change of the lattice perfection over a given diatom surface. The reason for this is that the threshold of x% deviation represents $t_\sigma * \sigma$ (see Section 6.2.4), and thus, for a larger standard deviation σ, the probability of corresponding to the "perfect" lattice increases (with t_σ decreasing). Thus, a lattice, which shows large variation within a valve, will have a greater probability of being well presented by a "perfect" non-oblique lattice compared to an example in which the lattice parameters are very well defined over the entire valve, and thus, the deviation becomes significant.

Table 6.5 A comparison between different 2D group arrays within the same micrograph of two selected examples.

Species	***Asteromphalus hookeri***	***Haslea* sp.**
First group array	**Figure 6.10 (1)**	**Figure 6.29c (1)**
Deviation of vector length b compared to vector length a (%)	6.3	23.6
Deviation of angle θ compared to closest perfect angle (%)	4.7 (compared to 60°)	0.42 (compared to 90°)
Lattice type with a deviation up to 5%	Oblique	Rectangular
Lattice type with a deviation up to 10%	Hexagonal	Rectangular
Lattice type with a deviation up to 20%	Hexagonal	Rectangular
Second group array	**Figure 6.10 (4)**	**Figure 6.29c (2)**
Deviation of vector length b compared to vector length a (%)	9.7	20.6
Deviation of angle θ compared to closest perfect angle (%)	6.4 (compared to 60°)	1.8 (compared to 90°)
Lattice type with a deviation up to 5%	Oblique	Rectangular
Lattice type with a deviation up to 10%	Hexagonal	Rectangular
Lattice type with a deviation up to 20%	Hexagonal	Rectangular
Third group array	**Figure 6.10 (6)**	
Deviation of vector length b compared to vector length a (%)	2.8	
Deviation of angle θ compared to closest perfect angle (%)	6.6 (compared to 60°)	
Lattice type with a deviation up to 5%	Centered rectangular	
Lattice type with a deviation up to 10%	Hexagonal	
Lattice type with a deviation up to 20%	Hexagonal	

6.3.4 Planar Symmetry Groups to Describe the Whole Diatom Valve Symmetries and Additionally Describe the Complicated 2D Periodic Pore Arrays' Symmetries

Besides, the overall symmetries that appear in the whole diatom valves [6.33], and the symmetry, between similar discrete pores, of different 2D lattices of diatom pore arrays, additional complicated symmetry relations between the pores of 2D periodic arrays can also frequently be observed in diatoms. To describe them, it is better to use the original description in translational geometry and related terms. Here, there are three kinds of planar symmetric groups defined (Rosette, Frieze, and wallpaper groups), which can be used to describe the overall symmetries within a given pattern, which in our case is either the whole diatom valve or frustule including the additional symmetries between given groups of pore arrays. Between these separate groups of pores, additional rotational, reflection, and/or glide reflection symmetries can exist, as clearly seen in Figures 6.12 and 6.27 to 6.29. In this study, only a brief overview of these symmetry relations is given (see below), as an inspiration for future research.

Moreover, we wish to understand what the presence of additional symmetry relations means for the complete description of the 2D periodicity of the pore arrays. For instance, in some cases the five 2D Bravais lattices may not be enough to describe the 2D symmetries of these observed 2D periodic pore arrays, because each point of the obtained lattice itself is composed of two or more pores (complicated basis) with additional symmetries (see Figures 6.12, 6.27, and 6.28 in contrast to Figures 6.11, 6.15, 6.17, 6.23, and 6.24, in which a single pore is found at each point).

6.3.4.1 *Rosette Groups*

The Rosette group is a group of geometric symmetries, which have rotational and sometimes reflectional symmetries of periodic patterns around a central point [6.55]. This group can be separated into two families; one with only rotational symmetry, the cyclic groups (C_n of n-fold), and the other with both rotational and reflectional symmetry, the dihedral groups (D_n of n-fold).

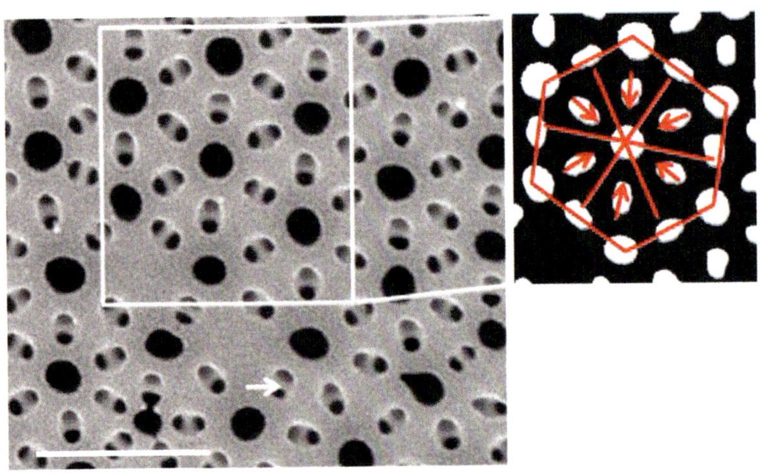

Figure 6.27 A zoom-in of the external view of a single slightly immature valve of *Asterolampra marylandica*. The white square defines the segmented cut part, which appears at the right illustrating a hexagonal shape composed of two kinds of pores. The round large pores and smaller ellipatical pores with a cirucular periodicity around each discrete round pore, which are defined as the C_3 cyclic symmetry group. It should be noted that the ultrastructure appearing at the white arrow illustrates the bottom layer of the areolae walls, does not belong to the upper silica layer, and was therefore excluded from the segmented micrograph. Scale bar is 2 μm.

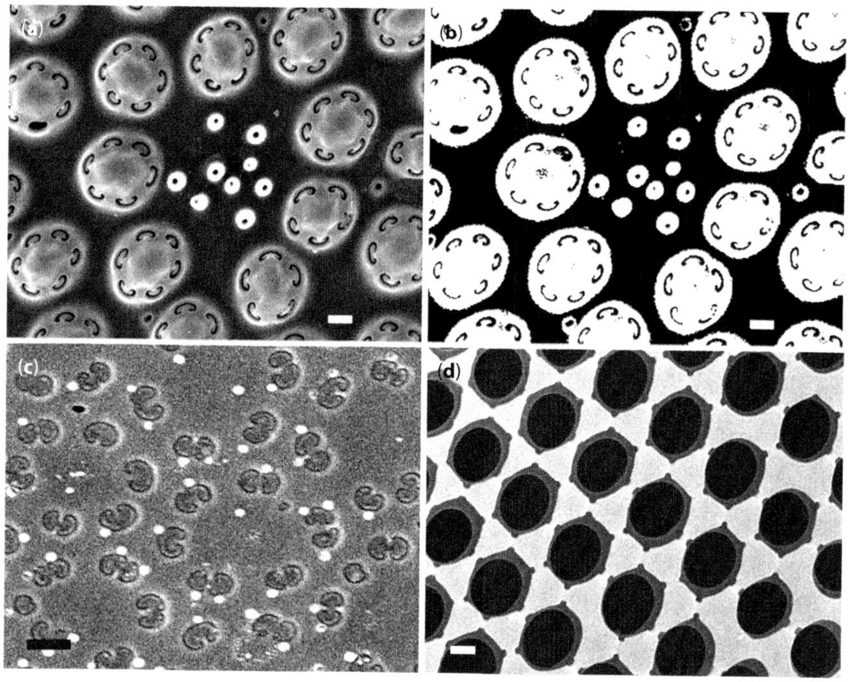

Figure 6.28 (a) SEM of an external view of the central area of a single valve of *Trigonium formosum* fo. *quadrata* showing mature cribra, and its corresponding thresholded micrograph (b), scale bar = 2 μm. (c) SEM of a close up of the external view of a single valve *Trigonium arcticum* showing mature cribra, Scale bar = 1 μm. (d) SEM of a close up of *Coscinodiscus* forming valve, scale bar = 2 μm.

Using these groups, we can thus describe the rotational symmetry of the whole centric valves (Figures 6.7 and 6.10), and possibly even the rotational and reflection symmetries that appear in pennates assigned to certain dihedral groups, for instance, D_2 dihedral symmetry between the four groups of 2D pore arrays appearing in *Cocconeis* sp. (Figure 6.29a).

In addition, these groups can also be used to describe the symmetries within specific parts in the 2D lattice structures which appeared in some diatom valves. As an example, the C_3 cyclic symmetry group is observed at the center of the hexagonal pattern that appears at the external view of a mature valve of *Asterolampra marylandica* (Figure 6.27), in which elliptical pores show a rotational symmetry around the central pore, while the whole pattern, including the round pores, could have a hexagonal 2D Bravais lattice arrangement if we accept deviation up to 10%, and assume that all the pores are equivalent bases (Figure 6.19a). Other examples were also found in *Trigonium* spp. in Figure 6.28. Furthermore, the 2D pore arrays at the internal view of *Pleurosigma* sp. (Figure 6.12) can be considered as an example of this phenomenon in pennates, where the unit cell of this lattice is 2D oblique (if we accepted deviation up to 5%) and acknowledge that each point in the lattice is composed of a structured basis, i.e., two pores with reflectional symmetry.

6.3.4.2 *Frieze Groups*

There are seven types of Frieze groups [6.48], which are related to reflection and translational symmetries in one direction along line. Many representative examples which can be

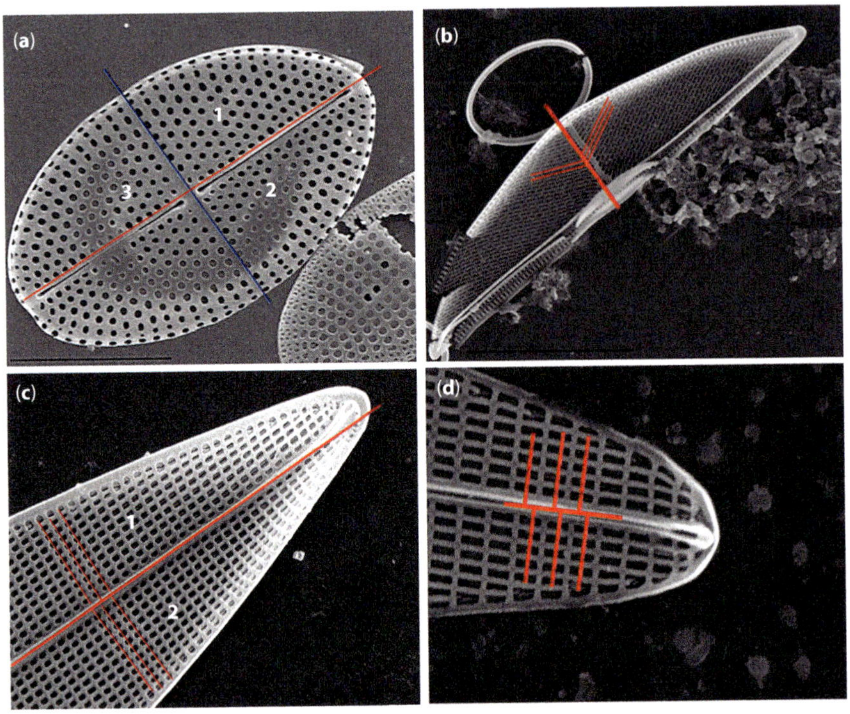

Figure 6.29 (a) SEM of the external view of a single valve of *Cocconeis* sp. illustrating the bilateral symmetry present within the valve around the perpendicular red and blue axes with four group pore arrays including 1, 2, and 3. This is a typical case for some pennates as each group of pore arrays is composed of curved striae and the discrete pores within the striae have a 2D lattice arrangement, which, in this case, is oblique (see Table 6.4). Note that along the red axis the striae might have a frieze symmetry group, and a possible additional dihedral rossete symmetry group D_2 around the center of the whole valve (nodule zone), scale bar = 10 µm. (b) SEM of the internal view of a single valve of *Haslea feriarum* illustrating a THG frieze group, scale bar = 10 µm. (c) SEM of the internal view of a single valve of *Haslea* sp. with a TRHVG frieze group, scale bar = 1 µm. (d) SEM of the internal view of a single valve of *Haslea tsukamotoi* with a TG frieze group, scale bar = 1 µm.

described using this group can be found in diatoms as most pennate valves exhibit reflection symmetry usually across the sternum (the raphe area in case of raphid pennates) (Figures 6.2 and 6.29). In pennates, pores are often arranged in linear or quasi-linear striae, which are sometimes curved, (e.g., *Cocconeis* in Figures 6.21 and 6.29a) or branched. These striae can exhibit symmetry relations between their discrete pores (2D lattice type) as discussed before and have additional symmetry relations across the sternum, if we consider each stria as a separate group of pores or a pattern.

We assume that three examples out of the seven types of frieze groups might be present in the current study, as following:

- THG type
  It includes the following symmetries; translations, horizontal reflections, and thus glide reflections [6.48], as shown in Figures 6.29a and b.
- TRHVG type

It includes the following symmetries; horizontal and vertical reflection lines, translations [6.56] and 180° rotations [6.48], as shown in Figure 6.29c.

- TG type
  It includes the following symmetries; translations and glide-reflections [6.48], as shown in Figure 6.29d.

6.3.4.3 *Wallpaper Groups*

Wallpaper groups can be separated into seventeen types [6.57] and are associated with the symmetries of the repetitive pattern in two directions (2D symmetries). The five known 2D Bravais lattices can describe the wallpaper groups, as the latter consider subgroups of the essential Bravais lattices [6.59]. Wallpaper groups may be present in diatoms if we consider Figures 6.27 and 6.28. However, the assignment of Wallpaper group types is more complex and needs more research in future work.

6.4 Conclusion

Diatoms create their siliceous frustule truly as an excellent engineer with the help of the suggested genetically controlled morphogenesis mechanisms, or possibly cytoplasmic inheritance [6.40, 6.43]. Although many living organisms show different degrees and types of periodicities and symmetries, diatom frustules get special attention because they are siliceous and have a multiscalar, multilevel porosity with an interesting range of scales [6.11] as target material for a number of applications. Understanding the arrangement of pore arrays within a single silica layer of a given valve or girdle band is one of the key tasks for diatom nanotechnology applications especially in the study of photonic properties of valves, girdle bands, whole frustules, or even linear colonies. Obtaining average lattice parameters using 2D ACF analysis of thresholded micrographs seems to be a precise and quick method to assign 2D lattice types within diatom valves. The careful selection of target pore array groups within the valve improves analysis. Following 2D ACF analysis, additional symmetry relations between distinct pore array groups or pores comprising the basis of the lattice points in the 2D lattice can be studied. Several interesting geometrical phenomena observed in diatoms deserve more attention in the future as many applications might emerge from such geometrical structures.

A few examples of irregular pore scattering, which do not have a definite periodicity between its pores, are also present in diatoms. Additionally, 1D periodicity within single stria is common in diatoms, as well as circular periodicity. The 2D lattices can be consider as a consequence of the distance and angle relations between the discrete pores within neighboring striae and might prove of taxonomic value, i.e., it could be a species-specific property. In this study, we could show that all five 2D Bravais lattices are present in diatoms (if we accept imperfection up to 20%). However, the results here are just examples of diatoms' ability to produce lattices and their description and do not consider the morphometric criteria of these taxa. The 2D hexagonal lattice might be the most common type in centric valves, possibly due to its rotational symmetry, while the 2D square lattice might be the most common in girdle bands. Further, the 2D rectangular lattice might be the most common one in pennates, possibly due to the 90 angle, commonly associated with the relation between striae and the sternum and its lateral symmetry. Interestingly, an example of a 2D non-Bravais lattice was described, when considering the symmetry of the silica walls between areaolae and not the pores themselves.

We could show that the structural perfection in distinct groups of 2D pore arrays could vary even within the same valve but especially between different valves of the same species. Such analysis provides information about how well diatoms can build their lattices, which will depend on the species. In future, the detailed understanding of the arrangement of the 2D pore arrays at the surface, and if possible at each layer, of diatom valves could help to understand their morphogenesis.

Some additional questions were only answered preliminary or not at all. We thus suggest the following points to be considered in future studies:

1. Extraction of the standard deviation of each lattice parameter using the 2D ACF analysis.
2. Extended studies on the presence of different types of 2D lattices in different diatom taxa.
3. Is the 2D lattice of a given species really a species-specific property, and how reproducible could this lattice be?
4. Does the 2D lattice type of a given species extend to other species in the same genus? (Is it a genus-specific property?)
5. The applicability of perfection calculations based on cross-correlation or difference analysis with artificially produced perfect lattices.
6. Generating an index of perfection for a biogenic inorganic 2D lattice structure that could be extended beyond diatoms to other organisms.
7. Extended studies on the presence of planar symmetry groups, i.e., rosette, frieze, and wallpaper groups, in diatoms.

Acknowledgment

This work was supported by the Deutsche Forschungsgemeinschaft (DFG) (grant number: RE-3109\2-1; project number: 422038271).

Glossary

Bravais lattice: the two or three-dimensional infinite periodic arrangements (array) of equivalent points with a structure similar to a mesh or net featuring translational symmetry. There exist five different types considering their symmetry in the 2D (Euclidean) space with distinct lattice parameters of their unit cell [6.4].
Cross-correlation: is a standard tool in (periodic) signal processing to study the coherence between two data sets depending on a temporal delay or spatial shift [6.42].
Crystal: is a description of the ordered arrangement of a certain structure or component in 1D, 2D, 3D, or 4D spaces featuring translational symmetry.
Crystallography: is the experimental science of studying crystal structures [6.50].
Diatom Nanotechnology: is the utilization of nanoporous silica microstructures created by diatoms in nanotechnological applications.

Euclidean plane: a two-dimensional geometry or space that extends infinitely.
Fourier analysis: the analysis, which converts a signal from its original domain (often time or space) to a representation in the frequency domain (Sin and Cos components) and vice versa.
Fourier domain: is a mathematical transform that decomposes a function (such as a signal) into its constituent frequencies. It is the result of Fourier analysis of this given signal [6.51].
Frieze group: is a design on a two-dimensional surface that is repetitive in one direction [6.48].
Image analysis: number of techniques used for extraction of information from a given image.
Lattice: is a pure mathematics concept, which can be defined as the two- or three-dimensional infinite periodic arrangements (array) of points connected by vectors resulting in a structure similar to a mesh or net [6.4].
Rosette groups (point groups): is a group of geometric symmetries, which have rotational and sometimes reflectional symmetries of the periodic patterns around a central point [6.55].
Spatial domain: The micrograph before doing FFT analysis.
Two-dimensional fast Fourier transform analysis (2D FFT): is an algorithm for Fourier analysis in Euclidean plane (see Fourier analysis).
Unit cell: is the smallest possible repetitive area or volume in a crystal structure, usually selected so that it features the characteristic symmetries of the crystal as well as occupies as little space and number of atoms as possible.
Wallpaper groups: is a mathematical classification of a two-dimensional repetitive pattern, based on the symmetries in the pattern. Such patterns occur frequently in architecture and decorative art, especially in textiles and tiles as well as wallpaper [6.57].

References

[6.1] Almeida, S.P. and Fujii, H., Fourier transform differences and averaged similarities in diatoms. *Appl. Opt.*, 18, 10, 1663–1667, 1979.

[6.2] B.W.W., Periodicity in weather and crops. *Nature*, 105, 370–371, 1920.

[6.3] Band, Y.B. and Avishai, Y., *Low-Dimensional Quantum Systems. In: Quantum Mechanics with Applications to Nanotechnology and Information Science*, Elsevier, pp. 749–823, 2013.

[6.4] Borchardt-Ott, W., *Crystallography: An Introduction*, Springer, Berlin, Germany, 2012.

[6.5] Cohoon, G.A., Alvarez, C.E., Meyers, K., Deheyn, D.D., Hildebrand, M., Kieu, K., Norwood, R.A., Analysis of quasi-periodic pore-network structure of centric marine diatom frustules. *Proc. SPIE*, 9341, 93410g, 2015.

[6.6] Cox, E.J., Diatom identification in the face of changing species concepts and evidence of phenotypic plasticity. *J. Micropalaeontol.*, 33, 111–120, 2014.

[6.7] De Stefano, L., Maddalena, P., Moretti, L., Rea, I., Rendina, I., De Tommasi, E., Mocella, V., De Stefano, M., Nano-biosilica from marine diatoms: a brand new material for photonic applications. *SuperlatticesMicrostruct.*, 46, 1–2, 84–89, 2009.

[6.8] de Wolff, P.M., Belov, N.V., Bertaut, E.F., Buerger, M.J., Donnay, J.D.H., Fischer, W., Hahn, T., Koptsik, V.A., Mackay, A.L., Wondratschek, H., Wilson, A.J.C., Abrahams, S.C., Nomenclature for crystal families, Bravais-lattice types and arithmetic classes report of the International Union of Crystallography Ad-Hoc Committee on the Nomenclature of Symmetry. *Acta Crystallogr. Sect. A*, A41, 278–280, 1985.

[6.9] Filipski, J., Leblanc, J., Youdale, T., Sikorska, M., Walker, P.R., Periodicity of DNA folding in higher order chromatin structures. *EMBO J.*, 9, 4, 1319–1327, 1990.

[6.10] Fuhrmann, T., Landwehr, S., El Rharbi-Kucki, M., Sumper, M., Diatoms as living photonic crystals. *Appl. Phys. B-Lasers Opt.*, 78, 3–4, 257–260, 2004.

[6.11] Ghobara, M.M., Mazumder, N., Vinayak, V., Reissig, L., Gebeshuber, I.C., Tiffany, M.A., Gordon, R., On light and diatoms: A photonics and photobiology review [Chapter 7], in: *Diatoms: Fundamentals & Applications [DIFA, Volume 1 in the series: Diatoms: Biology & Applications, series editors: Richard Gordon & Joseph Seckbach]*, J. Seckbach and R. Gordon (Eds.), pp. 129–190, Wiley-Scrivener, Beverly, MA, USA, 2019.

[6.12] Goessling, J.W., Frankenbach, S., Ribeiro, L., Serôdio, J., Kühl, M., Modulation of the light field by valve optical properties of raphid diatoms: Implications for niche differentiation in the microphytobenthos. *Mar. Ecol. Prog. Ser.*, 588, 29–42, 2018.

[6.13] Goessling, J.W., Gonzalez, A.S.A., Raj, V.S.P., Ashworth, M.P., Manning, S.R., Lopez-Garcia, M., Biosilica slab photonic crystals as an alternative to cleanroom nanofabrication? *FaradayDiscuss.*, 223, 0, 261–27, 2020.

[6.14] Goessling, J.W., Wardley, W.P., Lopez-Garcia, M., Highly reproducible, bio-based slab photonic crystals grown by diatoms. *Adv. Sci.*, 7, 10, 1903726, 2020.

[6.15] Gordon, N.K. and Gordon, R., *Embryogenesis Explained*, World Scientific Publishing, Singapore, 2016.

[6.16] Gordon, R., Diatoms and nanotechnology: Early history and imagined future as seen through patents, in: *The Diatoms: Applications for the Environmental and Earth Sciences*, 2nd, J.P. Smol and E.F. Stoermer (Eds.), pp. 585–602, Cambridge University Press, Cambridge, 2010.

[6.17] Gordon, R. and Drum, R.W., The chemical basis of diatom morphogenesis. *Int. Rev. Cytol.*, 150, 243–372, 421–422, 1994.

[6.18] Gordon, R., Losic, D., Tiffany, M.A., Nagy, S.S., Sterrenburg, F.A.S., The Glass Menagerie: Diatoms for novel applications in nanotechnology. *Trends Biotechnol.*, 27, 2, 116–127, 2009.

[6.19] Grimann, M. and Fuhrmann-Lieker, T., Biological photonic crystals, in: *Organic and Hybrid Photonic Crystals*, pp. 57–74, Springer, Switzerland, 2015.

[6.20] Iadonisi, G., Cantele, G., Chiofalo, M.L., Crystals and nanosystems structure, in: *Introduction to Solid State Physics and Crystalline Nanostructures*, pp. 1–96, Springer, Milano Italy, 2014.

[6.21] Joannopoulos, J.D., *Photonic Crystals: Molding the Flow of Light [+Errata online]*, 2nd, Princeton University Press, Princeton, NJ, USA, 2008.

[6.22] Kim, K., Choi, J., Park, J.B., Jeon, S.C., Kim, J.S., Lee, H.M., Lattice constant effect of photonic crystals on the light output of blue light-emitting diodes. *IEEE Photonics Technol. Lett.*, 20, 17–20, 1455–1457, 2008.

[6.23] Klapetek, P., Nečas, D., Anderson, C., Gwyddion user guide: Feature Measurement: Chapter 4, in: *Data Processing and Analysis*, 2019, http://gwyddion.net/documentation/user-guide-en/feature-measurement.html.

[6.24] Krumwiede, T.R., *Surface Energy in Bond-Counting Models on Bravais and Non-Bravais Lattices [PhD thesis]*, University of Tennessee, Knoxville, TN, USA, 2017.

]6.25] Longuet-Higgins, M.S., Geometrical constraints on the development of a diatom. *J. Theor. Biol.*, 210, 1, 101–105, 2001.

[6.26] Losic, D. (Ed.), *Diatom Nanotechnology: Progress and Emerging Applications*, Royal Society of Chemistry, London, 2018.

[6.27] Losic, D., Rosengarten, G., Mitchell, J.G., Voelcker, N.H., Pore architecture of diatom frustules: Potential nanostructured membranes for molecular and particle separations. *J. Nanosci.Nanotechnol.*, 6, 4, 982–989, 2006.

[6.28] Luo, Q., Gao, Y., Luo, J., Chen, C., Liang, J., Yang, C., Automatic identification of diatoms with circular shape using texture analysis. *J. Software*, 6, 3, 428–435, 2011.
[6.29] Lyakh, A.M., Bedoshvili, Y.D., Shikhat, O.V., New method of estimation of the relative area of perforations on valves of centric diatoms using SEM images on the example of *Minidisus vodyanitskiyi* Lyakh et Bedoshvili. *Ecol. Montenegrina*, 21, 46–52, 2019.
[6.30] Machán, R. and Hof, M., Fluorescence Correlation Spectroscopy (FCS), in: *Practical Manual for Fluorescence Microscopy Techniques*, S. Ahmad (Ed.), pp. 5-1 to 5-24, PicoQuant, Berlin, Germany, 2016.
[6.31] Nasri, A., Benslimane, R., Ouaazizi, A.E., A genetic based algorithm for automatic motif detection of periodic patterns, in: *2014 Tenth International Conference on Signal-Image Technology and Internet-Based Systems*, pp. 112–118, 2014.
[6.32] Noren, A.K., *Characterization of Structure and Optical Properties of Diatoms for improved Solar Cell Efficiency [Masters Thesis]*, Norwegian University of Science and Technology, Norwegian University of Science and Technology, Trondheim, 2011.
[6.33] Pappas, J.L., Tiffany, M.A., Gordon, R., The uncanny symmetry of some diatoms and not of others: A multi-scale morphological characteristic and a puzzle for morphogenesis, in: *Diatom Morphogenesis [DIMO, Volume in the series: Diatoms: Biology & Applications,* series editors: Richard Gordon & Joseph Seckbach], V. Annenkov, J. Seckbach, R. Gordon (Eds.), Wiley-Scrivener, Beverly, MA, USA, 2021.
[6.34] Parkinson, J., Brechet, Y., Gordon, R., Centric diatom morphogenesis: A model based on a DLA algorithm investigating the potential role of microtubules. *Biochim. Biophys. Acta - Mol. Cell Res.*, 1452, 1, 89–102, 1999.
[6.35] Pfleiderer, S., Ball, D.G.A., Bailey, R.C., AUTO: A computer program for the determination of the two-dimensional autocorrelation function of digital images. *Comput. Geosci.*, 19, 6, 825–829, 1993.
[6.36] Rasband, W., *NIH Image (ImageJ)*, U.S. National Institutes of Health, 2020, http://rsb.info.nih.gov/nih-image/.
[6.37] Round, F.E., Crawford, R.M., Mann, D.G., *The Diatoms, Biology & Morphology of the Genera*, Cambridge University Press, Cambridge, 1990.
[6.38] Schmid, A.M.M., Eberwein, R.K., Hesse, M., Pattern morphogenesis in cell walls of diatoms and pollen grains: A comparison. *Protoplasma*, 193, 1-4, 144–173, 1996.
[6.39] Seethalakshmi, N. and Selvakumar, R., Investigation of porous silica nanostructures in diatoms isolated from Kurichi and Sulur lakes of Coimbatore, India using field emission scanning electron microscopy. *Micron*, 79, 24–28, 2015.
[6.40] Shirokawa, Y. and Shimada, M., Cytoplasmic inheritance of parent-offspring cell structure in the clonal diatom *Cyclotella meneghiniana*. *Proc. R. Soc. B-Biol. Sci.*, 283, 1842, 20161632, 2016.
[6.41] Sinnott, E.W., *Plant morphogenesis*, Nabu Press, South Carolina, USA, 1960.
[6.42] Stein, J., *Digital Signal Processing: A Computer Science Perspective: Chapter 9: Correlation*, Wiley Online Library, USA, 2000.
[6.43] Sterrenburg, F.A.S., *Cytoplasmic inheritance in diatoms*, 1996.
[6.44] Sterrenburg, F.A.S., Gordon, R., Tiffany, M.A., Nagy, S.S., Diatoms: Living in a constructal environment, in: *Algae and Cyanobacteria in Extreme Environments. Series: Cellular Origin, Life in Extreme Habitats and Astrobiology*, vol. 11, J. Seckbach (Ed.), pp. 141–172, Springer, Dordrecht, The Netherlands, 2007.
[6.45] Taylor, J.R., *An Introduction to Error Analysis: The Study of Uncertainties in Physical Measurements*, University Science Books, Sausalito, California, USA, 1997.
[6.46] Tobin, P.C., Estimation of the spatial autocorrelation function: Consequences of sampling dynamic populations in space and time. *Ecography*, 27, 6, 767–775, 2004.

[6.47] Vukusic, P. and Sambles, J.R., Photonic structures in biology [Erratum: (2004) 42. *Nature*, 424, 6950, 852–855, 2003.

[6.48] Wikipedia, Frieze group, https://en.wikipedia.org/wiki/Frieze_group, 2019.

[6.49] Wikipedia, Gwyddion (software), https://en.wikipedia.org/wiki/Gwyddion (software), 2019.

[6.50] Wikipedia, Crystallography, https://en.wikipedia.org/wiki/Crystallography, 2020.

[6.51] Wikipedia, Fourier transform, https://en.wikipedia.org/wiki/Fourier_transform, 2020.

[6.52] Wikipedia, ImageJ, http://en.wikipedia.org/wiki/ImageJ, 2020.

[6.53] Wikipedia, List of planar symmetry groups, https://en.wikipedia.org/wiki/List_of_planar_symmetry_groups, 2020.

[6.54] Wikipedia, Penrose tiling, https://en.wikipedia.org/wiki/Penrose_tiling, 2020.

[6.55] Wikipedia, Point groups in two dimensions, https://en.wikipedia.org/wiki/Point_groups_in_two_dimensions, 2020.

[6.56] Wikipedia, Translational symmetry, https://en.wikipedia.org/wiki/Translational_symmetry, 2020.

[6.57] Wikipedia, Wallpaper group, https://en.wikipedia.org/wiki/Wallpaper_group, 2020.

[6.58] Xiong, C., Zhang, B., Kang, X.N., Fu, X.X., Zhang, G.Y., Lattice-parameter effects on diffracted transmission of GaN two-dimensional square-lattice photonic crystals. *Sci. China-Technol. Sci.*, 54, 1, 23–27, 2011.

[6.59] Wikipedia, Lattice (group), https://en.wikipedia.org/wiki/Lattice_(group)#Lattices_in_two_dimensions:_detailed_discussion, 2020.

[6.60] Losic, D., Mitchell, J. G., Lal, R., Voelcker, N. H., Rapid Fabrication of Micro and Nanoscale Patterns by Replica Molding from Diatom Biosilica. *Adv. Functional Mater.*, 17, 14, 2439–2446, 2007. https://doi.org/10.1002/adfm.200600872.

[6.61] Noyes, J., Sumper, M., Vukusic, P., Light manipulation in a marine diatom. *J. Mater. Res.*, 23, 12, 3229–3235, 2008. https://doi.org/10.1557/jmr.2008.0381.

[6.62] Kieu, K., Li, C., Fang, Y., Cohoon, G., Herrera, O. D., Hildebrand, M., Sandhage, K. H., Norwood, R. A., Structure-based optical filtering by the silica microshell of the centric marine diatom *Coscinodiscus wailesii*. *Opt. Express*, 22, 13, 15992, 2014. https://doi.org/10.1364/oe.22.015992.

7

Quantified Ensemble 3D Surface Features Modeled as a Window on Centric Diatom Valve Morphogenesis

Janice L. Pappas

Department of Mathematics, Michigan Math and Science Scholars, University of Michigan, Ann Arbor, MI, United States

Abstract

Morphological surface features are a record of genetic and developmental processes as well as environmental influences. The 3D geometric "terrain" of the surface consists of slopes via tangents, peaks and valleys via normals, smoothness of the transition between peaks and valleys, and point connections as flatness or curvature among all features. Such geometric quantities can be used to indicate morphological changes in valve formation over time. Quantified 3D surface features as geometric pattern ensembles may be representative of structural snapshots of the morphogenetic process.

For diatoms, valve formation and pattern morphogenesis reveal a layered valve face structure of areolae, cribra, and other morphological characters. Valve formation exhibits different combined geometries unfolding as 3D structural ensembles in particular spatial arrangements. Quantifying ensemble 3D surface geometries is attainable via models devised using parametric 3D equations and extracting surface features via partial derivatives for slopes, peaks and valleys, smoothness, and flatness as feature connectedness. Differences in ensemble 3D surface features may be used to assess structural differences among selected diatom genera as indicators of different valve formation sequences in surface generation and morphogenesis.

***Keywords*:** 3D surface, ensemble surface features, parametric equations, Jacobian, Hessian, Laplacian, Christoffel symbols, diatom morphogenesis

7.1 Introduction

As a guiding force historically in biology, the interweaving of evolution and genetics comprise the modern synthesis (e.g., [7.20, 7.59]) and the advent of mathematical population genetics showed that the application of quantitative methods was efficacious in understanding the role of genetics in evolution (e.g., [7.11]). To account for advances in genomics and molecular biology as well as developmental theory and morphology, and

Email: jlpappas@umich.edu

Vadim Annenkov, Joseph Seckback and Richard Gordon (eds.) Diatom Morphogenesis, (159–194) © 2022 Scrivener Publishing LLC

including abrupt timing events such as endosymbiosis (e.g., [7.27]), the extended evolutionary synthesis (e.g., [7.48]) and post-modern evolutionary synthesis (e.g., [7.25]) propose to remedy the shortcomings of the modern synthesis. Macroevolution across taxonomic levels going beyond species-specific biology is also a contributing factor in the impetus for a more accurate synthesis in understanding the biological world. The phenotype and its role in morphological evolution and evolutionary developmental biology (e.g., [7.12, 7.35]) are of particular interest in studying morphogenesis from a macroevolutionary viewpoint.

Developmental processes encompass information transfer over time. Some of the information transfer involves the phenotype of the organisms as changes in morphology and evidenced in morphological variation. The dynamics of form and structure comprising morphogenesis as an evolutionary process occurs via various mechanisms and at various spatial and temporal scales. Constraints on the kinds of phenotypes that evolve may be a result of the morphogenetic process, producing specific outcomes evidenced more broadly as morphological evolution. Morphogenesis is expressed externally in the morphological patterns of organisms as a result of internally occurring genetic, cytological, and developmental processes. Through developmental processes, morphogenetic outcomes may be the result of further refined evolutionary avenues so that such developmental constraints may produce morphological variation more specifically among taxa as phenotypic expression via epistasis and canalization (e.g., [7.67]).

Phenotype during development may be studied in unicellular organisms with regard to epistasis (e.g., [7.53]) and canalization (e.g., [7.71]). The processes of epistasis and canalization concerning the genotype may be extended to understand interactions among quantified characters of the phenotype in models of development [7.49]. Phenotypic characters are representations of morphological features, and their quantification enables a way to model inferences of their relation to information transfer during the life cycle and development of a unicellular organism.

As changes in phenotype occur, morphogenesis during unicell development may be quantifiable as three-dimensional (3D) surface features. Ensembles of these surface features may be used to track changes in phenotype over time using morphological 3D models. Documenting these phenotypic changes via 3D surface models enables studying the development of a whole unicellular organism more comprehensively in 3D rather than just looking at morphology as isolated or individual parts with no reasonable way to provide a cohesive view of the entire unicellular organism.

7.1.1 From 3D Surface Morphology to Morphogenesis

Morphology of an organism is what we perceive. We look at an organism to register its external features in our mind—its size, its color, and its shape. When describing these features, we often characterize them as measurable and countable attributes with boundaries. That is, we assign numerical descriptors about how much we see. For example, the length of a given feature, the number of repetitions of a given feature, and the amount of color and geometry of shape of a given feature are all observations that tell us something measurable about morphological attributes. All of these morphological measures are interpreted to have biological meaning.

Other aspects of morphology are evident, but not necessarily so easily summarized as a countable feature. How much does the surface slope? How do peaks and valleys vary on the surface? How smooth is the surface? Are some places on the surface much more pronounced than others? How flat or curved is the surface? How are surface attributes connected to one another? All of these morphological attributes may have biologically significant meaning that is just as important as countable features. Such surface features may be important in the dynamic growth of an organism and indicative of structural patterning. Morphological changes during growth potentially signifying phenotypic plasticity in which repeating patterned structures and their variation may be recorded. Such variation may be seen at the species or higher taxonomic levels via non-countable morphological surface features.

Morphological surface features are a record of the processes necessary to produce the way an organism looks and are identifiable from surface points to combinations of geometric arrangements of surface points and contours. The geometric "terrain" of the surface consists of slopes via tangents, the peaks and valleys via normals, smoothness at transitional zones between peaks and valleys, and point connections as curvature or flatness between tangents and normals spatially tying multiple features together. For diatoms, surface features may be functionally important in predation resistance or defense (peakedness), glide ability in water (smoothness), ability to be cloaked or disappear into the environment (flatness), or countershading and reflectability in the environment (slopeness). Quantified surface features may be used to represent structural features and changes during morphogenesis as ensembles. That is, the geometry of morphology is quantifiable as ensemble surface features in organisms. Because of their elaborate geometric morphology, diatoms would qualify as model systems [7.7] for study using ensemble surface features (Figure 7.1).

For diatoms, silica valve formation and pattern morphogenesis has been of interest and modeled using Turing-like [7.63] and similar algorithmic techniques (e.g., [7.13, 7.15, 7.44]). Models have been created to mimic the way in which diatoms exhibit the highly diverse patterns evident on their valve surfaces. Techniques including artificial neural networks via multi-agent algorithms or cellular automata systems have been used in various attempts to represent "mechanistically" the way the biological world works in all its complexity. Results from these techniques and algorithms have an intuitive appeal because of their resemblance to the consequents of biological processes. It is tempting to think of such algorithms as "biological mechanisms" when, in fact, they are not. They are only pattern generators. To date, such techniques have been useful in creating patterns but have not been integrated and matched at multiple scales with actual silica deposition events, developmental or life cycle changing events, or any

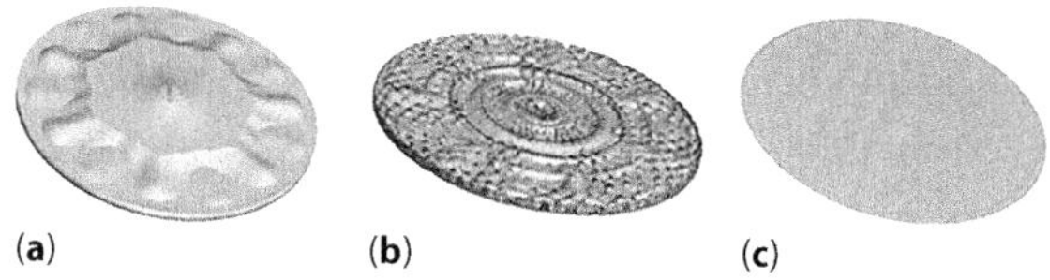

Figure 7.1 Ensemble surface features—examples: (a) smooth, but not flat; (b) flat, but not smooth; (c) smooth and flat. Smoothness is determined from the Laplacian, and flatness is determined from Christoffel symbols.

link to biologically evidenced morphogenetic steps in diatoms. Initiation and termination events of techniques such as those generating reaction-diffusion patterns, if they do occur in diatom valve formation, has yet to be addressed in more than an ad hoc way. Different applications of reactive-diffusion as different models may have multiple mechanistic explanations or make incongruent predictions (e.g., [7.5]). Despite the recent renewed interest in reaction-diffusion systems at the molecular and nanoscale levels (e.g., [7.16]), diatom valve patterning is a 3D process occurring during morphogenesis. Caution must be exercised concerning pattern-generating models as being equivalent to actual biological processes [7.70].

Having said this, the utility of modeling has been demonstrated time and time again as an aid to gain an understanding of the potential explanations for biological processes (e.g., [7.28]). Vegetative diatom cell morphology as a geometric construct can be used in a phenomenological approach and basis for modeling. From the whole organism cell pattern, the mature diatom vegetative cell is represented by valve face geometry, and changes in morphology during valve formation may be achieved via back-tracking from this cell to the start of silica deposition. Organism surface represents the boundaries of its 3D morphology, and as such, this surface geometry may be used as a proxy for the phenotype (e.g., [7.41]). Considering surfaces may enable a better understanding of potential function and evolution of biological structures and requires looking at 3D surface geometry and ensembles of those geometries.

For diatom valve formation, 3D surface geometry is necessary to consider the whole cell. Morphological changes over time occur in 3D, and all the information available on the surface is of importance concerning morphogenesis. Although quantitative techniques via a closed curve orthogonal polynomial or matching arbitrary boundary points among species-specific taxa via deformation techniques have been used to characterize valve shape in diatom identification and classification studies in taxonomic analysis (e.g., [7.42]), these techniques are inadequate in the assessment of surface morphology. Circular and oval centric valve outlines provide no differentiating information because of their shape. Even if differentiation of outline shapes is possible, this says nothing about the information contained in and on the valve surface. For diatom valve morphogenesis, 3D quantitative methods enable the inclusion of all boundary points on the 3D surface.

Diatom valve formation during morphogenesis may be "cataloged" as geometric surface changes. Absent direct matching with biological processes, a geometric interpretation can be qualitatively relational in terms of ensembles of morphological changes over time. With the striking geometry of the scaffolded structure of the diatom valve face of layered areolae, cribra, and other morphological characters, valve formation has different combined geometries from layer to layer, combining such structural elements in a spatial arrangement. In actual biological cells, the layering effect has been used as a means to examine structural morphology with imaging and microscopy techniques via *z*-stacking (e.g., [7.57]). Geometrically, a cell can be crudely thought of as a series of circles stacked one upon the other to form the 3D cylindrical shape. Its 3D reconstructed image is an analytic approximation of a real entity. Removal of each circular layer from a 3D cylinder would expose the relief of successive surfaces of a cell such as a centric diatom valve face. Back-tracking to the start of silica deposition, the whole diatom 3D model and its forming valves during morphogenesis may be devised as the reverse stepwise layering of successive surfaces of

z-stacking. Layered additions as a sequence of steps may be modeled with z-layers to produce the whole cell. A 3D surface model is an implicit approximation of a real entity.

7.1.2 Geometric Basis of 3D Surface Models and Analysis

Initially, a 3D surface model is created using parametric 3D equations. Geometric analysis of a 3D surface model necessarily means defining that surface as a tangent space. This space contains points in which first partial derivatives from the system of parametric 3D equations form the Jacobian matrix (i.e., the Jacobian) and represents the sloping of the surface. Connectivity of points on the surface necessitates the creation of ensembles of surface features because partial differentiation is insufficient in constituting connection of neighboring tangent spaces. To achieve connection of tangent spaces, Christoffel symbols are calculated as connectivity representing flatness or global curvature among points on the surface. In a tangent space, change in basis vectors relative to a basis forms the way in which to measure a differential change in coordinates in each coordinate direction [7.26] on the surface, and change in coordinates occurs via the Jacobian [7.22]. Connection coefficients are Christoffel symbols in which the basis vectors change linearly. As a result, Christoffel symbols are a linear connection among ensemble surface features as they give structure to a tangent space.

The second partial derivatives are the Hessian matrix (i.e., Hessian) and Laplacian (i.e., Laplace operator expressed via Laplace's equation) that are used to quantify the degree of local curvature of critical points and mapping of a one ensemble of surface features to another one on the surface, respectively. That is, the derivative of the slope of the tangent as peaks, valleys and saddle points (i.e., saddles) express maxima, minima and points of inflection on the surface, and the sum of the change among ensembles of these surface features express the degree of smoothness on the surface.

The Jacobian, Christoffel symbols, the Hessian, and the Laplacian express ensemble surface features as morphogenetic descriptors of each step of a diatom valve formation sequence as slopeness, connectivity via curvature or flatness, peak-to-valley changes, and smoothness, respectively, of the 3D surface. These quantities will characterize the valve surface as pointwise connectivity via degree of curvature or flatness and smoothness between elevations and depressions and sloping on a diatom valve surface.

7.1.3 Differential Geometry of 3D Surface

Each point on the surface is composed of a trihedron of vectors defined as a moving reference frame. For a curve, a Serret-Frenet frame consists of tangent (t), unit normal (n), and binormal (b) vectors, while for a surface, a Darboux frame consists of tangent (t), unit normal (principal normal) (u), and tangent normal vectors (v). At the tangent and tangent normal vectors of a point, normal planes at the unit normal can be used to cut the surface and define principal directions. Curvature, κ, at a point in relation to n and u is $\kappa n \cdot u$ [7.21].

Curvature is the rate of change of a tangent line to a curve [7.8]. The maximum and minimum curvatures at a point are the principal curvatures, κ_1 and κ_2. The product of the maximum and minimum curvatures is Gaussian curvature. Mean curvature is half the sum of the maximum and minimum curvatures. If both κ_1 and κ_2 are positive or negative, then the surface is locally convex. If κ_1 is positive and κ_2 is negative or vice versa, then the surface is locally a saddle. If κ_1 or κ_2 is equal to 0, then the surface is in between a convex and saddle. If κ_1 and κ_2 are both equal to 0, then the surface is a plane or a monkey saddle at a flat umbilic [7.21].

Rate of change of a tangent is a first partial derivative with respect to a given curve on a surface with arc length, *s*. All of the tangent lines at every point on the surface are summarized in a Jacobian. Numerical solution to the Jacobian characterizes all the tangent lines and planes on the whole organism surface. Tangent lines define an intrinsic metric via the arc length and are coefficients of the first fundamental form, which is an inner product on the tangent space of a surface. The first fundamental form enables the measurement of length, angle, curvature, surface area and any other metric on a surface in 3D space. The second fundamental form defines tangent planes in terms of change in surface shape of the normals with respect to the tangents [7.24] and is a quadratic form on the tangent space of a surface in 3D space. That is, the second fundamental form is a change of local shape over the surface.

For a two-dimensional (2D) patch on a 3D surface, tangent vectors span the tangent plane and serve as a basis in a matrix of coefficients for the differential of the tangent vectors. The matrix expressed in terms of the first and second fundamental forms represents Gaussian and mean curvature, respectively [7.8]. Using Gaussian curvature, shape can be determined as a general characterization of the surface [7.24].

There are normals to tangent vectors spanning a tangent plane. The differentiable field of the unit normal vectors on the surface is a Gauss map of the tangent space of a point. The differential of a Gauss map is expressed as the shape operator or a Weingarten map. The shape operator determines all the tangent planes in the neighborhood of a point and is the change in surface normals with tangent vectors [7.24].

Rate of change of a tangent line to a curve may be characterized by Gaussian curvature, $K = \kappa_1\kappa_2$, and mean curvature, $H = \frac{\kappa_1 + \kappa_2}{2}$. Gaussian curvature is an intrinsic property of a 3D surface object without reference to the geometric space in which the object resides, and mean curvature refers to the embedding of a 3D surface object in a 3D space and is its extrinsic property [7.24].

The first and second fundamental forms as well as Gaussian and mean curvature, respectively, provides a link from Jacobian to Gauss and Weingarten maps [7.8]. The Jacobian determinant of a Gauss map is Gaussian curvature. Principal curvatures of a Weingarten map are eigenvalues and are an indicator of how much the surface bends [7.24]. Principal directions are eigenvectors of a Weingarten map, and from the second fundamental form, rotation of the *x*, *y* tangent plane produces eigenvectors of the Hessian [7.8]. By projecting the covariant derivative of the Weingarten map onto the second fundamental form, the connection gradient, ∇, is obtained, which is also the derivative of the trihedron of vectors and represented by Christoffel symbols [7.8]. The connection matrix of Christoffel symbols represents the connection of points and the movement of a Darboux frame as it relates to a Serret-Frenet frame on a curve (i.e., trihedron of vectors) on the surface [7.24]. The Laplacian depends on connectivity via the divergence of the gradient on the surface and is the trace of the Hessian [7.8]. The Jacobian is related to the Christoffel symbols because each row of the Jacobian is a gradient [7.8, 7.22], and the

Jacobian determinant relates surface connections via coordinate transformation [7.60]. Gauss and Weingarten maps implicitly describe the functions that are used in characterizing 3D surface morphology.

7.1.4 3D Surface Feature Geometry and Morphological Attributes

Changes in the contours of the 3D surface or steepness or shallowness of slopes at a given u and v coordinate can be characterized in the x-, y-, and z-directions resulting in a Jacobian. The steepest and shallowest points as well as those points that indicate a transition from shallow to steep slopes (or vice versa) are calculated as second partial derivatives. These slopes are quantifications of local maxima, minima, and saddle points (i.e., saddles) with the highest maximum and lowest minimum as global values. The Hessian consists of second partial and mixed partial derivatives so that the Jacobian is transformed from a six element to a 12-element matrix.

When considering a patch as a tangent plane, the Hessian is much easier to calculate. Because x and y define a plane, only the z-direction needs to be considered, and in this case, the patch is a Monge patch [7.24]. The z-direction is the height at each point on a patch, and the partial derivatives of z with respect to u and v from the Jacobian are the maximum and minimum slopes that represent a morphology gradient of the direction of maximum height of the surface as normals to the tangent plane of the surface. The magnitude of maximum height is the square root of the sum of the second partial derivatives of z with respect to u and v. These second partial derivatives along with the mixed second partial derivatives comprise the Hessian for z. Mixed second partial derivatives of z represent a combination of slopes of the morphology at that point on the surface. Movement of the change in height along a curve on the surface determines the effect of a combination of slopes on the characterization of morphology.

From the Hessian, the sum of the elements of the diagonal is the Laplacian, ∇^2, and is the divergence of the gradient on the surface. Taylor expansion of the Laplacian around zero gives the state of the surface under conditions of no change in equilibrium and as a measure of stability that is solvable by finite differences [7.68]. Stability at equilibrium is also equivalent to being a measure of smoothness. Each of the terms from the finite differences approach with respect to u and v are eigenvectors, and the first term may be used to indicate approximate smoothness. Because the diagonal elements of the Hessian are used to calculate the Laplacian, it is these elements that are the eigenvalues of the eigenvectors. The variation in smoothness is evaluated by the sign of the sum of the eigenvalues or trace that is the Laplacian. That is, the Laplacian measures the degree to which perturbations on the surface reflect the smoothness of surface morphology.

Surface feature connections via Christoffel symbols are a measure of the degree of flatness or curvature of the surface [7.26]. For Christoffel symbols that are vanishing, the surface is flat. That is, for zero Gaussian curvature and Christoffel symbols at zero, the surface is flat [7.32, 7.60]. The connection between highly textured and smooth surfaces is summarized in the Christoffel symbols as an indicator of changing curvature and flatness both vertically and horizontally across a contiguous surface morphology.

7.1.5 Centric Diatom Taxa Used as Exemplars in 3D Surface Models for Morphogenetic Analysis

Centric diatom taxa chosen for study as exemplars were *Actinoptychus senarius*, *Arachnoidiscus ehrenbergii*, and *Cyclotella meneghiniana*. These taxa were readily modeled using parametric 3D equations and represent three widely different valve surfaces exhibiting distinctly different features associated with silica deposition during valve formation. Ensemble valve surface features are readily measured to illustrate quantitatively the distinctions among the taxa during valve formation as a part of the morphogenetic process.

Of the three exemplar centric diatom taxa, *Actinoptychus senarius* exhibits a circular buckling pattern as alternating undulations of sectors emanating from the center [7.14]. Such a pattern may be recoverable as a wave front [7.34]. Rimoportulae are curled, and the valve is divided into sectors by interior folds or walls [7.18, 7.29]. For *Actinoptychus senarius*, the areolae are a coarse or reticulate pattern on the valve surface, and there is a marginal ridge from which external projections of the rimoportulae protrude above the valve face [7.18]. *Actinoptychus senarius* has a narrow hyaline zone that extends from the central area to around halfway to the valve margin. Heterovalvy is not present in *Actinoptychus senarius* [7.14].

Arachnoidiscus ehrenbergii may exhibit heterovalvy during auxospore attachment to the epivalve of the parent cell [7.23, 7.54]. Different valve patterns may form as a result of differing attachment sites. Re-creation of a diatom valve was accomplished using digital holography, resulting in a model of *Arachnoidiscus* that was obtained via image reconstruction of an optical wave front [7.9, 7.10]. The model showed elongated central slits in a ring surrounding a small planar central area on the valve and pores diminishing in view of a planar field [7.9]. The ribs emanating from the central area are regularly spaced and may be bisected to look like windows on the pore structured pattern below [7.3].

Cyclotella meneghiniana has distinctive striations as ribs regularly placed at the valve margin covering about half of the valve face [7.17, 7.56]. These striations form a regular undulating pattern around the valve periphery giving *Cyclotella meneghiniana* its distinct appearance. Marginal rimoportulae are present as well [7.17], and the central area has fultoportulae. The clear central area exhibits a rather uniform surface [7.17, 7.56]. *Cyclotella meneghiniana* may exhibit heterovalvy [7.52].

7.1.6 Morphogenetic Descriptors of Centric Diatoms in Valve Formation as Sequential Change in 3D Surface Morphology

Diatom growth patterns follow two general tracks—one for centric diatoms, and the other for pennate diatoms [7.52]. Centric diatom valve formation may be characterized simply as growth from an annulus to the valve margin, while pennate diatoms exhibit growth from a sternum to the valve margin (e.g., [7.56]). Growth is horizontal across the valve face in 2D and horizontal and vertical in 3D. In the vertical aspect of the diatom growth system, the 3D surface may be characterized as layers of morphology. Quantitatively, as each layer is accounted for, vertical growth of the organism can be modeled using a series of systems of parametric 3D equations per taxon. The relation between developmental growth and 3D morphology is measurable as the change in whole surface morphology by additional surface morphologies, i.e., ensemble 3D surface morphologies. This change is more precisely a starting morphology plus an additional morphology. The addition is integrated into the original whole so that a new system of parametric 3D equations emerges from the original system. As subsequent additions are made, the cumulative result is modeled as the final valve produced after valve formation is completed.

Schmid and Volcani [7.56] determined that diatom valve formation roughly occurs as a three-stage process. For centric diatoms, the starting point is the annulus from which radial rows of silica are deposited in a horizontal fashion toward the valve margin in a crosswise pattern with connections so that a branching pattern materializes. This is the basal layer in which silica is deposited to initiate the rudimentary formation of internal rimoportulae as well, from which the next stage of valve formation occurs. This next stage involves the initiation of vertical silica deposition where the walls of the areolae increasingly emerge, and the round pores that may be formed in the first stage become hexagonal or more geometrically defined in shape as well as the formation of external rimoportulae occurring. Cribra and cribella mark their formation during the third stage in which completion of horizontal silica deposition occurs. The size and spacing of these structures are controlled by areolae [7.56] as the successive layers of silica are deposited to complete the valve structure to the margin [7.51].

Within the silicalemma and silica deposition vesicle (SDV), new wall formation occurs as silica is aggregated [7.51, 7.52, 7.56], and the plasmalemma molds cell wall shape prior to valve formation [7.55]. From the center to the valve margin, structural changes occur in a sequence over time. Common valve structures are created for a given species within a genus, and differences in those structures occur via silica deposition for multiple species within that genus in terms of the location and timing of structural elements emerging during valve formation.

7.1.7 Purposes of This Study

In this study, Schmid and Volcani's [7.56] schema is used to characterize ensemble 3D surface changes in the modeling of centric diatom valve formation stages during diatom morphogenesis. Models will be created using systems of parametric 3D equations for each sequence of valve formation steps. The ensemble 3D geometric surface properties are quantified as the Jacobian, Hessian, Laplacian, and Christoffel symbols from slopes, peaks-valleys-saddles, smoothness, and connectedness via flatness or curvature, respectively. Ensemble surface geometric features and the degree to which they change during valve formation is used to characterize a constructed diatom morphogenetic system with regard to silica deposition implicitly over time. Implications of ensemble surface features used in characterizing a process such as morphogenesis are explored for macroevolutionary processes such as epistasis and canalization.

7.2 Methods

Diatom valve formation is depicted using 3D surface models representing vegetative cells in 3D (x, y, and z) space. Parametric 3D equations are interpretable in terms of their utility in relating 3D surface geometry to 3D surface morphology. Variables x, y, and z parameterized by u and v are used in the characterization of a point anywhere on the surface as well as movement from point to point on the surface along a given curve on a given diatom model. Systems of parametric 3D equations of the general form $x = f(u, v)$, $y = g(u, v)$, $z = h(u, v)$ with Cartesian coordinates x, y, and z in parameters u and v and evaluated on the interval $[0, 2\pi]$ are used to create the 3D surface models [7.36–1.41]. Parametric 3D equations enable the quantification of points on curves and/or surfaces so that each point in x, y, and z Euclidean 3D space can be mapped to a new 2D u and v Euclidean space. Because surface morphology resides within a curved boundary (e.g., circle, oval or other polygonal 2D form), the new 2D space contains morphological information in u and v surface points. Each 2D surface point is a base for a vector in which the arrowhead of the vector points away from the surface. Using vectors that point perpendicularly from the surface at each point, maximum changes from point to point are recorded changes as one travels along the surface in any direction. The u and v parameters are 2D points extracted from a 3D x, y, and z diatom valve surface model that contain information on changes in the topography of the surface. These changes are measurable as the solution of parametric 3D equations that map the x, y, and z variables to the u and v parameters. In this way, 3D surface geometry is used in quantifying 3D surface morphology.

7.2.1 Measurement of Ensemble Surface Features and 3D Surface Morphology: Derivation and Solution of the Jacobian, Hessian, Laplacian, and Christoffel Symbols

7.2.1.1 The Jacobian of 3D Surface Morphology

From parametric 3D equations, first derivatives are expressed in terms of first partial derivatives with respect to arc length, s. The differential of s is $ds^2 = dx^2 + dy^2 + dz^2$ on the surface, and $dx = \frac{\partial x}{\partial u}du + \frac{\partial x}{\partial v}dv$, $dy = \frac{\partial y}{\partial u}du + \frac{\partial y}{\partial v}dv$, and $dz = \frac{\partial z}{\partial u}du + \frac{\partial z}{\partial v}dv$. The Jacobian is $J = \begin{bmatrix} \frac{\partial x}{\partial u} & \frac{\partial x}{\partial v} \\ \frac{\partial y}{\partial u} & \frac{\partial y}{\partial v} \\ \frac{\partial z}{\partial u} & \frac{\partial z}{\partial v} \end{bmatrix}$ as first partial derivative of x, y, and z with respect to u and v [7.22]. Numerical solution to the Jacobian characterizes all the tangent lines and planes on the whole organism surface. Eigenvalues, ψ, of the Jacobian may be negative, meaning a stable surface at extrema (i.e., maxima or minima), or positive, meaning an unstable surface at extrema (i.e., saddles or other extrema that are neither maxima nor minima).

7.2.1.2 Monge Patch

For a whole organism, 3D surface expressed parametrically as $x=f(u,v), y=g(u,v), z=h(u,v)$, $J=\begin{bmatrix} \frac{\partial x}{\partial u} & \frac{\partial x}{\partial v} \\ \frac{\partial y}{\partial u} & \frac{\partial y}{\partial v} \\ \frac{\partial z}{\partial u} & \frac{\partial z}{\partial v} \end{bmatrix}$. For a patch on or as that surface and redefining x, y, z, $x=g(u,v), y=h(u,v)$, $z=f(x,y)=f[g(u,v),h(u,v)]$, and $dx=g'(u,v)du\,dv=\frac{\partial x}{\partial u}du+\frac{\partial x}{\partial v}dv$, $dy=h'(u,v)du\,dv=\frac{\partial y}{\partial u}du+\frac{\partial y}{\partial v}dv$, $dz=f'[g(u,v),h(u,v)]\;dx\,dy=\frac{\partial z}{\partial x}\left(\frac{\partial x}{\partial u}du+\frac{\partial x}{\partial v}dv\right)+\frac{\partial z}{\partial y}\left(\frac{\partial y}{\partial u}du+\frac{\partial y}{\partial v}dv\right)$. For J, define each row as a linear combination of the other. At (u_0, v_0), the submatrices are $\begin{bmatrix} \frac{\partial x}{\partial u} & \frac{\partial x}{\partial v} \\ \frac{\partial z}{\partial u} & \frac{\partial z}{\partial v} \end{bmatrix}$, $\begin{bmatrix} \frac{\partial x}{\partial u} & \frac{\partial x}{\partial v} \\ \frac{\partial y}{\partial u} & \frac{\partial y}{\partial v} \end{bmatrix}$ and $\begin{bmatrix} \frac{\partial y}{\partial u} & \frac{\partial y}{\partial v} \\ \frac{\partial z}{\partial u} & \frac{\partial z}{\partial v} \end{bmatrix}$ for $(x, z) = (x(u, v), z(u, v))$, $(x, y) = (x(u, v), y(u, v))$, and $(y, z) = (y(u, v), z(u, v))$, respectively. By the inverse function theorem, the reparameterizations to Monge patches are $\begin{bmatrix} x \\ y(u(x,z),v(x,z)) \\ x \end{bmatrix}$, $\begin{bmatrix} x \\ y \\ z(u(x,z),v(x,z)) \end{bmatrix}$, and $\begin{bmatrix} x(u(x,z),v(x,z)) \\ y \\ z \end{bmatrix}$, respectively.

7.2.1.3 First and Second Fundamental Forms and Surface Characterization of the Monge Patch

Tangent lines define an intrinsic metric expressed as $I = Edu^2 + 2Fdudv + Gdv^2$ where E, F, G are coefficients of the first fundamental form [7.24]. The second fundamental form defines tangent planes as change in surface shape of the normals with respect to the tangents [7.8, 7.24] and are given as $II = Ldu^2 + 2Mdudv + Ndv^2$ [7.8].

For a Monge patch on a surface, tangent vectors, $r(u, v) = (r_u, r_v) = r_{uv}$, span the tangent plane and serve as a basis in a matrix of coefficients for the differential of the tangent vectors. For $r_{uv}(x) = u x_u + v x_v$, $r_{uv}(y) = u y_u + v y_v$, and $r_{uv}(z) = u z_u + v z_v$, coefficients of the first fundamental form are $E = r_u \cdot r_u$, $F = r_u \cdot r_v$, $G = r_v \cdot r_v$, and $||r_u \times r_v|| = (r_u \cdot r_u)(r_v \cdot r_v) - (r_u \cdot r_v)^2 = EG - F^2$. The normal to tangent vectors spanning a tangent plane is $n\,(u, v) = (n_u, n_v) = n_{uv}$ and the unit normal vector is $\frac{r_u \times r_v}{||r_u \times r_v||}$. The coefficients of the second fundamental form are $L = r_{uu} \cdot n_{uv}$, $M = r_{uv} \cdot n_{uv}$, $N = r_{vv} \cdot n_{uv}$ [7.24].

For the matrix of the differentials of the tangent vectors, the first and second fundamental forms have a determinant given as $K = \frac{LN - M^2}{EG - F^2}$, and half the trace is $H = \frac{LG - 2MF + NE}{2(EG - F^2)}$

which are Gaussian and mean curvature, respectively [7.8]. Using K, shape can be determined as a general characterization of the surface [7.24], and H is the overall shape of the surface as it is embedded in 3D space [7.8].

7.2.1.4 3D Surface Characterization via Gauss and Weingarten Maps and the Fundamental Forms

The differentiable field of the unit normal vectors, N, on the surface, S, is a Gauss map of the tangent space of a point and is given as $N: S \rightarrow S^2$ for $S \rightarrow R^3$ and $S^2 = \{(x, y, z) \in R^3; |x^2 + y^2 + z^2 = 1\}$ [7.8]. The differential of a Gauss map is expressed as the shape operator or a Weingarten map and given as $\nabla_{Dir} S_{uv} = (EG - F^2)^{-1} \begin{pmatrix} LG - MF & MG - NF \\ ME - LF & NE - MF \end{pmatrix}$, where $\nabla_{Dir} S_u = n_u = \frac{MF - LG}{EG - F^2} r_u + \frac{LF - ME}{EG - F^2} r_v$ and $\nabla_{Dir} S_v = n_v = \frac{NF - MG}{EG - F^2} r_u + \frac{MF - NE}{EG - F^2} r_v$. The shape operator determines all the tangent planes in the neighborhood of a point and is the change in surface normals with tangent vectors [7.24].

7.2.1.5 Peaks, Valleys, and Saddles of Surface Morphology and the Hessian

For a Monge patch with height function z, $\nabla z = \begin{bmatrix} \frac{\partial z}{\partial u} \\ \frac{\partial z}{\partial v} \end{bmatrix}$ is the maximum height, and the magnitude is $\|\nabla z\| = \sqrt{\left(\frac{\partial z}{\partial u}\right)^2 + \left(\frac{\partial z}{\partial v}\right)^2}$. The Hessian is used to express the to the maximum slope at each point on the surface and is given as $A = \begin{bmatrix} \frac{\partial^2 z}{\partial u^2} & \frac{\partial^2 z}{\partial u \partial v} \\ \frac{\partial^2 z}{\partial u \partial v} & \frac{\partial^2 z}{\partial v^2} \end{bmatrix}$.

The product of the maximum height and maximum slope is expressed as $\nabla z \cdot A = \begin{bmatrix} \frac{\partial z}{\partial u} \\ \frac{\partial z}{\partial v} \end{bmatrix} \begin{bmatrix} \frac{\partial^2 z}{\partial u^2} & \frac{\partial^2 z}{\partial u \partial v} \\ \frac{\partial^2 z}{\partial u \partial v} & \frac{\partial^2 z}{\partial v^2} \end{bmatrix}$. The characterization of extrema is accomplished by calculating eigenvalues of the Hessian, and the product of the eigenvalues is the Hessian determinant [7.24]. For each extremum on the surface of a Monge patch, the characteristic equation is $A\,h = \phi\,h$, where ϕ is the eigenvalue of the Hessian expressed in matrix format as A and the column vector h represents ∇z. For the purpose of characterizing surface morphology using $\nabla z \cdot A$, eigenvalues are measures of the degree of concavity and curvature at a given z-value on the surface.

7.2.1.6 Smoothness as a Characterization of Surface Morphology and the Laplacian

Characterizing the changes over the course of the surface—the ups and downs—is accomplished by the Laplacian [div (grad z)]. For a patch, the Laplacian is $\nabla^2 z = \frac{\partial^2 z}{\partial u^2} + \frac{\partial^2 z}{\partial v^2} = 0$ as the trace of the Hessian and a measure of the degree of average smoothness over the surface. The Laplacian is a harmonic function that is equal to the average of its nearby values on a surface. The Laplacian is also representative of steady-state or equilibrium conditions because the average smoothness is calculated at $\nabla^2 z = 0$ [7.22].

To solve this initial value problem with a homogeneous boundary for a patch, separation of variables is used and expansion of ∇^2 is a quadratic expressing each term as $\frac{\partial^2 z}{\partial u^2} = \frac{\partial z}{\partial x}\frac{\partial^2 x}{\partial u^2} + \left(\frac{\partial^2 z}{\partial x^2}\right)\left(\frac{\partial x}{\partial u}\right)^2 + 2\frac{\partial^2 z}{\partial x \partial y}\frac{\partial x}{\partial u}\frac{\partial y}{\partial u} + \left(\frac{\partial^2 z}{\partial y^2}\right)\left(\frac{\partial y}{\partial u}\right)^2 + \frac{\partial z}{\partial y}\frac{\partial^2 y}{\partial u^2}$ and $\frac{\partial^2 z}{\partial v^2} = \frac{\partial z}{\partial x}\frac{\partial^2 x}{\partial v^2} + \left(\frac{\partial^2 z}{\partial x^2}\right)\left(\frac{\partial x}{\partial u}\right)^2 + 2\frac{\partial^2 z}{\partial x \partial y}\frac{\partial x}{\partial v}\frac{\partial y}{\partial v} + \left(\frac{\partial^2 z}{\partial y^2}\right)\left(\frac{\partial y}{\partial u}\right)^2 + \frac{\partial z}{\partial y}\frac{\partial^2 y}{\partial v^2}$. Discretization of the second partial derivatives with difference quotients (finite differences) determined by Taylor expansion of the Laplacian is given as $\nabla^2 z_{i,j} \approx \frac{z_{i+1,j} - 2z_{i,j} + z_{i-1,j}}{\Delta u^2} + \frac{z_{i,j+1} - 2z_{i,j} + z_{i,j-1}}{\Delta v^2} + \cdots$.

Each term is approximately equal to the eigenvectors of ∇^2 in u and v so that $\frac{\partial^2 z}{\partial u^2} = \left(\frac{d^2}{du^2} \otimes I_v\right) z(u, v)$ and $\frac{\partial^2 z}{\partial v^2} = \left(I_u \otimes \frac{d^2}{dv^2}\right) z(u, v)$. For $\nabla^2 z_{i,j}$, the eigenvalue problem is $\frac{d^2 z}{du^2} = \lambda_u z(u)$ and $\frac{d^2 z}{dv^2} = \lambda_v z(v)$ so that $\nabla^2 z = \left[\left(\frac{d^2}{du^2} \otimes I_v\right) + \left(I_u \otimes \frac{d^2}{dv^2}\right)\right] z(x, y)$ to determine degree of smoothness. The eigenvectors $eig_{i,j}(u, v) = eig_u \otimes eig_v$ are $eig_{u_i}(u) = \sqrt{\frac{2}{L}} sin\left(\frac{i\pi u}{L}\right)$ and $eig_{v_j}(v) = \sqrt{\frac{2}{L}} sin\left(\frac{j\pi v}{L}\right)$ with Dirichlet boundary $\partial\Omega = 0$, and the eigenvalues are $\lambda_{i,j}(u,v) = \lambda_{u_i} + \lambda_{v_j}$, or more generally, $\lambda = \lambda_u + \lambda_v$ [7.68].

7.2.1.7 Point Connections of 3D Surface Morphology and Christoffel Symbols

The parameterization of a surface induces the connection between a 3D space and the embedding of 2D geometric structures in that 3D space. As an affine connection, a point is transported along a curve on the surface inducing a characterization of the geometry as tangent vectors of the surface in the neighborhood of that point in a tangent metric space. The derivative of tangent vectors of the surface induces a connection between tangent spaces that are nearby each other.

Christoffel symbols indicate where curvature is present in a locally flat neighborhood on a surface [7.24]. Values of Christoffel symbols that are non-zero are curvature indicators of degree of twisting, turning, contraction, or expansion of a surface locally about a given point; for vanishing or zero-valued Christoffel symbols, then flatness is defined in a local neighborhood of a given point [7.32].

The metric tensor as a representation of the first fundamental form [7.8] is a characteristic of differentiable inner products in tangent spaces, and because of this, Christoffel symbols can be expressed via the metric tensor components [7.22, 7.26]. Curvature is a characteristic of metric space as the metric tensor enables the calculation of distances [7.60] on space curves or surfaces embedded in 3D space. The metric tensor is $g_{ij} = e_i \cdot e_j$ where e_i and e_j are basis vectors in a coordinate system, and Christoffel symbols in basis k for these basis vectors are $\partial_j e_i = \Gamma^k_{ij} e_k$ [7.33]. Surface basis vectors are contravariant space vectors or covariant surface vectors [7.60]. A change in basis as a scalar product $\Gamma^k_{ij} e_k \cdot e_l$ and $g_{kl} = e_k \cdot e_l$ produces $\Gamma^k_{ij} g_{kl} + \Gamma^k_{lj} g_{ki} = \partial_j g_{il}$ [7.33]. Basis vectors can vary from point to point on a curve or surface, and permutated indices are used to generate multiple Christoffel symbols in terms of the metric tensor [7.22].

The addition of indices with respect to covariant and contravariant vectors undergoing differentiation will yield covariant and contravariant derivatives such as $\Gamma^k_{ij} = \dfrac{\partial^2 \xi^p}{\partial x^i \partial x^j} \dfrac{\partial xk}{\partial \xi p} = \Gamma^k_{ji}$ where ξ^p are standard coordinates from a fixed coordinated system, and the Jacobian is $J = \dfrac{\partial(x^i,\ldots,x^n)}{\partial(x^l,\ldots,x^m)} \neq 0$ for the two sets of coordinates in bases i and j [7.22] with $J^T J = g_{ij}$ as a contravariant component of the metric tensor. The Jacobian is related to Christoffel symbols as a generalized gradient [7.32]. While the Jacobian is undefined at zero [7.22], Christoffel symbols at zero are indicators of flatness of the surface [7.32].

The contravariant and covariant components of the metric tensor are expressed in terms of the standard coordinates as $g_{ij} = \dfrac{\partial \xi^p}{\partial x^i} \dfrac{\partial \xi^p}{\partial x^j}$ and $g_{lj} = \dfrac{\partial \xi^p}{\partial x^l} \dfrac{\partial \xi^p}{\partial x^j}$ so that $g_{ij} = \dfrac{\partial x^k}{\partial x^i} \dfrac{\partial x^l}{\partial x^j} g_{kl}$ and $g^{ij} = \dfrac{\partial x^i}{\partial \xi^k} \dfrac{\partial x^j}{\partial \xi^l} \delta_{kl} = \dfrac{\partial x^i}{\partial \xi^k} \dfrac{\partial x^j}{\partial \xi^k} = g^{ji}$, respectively, where δ is Kronecker delta [7.22]. Permutation of indices produces $\Gamma^k_{jlgki} + \Gamma^k_{ilgkj} = \Gamma^k_{ljgki} + \Gamma^k_{ilgkj} = \partial_l g_{ji}$ and $\Gamma^k_{li} g_{kj} + \Gamma^k_{jigkl} = \Gamma^k_{il} g_{kj} + \Gamma^k_{ilgkl} = \partial_l g_{lj}$ [7.6, 7.33]. Collecting metric tensor components from the Christoffel symbols, and because $g^{ij} g_{jk} = \delta^i_k$ [7.33], Christoffel symbols can be expressed in terms of the metric tensor components as $\Gamma^k_{ij} = \dfrac{1}{2} g^{kl} \left(\dfrac{\partial g_{il}}{\partial x^j} + \dfrac{\partial g_{lj}}{\partial x^i} - \dfrac{\partial g_{ji}}{\partial x^l} \right) = \dfrac{1}{2} g^{kl} (\partial_j g_{il} + \partial_i g_{lj} - \partial_l g_{ji})$ [7.22, 7.33].

The first fundamental form coefficients and partial derivatives that are related to the metric tensor are used to express the connections between tangent spaces and can be summarized in matrix form as $\Gamma^l \equiv \begin{bmatrix} \Gamma^l_{ii} & \Gamma^l_{ij} & \Gamma^l_{ik} \\ \Gamma^l_{ji} & \Gamma^l_{jj} & \Gamma^l_{jk} \\ \Gamma^l_{ki} & \Gamma^l_{kj} & \Gamma^l_{kk} \end{bmatrix}$ [7.69]. Christoffel symbols of the first kind via covariant surface basis vectors [7.60] are expressed in terms of the first fundamental form and its partial derivatives as $\Gamma^1_{11} E + \Gamma^2_{11} F = \frac{1}{2} E_u$, $\Gamma^1_{11} F + \Gamma^2_{11} G = F_u - \frac{1}{2} E_u$, $\Gamma^1_{12} E + \Gamma^2_{12} F = \frac{1}{2} E_v$, $\Gamma^1_{12} F + \Gamma^2_{12} G = \frac{1}{2} G_u$, $\Gamma^1_{22} E + \Gamma^2_{22} F = F_v - \frac{1}{2} G_u$, and $\Gamma^1_{22} F + \Gamma^2_{22} G = \frac{1}{2} G_v$ [7.8, 7.60]. Christoffel symbols of the second kind via contravariant surface basis vectors are given as $\Gamma^1_{11} = \dfrac{GE_u - 2FF_u + FEv}{2(EG - F^2)}$, $\Gamma^1_{12} = \dfrac{GE_v - FG_u}{2(EG - F^2)}$, $\Gamma^1_{22} = \dfrac{2EF_v - GG_u - FG_v}{2(EG - F^2)}$, $\Gamma^2_{11} = \dfrac{2EF_u - EE_v - FE_u}{2(EG - F^2)}$, $\Gamma^2_{12} = \dfrac{EG_u - FE_v}{2(EG - F^2)}$, and $\Gamma^2_{22} = \dfrac{EG_v - 2FF_v - FG_u}{2(EG - F^2)}$ with $\Gamma^1_{12} = \Gamma^1_{21}$ and $\Gamma^2_{12} = \Gamma^2_{21}$ resulting from a torsion-free orientation of a space curve with respect to a Serret-Frenet frame at a given point on the surface [7.8, 7.60]. The covariant surface metric tensor is $g_{ij} = \frac{1}{EG-F^2}\begin{bmatrix} G & -F \\ -F & E \end{bmatrix}$, and the contravariant surface metric tensor is $g^{ij} = \begin{bmatrix} E & F \\ F & G \end{bmatrix}$ [7.60].

For a parametric surface, $\mathrm{r}(u, v) = (x(u, v), y(u, v), z(u, v))$, $\mathrm{r}_{uu} = \Gamma^1_{11}\mathrm{r}_u + \Gamma^2_{11}\mathrm{r}_v + Ln$, $\mathrm{r}_{uv} = \Gamma^1_{12}\mathrm{r}_u + \Gamma^2_{12}\mathrm{r}_v + M\mathrm{n}$, and $\mathrm{r}_{vv} = \Gamma^1_{22}\mathrm{r}_u + \Gamma^2_{22}\mathrm{r}_v + N\mathrm{n}$ with $(\mathrm{r}_{uu})_v = (\mathrm{r}_{uv})_u$ and $(\mathrm{r}_{vv})_u = (\mathrm{r}_{uv})_v$ via Clairaut's Theorem, indicating the relation between Christoffel symbols and the second fundamental form [7.24]. Additionally, Peterson-Codazzi-Mainardi equations relate Christoffel symbols to the second fundamental form as $L_v - M_u = L\Gamma^1_{12} + M(\Gamma^2_{12} - \Gamma^1_{11}) - \mathrm{N}\Gamma^2_{11}$ and $M_v - N_u = L\Gamma^1_{22} + M(\Gamma^2_{22} - \Gamma^1_{12}) - \mathrm{N}\Gamma^2_{12}$ [7.24].

Monge patches are local surfaces in a 3D metric space [7.24]. To calculate Christoffel symbols for a Monge surface patch via the first fundamental form, a change of variables via the total derivative is necessary because of the parameterization of the surface [7.24]. From elements of the Jacobian, the first fundamental form coefficients are $E_u = \left|\frac{\partial x}{\partial u}\right|^2$, $F_u = \frac{\partial x}{\partial u} \cdot \frac{\partial x}{\partial v}$, and $G_u = \left|\frac{\partial x}{\partial v}\right|^2$ for parameter u, and $E_v = \left|\frac{\partial y}{\partial u}\right|^2$, $F_v = \frac{\partial y}{\partial u} \cdot \frac{\partial y}{\partial v}$, and $G_v = \left|\frac{\partial y}{\partial v}\right|^2$ for parameter v [7.22], and $E = 1 + z_u^2$, $F = z_u z_v$, and $G = 1 + z_v^2$ [7.24]. A Monge patch with height function z may be expressed in matrix form as $\begin{bmatrix} 1+z_u^2 & z_u z_v \\ z_u z_v & 1+z_v^2 \end{bmatrix}$, and $I_{Monge} = (1+z_u^2)du\,du + 2z_u z_v\; du\,dv + (1+z_v^2)dv\,dv$ is the first fundamental form [7.60]. For a Monge patch, the second fundamental form is $II_{Monge} = \frac{z_{uu}du\,du + 2z_{uv}du\,dv + z_{vv}dv\,dv}{\sqrt{1+z_u^2+z_v^2}}$ [7.60], and coefficients of the second fundamental form are expressed as $L = \frac{z_{uu}}{\sqrt{1+z_u^2+z_v^2}}$, $M = \frac{z_{uv}}{\sqrt{1+z_u^2+z_v^2}}$, and $N = \frac{z_{vv}}{\sqrt{1+z_u^2+z_v^2}}$ [7.24]. Christoffel symbols for a Monge patch may be calculated as $\Gamma^k_{ij} = \frac{z_{ij}z_k}{1+z_1^2+z_2^2}$ [7.69] and represent the intrinsic geometry locally of curves and surfaces and their measurable characteristics such as the distance metric [7.8, 7.24] to connect surface features quantifiable on a pointwise basis.

Christoffel symbols of the second kind are the 2D surface connection coefficients of a 3D space [7.60], and the coefficients are a quantification of the degree of twisting, turning, contraction, or expansion of the surface. That is, the Christoffel symbols as connection coefficients serve three purposes: they are a quantification of how fast the components of a vector turn, twist, contract, or expand to keep constant against the turning, twisting, contraction, or expansion of the basis vectors; they are a quantification of parallel transport of basis vectors along the components of a vector; they are a quantification of the flatness of a surface [7.32].

7.2.1.8 *Protocol for Using Centric Diatom 3D Surface Models and Their Ensemble Surface Features in Valve Formation Analysis*

Parametric 3D surface models of external and forming valves of *Actinoptychus senarius*, *Arachnoidiscus ehrenbergii*, and *Cyclotella meneghiniana* are used in analyses following

Schmid and Volcani's [7.56] schema for centric diatom valve formation. For the modeling process, the final external valve model has *n*-horizontal slices of its surface removed in a stepwise fashion so that precursor steps are created to comprise the valve formation sequence. The ornamentation of the external valves is removed numerically at each step to illustrate the changes in valve formation and surface features. Models of each step for each taxon are devised using systems of parametric 3D equations. The number of steps chosen to create back-tracked slices of the valve formation in 3D is a function of the numerical changes in coefficients in systems of parametric 3D equations. The number of parametric 3D equations in each system representing a single diatom taxon is arbitrary because numerical solution can be obtained for any incremental, even infinitesimal, numerical change as the geometric change of a 3D surface. That is, the number of modeled steps potentially may be ad infinitum, even though perceptible changes are not evident.

Changes in ensemble surface features are used to develop a valve formation sequence as the morphogenetic model. Changes are measured as differences from one step to the next, starting with a base layer and sequentially accumulating ensemble surface features to complete the valve morphology. The base layer will have the smallest values for ensemble surface features so that, theoretically, the start would be a blank disk. The subsequent layers representing different degrees of sloping, peaks, valleys and saddles, smoothness, and flatness or curvature increases to model the acquisition of surface features to arrive at the final valve morphology. The whole surface is treated as a Monge patch when calculating ensemble surface measures.

Specifically, ensemble surface features are measured via first partial derivatives as elements of the Jacobian of a whole parametric 3D surface (variables x, y, and z in parameters u and v), second partial derivatives as the elements of the x-, y-, and z-Hessians of a Monge patch (where height is z with respect to planar patch x-y), the Laplacians as the sums of the diagonal elements of the x-, y-, and z-Hessians, and Christoffel symbols as connection coefficients with contravariant indices $k = 1, 2,$ or 3 and covariant indices $i, j = 1, 2, 3$. That is, Christoffel symbols of the second kind are calculated.

For each taxon, resultant quantified ensemble surface measures are analyzed for their contribution to each valve formation step and depicted in 100% stacked area plots to infer trends over time. The 100% stacked area charts show the contribution of each ensemble surface feature to each valve formation step. Each ensemble surface measure is analyzed to determine its contribution to ensemble surface features for *Actinoptychus senarius*, *Arachnoidiscus ehrenbergii*, and *Cyclotella meneghiniana* as well as the role each surface measure contributed in a comparison among valve formation taxon sequences. Total contribution of all Christoffel symbols to combined valve formation steps for all taxa was analyzed, and ensemble surface features for all taxa were analyzed with regard to ensemble surface measures as surface descriptors. A morphospace was devised to illustrate the relation between ensemble surface measures of the centric diatom taxa *Actinoptychus senarius*, *Arachnoidiscus ehrenbergii*, and *Cyclotella meneghiniana* and changes during valve formation.

7.3 Results

Parametric 3D equations were used to construct models of *Actinoptychus senarius*, *Arachnoidiscus ehrenbergii*, and *Cyclotella meneghiniana* and their valve formation steps.

Five valve formation steps for each taxon were modeled to depict noticeable valve face changes to fit into Schmid and Volcani's [7.56] three stage schema and are depicted in Figures 7.2, 7.3, and 7.4. From single linkage cluster analysis using Hamming distance of non-zero values for ensemble measures, Schmid and Volcani's [7.56] stage 1 was covered by modeled valve formation steps 1 and 2, their stage 2 was covered by modeled valve formation steps 3 and 4, and their stage 3 was covered by the modeled final valve (Figure 7.5). The number of modeled steps is arbitrary because clustering enables the binning of such steps into Schmid and Volcani's tripartite schema. Five back-traced modeled morphologies were chosen only to enable perceptible change in valve formation steps. Ensemble surface features of the Jacobian, Hessian, Laplacian, and Christoffel symbols were calculated for all 3D surface models constructed from systems of parametric 3D equations. Christoffel symbols of the second kind are calculated [7.19] and are holonomic. For contravariant indices $k = 1, 2, 3$, covariant indices i and j are symmetric so that, $\Gamma^k_{12} = \Gamma^k_{21}$, $\Gamma^k_{13} = \Gamma^k_{31}$ and $\Gamma^k_{23} = \Gamma^k_{32}$.

For all models, the same x- and y-equations are used and are given as $16 \cos u \cos v (1 + \sin u)$ and $16 \cos u \sin v (1 + \sin u)$, respectively. Only the z-equation differs for each taxon and are given in Table 7.1. Because the x- and y-equations were the same for all 3D models,

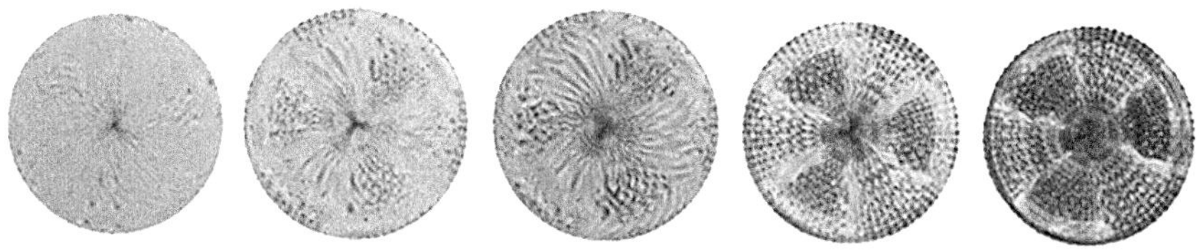

Figure 7.2 *Actinoptychus senarius* modeled valve formation sequence steps 1 through 5. From left to right, steps 1 and 2 represent stage 1, steps 3 and 4 represent stage 2, and step 5 as the finished valve represents stage 3 of Schmid and Volcani's [7.56] valve formation schema.

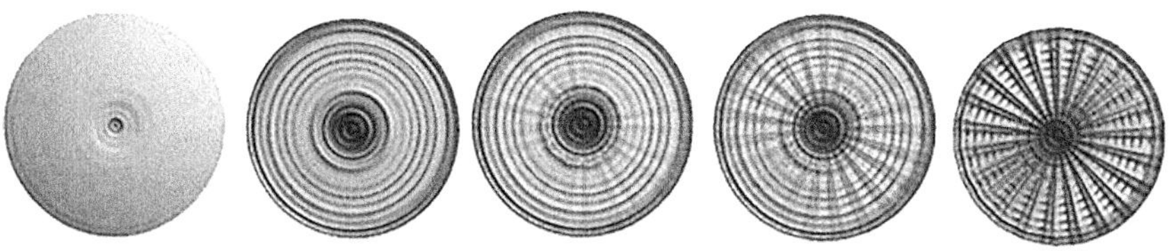

Figure 7.3 *Arachnoidiscus ehrenbergii* modeled valve formation sequence steps 1 through 5. From left to right, steps 1 and 2 represent stage 1, steps 3 and 4 represent stage 2, and step 5 as the finished valve represents stage 3 of Schmid and Volcani's [7.56] valve formation schema.

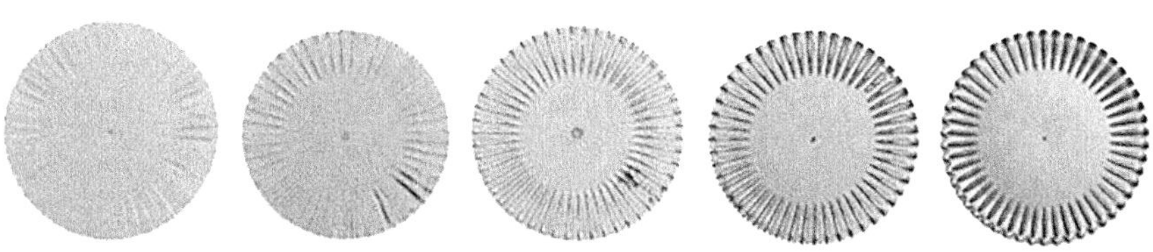

Figure 7.4 *Cyclotella meneghiniana* modeled valve formation sequence steps 1 through 5. From left to right, steps 1 and 2 represent stage 1, steps 3 and 4 represent stage 2, and step 5 as the finished valve represents stage 3 of Schmid and Volcani's [7.56] valve formation schema.

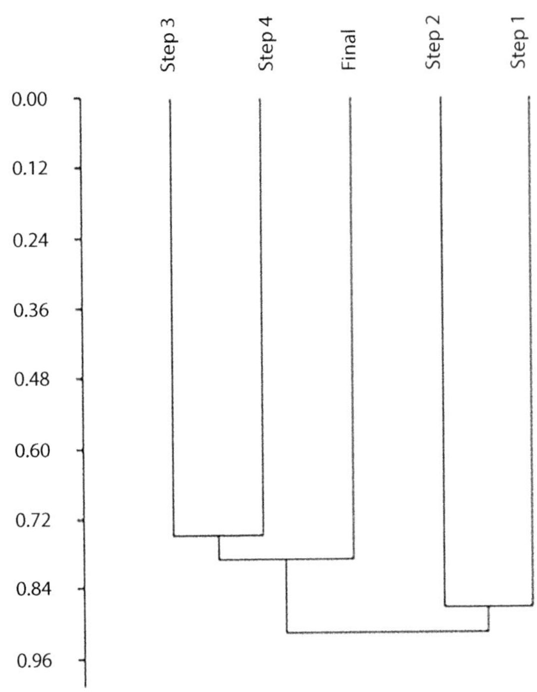

Figure 7.5 Single-linkage cluster analysis using Hamming distance for non-zero values of the Jacobian, Hessian, Laplacian, and Christoffel symbols.

Table 7.1 Terms with parameters and coefficients of the z-equations for centric diatom taxa and valve formation steps. For all parameters, $u, v \in [0, 2\pi]$.

Taxon	*z*-equation[†]
Actinoptychus senarius	$-\alpha u \cos 3u \cos 3v + \beta \sin 3u - \gamma \sin(6u)^5$
	$-\zeta u \cos(5 - 4u) \sin(3 - 4v)^4 + \mu \sin 120u \sin 6v$
	$-\varphi \cos(0.5u)^2 \cos(40u)^4 \cos 3v \sin 40u \sin(40v)^3$
	$-\psi u \sin 40u \sin 160v$
Arachnoidiscus ehrenbergii	$v \cos(0.5u^2) + \rho \cos(80u)^2 \sin(1.9u)^2 + \Xi \sin(1 + u)^3$
	$+\tau \cos(10 + 74u^2) \sin(0.5 + 0.8u)^3 \sin(2 + 2.9u)^3$
	$+\vartheta \cos u \sin(0.37u)^3 \sin(22v)^3$
Cyclotella meneghiniana	$\cos 1.5u[\xi \cos 2u \sin 0.5u \sin 2.9u + 0.14 \cos u \sin(0.5u)^4 \sin(7v)^4]$ $+\eta \cos(3.5u)^6 \cos 25v \sin 1.8u^6 \sin 25v$

[†] Coefficients: $0.0008 \leq \alpha \leq 0.64$, $0.01 \leq \beta \leq 0.1$, $0.001 \leq \gamma \leq 1$, $0.003 \leq \zeta \leq 0.09$, $0.0025 \leq \mu \leq 0.125$, $0.2 \leq \varphi \leq 10$, $0.0025 \leq \psi \leq 0.025$, $0.004 \leq \vartheta \leq 0.7$, $0.05 \leq v \leq 0.75$, $0.025 \leq \rho \leq 0.5$, $0.051 \leq \tau \leq 2.55$, $2 \leq \Xi \leq 3$, $0.01 \leq \xi \leq 0.5$, $0.18 \leq \eta \leq 3$.

each model can be treated as a Monge patch. As a result, the *z*-equation results are analyzed for the Jacobian in the *u*- and *v*-directions and the Hessian in the *u*-, *v*-, and *uv*-directions. Christoffel symbols with contravariant index $k = 3$ are used for comparison to the Jacobian and Hessian *z*-equation values and the Laplacian. All Christoffel symbols are used in further analyses.

A combination of all the ensemble measures of the Jacobian, Hessian, Laplacian, and Christoffel symbols produced the depiction of change from the start to each subsequent step of modeled valve formation with regard to the *z*-term and Christoffel symbols for contravariant index $k = 3$. The Jacobian in *u*- and *v*-parameters and the Hessian in the *u*-parameter were positive influences with increasingly positive contribution from start to final valve for *Actinoptychus senarius* (Figure 7.6). Negative influences were evident for the Hessian *v*- and *uv*-parameters, the Laplacian as well as Christoffel symbols Γ^3_{13} and Γ^3_{23},

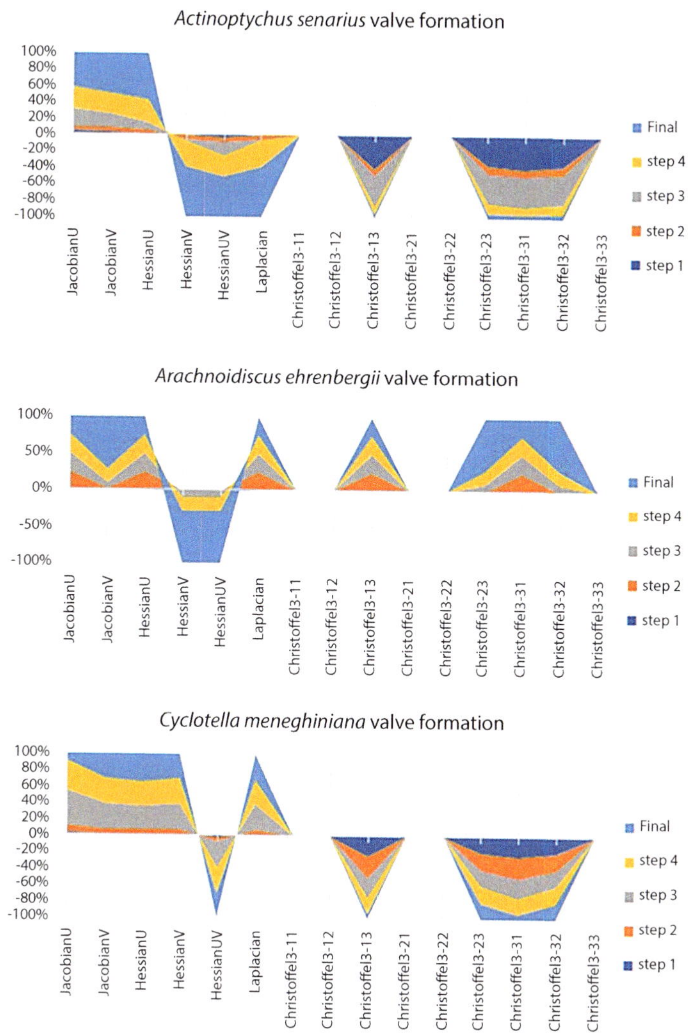

Figure 7.6 Contribution percentage of ensemble surface measures Jacobian, Hessian, Laplacian, and Christoffel symbols to each valve formation step.

Γ^3_{31}, Γ^3_{32}. Largest negative input was the Hessian *v*- and *uv*-parameters and Laplacian for step 4 and the final valve and Christoffel symbols Γ^3_{13} and Γ^3_{23}, Γ^3_{31}, Γ^3_{32} for steps 1 and 3 for *Actinoptychus senarius* (Figure 7.6).

For the Jacobian in the *u*-parameter and Hessian in all parameters, the scenario for *Arachnoidiscus ehrenbergii* was similar to that for *Actinoptychus senarius* (Figure 7.6). The Jacobian in the *v*-parameter had a near zero value for steps 1 thorough 4. Only the final valve had a positive value. The Laplacian and Christoffel symbols Γ^3_{13} and Γ^3_{23}, Γ^3_{31}, Γ^3_{32} were positive influences, contributing almost equally to steps 2 through the final valve for *Arachnoidiscus ehrenbergii*, with Christoffel symbols Γ^3_{23}, Γ^3_{31}, Γ^3_{32} contributing a larger portion to the final valve (Figure 7.6).

Cyclotella meneghiniana had positive influence from the Jacobian and Hessian in the *u*- and *v*-parameters and the Laplacian (Figure 7.6). The Hessian in the *uv*-parameter and Christoffel symbols Γ^3_{13} and Γ^3_{23}, Γ^3_{31}, Γ^3_{32} were negative influences in almost equal contribution for steps 1 through the final valve, with the exception of almost no contribution by Γ^3_{13} for the final valve of *Cyclotella meneghiniana* (Figure 7.6).

Modeled taxa were analyzed by their valve features using each ensemble surface measure. The ensemble surface measures that were most indicative of peak height changes in *Actinoptychus senarius* valve features were the Hessian in the *v*- and *uv*-directions from step 2 through the final valve, reflecting the changes in alternating undulations of the six valve surface sectors radiating outward from the center (Figures 7.7d and f). Only a slight uptick in value from step 3 to step 5 for the Jacobian in the *v*-direction was evident, indicating the change in sloping of undulations from sector to sector on the valve surface (Figure 7.7b). All other ensemble surface measures including the Jacobian and Hessian in the *u*-direction, Laplacian, and Christoffel symbols for contravariant index $k = 3$, indices $i = 1, 2$ and $j = 3$ were near zero, indicating an overall regular valve pattern of the smaller surface features in contrast to the larger valve features of the alternating undulations of the six sectors (Figures 7.7a, c, e, g, and h).

Arachnoidiscus ehrenbergii exhibited large surface changes from valve formation step 1 to step 2 as sloping and peakedness as the Jacobian and Hessian in the *u*-direction (Figures 7.7a and c). Regular undulations from the ribs to the concentric circles below the ribs is recovered by the Laplacian as "harmonics" (Figure 7.7e). While curvature changes in the ribs differing from the underlying concentric rings on the valve face are indicated by Christoffel symbols for contravariant index $k = 3$, indices $i = 1$ and $j = 3$ (Figure 7.7g), the Jacobian in the *v*-direction, Hessian in the *v*- and *uv*-directions, and Christoffel symbols for contravariant index $k = 3$, covariant indices $i = 2$ and $j = 3$ are near zero, indicating transitional or saddle areas of the valve surface that are flat (Figures 7.7d, f, and h). *Arachnoidiscus ehrenbergii* has a fairly even, regular patterned surface as measured by ensemble surface features.

The slightly undulating surface of *Cyclotella meneghiniana* was reflected in the Jacobian in the *v*-direction, increasingly from step 1 to step 3 during valve formation (Figure 7.7b), in contrast to the *u*-direction (Figure 7.7a). The same pattern held for the Hessian in the *v*-direction with positive values indicating valleys (Figure 7.7d) as well as Christoffel symbols for contravariant index $k = 3$, covariant indices $i = 1$ and $j = 3$ indicating greater curvature areas (Figure 7.7g). Sharp changes in peak height as an outer ring of "pleats" or plications at the valve margin of *Cyclotella meneghiniana* are recovered via these ensemble surface measures. A slight depression in the general valve surface was indicated by the

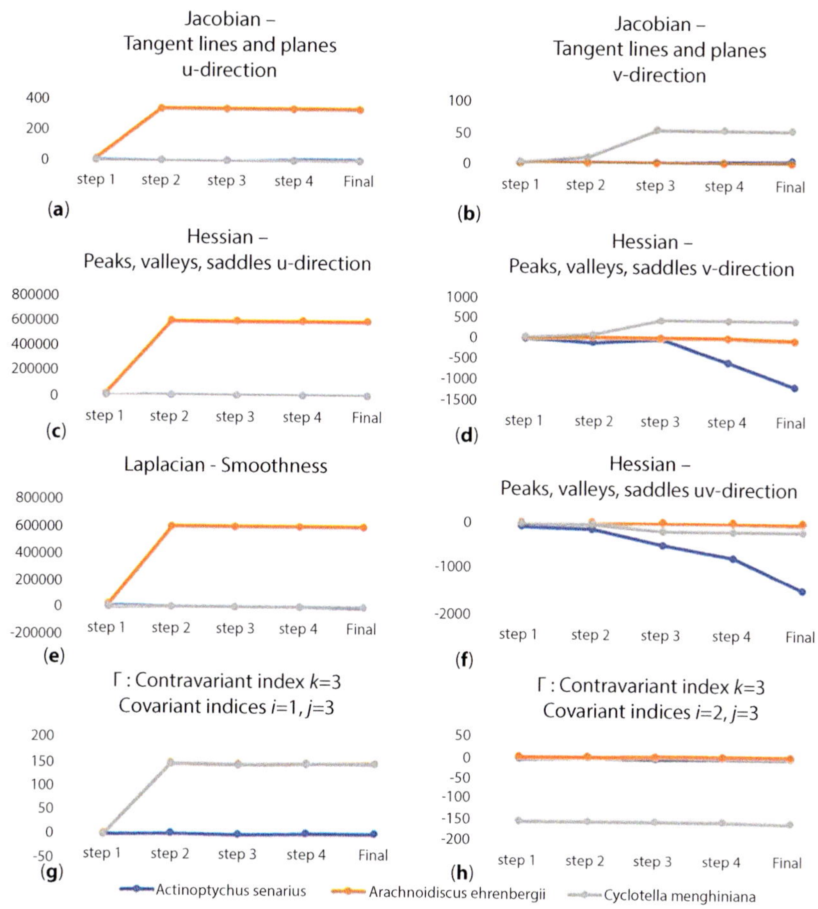

Figure 7.7 Jacobian, Hessian, and Laplacian from *z*-terms and Christoffel symbols Γ^3_{13} and Γ^3_{23} for modeled valve formation steps 1 through the final valve for *Actinoptychus senarius*, *Arachnoidiscus ehrenbergii*, and *Cyclotella meneghiniana*.

Hessian in the *uv*-direction (Figure 7.7f) representing the central area, and a transitional or saddle area between peaked and depressed areas along the marginal plications was indicated by the Hessian in the *u*-direction (Figure 7.7c). Valve surface smoothness is reflected in the very lightly radially streaked central area via the Laplacian (Figure 7.7e), while flatness of the overall valve face shape was indicated by Christoffel symbols for contravariant index $k = 3$, covariant indices $i = 2$ and $j = 3$ (Figure 7.7h).

The *u*- and *v*-parameters of the *z*-term in the Jacobian and Hessian and sum of the *z*-terms in the Laplacian contributed differing amounts to the outcome of valve formation for each taxon. For the Jacobian, all steps of *Cyclotella meneghiniana* were influenced by the *v*-parameter, and all steps of *Arachnoidiscus ehrenbergii* were influenced by the *u*-parameter as was the case for *Actinoptychus senarius*, but to a lesser degree (Figure 7.8, top). For the Hessian, the *uv*-parameter had a larger influence than the *v*-parameter on all steps of *Cyclotella meneghiniana* and *Actinoptychus senarius*. For *Cyclotella meneghiniana*, the influence was positive, while for *Actinoptychus senarius*, the influence was negative (Figure 7.8, middle). *Arachnoidiscus ehrenbergii* was influenced by the Hessian *z*-term in the

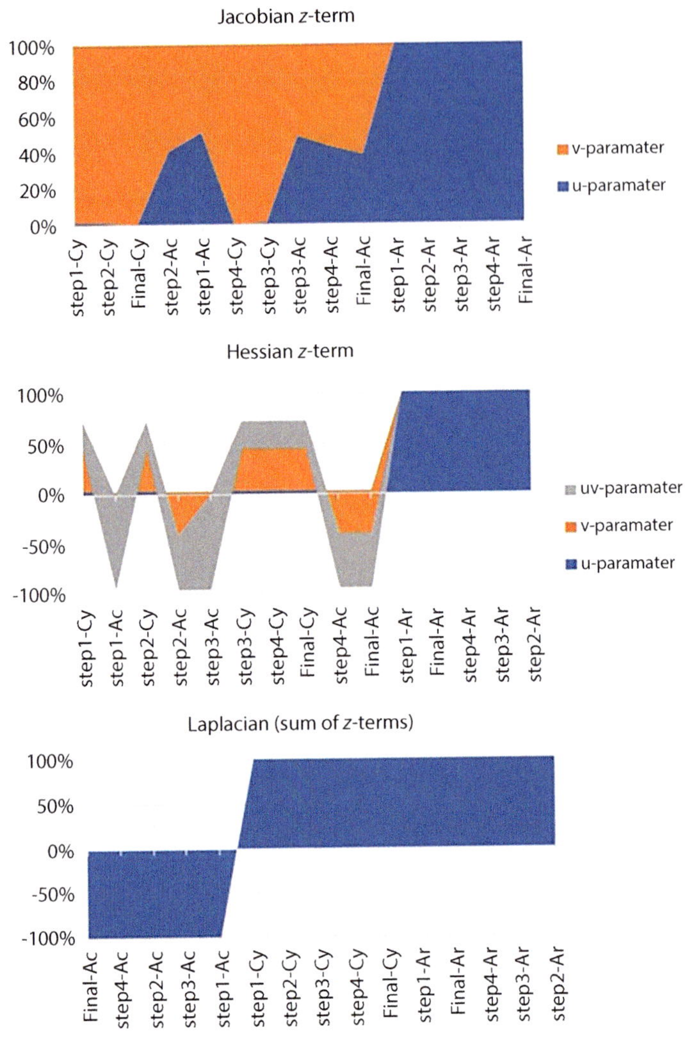

Figure 7.8 Contribution of the *u* and *v* parameters of the *z*-term of the Jacobian, Hessian, and Laplacian to each step of valve formation for *Actinoptychus senarius*, *Arachnoidiscus ehrenbergii*, and *Cyclotella meneghiniana* expressed as a percentage.

u-parameter (Figure 7.8, middle). The Laplacian was a negative influence for *Actinoptycus senarius* and a positive influence for *Cyclotella meneghiniana* and *Arachnoidiscus ehrenbergii* (Figure 7.8, bottom).

All Christoffel symbols were analyzed to determine degree of flatness and to determine their contribution to valve characteristics at each step of valve formation on the valve surface. The point at which the Christoffel symbols vanish are flat areas in a torsion-free or holonomic frame. The direction of change in a tangent space is symmetric in the covariant indices when Christoffel symbols equal zero. The non-zero values indicate distortion resulting from curvature as the connections on the valve surface change from point to point.

For contravariant index $k = 1$, steps 2, 3, 4, and the final valve were represented as a large percentage of negative values of Γ^1_{33}, then transitioning to step 1 as a narrower percentage of positive values for step 1 of *Arachnoidiscus ehrenbergii* (Figure 7.9, top). Step 1 through final valve for *Actinoptychus senarius* and *Cyclotella meneghiniana* were represented as a large percentage of positive values of Γ_{11}^{1} (Figure 7.9, top). All other values of Christoffel symbols for contravariant index $k = 1$ were equal to or near zero.

For contravariant index $k = 2$, step 1 through the final valve for *Arachnoidiscus ehrenbergii* had a large percentage of negative values for Γ_{22}^{2} (Figure 7.9, middle). For step 1 through the final valve of *Actinoptychus senarius* and steps 1 and 2 for *Cyclotella meneghiniana*, a larger percentage of negative values of Γ_{22}^{2} was indicated over a smaller negative percentage

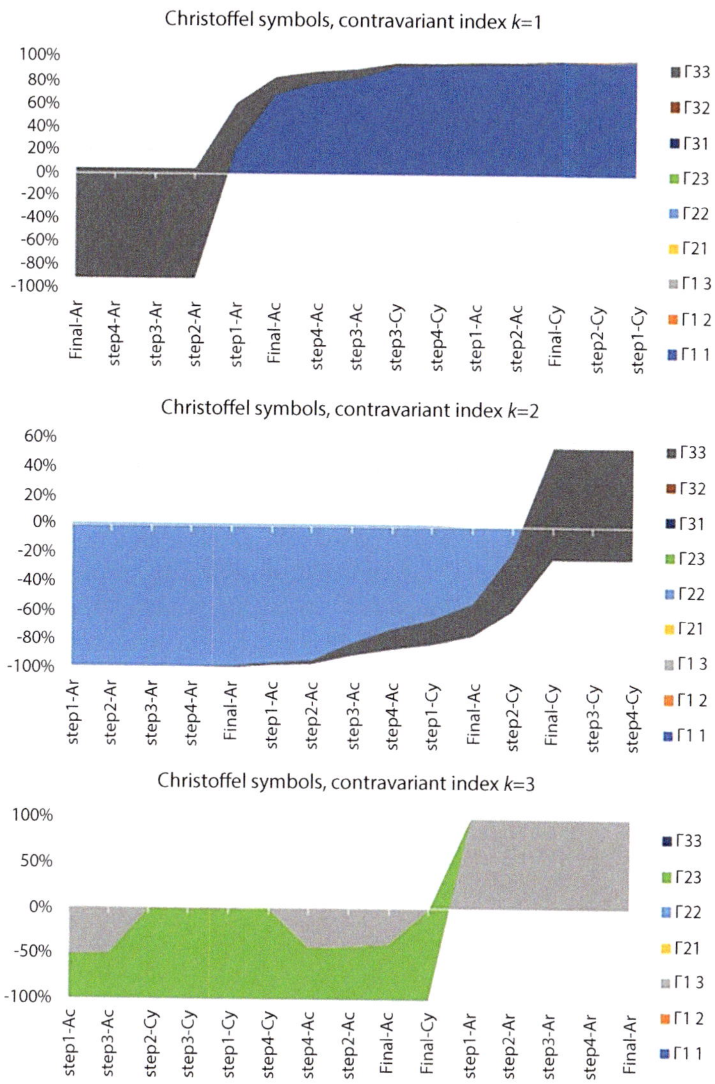

Figure 7.9 All Christoffel symbols for the three contravariant indices: top, $k = 1$; middle, $k = 2$; bottom, $k = 3$. Shaded areas indicate percentage that a given Christoffel symbol contributed to a valve formation step for each taxon. Because of symmetry, Christoffel symbols with contravariant index $k = 3$, covariant indices $i = 3, j = 1$, and $i = 3, j = 2$ are not reported.

for Γ_{33}^{2} (Figure 7.9, middle). For *Cyclotella meneghiniana* steps 3, 4 and the final valve, only Γ_{33}^{2} had a larger percentage of positive versus a lesser percentage of negative values (Figure 7.9, middle). All other values of Christoffel symbols for contravariant index $k = 2$ were equal to or near zero.

The general order of the taxa was switched for contravariant index $k = 3$ Christoffel symbols. All steps for *Actinoptychus senarius* and *Cyclotella meneghiniana* were large percentages of negative values for Γ_{13}^{3} and for Γ_{23}^{3}, where the latter indicated the bulk of the negative values (Figure 7.9, bottom). By contrast, all steps of *Arachnoidiscus ehrenbergii* were a large percentage of positive values for Γ_{13}^{3} and a smaller percentage of positive values for for Γ_{23}^{2} (Figure 7.9, bottom).

All Christoffel symbols for all three taxa were plotted for the sequence of steps in valve formation. Each step is covered by all contravariant indices $k = 1, 2, 3$. Of all the Christoffel symbols, those that were non-zero and had contributed to non-flat areas of the valve face were $\left(\Gamma_{32}^{1},\Gamma_{32}^{2},\Gamma_{32}^{3}\right)$, $\left(\Gamma_{22}^{1},\Gamma_{22}^{2},\Gamma_{22}^{3}\right)$, $\left(\Gamma_{33}^{1},\Gamma_{33}^{2},\Gamma_{33}^{3}\right)$, and $\left(\Gamma_{31}^{1},\Gamma_{31}^{2},\Gamma_{31}^{3}\right)$. The contribution of Christoffel symbols with covariant indices $i = 3, j = 2$ was a constant negative value for steps 1 through the final valve. Covariant indices $i = 2, j = 2$ had a similar contribution. On the positive side, covariant indices $i = 3, j = 1$ contributed to valve formation steps 2 through the final valve as a constant. Only step 1 had no contribution from Christoffel symbols with covariant indices $i = 3, j = 1$. For Christoffel symbols with covariant indices $i = 3, j = 3$, their contribution was exponential from step 1 to step 3 then levelled off as a maximum for steps 3, 4 and the final valve (Figure 7.10).

To aggregate ensemble surface measures and their contribution to valve formation steps for all taxa combined, a plot was devised to show the percentage of that contribution in terms of the descriptors flatness, peaked, sloped, and smoothed (Figure 7.11). Contraction described step 1, almost exclusively. Contraction to a lesser degree described step 2 with smaller contributions from almost flat to expansion, peaked and sloped features, and a slight contribution of dipped features. Valve formation step 3 included contraction, almost flat, expansion, dipped, peaked, and sloped descriptors along with the addition of the concave smoothness descriptor. Step 4 had elements of the same contributing descriptors with

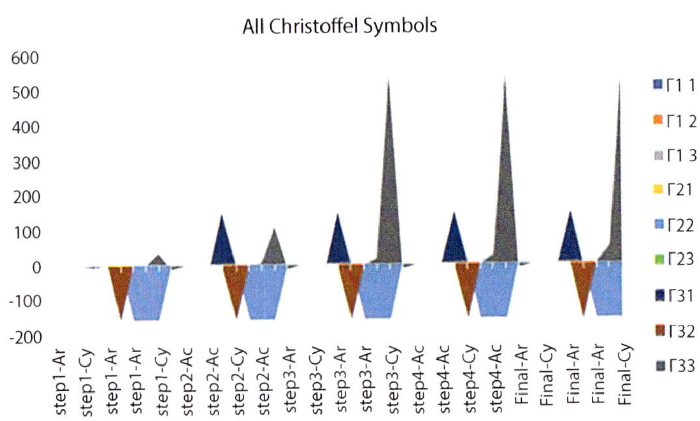

Figure 7.10 Plot of all Christoffel symbols shows change in contribution for $\left(\Gamma_{32}^{1},\Gamma_{32}^{2},\Gamma_{32}^{3}\right)$, $\left(\Gamma_{22}^{1},\Gamma_{22}^{2},\Gamma_{22}^{3}\right)$, $\left(\Gamma_{33}^{1},\Gamma_{33}^{2},\Gamma_{33}^{3}\right)$, and $\left(\Gamma_{31}^{1},\Gamma_{31}^{2},\Gamma_{31}^{3}\right)$ per valve formation step.

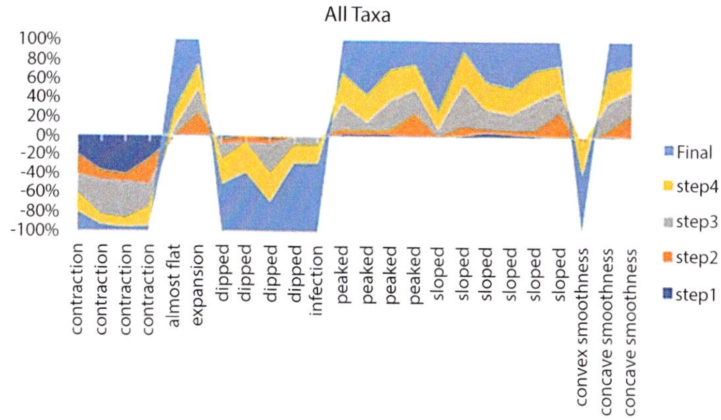

Figure 7.11 Ensemble surface measures as indicators of ensemble surface features for each valve formation step.

the addition of convex smoothness. The final valve was similar to step 4 but had the least contribution from the contractor descriptor compared to the previous four steps in the valve formation sequence (Figure 7.11).

A 3D morphospace was devised for *Actinoptychus senarius*, *Arachnoidiscus ehrenbergii*, and *Cyclotella meneghiniana* using ensemble surface measures of valve formation (Figure 7.12). Peakedness and smoothness were combined on one axis because the trace of the Hessian is the Laplacian. Slopeness and flatness were combined on another axis because differentiation of the Jacobian is a coordinate transformation of non-vanishing Christoffel symbols [7.2].

The morphospace shows a "cut-away" space to facilitate interpreting taxon ensemble surface feature results. The most pronounced changed in ensemble surface features occurred for *Arachnoidiscus ehrenbergii*. Emanating from its center, *Arachnoidiscus ehrenbergii* has alternating ribs and dips as its valve surface, exhibiting peaks and valleys. The ribs exhibit smoothness, while the branching or reticulated pattern beneath the ribs exhibits peakedness. The valve surface pattern exhibits a regular fluctuation in a harmonic fashion (Figure 7.12). During valve formation, *Arachnoidiscus ehrenbergii* proceeds with more changes in peakedness and smoothness and less so concerning slopeness and flatness (Figure 7.12).

Because the scale at which *Arachnoidiscus ehrenbergii* ensemble surface features differed was so great, a second morphospace "cut-away" was devised to enable the visualization of details for *Actinoptychus senarius* and *Cyclotella meneghiniana* (Figure 7.13). A marked change in peakedness and smoothness more so slopeness and flatness was evident for *Actinoptychus senarius*. A sharp dip proceeding from the first step to final valve in the valve formation sequence suggested the formation of the alternating undulations of the sectors on the valve surface (Figure 7.13).

For *Cyclotella meneghiniana*, the changes in peaks and valleys as well as slopeness and flatness varied in a shallow harmonic fashion (Figure 7.13). The shallow depth of the change in peakedness and smoothness suggested the finer undulation of the valve marginal plications in contrast to the regular harmonic undulating changes in the valve surface of *Arachnoidiscus ehrenbergii* or the large undulating sectors of *Actinoptychus senarius*. The slightly larger peak with respect to slopeness and flatness suggested the presence of the

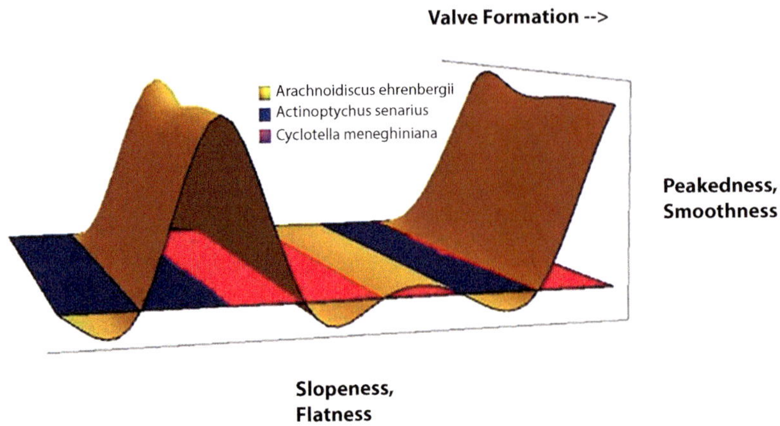

Figure 7.12 A 3D morphospace of ensemble surface measures of valve formation for *Actinoptychus senarius*, *Arachnoidiscus ehrenbergii*, and *Cyclotella meneghiniana*.

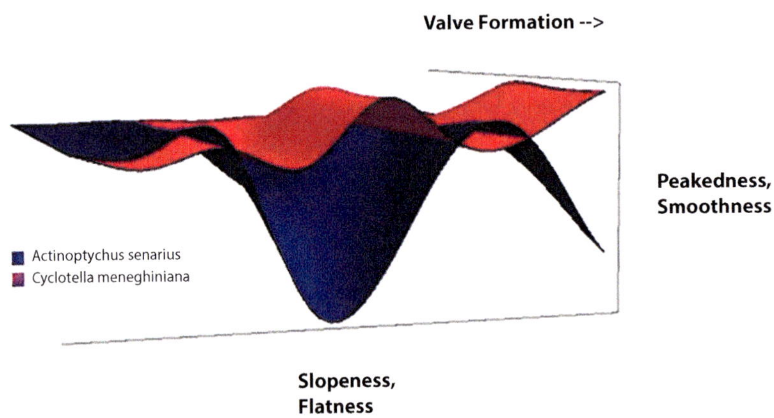

Figure 7.13 A 3D morphospace of ensemble surface measures of valve formation for *Actinoptychus senarius* and *Cyclotella meneghiniana*.

central area of *Cyclotella meneghiniana*. For the valve formation sequence, progression from more slopeness and flatness to more peakedness and smoothness indicated the final valve marginal surface of *Cyclotella meneghiniana* (Figure 7.13).

In the morphospace, the areas where smoothness equals flatness are inflection points that may be saddle points as well. That is, at the intersection of the areas for *Actinoptychus senarius*, *Arachnoidiscus ehrenbergii*, and *Cyclotella meneghiniana*, the value equals or is very close to zero (Figure 7.13).

7.4 Discussion

Diatom morphogenesis is a compilation of silica deposition, valve formation, mitotic growth, and other cytological, genetic, and epigenetic processes (e.g., [7.56]). Silica deposition during morphogenesis is not uniform [7.66]. Overall, valve surface pattern is maintained and reproducible, but diatom growth concerning silica deposition and the geometry of the surface occurs within finite spatial boundaries [7.66]. Regarding phenotypic plasticity as an influence on diatom valve patterns along with processes governing valve formation cytologically, 3D surface models are useful in studying potential modes of growth in morphogenesis.

Centric diatom valve morphogenesis was explored using 3D surface models and measurements of boundaryless, variable morphological characters combined as ensemble surface features. Studies were conducted with regard to stepwise ensemble surface changes as a sequence of events during valve formation. Each of the three general stages of valve morphogenesis as defined by Schmid and Volcani [7.56] were used as guidelines for the changes in ensemble surface features as slices in the height distribution of silica deposition. At the first stage of horizontal silica deposition and the presence of an annulus and branching pattern of silica strands with radial rows, this basal silica layer is the smoothest and flattest, having smaller values of slopes, peakedness, and saddles. At the second stage of vertical silica deposition, areolae walls increase in height that is reflected in larger values of peakedness and saddles. At the third stage of increasing horizontal and vertical silica deposition, peakedness and saddles become more differentiated as the size and spacing of such structures may become exaggerated with intervening areas of smoothness and flatness, as silica deposition extends to the valve margin.

At each point on the valve surface, ensemble surface features may have positive, negative, and zero values that are indicators of the degree of sloping, peakedness and saddles, smoothness, or flatness (Table 7.2). Positive values for elements of the Jacobian indicate increasing sloping or steepness of the surface, while negative values indicate decreasing sloping. Positive values for elements of the z-Hessian are peaks, negative values are valleys, and zero values are saddles. The transition from positive to negative values are indicators of changes in smoothness as harmonics, as smoothness is when the Laplacian is zero. Positive or negative Christoffel symbols are indicators of curvature, while flatness is recorded in zero values.

Table 7.2 Ensemble surface features, mathematical operator, and numerical results.

Surface feature	Mathematical operator	Analytical/Numerical results
Slopes	Jacobian	Positive or Negative
Peaks	Hessian	Maxima - Positive
Valleys	Hessian	Minima - Negative
Saddles	Hessian	Zero
Convex or Concave Smoothness	Laplacian	Positive or Negative
Smoothness	Laplacian	Zero
Curvature	Christoffel symbols	Positive or Negative
Flatness	Christoffel symbols	Zero

Ensemble surface measures from parametric 3D models were readily compiled and corresponded to Schmid and Volcani's [7.56] schema of diatom valve morphogenesis. Quantification of diatom surface morphology using such measures enabled the study of valve morphogenesis as well as showed the efficacy of using 3D surface models in valve formation studies. Such models can be used to extend the study of valve formation to include the initial cell and auxospore of the diatom life cycle. Systems of parametric 3D equations can be devised to create 3D surface models when initial cell and auxospore surface and shape morphology is known and documented.

7.4.1 Ensemble Surface Features and Physical Characteristics of Valve Morphogenesis

Ensemble surface features may be present indicating rotational, reflective, dihedral, or conformal symmetry. The same features may indicate degrees of complexity with regard to degree of sloping, peakedness, smoothness, or flatness. Increasing sloping and peakedness measured as the Jacobian and Hessians, respectively, means increasing complexity in contrast to increasing smoothness or flatness as measured from the Laplacian and Christoffel symbols, respectively. Symmetry and complexity assessments at each stage in valve formation may be used as composite descriptors of the changes in development, and such changes may be related to the changes occurring at the subcellular level with regard to organelles and their movement during mitosis and the size diminution cycle for many diatoms [7.43]. Ensemble surface features may indicate whether circular centric diatoms can become asymmetric at a given stage of growth and valve formation despite being formed from an annulus. From stepwise valve formation via composite ensemble surface features, a general ranking of most to least symmetric taxa are *Arachnoidiscus* to *Actinoptychus* to *Cyclotella*. A composite of ensemble surface features indicates that the highest to lowest degree of complexity in diatom valve surface formation patterns is *Arachnoidiscus* to *Actinoptychus* to *Cyclotella*.

Valve formation is directional with respect to valve characters and pattern, but at a local level. Initially, branching patterns may exhibit waviness, but spatial constraints on silica deposition occur to induce regularly spaced surface features, such as areolae, as wave front expansion from annulus to valve margin occurs in centric diatoms [7.66]. Expansion of a wave front with respect to the SDV may induce fractal-like structures at the valve margin with respect to growth [7.66].

Such growth may be indicative of the first stage of valve formation. At the second stage of valve formation, silica fills in the valve face producing another layer of patterning which occurs behind the initial wave front expansion. One possibility is that vertical silica deposition might occur stochastically, resulting in changing sizes and shapes of pores and rimoportulae in a random fashion over the valve face and in contrast to horizontal silica deposition that occurs chaotically resulting in fractal-like structures. At the third stage of valve formation, a combination of stochastic and directed processes might occur as cribra and cribella form within the constraints of the areolae, but silica filling into gaps in the valve face structures may do so in a random fashion. This represents growth where cell division and symmetry occur as a propagation of wave fronts [7.34].

7.4.2 Factors Affecting Valve Formation

Size and geometry of a morphological character may be affected by enzymatic inhibitors in contrast to the spatial distribution of those characters on the valve surface [7.66]. At the first stage of valve formation, constraints on the valve surface may be laid down with regard to transapical costae and at the second stage of valve formation with the formation of cross costae in some centric diatom taxa [7.66]. Sequential deposition with respect to silica height on the valve surface records pattern formation over time and may be affected by inhibitors. Low concentrations of inhibitor may induce change in silica deposition and a change in development of morphological characters on the valve. High concentration of inhibitors may induce aberrant valves [7.66].

Development of a morphological character is affected by inhibitors in contrast to the spatial relation among morphological character on the valve surface of a diatom [7.66]. Fluctuations in processes at the molecular level may induce slight differences in silica deposition on different parts of the valve surface. Physically, exponential cell growth may produce excess silica material generation to induce buckling of the domain boundary [7.34] or buckling on the surface. Structural anomalies such as areolar row disordering or distorted valve face shape may be induced via microtubule inhibitors because microtubule dissolution occurs or chromosome separation does not [7.1]. In either case, buckled valve structures may result as valve morphogenesis occurs.

Because valve morphogenesis commences at the microtubule center, and subsequent development of large valve structures (e.g., [7.47, 7.61, 7.62, 7.64, 7.65] occurs in conjunction with microtubule activity, morphogenetic stages that are interrupted or changed by microtubule inhibitors may produce abnormal valve structures or valve shape [7.1] as buckled surfaces or marginal valve margins. Such differences may be modeled using ensemble surface features with the potential of matching numerical results with the associated positioning of microtubules and possible valve surface structural or marginal buckling. The appropriate combination of such features has yet to be discerned and how such combinations might be associated to particular steps in the silica deposition process, and how this represents valve formation at given times during this process remains to be studied.

7.4.3 Diatom Growth Patterns—Buckling and Wave Fronts

Changes in early growth during diatom morphogenesis may resemble the propagation of an optical ray. Changes in the trajectory of the ray may resemble changes in the growth path during valve formation of diatom morphogenesis. For a homogeneous medium, that path may be linear and constant. Otherwise, the path may be non-linear but follow an exponential or logarithmic path. Rays normal to the surface are wave fronts [7.50] and may represent buckling patterns on diatom surfaces. A buckled surface is a patterned surface [7.4, 7.30].

A wave front or buckling may be interpreted to occur on the surface of *Cyclotella meneghiniana* with regard to the changes from the central area to the margin of the entire valve. Another aspect of a wave front property may be construed to be in the pleating or plications near the valve margin with regard to the areolae structure in a ring. The undulating sectors of *Actinoptychus senarius* or the height changes between ribs and underlying structural branching pattern of *Arachnoidiscus ehrenbergii* may be representational of buckling that has wave front properties.

The Jacobian and Hessian elements of first and second partial derivatives, respectively, record the indications of buckling on the surface of a diatom valve. By contrast, the Laplacian and Christoffel symbols record the smoothness and flatness, respectively, of the diatom valve surface which indicates lack of buckling. As a physical phenomenon, buckling is non-uniform compression given as differential growth [7.34]. For isometric planar growth, buckling does not exist. That is, early growth from a conical surface above a disk corresponding to isometry is a constant growing surface where growth is low and the surface is flat, and buckling does not occur [7.34]. Ensemble surface features and the composite change in these features across a surface may be a variant of wave front propagation. The Laplacian has solutions that are representative of harmonics, and the combinations of first and second partial derivatives are extractable as variants of wave front solutions.

For an exponentially growing surface embedded in a plane, the Jacobian of a conformal mapping is dependent on angular symmetry and is recoverable in terms of hyperbolic metrics of the surface. This growth is a type of motion in which energy is propagated from one stage to another along a continuum from nascent cell to mature organism. Buckling as a wave front is a natural phenomenon of cell growth at different scales [7.34].

Buckling on the surface may be determined by the metric tensor. Geodesics defining paths on the surface are parameterized and determine the Christoffel symbols. Geodesics along a surface are associated with the motion of silica deposition at regular intervals producing a characteristic valve pattern. Movement of silica with regard to a valve surface may be indicative of cell division and symmetry changes as a propagation of wave fronts. Surface height above the domain is characterizable by such wave front movement [7.34]. Degrees and kinds of surface buckling may be indicators of different rigidities of surface features [7.14].

At which junctures buckling occurs during diatom valve morphogenesis has yet to be determined [7.15]. One hypothesis is that buckling occurs when frustule structures are thin [7.14]. Another hypothesis is that via wave front analysis, buckling may happen at any time during valve formation as a natural phenomenon that is genetically, environmentally, or epistatically controlled. This may induce plasticity in the phenotype. Epistasis may be instrumental in buckling as a more complicated, unpredictable, chaotic phenomenon. Ensemble surface measures may be used to characterize phenotypic buckling during morphogenesis because a 3D surface is a proxy for the phenotype [7.41].

At some critical values, the Jacobian may become negative and the resultant growth may be squeezed in rather than buckling out. Squeezing in may be a regulation of the rigidity or stiffness of the surface [7.34] as in buckling, which is constrained by surface features that are ridged or stiff [7.14]. Growth may be geometrically hyperbolic in which there is zero Gaussian curvature [7.34]. At constant negative curvature, the surface bending may be similar to a pseudosphere. Cascades of pseudospheres may resemble peaks and saddles, exhibiting buckling in this way. High positive values of the Hessian may measure buckling with respect to silica height off the valve surface, while the high positive values for the Laplacian and zero values for Christoffel symbols may be used to measure lack of buckling, or possibly, transitions between buckles that are not saddles on valve surfaces with few silica elevations as morphological characters. Locally on a valve surface, negative values of Hessians, Laplacians and Christoffel symbols may be indicators of squeezing out rather than buckling in. With the potential for a hierarchical, nested or cascading structure in terms of the relation among ensemble surface features, repetition of such features may be present at multiple scales.

Early in growth, there are buckling instabilities on the circumference as negative values. At a critical point on the surface, buckling proliferates in the direction of growth where peaks and valleys multiply. A hierarchy in peak size occurs related to degree of buckling. At later stages in growth, buckling instabilities subside as the limits of growth occur. At the circumference, a self-similar buckling profile may emerge [7.34].

7.4.4 Valve Formation, Ensemble Surface Features, and Self-Similarity

Self-similarity may be evident during valve formation, but such a determination requires special testing. One possible test may be to detect the smallest changes among pairwise smallest z-Hessian elements and using ultrametricity [7.34], find the smallest possible scale at which a given feature is measurable, and find the next larger scale at which that feature is manifested which may be present on the same valve face or on a valve face generated during valve formation. If a given feature is found at different scales, then there is self-similarity on the diatom valve surface at times when growth or silica deposition occurs. Self-similarity may be at work during growth at different scales and may be evident and measurable as scale symmetry. The limits to growth and the MacDonald-Pfitzer rule [7.31, 7.45, 7.46] may determine the threshold at which self-similarity may or may not be evident.

7.4.5 Diatom Morphogenesis: Cytoplasmic Inheritance and Phenotypic Plasticity

Cytoplasmic inheritance is non-genetic inheritance in which there is a division of cell states between cell lineages [7.58]. Structure of the parental cell can be expressed by the daughter cell through direct descendance or there could be a pre-existing cell membrane that functions as a template for development of the daughter cell. Diatoms reproduce via daughter cells forming within the SDV that expands within the parent cell, decreasing the size of the offspring. Daughter cell microscale surface structure patterns (i.e., morphological characters such as central fultoportulae, striae, central area, and cell diameter) correlate with parental organelles [7.58].

Cell structure traits are quantitatively different from structural traits inherited by genetic or non-genetic factors. In diatoms, cell structure variation results from the new formation of daughter cells within the parent as incomplete inheritance [7.58]. Microscale structural patterns on the daughter cell valve reflect not only the environmental conditions the parent cell was subjected to but also the location of organelles within the parental cell [7.58]. As a result, diatoms exhibit phenotypic plasticity, and structural variation in shape and pattern of a cell are reflected during size diminution during the diatom life cycle. Size reduction may occur over many generations, and partial renewal inheritance occurs such that one valve is always a parental cell in contrast to the new cell generated [7.58]. Ensemble surface features could be measured for parent and daughter cells, and values for parent cells could be plotted and regressed on values for daughter cells to determine a regression coefficient. Potentially, hypotheses concerning cytoplasmic heritability [7.58] could be devised and tested using ensemble surface measures.

Non-heritable factors could be studied using ensemble surface features. The combinations of such features that indicate buckling may be related to degree of phenotypic plasticity in diatom valve surfaces. Degree of buckling at the subcellular level may be influenced by environmental factors and indicative of degree of plasticity of a given diatom phenotype, and in turn, be indicative of the degree of phenotypic variation for a given diatom morphology.

7.4.6 Phenotypic Variation and Ensemble Surface Features: Epistasis and Canalization

Diatom phenotype is affected by developmental changes. Such changes during development including valve formation may include epistasis and canalization [7.67]. Phenotypic variation or the amount of variation in developmental factors is measurable as degree of canalization via the slope in a curved surface [7.49]. That is, minimum canalization (i.e., decanalization) is an increasing or maximum slope, while maximum canalization is a decreasing or minimum slope along a curve on a surface. Measurement of change phenotypically may be compared by the amount of developmental change as the phenotypic character becomes canalized [7.49].

By contrast, epistasis represents distinct interactions among developmental states that are not additive. A phenotype gradient is the maximum slope of the curve on a surface and is comprised of first partial derivatives of a phenotype measure and are elements of the Jacobian. Eigenvectors of the Jacobian are phenotype gradients. An epistasis matrix has elements of second partial derivatives, and developmental epistasis is represented by the off-diagonal elements of the matrix, while the diagonal elements represent dominance and are eigenvalues [7.49]. This matrix is the Hessian, and the sum of the diagonal elements is the Laplacian. For additive effects given as zero valued off-diagonal elements and the same value for diagonal elements of the epistasis matrix, maximum slope is the same on a given surface and may be indicative of drift [7.49]. Ensemble surface measures of slopeness, peakedness, and smoothness may be useful in constructing possible scenarios on the relation between morphogenesis, canalization and epistasis in diatom evolution.

7.5 Conclusions

Quantifying diatom surface features from parametric 3D models enabled the analysis of multiple changes in morphology during a valve formation sequence. The analyses were centered on the combination of surface descriptors called ensemble surface features—slopeness, peakedness, smoothness, and flatness—as quantitative changes that could be compared among taxa. Three exemplars were used. *Actinoptychus senarius* was characterized by a combination of peakedness and smoothness, *Arachnoidiscus ehrenbergii* was characterized by a different combination of smoothness and peakedness, and *Cyclotella meneghiniana* was characterized by a varying combination of all the ensemble surface features measured.

For valve formation, the degree to which each taxon valve surface changed was measurable and elucidated for steps that were clustered and matched to Schmid and Volcani's [7.56] schema. Ensemble surface measures were used to determine that in stage one of valve formation, early change in the valve was horizontally flatter but sloped in terms of vertical changes. Midway through valve formation as stage two, horizontal and vertical surface changes increased in peakedness although smoothness of areas was also evident. Finally, toward completion of the valve at stage three, all ensemble surface measures contributed in varying degrees to the final surface "terrain" in terms of flat, smooth, sloped, and peaked features.

Matching ensemble surface features to definable stages in cytological processes, such as the position of organelles during mitosis or determining the degree to which ensemble surface features are a result of epistasis, is potentially in the offing and of interest in gaining an understanding of the role that cytology, inheritance and evolution have in diatom morphogenesis. Ensemble surface features have been demonstrated to be a quantifiable basis in morphological changes during diatom morphogenesis. Such features have potential application in studies of diatom valve formation with regard to phenotypic plasticity, epistasis, and canalization.

Acknowledgment

I would like to dedicate this chapter to Norman A. Andresen. Norman was a friend and colleague who died unexpectedly on August 7, 2019. He was a diatom researcher for most of his life, being dedicated to doing the very best in studies concerning diatom taxonomy, morphology, and limnology, and understood the importance of morphogenesis in all of these areas of diatom research. A tireless worker, Norman was always willing to lend a helping hand, being instructive and insightful in all things concerning diatoms. Those of us who got to know Norman over the years will miss him, but his contributions to diatom research will live on.

References

[7.1] Bedoshvili, Y., Gneusheva, K., Popova, M., Morozov, A., Likhoshway, Y., Anomalies in the valve morphogenesis of the centric diatom alga *Aulacoseira islandica* caused by microtubule inhibitors. *Biol. Open*, 7, 8, bio035519, 2018.

[7.2] Bertschinger, E., *Introduction to tensor calculus for general relativity*, pp. 1–34, Massachusetts Institute of Technology, Department of Physics, Cambridge, Massachusetts, Physics 8.962, 1999.

[7.3] Brown, N.E., *Arachnoidiscus*, W. Watson and Sons, Ltd., London, United Kingdom, 1933.

[7.4] Chen, C.S., Mrksich, M., Huang, S., Whitesides, G.M., Ingber, D.E., Micropatterned surfaces for control of cell shape, position, and function. *Biotechnol. Progr.*, 14, 3, 356–363, 1998.

[7.5] Cotterell, J., Robert-Morena, A., Sharpe, J., A local, self-organizing reaction-diffusion model can explain somite patterning in embryos. *Cell Syst.*, 1, 257–269, 2015.

[7.6] Deserno, M., *Notes on differential geometry with special emphasis on surfaces in* $\mathbb{R}^3$, pp. 1–64, UCLA, Los Angeles, California and Max Planck Institute, Mainz, Germany, 2004.

[7.7] De Tommasi, E., Gielis, J., Rogato, A., Diatom frustule morphogenesis and function: a multidisciplinary survey. *Mar. Genomics*, 35, 1–18, 2017.

[7.8] do Carmo, M.P., *Differential Geometry of Curves and Surfaces*, Prentice-Hall, Englewood Cliffs, New Jersey, USA, 1976.

[7.9] Ferrara, M.A., Dardano, P., De Stefano, L., Rea, I., Coppola, G., Redina, I., Congestri, R., Atonucci, A., De Stefano, M., De Tommasi, E., Optical properties of diatom nanostructured biosilica in *Arachnoidiscus* sp.: micro-optics from mother nature. *PLoS One*, 9, 7, e103750, 2014.

[7.10] Ferrara, M.A., De Tommasi, E., Coppola, G., De Stefano, L., Rea, I., Dardano, P., Diatom valve three-dimensional representation: a new imaging method based on combined microscopies. *Int. J. Mol. Sci.*, 17, 10, 1645, 2016.

[7.11] Fisher, R.A., The correlation between relatives on the supposition of Mendelian inheritance. *Trans. R. Soc. Edinburgh*, 52, 2, 399–433, 1918.

[7.12] Gilbert, S.F., The morphogenesis of evolutionary developmental biology. *Int. J. Dev. Biol.*, 47, 467–477, 2003.

[7.13] Gordon, R. and Drum, R.W., The chemical basis for diatom morphogenesis. *Int. Rev. Cytol.*, 150, 243–372, 421–422, 1964.

[7.14] Gordon, R. and Tiffany, M.A., Possible buckling phenomena in diatom morphogenesis, in: *The Diatom World*, J. Seckbach and J.P. Kociolek (Eds.), pp. 245–272, Springer, Dordrecht, The Netherlands, 2011.

[7.15] Gordon, R., Losic, D., Tiffany, M.A., Nagy, S.S., Sterrenburg, F.A.S., The glass menagerie: diatoms for novel applications in nanotechnology [invited]. *Trends Biotechnol.*, 27, 2, 116–127, 2009.

[7.16] Green, J.B.A. and Sharpe, J., Positional information and reaction-diffusion: two big ideas in developmental biology combine. *Development*, 142, 1203–1211, 2015.

[7.17] Håkansson, H. and Chepurnov, V., A study of variation in valve morphology of the diatom *Cyclotella meneghiniana* in monoclonal cultures: effect of auxospore formation and different salinity conditions. *Diatom Res.*, 14, 2, 251–272, 1999.

[7.18] Hasle, G.R. and Sims, P.A., The diatom genera *Stellarima* and *Symbolophora* with comments on the genus *Actinoptychus*. *Br. Phycol. J.*, 21, 1, 97–114, 1986.

[7.19] Hartle, J.B., http://web.physics.ucsb.edu/~gravitybook/mathematica.html, Accessed on 31 Aug 2018.

[7.20] Huxley, J., *Evolution: The Modern Synthesis*, George Allen & Unwin Ltd, London UK, 1942.

[7.21] Kaplan, W., *Maxima and Minima with Applications – Practical Optimization and Duality*, John Wiley & Sons, Inc., New York, USA, 1999.

[7.22] Kaplan, W., *Advanced Calculus*, 5th edition, Addison-Wesley, Reading, Massachusetts, 2003.

[7.23] Kobayashi, A., Osada, K., Nagumo, T., Tanaka, J., An auxospore of Arachnoidiscus ornatus Ehrenberg, in: *Proceedings of the 16th International Diatom Symposium*, Amvrosiou Press, University of Athens, Athens, Greece, pp. 197–204, 2001.

[7.24] Koenderink, J.J., *Solid Shape*, MIT Press, Cambridge, Massachusetts USA, 1990.

[7.25] Koonin, E.V., The Origin at 150: is a new evolutionary synthesis in sight? *Trends Genet.*, 25, 11, 473–475, 2009.

[7.26] Koop, R., *Global Gravity Field Modelling Using Satellite Gravity Gradiometry*, W.D. Meinema B.V., Delft, The Netherlands, 1993.

[7.27] Kutschera, U. and Niklas, K., The modern theory of biological evolution: an expanded synthesis. *Naturwissenschaften*, 91, 255–276, 2004.

[7.28] Laubichler, M.D. and Müller, G.B., Models in theoretical biology, in: *Modeling Biology. Structures, Behaviors, Evolution*, M.D. Laubichler and G.B. Müller (Eds.), pp. 1–10, MIT Press, Cambridge, Massachusetts, USA, 2007.

[7.29] Lee, J.H. and Chang, M., Morphological variations of the marine diatom genus *Actinoptychus* in the coastal waters of Korea. *Algae*, 11, 4, 365–374, 1996.

[7.30] Lin, K.H., Crocker, J.C., Prasad, V.V., Schofield, A., Weitz, D.A., Lubensky, T.C., Yodh, A.G., Entropically driven colloidal crystallization on patterned surfaces. *Phys. Rev. Lett.*, 85, 8, 1770–1773, 2000.

[7.31] MacDonald, J.D., On the structure of the diatomaceous frustule, and its genetic cycle. *Ann. Mag. Nat. Hist.*, 4, 1, 1869.

[7.32] Misner, C.W., Thorne, K.S., Wheeler, J.A., *Gravitation*, W.H. Freeman and Company, San Francisco, California, USA, 1973.

[7.33] Moore, T.A., Chapter 17, Box 17.4: Christoffel symbols in terms of the metric tensor, in: *A General Relativity Workbook*, University Science Books, Mill Valey, California, 2013.

[7.34] Nechaev, S., Non-Euclidean geometry in nature. arXiv:1705.08013v2 [cond-mat.soft] 27May, 2017.

[7.35] Organ, C.L., Cooper, L.N., Hieronymus, T.L., Macroevolutionary developmental biology: embryos, fossils, and phylogenies. *Dev. Dyn.*, 244, 1184–1192, 2015.

[7.36] Pappas, J.L., Geometry and topology of diatom shape and surface morphogenesis for use in applications of nanotechnology. *J. Nanosci. Nanotechnol.*, 5, 1, 120–130, 2005a.

[7.37] Pappas, J.L., Theoretical morphospace and its relation to freshwater gomphonemoid-cymbelloid diatom (Bacillariophyta) lineages. *J. Biol. Syst.*, 13, 4, 385–398, 2005b.

[7.38] Pappas, J.L., More on theoretical morphospace and its relation to freshwater gomphonemoid-cymbelloid diatom (Bacillariophyta) lineages. *J. Biol. Syst.*, 16, 1, 119–137, 2008.

[7.39] Pappas, J.L., Graph matching a skeletonized theoretical morphospace with a cladogram for gomphonemoid-cymbelloid diatoms (Bacillariophyta). *J. Biol. Syst.*, 19, 1, 47–70, 2011.

[7.40] Pappas, J.L., Multivariate complexity analysis of 3D surface form and function of centric diatoms at the Eocene-Oligocene transition. *Mar. Micropaleontol.*, 122, 67–86, 2016.

[7.41] Pappas, J.L. and Miller, D.J., A generalized approach to the modeling and analysis of 3D surface morphology in organisms. *PLoS One*, 8, 10, e77551, 2013.

[7.42] Pappas, J.L., Kociolek, J.P., Stoermer, E.F., Quantitative morphometric methods in diatom research, in: *Nina Strelnikova Festschrift*, vol. 143, Kociolek, J.P., Kulivoskiy, M., Witkowski, J., Harwood, D.M. (Eds.), *Nova Hedwigia, Beihefte*, pp. 281–306, 2014.

[7.43] Pappas, J.L., Tiffany, M.A., Gordon, R., The uncanny symmetry of some diatoms and not of others: a new multi-scale morphological characteristic and a puzzle for morphogenesis, in: *Diatom Morphogenesis [DIMO, Volume 2 in the series: Diatoms: Biology & Applications, series editors: Richard Gordon & Joseph Seckbach]*, V. Annenkov, J. Seckbach, R. Gordon (Eds.), Wiley-Scrivener, Beverly, Massachusetts, USA, 2021.

[7.44] Parkinson, J., Brechet, Y., Gordon, R., Centric diatom morphogenesis: a model based on a DLA algorithm investigating the potential role of microtubules. *Biochim. Biophys. Acta Mol. Cell Res.*, 1452, 1, 89–102, 1999.

[7.45] Pfitzer, E., Über den bau and zellteilung der diatomeen. *Bot. Ztg.*, 27, 774–776, 1869.

[7.46] Pfitzer, E., Untersuchungen über Bau und Entwicklung der Bacillariaceen (Diatomeen), in: *Botanische Abhandlungen aus dem Gebiet der Morphologie und Physiologie*, vol. 1, Hanstein, J. (Ed.), pp. 1–189, 1871.

[7.47] Pickett-Heaps, J.D., Cell division and morphogenesis of the centric diatom *Chaetoceros decipiens* (Bacillariophyceae): II. Electron microscopy and a new paradigm for tip growth. *J. Phycol.*, 34, 995–1004, 1998.

[7.48] Pigliucci, M., Murren, C.J., Schlichting, C.D., Phenotypic plasticity and evolution by genetic assimilation. *J. Exp. Biol.*, 209, 2362–2367, 2006.

[7.49] Rice, S.H., The evolution of canalization and the breaking of Von Baer's laws: modeling the evolution of development with epistasis. *Evolution*, 52, 3, 647–656, 1998.

[7.50] Ries, H. and Muschaweck, J., Tailored freeform optical surfaces. *J. Opt. Soc. Am. A*, 19, 3, 590–595, 2002.

[7.51] Rogerson, A., DeFreitas, A.S.W., McInnes, A.G., Observations on wall morphogenesis in *Coscinodiscus asteromphalus* (Bacillariophyceae). *Trans. Am. Microsc. Soc.*, 105, 1, 59–67, 1986.

[7.52] Round, F.E., Crawford, R.M., Mann, D.G., *The Diatoms – Biology and Morphology of the Genera*, Cambridge University Press, Cambridge, United Kingdom, 1990.

[7.53] Sanjuán, R. and Elena, S.F., Epistasis correlates to genomic complexity. *Proc. R. Soc.*, 103, 39, 14402–14405, 2006.

[7.54] Sato, S., Nagumo, T., Tanaka, J., Auxospore formation and the morphology of the initial cell of the marine araphid diatom *Gephyria media* (Bacillariophyceae). *J. Phycol.*, 40, 684–691, 2004.

[7.55] Schmid, A.-M.M., III.2 Morphogenetic forces in diatom cell wall formation, in: *Cytomechanics: The Mechanical Basis of Cell Form and Structure*, J. Bereiter-Hahn, O.R. Anderson, W.-E. Reif (Eds.), pp. 183–199, Springer-Verlag, Berlin Heidelberg, Germany, 1987.

[7.56] Schmid, A.-M.M. and Volcani, B.E., Wall morphogenesis in *Coscinodiscus wailesii* I. valve morphology and development of its architecture. *J. Phycol.*, 19, 4, 387–402, 1983.

[7.57] Shihavuddin, A., Basu, S., Rexhapaj, E., Delestro, F., Menezes, N., Sigoillot, S.M., Nery, E.D., Selimi, F., Spassky, N., Genovesio, A., Smooth 2D manifold extraction from 3D image stack. *Nat. Commun.*, 8, 15554, 2017.

[7.58] Shirokawa, Y. and Shimada, M., Cytoplasmic inheritance of parent-offspring cell structure in the clonal diatom *Cyclotella meneghiniana*. *Proc. R. Soc. B*, 283, 20161632, 2016.

[7.59] Simpson, G.G., *Tempo and Mode in Evolution*, Columbia University Press, New York, USA, 1944.

[7.60] Sochi, T., *Principles of Differential Geometry* arXiv:1609.02868v1 [math.HO] 9 Sep 2016, 2016.

[7.61] Tesson, B. and Hildebrand, M., Extensive and intimate association of the cytoskeleton with forming silica in diatoms: control over patterning on the meso- and micro-scale. *PLoS One*, 5, 12, e14300, 2010.

[7.62] Tesson, B. and Hildebrand, M., Characterization and localization of insoluble organic matrices associated with diatom cell walls: insight into their roles during cell wall formation. *PLoS One*, 8, 4, e61675, 2013.

[7.63] Turing, A.M., The chemical basis of morphogenesis. *Philos. Trans. R. Soc. Lond. B*, 237, 37–72, 1952.

[7.64] Van de Meene, A.M.L. and Pickett-Heaps, J.D., Valve morphogenesis in the centric diatom *Proboscia alata* Sundstrom. *J. Phycol.*, 38, 2, 351–363, 2002.

[7.65] Van de Meene, A.M.L. and Pickett-Heaps, J.D., Valve morphogenesis in the centric diatom *Rhizosolenia setigera* (Bacillariophpyceae, Centrales) and its taxonomic implications. *Eur. J. Phycol.*, 39, 1, 93–104, 2004.

[7.66] Vartanian, M., Desclés, Quinet, M., Douady, S., Lopez, P.J., Plasticity and robustness of pattern formation in the model diatom *Phaeodactylum tricornutum*. *New Phytol.*, 182, 2, 429–442, 2009.

[7.67] Waddington, C.H., Canalization of development and the inheritance of acquired characters. *Nature*, 150, 563–565, 1942.

[7.68] Weinberger, H.F., *A First Course in Partial Differential Equations with Complex Variables and Transform Methods*, Dover, Publications, Inc., New York, USA, 1965.

[7.69] Weisstein, E.W., *CRC Concise Encyclopedia of Mathematics*, 2nd ed., Chapman & Hall/CRC, London, United Kingdom, 2002.

[7.70] Wimsatt, W.C., False models as means to truer theories, in: *Neural Models in Biology*, M.H. Nitecki and A. Hoffman (Eds.), pp. 23–55, Oxford University Press, New York, USA, 1987.

[7.71] Yadav, A., Dhole, K., Sinha, H., Genetic regulation of phenotypic plasticity and canalization in yeast growth. *PLoS One*, 11, 9, e0162326, 2016.

8

Buckling: A Geometric and Biophysical Multiscale Feature of Centric Diatom Valve Morphogenesis*

Janice L. Pappas[1†] **and Richard Gordon**[2,3]

[1]*Department of Mathematics, Michigan Math and Science Scholars, University of Michigan, Ann Arbor, MI, United States*
[2]*Gulf Specimen Marine Laboratory & Aquarium, Panacea, FL, United States*
[3]*C.S. Mott Center for Human Growth & Development, Department of Obstetrics & Gynecology, Wayne State University, Detroit, MI, United States*

Abstract
Buckling is a ubiquitous phenomenon in living organisms that occurs often throughout their life cycles. For diatoms, buckling occurs as a micro-, meso-, and nanoscale feature of morphogenesis. From silica deposition of valve characters to cytoplasmic structure behavior and position during morphogenesis, buckling is evident as a three-dimensional (3D) consequence. While it is currently not possible to match cytoplasmic structural behavior with the onset of valve surface features at distinct morphogenetic stages, it is possible to determine the association and influence that cytoplasmic structural behavior has on the consequent morphological features formed via buckling during morphogenesis. We characterize three exemplar centric diatom taxa via their 3D surface morphology and associate geometric and biophysical buckling of the valve surface to microtubule filament and bundle and actin filament (microfilament) ring geometric buckling during morphogenesis. Analytically, these changes are characterized via properties of buckling including deformation and displacement. The resultant association indicates the magnitude of buckling by cytoplasmic structures that influences the degree of buckling on the valve surface during morphogenesis.

Keywords: 3D surface, buckling, microtubule, actin, wave front, eikonal equation, Hamiltonian mechanics, Krylov methods

*We dedicate this chapter in memory of Mark Hildebrand. His pioneering work is in diatom morphogenesis, especially concerning the regularity of nanoscale patterning of silica in diatom frustules. Mark's dedication and ability to push the envelope concerning genomics, biochemistry, and molecular biology with regard to algae, in particular, to diatoms has advanced our understanding at the micro-, meso-, and nanoscale levels. We are grateful for his foresight and fortitude in life and diatom research and will be forever bereft of his potential insightful research that we will never know.

†*Corresponding author*: jlpappas@umich.edu

Vadim Annenkov, Joseph Seckback and Richard Gordon (eds.) Diatom Morphogenesis, (195–230) © 2022 Scrivener Publishing LLC

8.1 Introduction

Diatom morphogenesis of a vegetative cell involves changes in the three-dimensional (3D) valve surface from an annulus to the mantle in centric diatoms and from the sternum to mantle in pennate diatoms [8.94]. Each valve and cingula grows inside a flat, membrane bag where silica condensation occurs, called the silica deposition vesicle (SDV), attached to the inside of the plasmalemma. The membrane surrounding the SDV is called the silicalemma. Forces impinging on the cell are the result of turgor internally [8.100] and the external environment, which more often than not, involves water [8.24]. At a smaller scale, silica deposition involves forces that occur during the initiation and termination of valve formation as uptake of silicic acid is necessary for silica polymerization to commence [8.108]. External forces influence the shape of the cell ecophenotypically, while internal forces are the result of genetic, molecular, and cellular processes that affect the cytoskeleton [8.125].

During the past few decades, since our review of diatom valve morphogenesis [8.32], much has been learned about the process with respect to cytological [8.96, 8.114], chemical [8.108], molecular [8.117], and biomechanical [8.41] experimental evidence. The earliest studies on diatom morphogenesis showed that silica formation occurs within cytological structures [8.21], and these studies were accomplished using transmission electron or fluorescence light microscopical imaging techniques [8.87, 8.99, 8.102, 8.120]. Experimentally, cytological and proteomic evidence has been accumulating showing that microtubules [8.15, 8.96, 8.107] and transmembrane proteins [8.23, 8.110, 8.111, 8.112] play vital roles in silica structural formation in concert with the SDV and silicalemma [8.99, 8.114]. The role of silica biochemistry in diatom biomineralization [8.3, 8.4, 8.5, 8.65, 8.108] involves soluble biomolecules including silaffins [8.108], frustulins [8.64], cingulins [8.97], silacidins [8.124], silicanins [8.62], and long-chain polyamines (LCPAs) [8.108]. These are associated with silica formation on a nanoscale, while insoluble organic molecules and matrices [8.63] are associated with meso- to microscale silica organization [8.46]. Cause and effect are difficult to unravel here. Suggestions of prepatterns, reviewed in [8.32], only beg the question of how the prepattern forms. Hildebrand has suggested top down causation RFR, but how this occurs through the sexual cycle has yet to be addressed. Buckling, which occurs over various spatial and time scales, may be one of the keys to understanding causation in morphogenesis.

The microscale includes the 3D siliceous diatom structure, the mesoscale includes assemblies of silicified valve structural elements in the range of hundreds of nanometers, and the nanoscale includes initial silica polymerization structures in the range of tens of nanometers [8.110]. The full range of silica structures in diatoms spans 8 orders of magnitude [8.25]. While the above-mentioned biomolecules are associated with the SDV, SAP3 [8.112] and Sin1 [8.62], which are associated with the plasma membrane and an intracellular vesicle, are important in relation to the silicalemma [8.46]. Transcripts for silicification have been found in *Thalassiosira pseudonana*, such as the gene encoding protein, TpSil3 [8.37], which provide markers for the timing of particular structures formed during valve formation [8.23].

Matching of interior cellular processes to the production of diatom valve surface features has yet to be accomplished. Thus far, there are hints that on a macromolecular and subcellular structural level, such matching to some structural features may be in the offing [8.46]. Exterior valve structural formation during cell growth along with the morphology of cytological structures may be associated through a commonality of the morphogenetic process on a nano- to meso- to micro-scale [8.110]. In this regard, computational and modeling techniques are a useful aid in explicating how disparate pieces of the known experimental and empirical evidence may be used to understand the multiscale nature of valve morphogenesis [8.10, 8.11, 8.18, 8.32, 8.70, 8.86, 8.126]. Valve morphological properties as a result of silica deposition via precipitation during valve formation and subcellular structural morphological changes during this process may be modeled with regard to buckling. Buckling is a natural phenomenon of cell growth at different scales [8.78] that provides strength [8.39] and occurs during valve morphogenesis [8.35].

8.2 Purpose of Study

The purpose of our study is the explication and evaluation of the geometrical and biophysical understanding of buckling during valve morphogenesis as a multiscale property. Specifically, we use centric diatoms as the simplest case in which to perform our study. As a guide, the valve morphogenetic model of Schmid and Volcani [8.102] is used as well as the condensed version of the same model by Hildebrand, Lerch, and Shrestha [8.46]. Computational modeling will be used to show how buckling at multiple scales facilitates associating cellular components to valve surface features during centric diatom morphogenesis. We expect to find, in accordance with the schema of Schmid and Volcani [8.102], that buckling occurs in the valve face plane early in morphogenesis as the horizontal base layer forms, while the vertical height of the valve features forms via buckling out of the valve face late in morphogenesis.

We take the approach of viewing valve formation being influenced by biophysical processes, from initiation of the valve face pattern to its completion on a micro- to meso- to nanoscale during valve morphogenesis. Along the way, the state of the materials involved—amorphous silica, silica-associated proteins, and insoluble organic materials—affect valve formation at multiple scales and may be viewed as a tiered non-linear 4D process. Geometrically, valve face changes in the z-direction, progressing along linear paths in the radial direction, may characterize the progression of morphological changes where the flow of valve formation commences at an "annulus" [8.87, 8.121] ("pattern center" [8.87, 8.92]; "primary silicification site" [8.87, 8.98]) in centric diatoms to the valve margin culminating at the mantle-girdle band interface.

On a biophysical basis, the production and progression of valve face feature formation may be viewed as the buckling of amorphous silica as a concomitant result of forces generated by the cytological elements in the diatom cell. Geometrically, a circular centric diatom provides the simplest exemplar case where buckling may be found. The points at which buckling occurs on the structural features of the external cell as well as the internal morphology of cell elements may be associated with one another via measurement of buckling. To start, 3D surface centric diatom models are constructed and analyzed for use in measurement of valve surface buckling. Modeling of cellular components that have a role in valve surface morphological feature construction is also used. The geometrical and biophysical results from modeling are used in determining the association of cellular components to valve feature appearance via buckling in the valve morphogenetic process.

8.3 Background: Multiscale Diatom Morphogenesis

8.3.1 Valve Morphogenesis—Schemata of Schmid and Volcani and of Hildebrand, Lerch, and Shrestha

Diatom valve formation is a directional growth process with respect to silica deposition and pattern formation [8.117]. Evidence has accumulated that given the directionality of the process, the cardinality of structural features does not exhibit a one-to-one association between first appearance of a given valve face structural feature, the underlying cell element morphology and positional state influencing that structural feature [8.15, 8.46, 8.107, 8.117]. For example, silica deposition in centric diatoms within the annulus occurs in spurts, resulting in an irregular structural pattern [8.96] (cf. [8.32, 8.126]). The valve mantle thickens via silica deposition in which processes such as rimoportulae may form prior to the completion of the valve mantle [8.96]. This association between structural valve patterning and cell element involvement is traceable from the initiation of valve formation centrally in the cell to formation of the valve mantle edge. Valve face pattern and cell shape formation is globally a directional flow from valve center to periphery [8.32, 8.102].

Valve formation follows a process that is delineated into three stages [8.102]. (For a finer dissection, see [8.87].) The two-stage process of Hildebrand, Lerch, and Shrestha [8.46] combines the second and third stages of Schmid and Volcani [8.102]. In both cases, the first stage encompasses valve formation in the x, y-plane as proximal or horizontally as a base layer, and the second or combined second and third stages encompass the z-direction changes in valve formation as distal or vertical changes in silica deposition with secondary or outer additional silicification horizontally over the valve surface. Growth in the x, y-plane is reported to take 10 minutes while thickening in the z-direction takes 6 hours [8.32], although this timetable is not necessarily applicable to all diatoms. “The first feature that resembles a valve component is called the base layer, which will form the proximal surface of the completed valve. The base layer is a footprint of the valve which includes the ribs and precursors to the fultoportulae, defining structure in the x and y plane. The structure is very flexible at this stage [8.44]” [8.38] and thus readily subject to buckling:

> “This has two mechanical consequences. First, a thin layer of silica is likely to be easy to bend and buckle, just the way a glass capillary bends more readily the thinner it is. Second, diatoms can do something we rarely indulge in to increase mechanical strength of a manufactured structure: the valve thickens after it is shaped. Corrugation also increases strength; thickening after corrugation would do so that much more” [8.35].

The base layer includes the formation of ribs, cross extension of silica deposition between radial silica strings of the areolae, and compaction, to form a nanoscale thick networked template for the next steps in distal growth [8.46, 8.102]. Initial traces of the fultoportulae and/or internal rimoportulae tubes also are formed [8.102]. Next, areolae walls increase in height as silica is deposited vertically, which can occur in both vertical directions [8.87]. Hexagonal pores are formed from round ones, and rimoportulae are completed [8.102]. Growth via silica deposition occurs distally [8.32, 8.46]. Finally, cribra and cribella are completed during additional horizontal silica deposition with regard to areolar structural constraints as to size and spacing [8.102]. The valve and its structures extend to the margin as the final sequential vertical silica layers are deposited [8.93].

8.3.2 Valve Formation—An Overview at the Microscale

Silica depositional and associated proteins are associated with cell structures that influence surface features during the valve formation process. The SDV and the silicalemma play a large role in cell formation and shape [8.67]. The SDV is an organelle in which silicification occurs and new valves of daughter cells form within the mother cell at interphase [8.114]. The silicalemma is a membrane surrounding the SDV, and its plasticity influences cell shape [8.46].

Initial silica deposition occurs before the SDV is fully formed. A shift to low pH in the SDV aids or triggers silicification [8.122]; lack of pH change prevents silicification [8.114, 8.115]. Expansion in the z-direction of the base layer may be dependent on a change in pH triggering the SDV to commence silica deposition [8.102, 8.46]. Low pH is induced by the presence of the soluble proteins, silaffins [8.63].

Cytoskeleton positions mesoscale structures via microtubules, and actin is involved in the overall surface features pattern [8.22, 8.110]. The cytoskeleton may interact with the SDV [8.92] facilitated by proteins associated with the silicalemma; i.e., it is speculated that cytoskeleton and SDV assemble is such a way as to enable the cytoskeleton to position and organize components for base layer formation. Microtubules and microfilaments are involved in the shaping and expansion of the SDV during valve formation [8.57, 8.62]. Within the SDV, silica is polymerized, and as the SDV interacts with its membrane, the silicalemma, the cytoskeleton is involved in silica patterning via the silicalemma and transmembrane proteins [8.112].

Microtubules and actin filaments correlate with the position of the SDV [8.46, 8.114]. Centrally located, the SDV expands with respect to radial microtubules potentially held rigid by intermediate filaments [8.16, 8.56]. The microtubule organizing center (MTOC) is located adjacent to the annulus, at the cytoplasmic surface of the silicalemma closest to the cell nucleus [8.34]. Silica transport vesicles, each containing a colloidal silica sphere [8.32, 8.101], may move along the microtubules and fuse with the silicalemma at its perimeter [8.3, 8.10, 8.11, 8.18, 8.32, 8.34, 8.70, 8.86, 8.126]. The potentially contractile and probably pliable actin ring defines the perimeter of the SDV [8.34], and as the ring expands, the SDV expands [8.110, 8.114], though it may be in mechanical opposition to the microtubules [8.34]. As we have noted: "...diatom shells [8.29] are formed inside a membrane bag (silicalemma) that is apparently encircled by a microfilament bundle [actin ring] and spanned by spokes of microtubules emanating from a microtubule-organizing center [8.88, 8.89]" [8.30]. Actin filament assembly occurs at the external perimeter of the SDV [8.32, 8.89] and is presumed to affect silica patterning [8.46, 8.92]. Pliable intermediate filaments provide connections among actin and microtubule both cytoskeletally and biomechanically [8.123] and along with actin filaments are offset by microtubules that contribute to cell shape stabilization [8.68]. While microtubules and actin filaments have polar ends, intermediate filaments contain neutral protein subunits and provide stability to the cytoskeleton [8.123]. However, while microtubules and actin are associated with silica valve structure, intermediate filaments are not [8.62, 8.110].

8.3.3 Valve Formation—An Overview at the Meso- and Microscale

Microtubules are associated with the forming valve [8.110]. Radiating microtubules from the microtubule center to the periphery occur very early in the valve formation process, indicating an association with the Schmid and Volcani [8.102] first morphogenesis stage of the base layer proximal surface. Microtubule bundles are associated with indentations on the valve surface, being in direct contact with the valve and the SDV. Portulae location was also associated with the tips of microtubules, and initial positioning of silica structures of the forming valve has been postulated to be associated with microtubule function. Microtubules are associated with microscale structure positioning [8.110].

Actin filaments are also associated with the forming valve [8.110]. Although they, like microtubules, occur outside the SDV, actin filaments are associated with the structure and positioning of the ribs and foramen, forming in the base layer during the first morphogenetic stage of Schmid and Volcani [8.102]. Actin has been postulated to be associated with areolae structure and cribrum and cribellum positioning, which suggests that these filaments are involved in distal growth [8.110]. As such, actin filaments may be instrumental in the Schmid and Volcani [8.102] second stage of valve morphogenesis. Actin filaments are associated with micro- and mesoscale structure, and actin rings surround the SDV perimeter and may be a controlling factor in SDV expansion [8.34, 8.110].

Because actin occurs outside the SDV, it is also involved in shaping the silicalemma. The implication is that actin provides a constraint to the topography of silicification so that templating is not necessary in the valve formation process [8.45, 8.110], perhaps by being in a radial tug-of-war with the radial microtubules [8.16, 8.34]. Actin and microtubules are limited mostly in affecting silicification of the base layer as a result of the degree of foramen present, which varies for centric diatoms. That is, while the base layer may be similar across centric diatoms, the distal layer varies widely as exhibited by the resultant morphologies [8.110].

8.3.4 Valve Formation—An Overview at the Meso- and Nanoscale

Early silica deposition enables flexible structural features of the base layer which is positioned near the edge of one cell [8.45]. In the base layer, silaffins, which are soluble, short proteins, are involved in silica sphere deposition and resultant disordered nanoscale valve features [8.46]. LCPAs, silacidins, and TpSil 1/2 L generate randomly-fused, disordered silica sphere depositional structures as well. When all of these macromolecules combine, a more ordered silica valve structure emerges as a vertical expansion of the valve face. Insoluble organic proteins, including ammonium fluoride insoluble material (AFIM) [8.97, 8.111], are involved in base layer valve features such as pore formation of projections and occlusions and may be associated with *z*-direction valve thickening [8.46]. Self-assembly of silaffins, silacidins, and LCPAs act as catalysts for silica polymerization, and insoluble organic materials, including AFIMs form the same fine structural details evident on the siliceous valve surface, suggesting their role in producing valve features [8.111, 8.112].

Base layer cytoskeletal positioning occurs in association with the silicalemma-associated protein (SAP), Sin1 [8.46, 8.62] which was also extracted from AFIM as SiMat7 [8.46, 8.63], and SAP1 [8.46]. Positioning of the microtubule center [8.87] and SDV constituents is facilitated by SDV-associated proteins, and actin filaments contribute to the extent of silica polymerization and fine structural base layer valve features at a meso- to microscale [8.46]. SDV expansion via actin filaments and AFIM accompanies base layer completion [8.111], with the assembly of actin filaments externally to the SDV corresponding to the overall silica structure of the valve [8.110]. Completed base layer formation is accompanied by expansion of the SDV [8.110] involving the actin ring, the radial microtubules, and perhaps insoluble organic components [8.111].

Silicalemma associated proteins (SAPs), such as TpSAPs for the model organism, *Thalassiosira pseudonana*, have been found to play a role in frustule morphogenesis [8.112]. SAPs are embedded in the silica [8.112] and may be associated with specific valve surface features [8.46]. SAP3 [8.47], Sin1 [8.62], and TpSil3 [8.23] may have specific functions with regard to particular valve or girdle band silicification processes [8.46]. SAPs and Sin1 are embedded in silica and AFIM, and SAPs are associated with proteins involved in silica deposition, including silaffins, silacidin, cingulins, and SiMat [8.46]. SAP1 and SAP3 are associated with phenotypic silica structural changes via valve formation. TpSAP1 is associated with location of the pattern forming center and may interact with the microtubule network, while TpSAP3 is associated with location of portulae and forming silica ridges on base layer ribs [8.112].

8.4 Biophysics of Diatom Valve Formation and Buckling

8.4.1 Buckling as a Multiscale Measure of Valve Formation

One characterization of buckling is as deformation or displacement of materials [8.90]. In valve formation, morphological features form as silica is deposited while other processes occur involving subcellular structures. Buckling as a biophysical influence in valve formation includes forces that may induce elasticity, plasticity, or shear [8.1]. For diatom valve formation, plasticity, and shear are not necessarily involved in deformation or displacement to the degree that such quantities affect overall buckling behavior [8.1]. Compressive axial loads may induce buckling in which an elastic response is invoked, and therefore, elasticity is important in buckling during valve formation [8.55].

Buckling may occur at multiple scales. Valve surface formation, silica deposition, and behavior of cytoplasmic structures in positioning and shape change may all undergo buckling. At the microscale, buckling is an indicator of morphological feature formation at the locations on the valve surface as indicative of a particular species [8.46, 8.110]. Silica deposition produces different thickness of material where there is a change in elevation on the valve surface which may be characterizable as buckling. When considering the meso- and nanoscales, microtubules and actin filaments undergo movement and shape changes. Such changes may coincide with morphological character formation on the valve surface [8.46] and may be indicative of buckling. Associating buckling occurrences at the micro-, meso-, and nanoscales may enable the compilation of a multiscale process involving silica deposition and subcellular biophysics during valve formation.

8.4.2 Valve Formation—Cytoplasmic Features and Buckling

Eukaryotic cytological systems are composed of microtubules, intermediate, and actin filaments [8.46, 8.75]. Each of these elements has a role in determining cell shape and mechanical properties of the cell. Diatom valve formation involves changes in compressive and tensile forces throughout the morphogenetic process [8.99]. These forces are dependent on whether a given valve formation stage involves viscosity, elasticity, viscoelasticity, or plasticity of the materials and their depositional attributes as well as surrounding cellular content. At the subcellular level, some structures may form and induce compressive forces. Microtubule formation via tubulin polymerization has been shown to influence diatom cell shape and mechanics via compression resistance to the contraction of the surrounding cytoskeleton [8.15]. Microtubule buckling has been shown to occur in different arrangements of bundles [8.107].

Cell shape and biomechanics are influenced by microtubules which withstand compressive forces as the cytoskeleton contracts and buckles. The cytoskeleton is elastic which aids in relieving buckling so that the ultimate instability of breakage does not occur [8.15, 8.107]. As a result, microtubule polymerization at the cell mantle exhibits buckling [8.15] which contributes to valve mantle and cell shape morphology [8.107].

Compressive forces of rigid microtubules are offset by tensile forces of the more flexible intermediate and actin filaments. Tensegrity can produce a stabilized cell shape [8.28, 8.51, 8.68]. Microtubule buckling occurs as short-wavelength phenomena [8.16, 8.55, 8.56, 8.57] as the microtubule tips on the growing surfaces polymerize [8.15]. Microtubules undergo buckling when exposed to compressive forces [8.1, 8.16, 8.57]. Proteins associated with microtubules play an important role in compressive induced buckling [8.1, 8.57] so that as compression increases, buckling wavelength decreases, and buckling amplitude increases [8.57]. The radius of curvature of buckled microtubules also changes with changes in the concentration of associated proteins [8.57].

Polymerizing tubulin lengthens microtubules, alters their positioning [8.114], and can eventually lead to buckling [8.50, 8.69, 8.107]. Microtubules can self-organize into aligned parallel bundles forming a striped pattern that is detectable via birefringence [8.69]. Then, the bundles form a wave-like structure. Such microtubule bundle buckling acts as a release from polymerization compressional forces [8.69, 8.107].

Polymerization of tubulin influences the pushing forces of microtubule arrays that are expanding or self-assembling [8.75]. Pulling forces are induced by contraction of actin filaments in cross-linked bundles to the plasma membrane. Actin filaments are found in disordered gels as well. Microtubule and actin network assembly and disassembly can produce changing cell shape in response to external and internal forces. When growing, microtubules can lengthen and shorten, while actin filaments can lengthen. Microtubule radial array networks have polarized ends within a cell at the center and edge [8.75]. Persistent cytological structures may influence cell shape epigenetically [8.22].

Microtubule, intermediate, and actin filament networks have viscous, elastic, or viscoelastic properties because of cross-linked polymers [8.6, 8.75]. External forces as well as internal forces by molecular motors can induce different responses by cytoskeletal elements [8.75]. The rigidity of microtubules prevents buckling as they are moved by kinesin [8.58] and dynein [8.73], molecular motors [8.2, 8.12]; however, microtubule buckling may occur in response to molecular motors as well [8.57]. Myosins act as molecular motors for actin filaments [8.43]. Motor proteins at the cell edge enable self-centering [8.127] of radial arrays of microtubule networks [8.75]. Cross-linked actin filaments determine the capacity of the network to expand and contract and withstand tensive forces [8.75].

Microtubule and actin networks are stabilized by intermediate filaments against compressive forces [8.13, 8.123]. In turn, stabilization by intermediate filaments enables microtubule bundles to withstand increasing compressive forces. Intermediate filaments are highly flexible enabling high resistance to deformation and high recovery from long periods of strain [8.16]. However, breakdown of intermediate filaments occurs if deformation exceeds 20% or the capacity of microtubules and actin filaments in the cytoskeleton is exceeded to withstand compression [8.123]. Intermediate filaments prevent overextension of the cell which results in net added stiffness [8.13, 8.123].

Microtubules as well as actin and intermediate filaments are contained in cytoplasm, which has an effect on the biomechanical tolerances of compressive forces applied to these structures. Cytoplasm may be characterized as viscous, elastic, or viscoelastic and is a supportive factor in the ability of microtubules to sustain compressive forces as buckling occurs [8.16, 8.55]. Cytoplasm may account for shear forces so that microtubule buckling in the cytoskeleton occurs linearly with strain displacement [8.55] via the strain rate tensor.

8.4.3 Buckling: Microtubule Filaments and Bundles

Biophysical changes to the diatom cell occur during morphogenesis at the micro-, meso-, and nano-scales. Deformation of the diatom valve occurs as silica is deposited, and this, in turn, is affected by microtubule polymerization and bundling as well as SDV and silicalemma expansion. Microtubules can undergo deformation in which buckling occurs at the tip only while the remainder of the structure is mostly unaffected [8.15, 8.107]. Buckling occurs at short wave-lengths for individual and bundled microtubules [8.107]. In general, microtubules consist of 9–16 tubulin protofilaments, and the orientation of the filaments does not affect the biomechanical response of microtubules during buckling [8.72]. Geometry of a microtubule bundle consists of a hexagonal [8.107] to cylindrical [8.72] arrangement. Axial compression may induce buckling, while shear deformation is considered to be negligible by comparison [8.1]. Circumferential compression is non-uniform and dependent on microtubule wall thickness so that modeling a microtubule as a thin shell is only approximates the buckling response [8.72]. Modeling of microtubule buckling as a result of axial compression is approximately the same whether considering an individual filament or bundle.

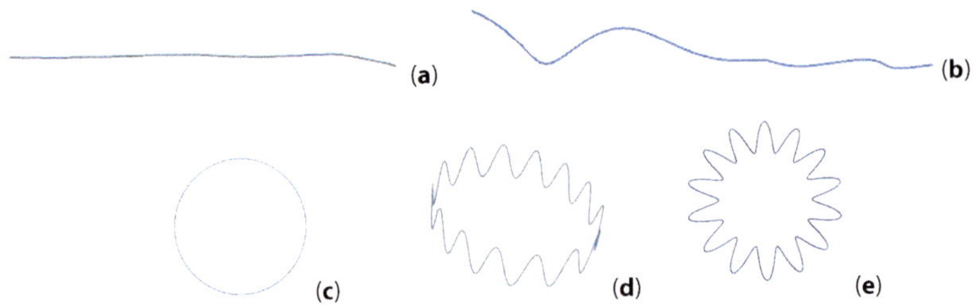

Figure 8.1 Top row: A non-buckled (a) and buckled (b) microtubule filament or bundle (modeled after [8.15], Figure 6a). Bottom row: a non-buckled (c), buckled in the z-direction (d), and buckled in the *x*, *y*-plane (e) actin filament ring.

A microtubule may be thought of as a stiff spring where an optical "bead" travels laterally as the microtubule buckles [8.26]. This is a form of ray tracing. Stretching and bending occur along the microtubule until buckling at the tip [8.107] peaks as a reaction to axial compression. A spring system models the relaxation stage of the microtubule after buckling [8.72]. The buckling event may be hysteretic via damping effects [8.72]. A microtubule bundle can be modeled as a unit undergoing deformation via individual microtubule filaments moving in a non-simultaneous fashion [8.107]. Experimental outcomes have yielded buckling occurring at the microtubule tips, and the response decays along and approaching the opposing end of the bundle [8.15, 8.107]. If microtubules in a bundle are not cross-linked, each exhibits an equivalent compressive response. If cross-linking is present, different buckling times have been found for each microtubule and this enables the bundle to withstand increased compressive forces [8.107].

The path from start to end of the microtubule bundle is linear in the *x*, *y*-plane, while buckling in the *z*-direction is non-linear (Figure 8.1b). A ray may be used to trace the planar path, while an optical "bead" may be used to trace the buckling event. The ray may be modeled as a parametric 2D phenomenon.

8.4.4 Buckling: Actin Filament Ring

In centric diatoms, the perimeter of the SDV is encircled by an actin filament ring, and outside of the SDV, the actin filament ring expands as the SDV expands [8.46, 8.110, 8.114]. As with microtubules, actin is involved in valve morphogenesis at the meso- and microscale levels [8.46]. Actin polymerization is regulated by ring contraction via myosin motors, and constriction is evident during cytokinesis [8.131, 8.132]. Actin filaments shorten as buckling occurs during constriction, and cell shape changes result [8.76]. Buckling serves as the connection between compressive and tensile forces so that mesoscale contraction occurs inducing deformation [8.76]. Although the position of actin is different from microtubules concerning the SDV, actin behaves similarly to microtubules to affect valve surface changes. An actin filament ring is a connected path so that maximum buckling may occur in the *x*, *y*-plane of the valve surface as well as in the *z*-direction out of the valve face plane (Figures 8.1d and e). The *z*-direction of buckling by an actin filament ring affects microscale deformation more readily in contrast to *x*, *y*-planar buckling which squeezes in [8.78] rather than a buckling out. Ray tracings of either path are modeled as closed parametric 2D curves.

8.5 Geometrical and Biophysical Aspects of Buckling and Valve Formation

8.5.1 Buckling: Geometry of Valve Formation as a Multiscale Wave Front

To emulate the valve morphogenesis process from annulus to valve mantle, let the directional growth be modeled as a linear optical ray propagated in an approximately homogeneous medium, amorphous silica. Define an optical ray emanating from the annulus where valve formation propagates from the center to the periphery where the valve mantle forms. Ray propagation is linear indicating directional growth but follows the topography of the formed valve surface that may cover the possibly non-linear surface. Geometrically, the propagation of rays is a form of ray tracing. Multiple rays propagating from the center collectively follow the topography of the surface as a wave front. Growth of the valve surface is a propagation of wave fronts. The surface over which the optical ray travels is non-linear and represented by the eikonal equation [8.78, 8.118, 8.119]. For the special case $f = 1$ (constant "buckling profile" [8.78], i.e., no buckling), then the eikonal equation equals the Euclidean distance function. Overall, valve morphogenetic growth is modeled as directional along a linear ray in the x, y-plane that reveals the non-linear surface being created in the z-direction during valve morphogenesis.

The 3D diatom valve surface is definable via geodesics such that multiple rays from a common center (the center of the annulus) are propagated in all directions from source points along the annulus, to the mantle as wave fronts [8.9, 8.78]. Rays may cross or not along multiple wave fronts indicating the structural surface pattern. The diatom valve surface as a whole cell may exhibit positive curvature, while the surface at a micro- to nanoscale exhibits negative curvature [8.78].

Waviness along an optical ray from annulus to valve margin represents stages of valve formation when buckling occurs. Each optical ray propagation traces the geometric path of buckling as a cross-section of the valve surface. Multiple propagated rays traced similarly record valve formation with respect to buckling. The average difference of waviness as buckling from a flat surface is deformation and recovered via the Jacobian matrix (i.e., Jacobian). The Jacobian determinant is a measure of maximum deformation. For deformation as displacement, the eikonal equation is related to the Jacobian via conformal mapping [8.78]. For an early application of these ideas to morphogenesis, see [8.54].

The buckling motion propagated along a given ray is a displacement via the eikonal equation. From geometrical optics, the eikonal equation provides a connection among propagated rays on a surface from annulus to valve margin for circular centric diatoms (Figure 8.2). The eikonal equation incorporates the Jacobian representation of deformation via conformal mapping so that valve surface area is proportionately preserved [8.78]. As valve formation occurs from annulus to margin, the surface may expand in an exponential manner because of buckling (cf. buckling of cylinders [8.33, 8.91]), and this deformation may be representative of displacement via the eikonal equation and representative of Hamiltonian mechanics [8.118, 8.119]. The degree of buckling on the surface may occur

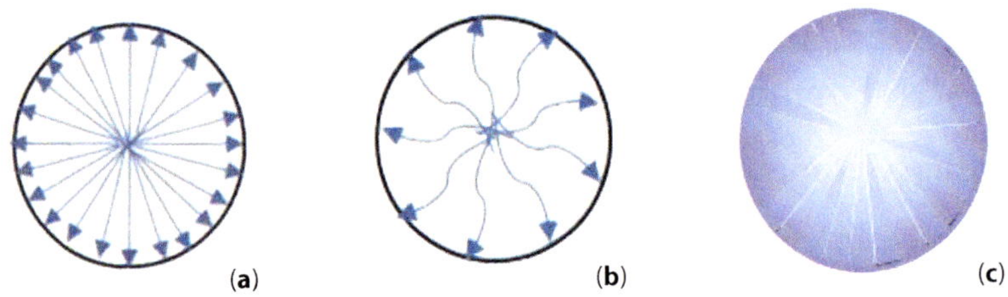

Figure 8.2 (a) Ray tracing valve formation on the valve face from annulus to margin; (b) buckling in the x, y-plane; (c) buckling in the z-direction, where the darker the color, the more buckling on the valve surface.

not only exponentially but also factorially or quadratically compared to a flat surface so that surface buckling is geometrically measurable in modeling valve morphogenesis.

Differential growth is necessary as a precursor to a non-flat surface and is an indicator of buckling. A flat rather than buckled surface may exist very early in growth. Concavity or convexity may not be evident initially. As curving of the surface occurs, a negatively curved, angularity may appear which is the first indication of buckling. At later stages of growth, the surface may exhibit an increasingly negative curvature and an increase in buckling as finite bending occurs. Locally, flat areas of the surface may be apparent, but the global valve surface may exhibit buckling. At the valve mantle, there is the potential of an increase in peaks, saddles and valleys resulting in buckling. If in-plane deformation occurs, squeezing in occurs rather than buckling out [8.78]. As growth proceeds from annulus to valve mantle, silica deposition may induce buckling. As growth is completed, buckling is quenched so that structural instability and failure does not occur [8.78] at the valve mantle as a result of growth.

Amorphous silica, while having elastic properties [8.104], is generally considered to be a stiff material [8.53, 8.104]. Plastic deformation may occur if the compressive stresses on the valve mantle are sufficient to cause breakage [8.60]; however, mature diatom valves have not been found to exhibit plastic properties when tested with nanoindentation techniques [8.1]. SEMs of cleaned, mature diatom valves often contain internal cracks, but these may be caused by internal stresses due to dehydration. Loads to the valve margin may be tangential or perpendicular (normal) to the surface [8.60]. Tangential loading may be frictional, while perpendicular loading may be adhesive [8.60]. In either case, sliding may occur at the valve margin where the tangential and perpendicular loads are offset causing suboptimal compressive stresses or deformation [8.53]. Shear stress is also a consideration, although the dependence of shear stress on sliding has yet to be exhibited at the micro- and nanoscale [8.53]. Lateral stiffness in contrast to contact stiffness is also a factor in shear stress [8.53]. The eikonal equation measures stiffness of silica deposition materials via the Jacobian and bending energy. All of these stresses may occur at the valve mantle inducing buckling.

While diatom silica deposition occurs at the nanoscale, the potential wave front properties of amorphous silica as silica nanospheres being deposited may be characterized as a continuous phenomenon during valve formation. The eikonal equation recovers this continuity, and the mechanics concomitant with this phenomenon are obtained to bridge continuity with molecular dynamics [8.128]. Propagating wave fronts of siliceous material arrive at the valve margin and are halted without reflection, indicating that Hamiltonian mechanics at the nanoscale may characterize continuous silica deposition and the overall approximate molecular dynamics of discrete silica nanosphere deposition. This coupling of mechanics covers the continuity and nanoscale [8.128] nature of silica deposition during valve formation.

8.5.2 Buckling: Valve Formation and Hamiltonian Biophysics

The amorphous silica of diatom frustules is considered to be an approximately homogeneous, isotropic material that co-occurs with other organic biomolecules as valve formation of silica structures occurs. Hamiltonian mechanics is useful with respect to characterizing flow of amorphous silica in valve formation from the annulus to the valve mantle. That is, Hamiltonian mechanics, enabling modeling of silica deposition, is a wave-like process, representing radially occurring propagation of waves in centric diatom valve formation. Silica spheres are deposited and comprise the majority and most important of the materials present, even though organic materials are also present and involved in valve structural features. If extracellular "elastic round microgels" [8.74] could be miniaturized and introduced inside the SDV, say of giant Antarctic diatoms [8.36], then we could possibly directly observe compressive and stress forces within the SDV as morphogenesis proceeds. The inferred appearance of salt crystals inside the SDV in some cases [8.31] suggests that such interventions may be possible. The SDV is accessible to molecular pH probes [8.122], so perhaps larger objects could indeed be inserted.

Silica deposition from annulus to mantle is a dynamic process [8.117] and may be modeled as a time evolution of the valve formation system, starting at the change in state from the annulus to state-changes along a path to the valve mantle. In a Hamiltonian system, equations of motion are based on an action functional via the Lagrangian, which represents all the forces acting on the valve formation system. Advancing from the annulus to mantle monotonically, the action functional is $S(\mathbf{q},t)=\int_{t_1}^{t_2} L[\mathbf{q}(t),\dot{\mathbf{q}}(t),t]dt$ where $\mathbf{q}_1=\mathbf{q}(t_1)$ and $\mathbf{q}_2=\mathbf{q}(t_2)$ are fixed points at fixed times, and L is the Lagrangian. More expansively, the action functional at a fixed initial point and time, $\mathbf{q}_0$ and t_0, respectively, for silica deposition from the annulus to variational endpoint and end time at the mantle is represented by $S_{\mathbf{q}_0,t_0}(\mathbf{q},t)=\int_{\gamma} L\,dt$ the Hamiltonian system for extremal curve γ [8.103] as a minimum with respect to Lagrangian L [8.17]. If the curve is straight, and therefore linear, the action functional records Euclidean length, and Euclidean distance may be used. For a smooth surface, the magnitude of the gradient is $||\nabla S||=1$. Speed of silica deposition along the curve and the amount of time it takes for different structures to appear during valve formation is assumed to be constant.

More formally, the Hamiltonian is $H:\mathbb{R}^n_{\mathbf{p}}\times\mathbb{R}^n_{\mathbf{x}}\to\mathbb{R}$ where $(\mathbf{p},\mathbf{x})\in\mathbb{R}^n_{\mathbf{p}}\times\mathbb{R}^n_{\mathbf{x}}$ and $\mathbf{p}=(p_1,\ldots,p_n)$ and $\mathbf{x}=(x_1,\ldots,x_n)$ are the generalized momentum and position, respectively. Hamilton's equations are $\dot{\mathbf{p}}(t)=-\nabla_x H(\mathbf{p}(t),\mathbf{x}(t))$ and $\dot{\mathbf{x}}(t)=\nabla_{\mathbf{p}}H(\mathbf{p}(t),\mathbf{x}(t))$, where the dot connotes the time derivative. The Lagrangian is $L:\mathbb{R}^n_{\mathbf{v}}\times\mathbb{R}^n_{\mathbf{x}}\to\mathbb{R}$ where $\mathbf{v}$ is the generalized velocity, and L has parameters $L(\mathbf{p},\dot{\mathbf{p}})$ [8.17, 8.103].Using $\mathbf{p}=\nabla_{\mathrm{v}}L(\mathbf{v},\mathbf{x})$, mapping to the Hamiltonian occurs via the Legendre transform as $H(\mathbf{p},\mathbf{x})=\sup_{\mathbf{v}\in\mathbb{R}^n}\{\mathbf{p}\cdot\mathbf{v}-L(\mathbf{v},\mathbf{x})\}$ [8.17] as the time-independent version of the Hamiltonian. For $H=\sum_{i=1}^{n}\dot{q}_i\frac{\partial L}{\partial \dot{q}_i}-L=\sum_i \dot{q}_i\,p_i-L$ the Hamiltonian involves first-order partial differential equations [8.17], while the Lagrangian involves second-order partial differential equations [8.103].

The Hamiltonian and Lagrangian are also applicable to geometrical optics [8.78]. For an optical ray representing the directional linear path that silica non-linearly deposits from annulus to mantle, the optical Lagrangian with respect to the parameterized path is, $L(\mathbf{x},\dot{\mathbf{x}}) = n\frac{ds}{d\mathbf{x}} = n(\mathbf{x}(s))\sqrt{\dot{\mathbf{x}}(s)\dot{\mathbf{x}}(s)}$ where s is the Euclidean distance between two points on the path, ds is the infinitesimal change in distance, and $\dot{\mathbf{x}}(s)^2 = \sum_{i=1}^{D}\left(\frac{dx^i}{ds}\right)^2$ [8.78]. The optical momentum is $p_i = \frac{\partial L}{\partial \dot{x}_i}$ and the optical path length is $s_{optical} = \int_a^b n\, ds = \int \mathbf{p}\cdot ds$ [8.17].

The eikonal equation may be written in the form $(\nabla S(x))^2 = n^2(x))$, which is the same form as the Hamilton-Jacobi equation as a special case in Hamiltonian mechanics for an n-dimensional space [8.78], and may be given as $\nabla S(X) = f(X)$, $X \in \Omega$ with boundary condition $S|_{\partial\Omega} = U(X)$, where Ω is an open subset of $\mathbb{R}^D$, $f(X)$ is a positive valued forcing function, and ∇ is a gradient operator. Hamiltonian mechanics is used to associate propagation of optical rays with the motion of particles. Propagation as a wave front and determination of geodesics as a non-linear first-order partial differential equation is determined by a Hamilton-Jacobi equation as an approximation to the eikonal equation.

Valve formation as a time-evolved event involves the geodesics on the surface, indicating the monotonic motion of an advancing wave front of silica depositional particles from annulus to mantle following variational optical paths. That is, a wave front is an evolving front [8.103]. The optical length also is expressible as $S_{optical} = \int_x^{x+dx} L(\mathbf{x},\dot{\mathbf{x}})\, ds$ and this is the eikonal equation [8.78]. The traveling motion of a wave front is representationally equivalent to geometrical optical rays and is a minimization of the action functional represented by the eikonal equation. Minimization of the Lagrangian is recovered via the eikonal equation, and the optimal buckling surface is determined via the metric tensor [8.78]. Optimal ray propagation is a Hamiltonian flow during valve morphogenesis. The eikonal equation measures the surface height action of a particle moving in the direction of valve morphogenesis of a centric diatom, that is, buckling, as a geometric displacement.

8.5.3 Buckling: Valve Formation and Deformation Gradients

Bucking is evidenced by deformation and is measurable as strain [8.55]. Moment of inertia is a measure of resistance to buckling [8.55]. For a diatom valve surface, the change in height of silica as deposition occurs during valve formation is the strain rate, and the change in the rate at which silica deposition occurs is the strain rate tensor. This tensor measures the rate of deformation and is an indicator of buckling. As silica is deposited during valve formation, a rate of rotation may occur as well and is measured as the spin tensor (anti-symmetric), but this component does not contribute to deformation. The strain rate and spin tensors may be defined with regard to the velocity gradient, ∇v, at relatively small changes in position of materials elements and are measured as $E_{strain\ rate} = \frac{1}{2}[(\nabla v)+(\nabla v)^{\mathrm{T}}]$

and $W_{rotation\ rate} = \frac{1}{2}[(\nabla v) - (\nabla v)^T]$, respectively. The velocity gradient is a measure of the rate of deformation.

Deformation is a change in shape, while displacement is change in position. Shape change may be construed to be a spatial change, and therefore a position change. A deformation gradient is a displacement gradient. The deformation gradient is a change in length and representative of buckling of solid materials such as silica. The Jacobian is a measure of deformation and is the generalized gradient [8.59]. The Jacobian determinant is a measure of maximum deformation. Jacobian deformation is the determinant of the deformation gradient tensor which is the ratio of deformed to undeformed silica elements of which the diatom valve consists. The deformation gradient tensor may be equilibrated to the strain rate tensor as $E_{Jacobian} = \frac{1}{2}[(J) + (J)^T]$, where J is the Jacobian. For buckling derived from a model with parameters u, v, $J^2(u, v) = n^2(u, v) + 1$ is a representation of a uniformly compressive surface that is flat, i.e., unbuckled. A negative Jacobian indicates squeezing in as buckling in the plane rather than buckling out [8.78]. The Jacobian may be applied to displacement as a deformation gradient concerning buckling microtubules and microtubule bundles or actin filaments. As such, the strain rate tensor is calculable from a parametric representation of these cytoskeletal elements in the same manner as it is for buckling on the siliceous valve surface. Strain is expressible as displacement.

Deformation is also involved with respect to bending energy. The Helfrich Hamiltonian is a gradient approximation that represents the total elastic energy as $F_{Helfrich-Hamiltonian} = \frac{k_c}{2}\oint(2H - c_0)^2\, dA$ and the elastic energy of curvature is, $F_c = \frac{k_c}{2}(2H - c_0)^2 + k\,K$, where k_c is the bending modulus, c_0 is the mean curvature with no deformations, dA is the area element, H is mean curvature, and K is Gaussian curvature [8.14, 8.106]. Total bending energy is $E_{bending} = \int dA\left(\frac{1}{2}\kappa\, H^2 + \bar{\kappa}K\right)$ [8.19]. The gradient of the bending angle is represented by the rows of the Jacobian [8.38]. The Jacobian with respect to bending energy on the parametric surface is, $J(u,v) = \sqrt{1 + (\partial_u f)^2 + (\partial_v f)^2}$ and more generally, [8.77, 8.78], $J = \sqrt{EG - F^2}$ where E, F, and G are coefficients of the first fundamental form and $g = \begin{bmatrix} E & F \\ F & G \end{bmatrix} = \mathbf{J}^T\mathbf{J}$ relates the metric tensor, g, with the Jacobian [8.20]. The Jacobian of the bending forces is $\nabla\left[(-\nabla E_{bending})^T\right]$ as the gradient of the bending energy [8.109]. While we expect the first phase of valve morphogenesis, with a thin, perhaps buckled, nascent valve, to exhibit bending energy, its subsequent thickening may mask or even eliminate this energy, stabilizing the buckled shape (cf. [8.35]).

The Jacobian is symplectic as $J^{-1} = -\omega J^T\,\omega$, where $\omega = \begin{bmatrix} 0 & \mathbf{I} \\ -\mathbf{I} & 0 \end{bmatrix}$ and is a linearization of a Hamiltonian flow [8.38]. As a result, a Hamiltonian flow is a canonical transformation as a time evolution event where the flow is the wave front in motion. That is, a canonical transformation as a linearization of a Hamiltonian flow is a time evolution that is a symplectic transformation [8.38, 8.109, 8.129]. Thus, bending and deformation that occur as a Hamiltonian are reflected via the Jacobian so that the Jacobian with respect to a Hamiltonian flow is a measure of buckling during valve morphogenesis.

8.5.4 Buckling: Multiscale Measurement With Respect to Valve Formation

The model of optical ray propagation as a Hamiltonian system of deformation and bending during valve surface formation of morphogenesis may be summarized in the Jacobian as a common feature of each mode of buckling. The Jacobian can be used to assess 3D centric diatom surfaces with respect to deformation and their strain rate tensor, as buckling occurs in the valve surface plane as well as out of the valve surface plane. First partial derivatives such as vector quantities of microtubule filaments or bundles and actin filament ring behavior during valve morphogenesis may be used as geometric representatives of their buckling behavior.

The Jacobian as a generalized gradient is a measure of the changes in tangent lines and planes on a 3D surface [8.59]. As such, the Jacobian is a measure of deformation and therefore a geometric measure of buckling. Valve surface buckling is represented by the strain rate tensor as $E_{Jacobian} = \frac{1}{2}[(J)+(J)^{T}]$ where J is the Jacobian. Buckling of microtubules and the actin filament ring are displacements and relatable to a buckled valve surface via a linear system [8.113]. Again, the Jacobian is a biophysical measure of buckling.

8.5.5 Buckling: Krylov Methods and Association of Valve Surface Buckling With Microtubule and Actin Buckling

To associate buckled microtubules and actin filaments to buckled 3D centric diatom valves, an iterative method is necessary to solve this nonsymmetric problem involving tensors, matrices, and vectors. To facilitate convergence, preconditioning of the nonsymmetric system may be necessary. By treating the problem via successive approximations, an approximate linear solution may be achieved. Krylov methods [8.66] are useful with dense non-symmetric [8.116] or sparse matrices [8.105] and enable a solution in a least squares sense [8.52, 8.105]. Iteration occurs such that a higher-dimensional problem is projected into a lower-dimensional subspace. That is, Krylov subspaces solve the original problem as a series of smaller matrix problems [8.130]. A Krylov subspace is $\mathcal{K}_k(A,c) \equiv span\{c, Ac, \ldots, A^{k-1}c\}$, constructed from the linear system, $Ax = b$, with an approximate solution, x_k, via vector, c [8.52, 8.95, 8.105].

Via Arnoldi iteration, QR-factorization is used to produce upper Hessenberg matrices via orthogonalization and orthonormalization [8.52, 8.105]. Arnoldi iteration is a form of a modified Gram-Schmidt process, and the vectors formed are the bases of the successive Krylov subspaces generated [8.95]. The Arnoldi process uses a generalized minimal residual [8.95] or biconjugate gradient (BiCGSTAB) stabilized [8.116] linear systems to form Krylov subspaces. GMRES is achieved via solving the least squares problem $\min_{z \in \mathcal{K}_k(A,\,b)} \|b - Az\|$ where the residual norm is small enough [8.52]. Orthogonalization and orthonormalization of vectors $\{v_1, v_2, \ldots, v_k\}$ for $\mathcal{K}_k(A,b)$ produce new basis vectors, $z = AV_j$ [8.105], so that $AV_j = V_{j+1}H_j$, where H_j is a $(j + 1) \times j$ upper Hessenberg matrix [8.52]. The conjugate transpose is used in place of the transpose for complex values. BiCGSTAB is a method of alternating biconjugate gradient and GMRES steps in the algorithm [8.7, 8.116]. Krylov subspaces are linear combinations of the residual vectors [8.40].

The upper Hessenberg matrices are the corresponding projections into the Krylov subspaces [8.130]. The diagonals of the upper Hessenberg matrices usually contain extreme eigenvalues of the spectrum [8.130]. The eigenvalues may be evaluated to indicate growing, decaying or oscillating buckling for positive, negative, or imaginary values, respectively; eigenvalues connote expansion or contraction [8.42].

For valve formation, Krylov methods are used to produce an association among 3D valve surfaces, microtubule and actin filament buckling. The matrix representing valve buckling is the Jacobian, and the vectors representing microtubule and actin filament buckling consist of first derivatives. To determine the vector **x** as the association of microtubule and actin filament buckling with valve surface buckling, we apply the Jacobian, **J**, and the linear system, **Ae** = **x** is solved by $\mathbf{A} = \mathbf{J}^{-1}$ [8.113]. The resultant upper Hessenberg matrices are evaluated via the eigenvalues. In this regard, the eigenvalues characterize buckling.

Similarly, the strain rate tensor is used in place of the Jacobian to evaluate buckling associated with the valve surface, microtubule, and actin filaments. While the evaluation of buckling with the Jacobian is geometric, the evaluation of buckling with the strain rate tensor is a measure of biophysical buckling.

8.6 Methods

8.6.1 Constructing and Analyzing 3D Valve Surface and 2D Microtubule and Actin Filament Models

Systems of parametric 3D equations are used to construct 3D surface models of centric diatom valves with boundary conditions $u, v \in [0,2\pi]$ [8.79, 8.80, 8.81, 8.82, 8.84]. For a 3D valve surface model, the 3×2 Jacobian is made square by using the magnitude of the cross product of row elements to form the third column [8.82], which is a deformation gradient. The Jacobian is a displacement gradient as well, showing the change in length and angles of deformable materials of the valve surface relative to a smooth surface. From this, Jacobian deformation as the determinant of the deformation gradient tensor represents the ratio of deformed to undeformed infinitesimal volume elements of the materials. The Jacobian determinant is a coefficient of maximum deformation and a summary value representing buckling.

The Jacobian is used as a geometric representation of buckling for three centric diatom exemplar taxon 3D models: *Actinoptychus*, *Arachnoidiscus*, and *Cyclotella*. Cylindrical centric diatom models provide the simplest case because there is no need to consider valve perimeter shape change. The square non-symmetric Jacobian for each model is used in calculating the strain rate tensor for each exemplar. The strain rate tensor is a biophysical measure of buckling for each exemplar taxon. Evaluation for each exemplar taxon and comparisons among taxa are made to determine the contribution of buckling to valve formation.

For microtubules and microtubule bundles, displacement is shortening of the microtubule bundle as recovered via the first derivative of the curve fit to its buckled structure. A microtubule bundle buckles similarly to a single microtubule as deformation at the tip [8.15, 8.107]. Experimental results of a buckled microtubule filament from Brangwynne *et al.* [8.15] as depicted in their Figure 6A are used to determine a curve fit using a system of parametric 2D equations. Similarly, a curve fit of actin filaments in a ring is determined via a system of parametric 2D equations of rippling as displacement occurring perpendicular to a planar circular ring. The microtubule filament or bundle and actin filament ring in a buckled state represent a given stage of polymerization during valve formation. Conceptually, the curve fits for a microtubule and actin filament ring represent displacement such that each model consists of motion of a particle along the microtubule or actin filament ring as a wave and represents contraction of each structure during buckling. Elastic displacement is the general response that occurs in each of these structures when buckling occurs [8.72]. For microtubule filaments or bundles, and actin filament rings, first-order derivatives are obtained from systems of parametric 2D equations with boundary conditions $u \in [0, 2\pi]$, and these results form the elements of displacement vectors representing buckling.

8.6.2 Krylov Methods: Associating Valve Surface With Microtubule and Actin Filament Buckling

To associate valve surface buckling with microtubule and actin filament buckling, Krylov methods were instituted. For each 3D valve surface model exemplar taxon—*Actinoptychus*, *Arachnoidiscus*, and *Cyclotella*—the Jacobian 3×3 matrix was used in conjunction with displacement vectors for microtubules. The same procedure was used with displacement vectors for the actin filament ring. Comparisons were made between non-buckled and buckled microtubules and actin filament rings. Krylov methods were also used in conjunction with the strain rate tensor compiled in matrix form for each exemplar taxon in the same way as was calculated for the Jacobian. BiCGSTAB was the methodology used in Krylov iteration. An upper Hessenberg matrix was determined for each exemplar taxon via combination of the valve surface, microtubule and actin filament ring buckling. Both the Jacobian and strain rate tensor were used. Non-buckling among valve surface, microtubule and actin filament ring association was used as a comparative starting point in valve morphogenesis for each exemplar taxon.

8.7 Results

For *Actinoptychus*, *Arachnoidiscus*, and *Cyclotella*, 3D surface models were constructed using parametric equations (Figure 8.3 and Table 8.1). Jacobians and strain rate tensors were calculated. For the Jacobian, the changes in tangent lines and planes from parameters u and v indicate that the base layer horizontal buckling is most prevalent in *Actinoptychus*,

Figure 8.3 From left to right, centric diatom 3D surface models of *Actinoptychus*, *Arachnoidiscus*, and *Cyclotella*.

Table 8.1 Parametric 3D equations on the interval $u, v \in [0, 2\pi]$ for a non-buckled disk and exemplar centric diatom taxa *Actinoptychus, Arachnoidiscus,* and *Cyclotella.* The x and y equations for all taxa are $80 \cos v(1 + \sin u)\, 0.2 \cos u$ and $80 \sin v(1 + \sin u)\, 0.2 \cos u$, respectively, and the z equation is given below. The *x, y* and *z* equations for the disk are given in the table.

Taxon	**z equation**[†]
Disk	$x = \cos v + \sin u,\ y = \sin v + \cos u,\ z = 0.5 \cos v + 0.5 \sin u$
Actinoptychus[†]	$-\alpha u \cos 3u \cos 3v + \beta \sin 3u - \gamma \sin(6u)^5$ $-\zeta u \cos(5 - 4u)\sin(3 - 4v)^4 + \mu \sin 120u \sin 6v$ $-\varphi \cos(0.5u)^2 \cos(40u)^4 \cos 3v \sin 40u \sin(40v)^3$ $-\psi u \sin 40u \sin 160v$
Arachnoidiscus[†]	$\nu \cos(0.5u^2) + \rho \cos(80u)^2 \sin(1.9u)^2 + \Xi \sin(1 + u)^3$ $+ \tau \cos(10 + 74u^2)\sin(0.5 + 0.8u)^3 \sin(2 + 2.9u)^3$ $+ \vartheta \cos u \sin(0.37u)^3 \sin(22v)^3$
Cyclotella[†]	$\cos 1.5u[\xi \cos 2u \sin 0.5u \sin 2.9u + 0.14 \cos u \sin(0.5u)^4 \sin(7v)^4]$ $+\eta \cos(3.5u)^6 \cos 25v \sin 1.8u^6 \sin 25v$

[†] z-coefficients: $0.0008 \le \alpha \le 0.64$, $0.01 \le \beta \le 0.1$, $0.001 \le \gamma \le 1$, $0.003 \le \zeta \le 0.09$, $0.0025 \le \mu \le 0.125$, $0.2 \le \varphi \le 10$, $0.0025 \le \psi \le 0.025$, $0.004 \le \vartheta \le 0.7$, $0.05 \le \nu \le 0.75$, $0.025 \le \rho \le 0.5$, $0.051 \le \tau \le 2.55$, $2 \le \Xi \le 3$, $0.01 \le \xi \le 0.5$, $0.18 \le \eta \le 3$.

followed by *Arachnoidiscus*, then *Cyclotella* (Figure 8.4). From the schema of Schmid and Volcani [8.102], a large change in buckling occurs between stages 1 and 2 in *Arachnoidiscus*, while buckling is most prevalent in stage 3 for *Cyclotella*. Large sector undulations in *Actinoptychus* are indicated by buckling throughout valve formation in contrast to selected buckling such as the ribs in *Arachnoidiscus* and marginal finer undulations in *Cyclotella* (Figure 8.4).

The Jacobian determinant as coefficients of deformation representing maximum buckling was calculated for each exemplar taxon. A best-fit curve was determined exemplifying different modes of buckling during valve formation. For *Actinoptychus*, a positive quadratic curve fit resulted (Figure 8.5a). A logarithmic curve fit was found for *Arachnoidiscus* (Figure 8.5b). A negative quadratic curve fit resulted for *Cyclotella* (Figure 8.5c). The *Actinoptychus* valve surface buckling grew throughout valve formation, while it leveled off almost asymptotically for *Arachnoidiscus* and *Cyclotella*.

Using the Jacobian determinant, comparative valve surface buckling during formation was determined in a combined plot for the exemplar taxa. In terms of magnitude, *Arachnoidiscus* had the most buckling early in valve formation (Figure 8.6). For *Actinoptychus* and *Cyclotella*, buckling was slight and gradually increased to almost steady during valve formation (Figure 8.6).

Looking at strain rate tensors for the three exemplar taxa, we see that buckling was mostly in the *z*-direction (Figure 8.7). These results reinforce the findings from comparative valve surface buckling via the Jacobian determinant (Figure 8.6). For the most part, buckling in the *x, y* valve plane is similar in *Arachnoidiscus, Actinoptychus,* and *Cyclotella* where there is a steady indication of buckling throughout valve formation (Figure 8.7). That is, planar buckling is not very different among the exemplar taxa. The ribs of *Arachnoidiscus*, the large sector undulations of *Actinoptychus* valve, and the finer marginal undulations of *Cyclotella* reflect the approximately equidistant spacing of these valve surface features. Comparing

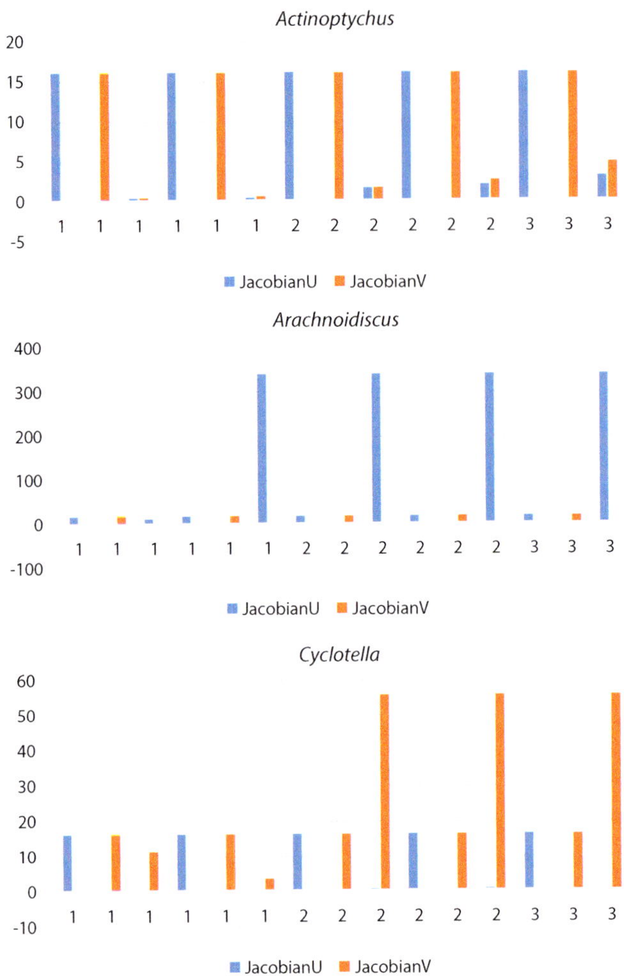

Figure 8.4 Change in tangent lines and planes during valve formation indicate that base layer horizontal changes are more prevalent in *Actinoptychus* than in *Arachnoidiscus* and *Cyclotella*. The latter two plots indicate more changes vertically in valve formation.

exemplar taxa via the strain rate tensor, less buckling occurred during base layer formation than during later valve formation where there was more buckling in the *z*-direction (Figure 8.8).

Non-buckled and buckled microtubule and actin ring filaments were simulated using parametric 2D equations (Figure 8.1 and Table 8.2). Non-buckled microtubules were approximately straight lines (Figure 8.1a). The buckled form was based on a model of Brangwynne *et al.* [8.15], their Figure 6A (Figure 8.1b). Actin filament ring models were based on a circle for non-buckling, buckling in the *z*-direction and *x*, *y*-plane (Figures 8.1c–e).

Microtubule and actin filament ring models were used in combined buckling analysis with exemplar taxa. To set a starting base of no buckling with respect to the valve surface, non-buckling microtubule and actin filament ring vectors, **x**, were associated with a flat

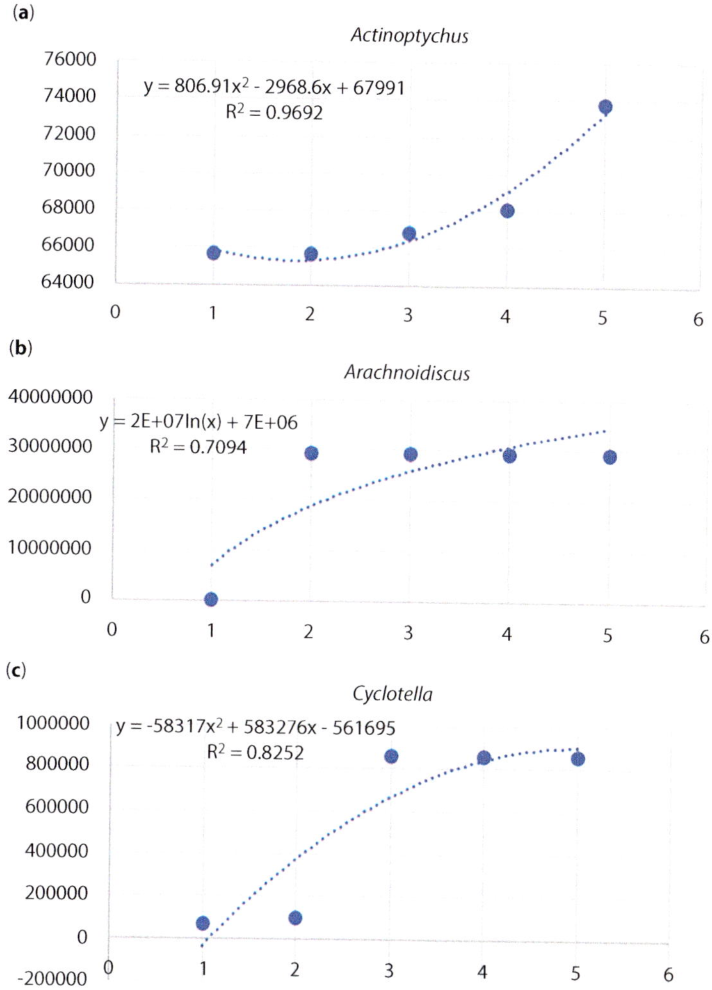

Figure 8.5 Plots of *Actinoptychus*, *Arachnoidiscus*, and *Cyclotella* buckling during valve formation. A quadratic best-fit curve is indicated for *Actinoptychus* as positive growth and for *Cyclotella* as asymptotic. E + 07 = 10^7, etc.

non-buckled disk using Krylov methods. The Jacobian and strain rate tensor were calculated for the disk as the input matrix, *A*.

For buckling, Krylov methods were used to associate each exemplar taxon with buckled microtubule and actin filament ring vectors. The Jacobian and strain rate tensors were used as the input matrix, A. Buckling vector, *b*, was determined via Arnoldi iteration, producing orthogonalized and orthonormalized results. The residual norm for all runs achieved convergence to zero.

Upper Hessenberg matrix plots for each of the exemplar taxa were used to interpret non-buckling vs. buckling associated with microtubules and the actin filament ring. The values used for the plots are the non-zero values of the upper Hessenberg matrices. Eigenvalues are reported in Table 8.3 for non-buckling and buckling (Figure 8.9).

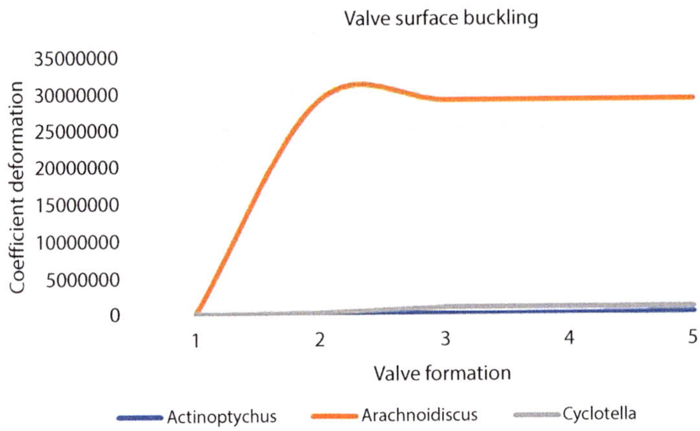

Figure 8.6 The Jacobian determinant as the coefficient of deformation during valve formation. The higher the coefficient of deformation, the more buckling on the surface. *Arachnoidiscus* has higher deformation than *Actinoptychus* or *Cyclotella*. For *Arachnoidiscus*, buckling is greatest in the base layer during valve formation. For *Actinoptychus* and *Cyclotella*, buckling gradually occurs during valve formation.

Based on the Jacobian, all eigenvalues are negative for all taxa, while for the strain rate tensor, eigenvalues are positive for *Actinoptychus* and *Cyclotella* and negative for *Arachnoidiscus* (Table 8.3). The difference between non-buckling and buckling is positive for *Actinoptychus*, indicating an increase in buckling during valve formation. A negative value for Arachnoidiscus indicates decay in buckling during valve formation. These results are in agreement between Jacobian and strain rate tensor evaluation. For *Cyclotella*, geometrical buckling via the Jacobian is decaying, but biophysical buckling via the strain rate tensor is growing (Table 8.3).

From the upper Hessenberg matrices, curves with an almost flat slope represent no buckling. For *Actinoptychus*, *Arachnoidiscus*, and *Cyclotella*, the Jacobian buckling changes as a peak relative to a non-buckling curve (Figures 8.10a, c, and e). The plot with the strain rate tensor accentuates this result for *Actinoptychus* and *Cyclotella* (Figures 8.10b and f). However, for *Arachnoidiscus*, buckling occurs asymptotically in contrast to a flat line for non-buckling (Figure 8.10d).

Actinoptychus and *Cyclotella* register more buckling via the Jacobian in Krylov analysis relative to *Arachnoidiscus* when all taxa are compared relative to each other (Figure 8.11). A greater change in buckling from early to later stage morphogenesis may be inferred for *Arachnoidiscus* when comparing all taxa with regard to Krylov analysis using the strain rate tensor (Figure 8.11b).

8.8 Conclusion

We presented a computational modeling approach for evaluating buckling during centric diatom valve morphogenesis. This approach included a geometrical as well as a biophysical aspect. While we cannot match specific buckling events by microtubules or actin filament rings to valve formation events such as the creation of rimoportulae or cribra, we can evaluate buckling as it is associated to valve morphogenetic stages as delineated by [8.46, 8.102].

The exemplar centric diatom taxa modeled were indicative of different types of buckling evident on their valve face. *Actinoptychus*, *Arachnoidiscus*, and *Cyclotella* exhibit geometrical buckling as large undulating sectors, radiating undulating ribs and fine undulations

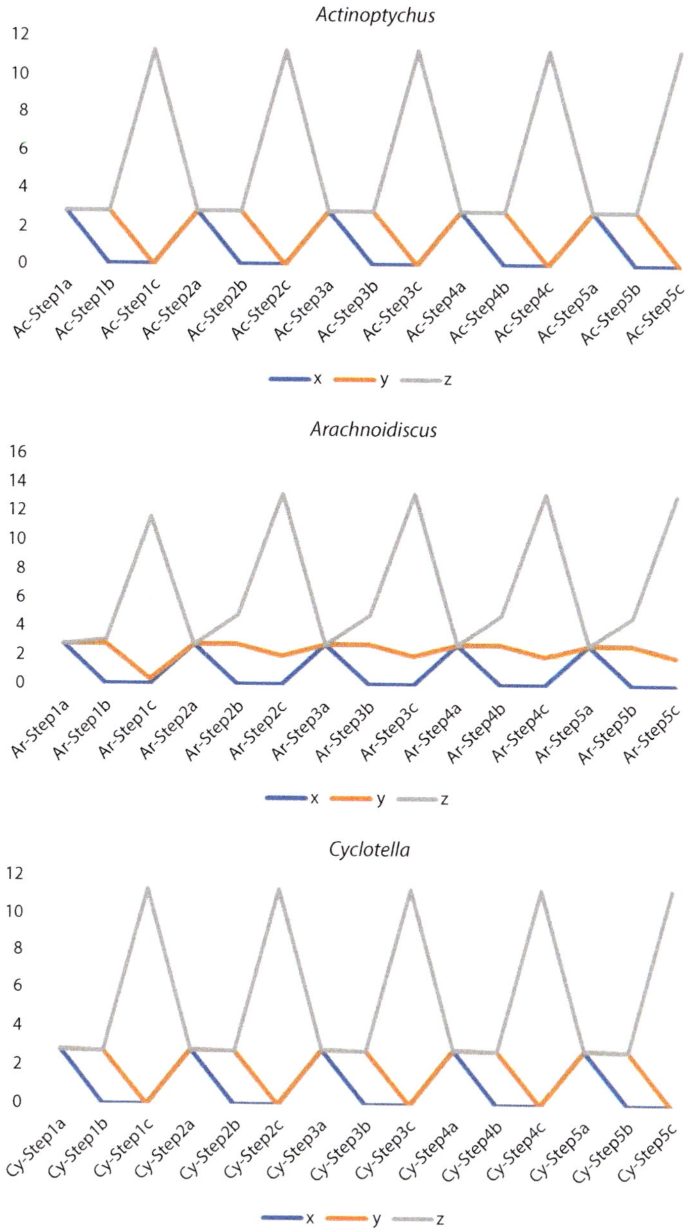

Figure 8.7 Strain rate tensor deformation for *Actinoptychus*, *Arachnoidiscus*, and *Cyclotella*. For *Actinoptychus*, deformation is regular during valve formation, reflecting the large undulating sectors of the valve surface. For *Arachnoidiscus*, deformation is less pronounced early in valve formation. For *Cyclotella*, deformation is regular during valve formation because of the regular undulations near the valve margin.

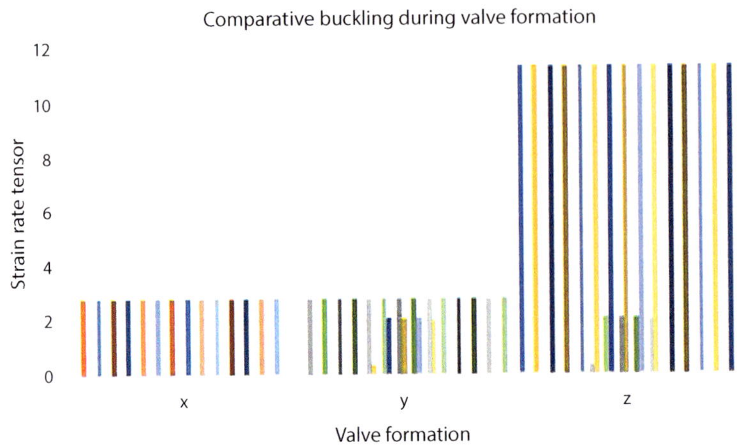

Figure 8.8 Combined strain rate tensor deformation for *Actinoptychus, Arachnoidiscus,* and *Cyclotella.* Deformation in the x, y-plane indicates less buckling during base layer formation in contrast to vertical buckling in the z-direction.

Table 8.2 Parametric 2D equations on the interval $u \in [0, 2\pi]$ for non-buckled and buckled microtubule and actin filament rings.

Taxon	x, y, z equations
Non-buckled microtubule filament or bundle	$1.1u \sin(2.6u + 2.5) - 2.2u^2 - 7u - 0.5 \cos(0.1u + 1)$, $0.5u \cos(0.9u) - 0.5u^2 + 2u$, $0.5u \cos(2.5u + 1.5) - 0.8u^2 - u + 0.5 \cos(0.1 + 5)$
Buckled microtubule filament or bundle	$1.2u \sin(3u + 3) - 2.5u^2 - 9u$, $u \cos(4u) - 0.5u^2$, $u \cos(2.5u + 2) - 4\mathrm{u}$
Non-buckled actin filament ring	$4 + 0.2 \cos(2u) \cos u$, $4 + 0.2 \cos(2u) \sin u$, $\sin u$
Buckled actin filament ring	$4 + 0.2 \cos(2u) \cos u$, $4 + 0.2 \cos(2u) \sin u$, $\sin 14\, u$

close to the valve margin, respectively. The utility of 3D surface models has been evident in diatom studies of complexity [8.82], symmetry [8.85], and 3D morphological changes during morphogenesis [8.83]. Our computational modeling of these taxa and their concomitant buckling scenarios with respect to valve formation were accurately determined and differentiated from one another. The utility of Krylov methods is evident when combining disparate pieces of data into a least-squares analysis, so that the association among valve face, microtubule, and actin filament ring buckling could be achieved.

At early growth during valve morphogenesis [8.102] (stage 1), buckling may be absent as the valve surface is nearly flat. Instability may be present. A zero Jacobian (= flatness) approaching a negative Jacobian indicates an increase in stiffness because of an increase in

Table 8.3 Eigenvalues from upper Hessenberg matrices for centric diatom exemplar taxa *Actinoptychus*, *Arachnoidiscus*, and *Cyclotella* associated with non-buckled and buckled microtubule and actin filament rings.

Jacobian			
	non-buckling	**buckling**	**difference**
Actinoptychus	−0.676118	−0.49697	0.179147
Arachnoidiscus	−0.704868	−1.05503	−0.35016
Cyclotella	−0.331786	−0.48549	−0.1537
Strain Rate Tensor			
	non-buckling	**buckling**	**difference**
Actinoptychus	3.86051	12.2307	8.37019
Arachnoidiscus	−3.90363	−645995	−645991
Cyclotella	1.20315	3.5087	2.30555

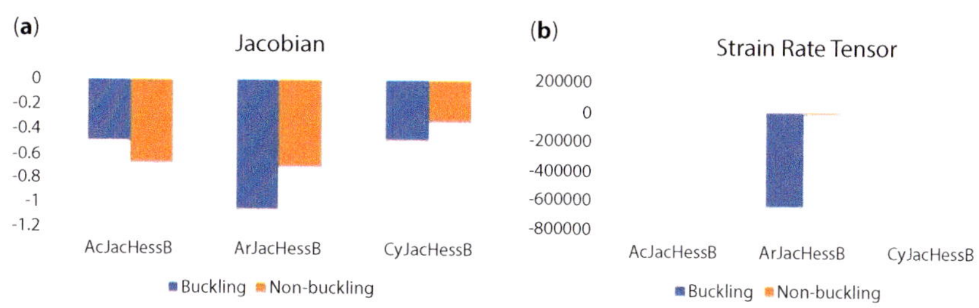

Figure 8.9 Upper Hessenberg matrix eigenvalues for *Actinoptychus, Arachnoidiscus* and *Cyclotella* via the Jacobian (a) and the strain rate tensor (b).

buckling [8.78]. As buckling increases toward the valve mantle in conjunction with Schmid and Volcani [8.102] stage 3 of valve morphogenesis, the diatom cell structure becomes more stable. Proliferation in buckling as late stage growth slows to complete the valve formation process, and this provides stability for the diatom cell. At the valve mantle, buckling is quenched so that instability and breakage do not occur as a result of growth; that is, structural disorder does not occur at the valve mantle [8.1]. Buckling is an integral part of valve morphogenesis.

Geometrically, buckling out means constant negative curvature of the surface [8.78]. The metric tensor encompasses parameterized geodesics defining the paths of ray propagation. Christoffel symbols are a measure of the parameterized ray paths. Interpreting the eikonal equation in terms of buckling, the metric tensor can be obtained in Cartesian coordinates as $g^{ik} = \begin{pmatrix} n^2 & 0 & 0 \\ 0 & 1 & 0 \\ 0 & 0 & 1 \end{pmatrix}$. From this, the geodesics that define the optical rays (propagations) on the surface are derived from $\frac{d^2x^\lambda}{d\tau^2} + \Gamma^\lambda_{ij}\frac{dx^i}{d\tau}\frac{dx^j}{d\tau} = 0$ where

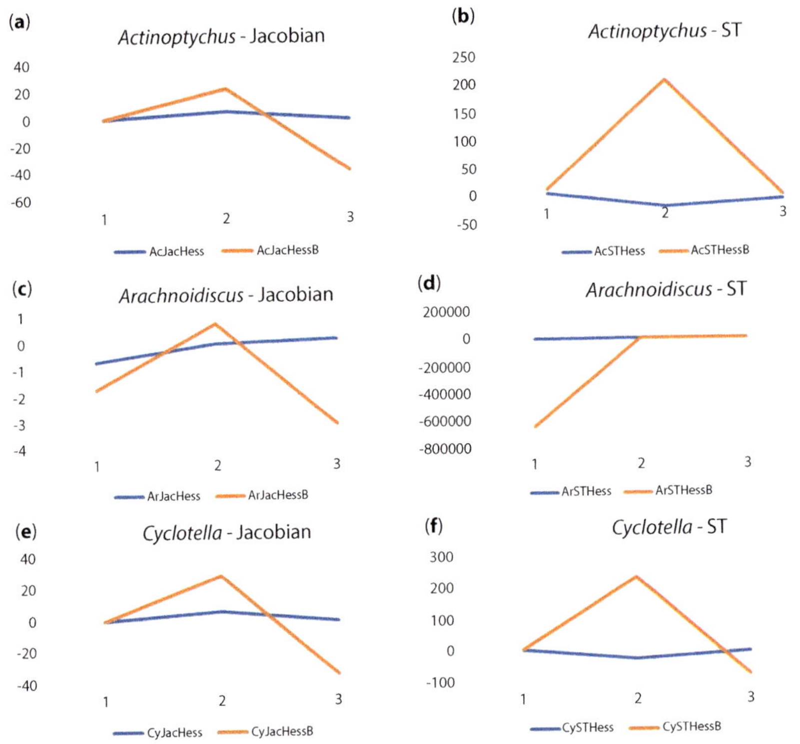

Figure 8.10 Non-buckling (blue curve) and buckling (orange curve) for *Actinoptychus* (a, b), *Arachnoidiscus* (c, d), and *Cyclotella* (e, f) via the Jacobian (first column) and the strain rate tensor (second column).

$\Gamma^i_{kl} = \frac{1}{2} g^{im} \left(\frac{\partial g_{mk}}{\partial x^l} + \frac{\partial g_{ml}}{\partial x^k} - \frac{\partial g_{kl}}{\partial x^m} \right)$ are Christoffel symbols with g_{ij} covariant and $g_{ij} g^{jk} = \delta_i^k$, where δ is the Kronecker delta [8.59]. Christoffel symbols are connection coefficients of points on the surface, and the Jacobian is the gradient of those connection points [8.20]. Surface buckling is extractable from valve morphology via its relation to valve surface geometrical characteristics as grouped composites of varying sections of valve surface heights [8.83].

Biophysically, buckling during morphogenesis also may be viewed with respect to the whole diatom cell and the effects of the surrounding environment. For example, at the latest stage of valve formation in *Coscinodiscus wailesii*, turgor increases and plasma membranes may fold outward at the mantle at the time interconnected daughter cells are adhered to the mother cell's girdle bands and are about to be released [8.99]. Instantly thereafter, these folded out mantles are retracted because of tensile forces as the plasma membranes are released and exocytosis of the valve occurs [8.99]. Compressive forces between daughter cells influence silica structure and ultimately cell shape as well [8.114]. Compression is offset by an exterior mucilage coating of the cell [8.99]. The folding out of plasma membranes is a kind of buckling as the final stage of silicification occurs and the valve surface and shape are complete.

Convexity or concavity of the valve face may affect valve mantle formation with respect to buckling. As buckling increases in the z-direction on the valve face at the valve mantle,

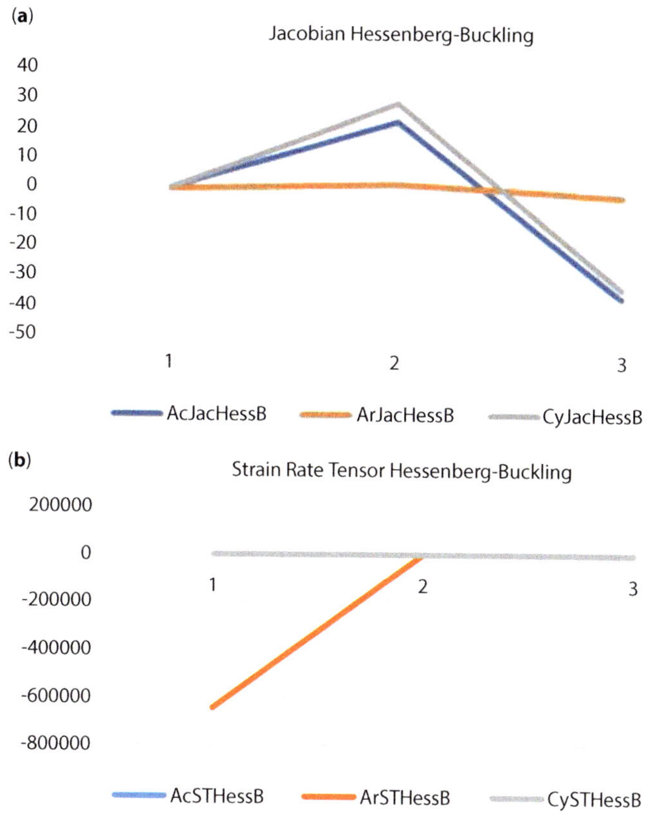

Figure 8.11 Upper Hessenberg matrices for all taxa based on the Jacobian (a) and the strain rate tensor (b).

the x, y-plane of the diatom valve face may bend upward or downward, reflecting the overall cell shape. As a measurable quantity, the valve mantle buckling profile with respect to valve face convexity or concavity may be represented by an expansion of a polynomial, and factorization of the polynomial would produce coefficients representative of buckling extent. Orthogonal transformation would produce coefficients that are independent descriptors of valve mantle buckling with respect to overall valve shape convexity and/or concavity. The same principal could be used with regard to pre-buckling and post-buckling behavior of the diatom cell as valve morphogenesis occurs. Comparison of sets of coefficients may be used as an overall indicator of degree of buckling of a given diatom cell, for example, as radial buckling [8.35].

Focusing on the valve mantle, a diatom cell has distinct characteristics that may be used to understand potential compressive or elastic forces as a result of cytoskeleton formation during silica deposition. The concomitant behavior of cytoplasmic structures and valve face patterning during morphogenesis influences the resultant valve mantle structure. The valve mantle may undergo buckling in centric diatoms recovering compressive resistance to tensile forces occurring via turgor within the diatom cell and externally as a result of exocytosis at the end of the valve formation process during morphogenesis.

Purely viscous materials may be modeled as well. Local minima are optima found as viscosity solutions to the Hamilton-Jacobi equation, and these solutions are the discretized solutions to a continuous eikonal equation [8.9]. With viscosity solutions, diffraction may be used to explain the valve structural surface pattern from annulus to valve mantle as multiscale wave fronts of buckling. Arrival at the valve mantle results in an abrupt halt and deflection downward, rather than reflection of the multiple rays back on themselves. This is perhaps accompanied by loss of compression of the actin ring against the radial microtubules [8.34]. At this point, mantle formation is a cascade of descending silica depositions that may result in distorted areolae or pores [8.8] or undulations of parallel rows of areolae pores [8.71], or irregularly spaced areolae or pores [8.8]. Viscoelastic as well as viscous responses may occur during silica deposition where moduli vary with varying direction at the valve mantle. Time dependence of mechanical responses to compressive stresses may result in a lag time in strain deformation responses [8.53] at the valve mantle when buckling occurs.

In valve morphogenesis, buckling behavior may be different among vegetative, initial, and auxospore cells, possibly explaining the sequence of forms as vegetative morphology is restored. Buckling may be an avenue to distinguish biophysical as well as geometrical differences in formation of different cell types. The same may be determined for teratogenic or other aberrant cells formed during morphogenesis. Experimental evidence suggests that microtubule inhibitors may affect valve formation and the production of teratogenic forms [8.15, 8.107].

Propagation of rays as ray tracing is a useful way to model valve morphogenesis in order to determine and evaluate buckling. This same modeling technique is applicable to pennate diatom morphogenesis using a central line representing the sternum where the raphe would form (or not) as the starting point of valve formation. Valve surface features and associated subcellular structural elements using a pennate diatom morphogenetic schema [8.61] may be used to represent multi-scale buckling. A line parallel or at an angle from each other [8.17] could be added in sequence as striae and as other structural features form. Interruptions in striae formation, such as Voigt discontinuities, could be included as a step-loading buckling process in the geometric model and recovered via the discrete eikonal equation. More expansively, teratogenic forms may have buckling occurrences that

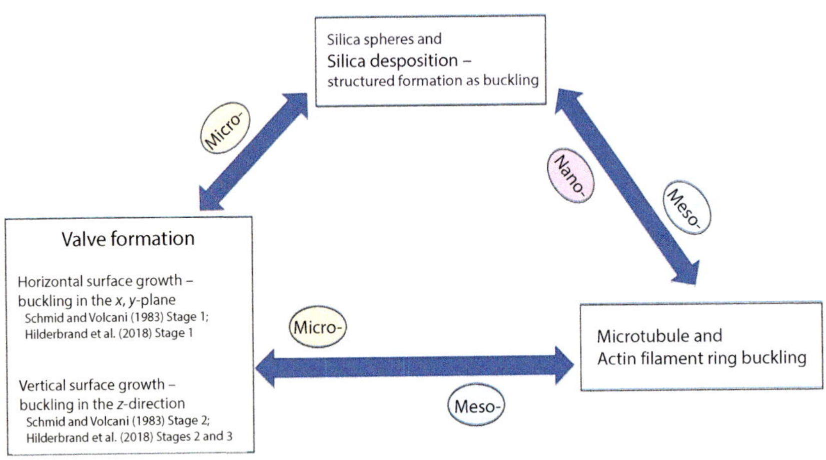

Figure 8.12 Multiscale buckling with respect to association of valve surface, microtubule, and actin filament ring.

interrupt valve formation at the subcellular level. Interruptions during valve formation may be ergodic or stochastic. The multiscale approach to understanding buckling enables association of nano- to meso- to microscale structures, regardless of cell shape or aberrant structures or forms.

The two silicalemmas of adjacent daughter cells, before they separate, can be quite close to one another so that their buckling patterns are complementary (Figure 27 in [8.94]; cf. [8.35]). As such they form a buckling laminate [8.27, 8.48, 8.49], and the forces between them, for example, sliding friction as between sorting cells [8.33], deserve further analysis, taking their close proximity into account.

Multiscale buckling may be summarized pictorially to illustrate the associations of multiscale structural features (Figure 8.12). These associations are a general framework to induce potential hypothesis creation and testing in terms of diatom morphogenesis. As new experimental evidence becomes available, computational modeling techniques can serve as an aid in understanding valve formation on a multiscale basis.

References

[8.1] Aitken, Z.H., Luo, S., Reynolds, S.N., Thaulow, C., Greer, J.R., Microstructure provides insights into evolutionary design and resilience of *Coscinodiscus* sp. frustule. *Proc. Natl. Acad. Sci. U. S. A.*, 113, 8, 2017–2022, 2016.

[8.2] Alberts, B., Bray, D., Lewis, J., Raff, M., Roberts, K., Watson, J.D., *Molecular Biology of the Cell*, 4th, Garland Publishing, New York, NY, USA, 2002.

[8.3] Annenkov, V.V., Basharina, T.N., Danilovtseva, E.N., Grachev, M.A., Putative silicon transport vesicles in the cytoplasm of the diatom *Synedra acus* during surge uptake of silicon. *Protoplasma*, 250, 5, 1147–1155, 2013.

[8.4] Annenkov, V.V., Danilovtseva, E.N., Gordon, R., Steps of silicic acid transformation to silica frustules: Main hypothesis and discoveries [SIFR], in: *Diatom Morphogenesis [DIMO, Volume 2 in the series: Diatoms: Biology & Applications, series editors: Richard Gordon & Joseph Seckbach]*, V. Annenkov, J. Seckbach, R. Gordon (Eds.), Wiley-Scrivener, Beverly, MA, USA, 2019, In preparation.

[8.5] Annenkov, V.V., Patwardhan, S.V., Belton, D., Danilovtseva, E.N., Perry, C.C., A new stepwise synthesis of a family of propylamines derived from diatom silaffins and their activity in silicification. *Chem. Commun. (Camb)*, 14, 1521–1523, 2006.

[8.6] Banigan, E.J., Stephens, A.D., Marko, J.F., Mechanics and buckling of biopolymeric shells and cell nuclei. *Biophys. J.*, 113, 8, 1654–1663, 2017.

[8.7] Barrett, R., Berry, M.W., Chan, T.F., Demmel, J., Donato, J., Dongarra, J., Eijkhout, V., Pozo, R., Romine, C., van der Vorst, H., *Templates for the Solution of Linear Systems: Building Blocks for Iterative Methods*, Society for Industrial and Applied Mathematics, SIAM, 3600 Market Street, Floor 6, Philadelphia, PA 19104, 1994.

[8.8] Bedoshvili, Y., Gneusheva, K., Popova, M., Morozov, A., Likhoshway, Y., Anomalies in the valve morphogenesis of the centric diatom alga *Aulacoseira islandica* caused by microtubule inhibitors. *Biol. Open*, 7, 8, 2018.

[8.9] Benamou, J.-D., Multivalued solution and viscosity solutions of the Eikonal equation, in: *Thème 4, Simulation et optimization de systems complexes, Projet Ondes, Rapport de recherche* no *3281*, pp. 1–40, Unité de recherche INRIA Rocquencourt, Domaine de Voluceau, Rocquencourt, Le Chesnay, Cedex, France, 1997.

[8.10] Bentley, K., Clack, C., Cox, E.J., Diatom colony formation: A computational study predicts a single mechanism can produce both linkage and separation valves due to an environmental switch. *J. Phycol.*, 48, 3, 716–728, 2012.

[8.11] Bentley, K., Cox, E.J., Bentley, P.J., Nature's batik: A computer evolution model of diatom valve morphogenesis. *J. Nanosci. Nanotechnol.*, 5, 1, 25–34, 2005.

[8.12] Berg, J.M., Tymoczko, J.L., Stryer, L., *Biochemistry*, 5th, W. H. Freeman, New York, 2002.

[8.13] Bertaud, J., Qin, Z., Buehler, M.J., Intermediate filament-deficient cells are mechanically softer at large deformation: A multi-scale simulation study. *Acta Biomater.*, 6, 7, 2457–2466, 2010.

[8.14] Black, E., Mathematical models of membranes, University of Chicago, http://math.uchicago.edu/~may/REU2013/REUPapers/Black.pdf, 2013.

[8.15] Brangwynne, C.P., MacKintosh, F.C., Kumar, S., Geisse, N.A., Talbot, J., Mahadevan, L., Parker, K.K., Ingber, D.E., Weitz, D.A., Microtubules can bear enhanced compressive loads in living cells because of lateral reinforcement. *J. Cell Biol.*, 173, 5, 733–741, 2006.

[8.16] Brodland, G.W. and Gordon, R., Intermediate filaments may prevent buckling of compressively-loaded microtubules. *J. Biomech. Eng.*, 112, 3, 319–321, 1990.

[8.17] Chang, A., An overview of the Hamilton-Jacobi equation, University of Chicago, http://math.uchicago.edu/~ac/hje.pdf, 2013.

[8.18] Cox, E.J., Willis, L., Bentley, K., Integrated simulation with experimentation is a powerful tool for understanding diatom valve morphogenesis. *BioSystems*, 109, 3, Special Issue on Biological Morphogenesis, 450–459, 2012.

[8.19] Deserno, M., Kremer, K., Paulsen, H., Peter, C., Schmid, F., Computational studies of biomembrane systems: Theoretical considerations, simulation models, and applications, in: *From Single Molecules to Nanoscopically Structured Materials*, vol. 260, T. Basche, K. Mullen, M. Schmidt (Eds.), pp. 237–283, Springer-Verlag Berlin, Berlin, 2014.

[8.20] do Carmo, M.P., *Differential Geometry of Curves and Surfaces*, 2nd, Dover Publications, Mineola, New York, 2016.

[8.21] Drum, R.W. and Pankratz, H.S., Post mitotic fine structure of *Gomphonema parvulum*. *J. Ultrastruct. Res.*, 10, 217–223, 1964.

[8.22] Fletcher, D.A. and Mullins, D., Cell mechanics and the cytoskeleton. *Nature*, 463, 7280, 485–492, 2010.

[8.23] Frigeri, L.G., Radabaugh, T.R., Haynes, P.A., Hildebrand, M., Identification of proteins from a cell wall fraction of the diatom *Thalassiosira pseudonana*: Insights into silica structure formation. *Mol. Cell. Proteomics*, 5, 1, 182–193, 2006.

[8.24] Geitmann, A., Experimental approaches used to quantify physical parameters at cellular and subcellular levels. *Am. J. Bot.*, 93, 10, 1380–1390, 2006.

[8.25] Ghobara, M.M., Mazumder, N., Vinayak, V., Reissig, L., Gebeshuber, I.C., Tiffany, M.A., Gordon, R., On light and diatoms: A photonics and photobiology review [Chapter 7], in: *Diatoms: Fundamentals & Applications [DIFA, Volume 1 in the series: Diatoms: Biology & Applications, series editors: Richard Gordon & Joseph Seckbach]*, J. Seckbach and R. Gordon (Eds.), pp. 129–190, Wiley-Scrivener, Beverly, MA, USA, 2019.

[8.26] Gittes, F., Mickey, B., Nettleton, J., Howard, J., Flexural rigidity of microtubules and actin filaments measured from thermal fluctuations in shape. *J. Cell Biol.*, 120, 4, 923–934, 1993.

[8.27] Gordon, R., A review of the theories of vertebrate neurulation and their relationship to the mechanics of neural tube birth defects. *J. Embryol. Exp. Morphol.*, 89, Suppl., 229–255, 1985.

[8.28] Gordon, R., *The Hierarchical Genome and Differentiation Waves: Novel Unification of Development, Genetics and Evolution*, World Scientific & Imperial College Press, Singapore & London, 1999.

[8.29] Gordon, R. and Aguda, B.D., Diatom morphogenesis: Natural fractal fabrication of a complex microstructure, in: *Proceedings of the Annual International Conference of the IEEE Engineering in Medicine and Biology Society, Part 1/4: Cardiology and Imaging*, New Orleans, LA, USA, 4-7 Nov. 1988 Institute of Electrical and Electronics Engineers, New York, pp. 273–274, 1988.
[8.30] Gordon, R. and Brodland, G.W., The cytoskeletal mechanics of brain morphogenesis. Cell state splitters cause primary neural induction. *Cell Biophys.*, 11, 1, 177–238, 1987.
[8.31] Gordon, R. and Brodland, G.W., On square holes in pennate diatoms. *Diatom Res.*, 5, 2, 409–413, 1990.
[8.32] Gordon, R. and Drum, R.W., The chemical basis of diatom morphogenesis. *Int. Rev. Cytol.*, 150, 243–372421–422, 1994.
[8.33] Gordon, R., Goel, N.S., Steinberg, M.S., Wiseman, L.L., A rheological mechanism sufficient to explain the kinetics of cell sorting. *J. Theor. Biol.*, 37, 1, 43–73, 1972.
[8.34] Gordon, R., Losic, D., Tiffany, M.A., Nagy, S.S., Sterrenburg, F.A.S., The Glass Menagerie: Diatoms for novel applications in nanotechnology. *Trends Biotechnol.*, 27, 2, 116–127, 2009.
[8.35] Gordon, R. and Tiffany, M.A., Possible buckling phenomena in diatom morphogenesis, in: *The Diatom World*, J. Seckbach and J.P. Kociolek (Eds.), pp. 245–272, Springer, Dordrecht, The Netherlands, 2011.
[8.36] Gordon, R., Witkowski, A., Gebeshuber, I.C., Allen, C.S., The diatoms of Antarctica and their potential roles in nanotechnology, in: *Antarctica: Time of Change*, M. Masó (Ed.), pp. 84–95, Editions ACTAR, Barcelona, 2010.
[8.37] Gröger, P., Poulsen, N., Klemm, J., Kröger, N., Schlierf, M., Establishing super-resolution imaging for proteins in diatom biosilica. *Sci. Rep.*, 6, 36824, 2016.
[8.38] Guckenheimer, J. and Osinga, H.M., The singular limit of a Hopf bifurcation. *Discrete Contin. Dyn. Syst.*, 32, 8, 2805–2823, 2012.
[8.39] Gutiérrez, A., Gordon, R., Dávila, L.P., Deformation modes and structural response of diatom shells. *J. Mater. Sci. Eng. Adv. Technol.*, 15, 2, 105–134, 2017.
[8.40] Gutknecht, M.H., A brief introduction to Krylov space methods for solving linear systems, in: *Frontiers of Computational Science*, Y. Kaneda, H. Kawamura, M. Sasai (Eds.), pp. 53–62, Springer-Verlag Berlin, Berlin, Germany, 2007.
[8.41] Hamm, C.E., Interactions between biomineralization and function of diatom frustules, in: *Handbook of Biomineralization: Biological Aspects and Structure Formation*, E. Bäuerlein, P. Behrens, M. Epple (Eds.), pp. 83–94, Wiley-VCH Verlag GmbH, Weinheim, Germany, 2007.
[8.42] Heath, M.T., *Scientific Computing: An Introductory Survey*, 2nd, Society for Industrial and Applied Mathematics Philadelphia, Pennsylvania, USA, 2018.
[8.43] Heisenberg, C.P. and Bellaïche, Y., Forces in tissue morphogenesis and patterning. *Cell*, 153, 5, 948–962, 2013.
[8.44] Hildebrand, M., Frigeri, L.G., Davis, A.K., Synchronized growth of *Thalassiosira pseudonana* (Bacillariophyceae) provides novel insights into cell-wall synthesis processes in relation to the cell cycle. *J. Phycol.*, 43, 4, 730–740, 2007.
[8.45] Hildebrand, M., Kim, S., Shi, D., Scott, K., Subramaniam, S., 3D imaging of diatoms with ion-abrasion scanning electron microscopy. *J. Struct. Biol.*, 166, 3, 316–328, 2009.
[8.46] Hildebrand, M., Lerch, S.J.L., Shrestha, R.P., Understanding diatom cell wall silicification—moving forward. *Front. Mar. Sci.*, 5, 125, 2018.
[8.47] Hildebrand, M., York, E., Kelz, J.I., Davis, A.K., Frigeri, L.G., Allison, D.P., Doktycz, M.J., Nanoscale control of silica morphology and three-dimensional structure during diatom cell wall formation. *J. Mater. Res.*, 21, 10, 2689–2698, 2006.

[8.48] His, W., *Unsere Körperform und das Problem ihrer Entstehung, Briefe an einen befreundeten Naturforscher/Our Body Form and the Problem of its Emergence, Letters to a Friendly Natural Scientist*, F.C.W. Vogel, Leipzig, Germany, 1874.
[8.49] His, W., On the principles of animal morphology. *R. Soc. Edinb. Proc.*, 15, 287–298, 1888.
[8.50] Hitt, A.L., Cross, A.R., Williams, R.C., Microtubule solutions display nematic liquid crystalline structure. *J. Biol. Chem.*, 265, 3, 1639–1647, 1990.
[8.51] Ingber, D.E., Cellular tensegrity: Defining new rules of biological design that govern the cytoskeleton. *J. Cell Sci.*, 104, 613–627, 1993.
[8.52] Ipsen, I.C.F. and Meyer, C.D., The idea behind Krylov methods. *Am. Math. Mon.*, 105, 10, 889–899, 1998.
[8.53] Jacobs, T.D.B., Mate, C.M., Turner, K.T., Carpick, R.W., Understanding the tip-sample contact: An overview of contact mechanics from the macro-to the nanoscale, in: *Scanning Probe Microscopy in Industiral Applications: Nanomechanical Characterization*, D.B. Yablon (Ed.), pp. 15–48, John Wiley & Sons, Hoboken, New Jersey, USA, 2014.
[8.54] Jacobson, A.G. and Gordon, R., Changes in the shape of the developing vertebrate nervous system analyzed experimentally, mathematically and by computer simulation. *J. Exp. Zool.*, 197, 2, 191–246, 1976.
[8.55] Jiang, H.Q. and Zhang, J.P., Mechanics of microtubule buckling supported by cytoplasm. *J. Appl. Mech.-T. ASME*, 75, 6, 061019, 2008.
[8.56] Jin, M.Z. and Ru, C.Q., Localized buckling of a microtubule surrounded by randomly distributed cross linkers. *Phys. Rev. E*, 88, 1, 012701, 2013.
[8.57] Kabir, A.M.R., Inoue, D., Afrin, T., Mayama, H., Sada, K., Kakugo, A., Buckling of microtubules on a 2D elastic medium. *Sci. Rep.*, 5, 17222, 2015.
[8.58] Kapitein, L.C. and Peterman, E.J.G., *Single molecule experiments and the kinesin motor protein superfamily: Walking hand in hand*, Elsevier Science Bv, Amsterdam, 2009.
[8.59] Kaplan, W., *Advanced Calculus*, 5th, Addison-Wesley, Reading, Massachusetts, USA, 2002.
[8.60] Karam, G.N. and Gibson, L.J., Elastic buckling of cylindrical shells with elastic cores. I. Analysis. *Int. J. Solids Struct.*, 32, 8–9, 1259–1283, 1995.
[8.61] Kharitonenko, K.V., Bedoshvili, Y.D., Likhoshway, Y.V., Changes in the micro- and nanostructure of siliceous valves in the diatom *Synedra acus* under the effect of colchicine treatment at different stages of the cell cycle. *J. Struct. Biol.*, 190, 1, 73–80, 2015.
[8.62] Kotzsch, A., Gröger, P., Pawolski, D., Bomans, P.H.H., Sommerdijk, N., Schlierf, M., Kröger, N., Silicanin-1 is a conserved diatom membrane protein involved in silica biomineralization. *BMC Biol.*, 15, 65, 2017.
[8.63] Kotzsch, A., Pawolski, D., Milentyev, A., Shevchenko, A., Scheffel, A., Poulsen, N., Shevchenko, A., Kröger, N., Biochemical composition and assembly of biosilica-associated insoluble organic matrices from the diatom *Thalassiosira pseudonana*. *J. Biol. Chem.*, 291, 10, 4982–4997, 2016.
[8.64] Kröger, N., Bergsdorf, C., Sumper, M., Frustulins: Domain conservation in a protein family associated with diatom cell walls. *Eur. J. Biochem.*, 239, 2, 259–264, 1996.
[8.65] Kröger, N. and Poulsen, N., Diatoms: From cell wall biogenesis to nanotechnology. *Annu. Rev. Genet.*, 42, 83–107, 2008.
[8.66] Krylov, A., On the numerical solution of the equation by which in technical questions frequencies of small oscillations of material systems are determined [Russian]. *Izvestija AN SSSR (News Acad. Sci. USSR) Otdel. Mat. Estest. Nauk.*, 7, 4, 491–539, 1931.
[8.67] Li, C.W., Chu, S., Lee, M., Characterizing the silica deposition vesicle of diatoms. *Protoplasma*, 151, 2–3, 158–163, 1989.
[8.68] Li, T., A mechanics model of microtubule buckling in living cells. *J. Biomech.*, 41, 8, 1722–1729, 2008.

[8.69] Liu, Y., Guo, Y., Valles Jr., J.M., Tang, J.X., Microtubule bundling and nested buckling drive stripe formation in polymerizing tubulin solutions. *Proc. Natl. Acad. Sci. U.S.A.*, 103, 28, 10654–10659, 2006.
[8.70] Lobel, K.D., West, J.K., Hench, L.L., Computational model for protein mediated biomineralization of the diatom frustule. *Mar. Biol.*, 126, 3, 353–360, 1996.
[8.71] Maidana, N.I., Morales, E.A., Bradbury, J.P., Schabitz, F., Houk, V., A new order and family of diatoms: Arcanodiscales, Arcanodiscaceae (Bacillariophyta) to accommodate Arcanodiscus platti gen. nov. et sp. nov. from the Argentinian Patagonia, in: *Diatom Taxonomy and Ecology: From France to the Sub-Antarctic Islands*, vol. 146, B. VanDeVijver, L. Tudesque, L. Ector (Eds.), pp. 63–72, J. Cramer, Stuttgart, Germany, 2017.
[8.72] Memet, E., Hilitski, F., Morris, M.A., Schwenger, W.J., Dogic, Z., Mahadevan, L., Microtubules soften due to cross-sectional flattening. *Elife*, 7, e34695, 2018.
[8.73] Mizuno, N., Narita, A., Kon, T., Sutoh, K., Kikkawa, M., Three-dimensional structure of cytoplasmic dynein bound to microtubules. *Proc. Natl. Acad. Sci. U. S. A.*, 104, 52, 20832–20837, 2007.
[8.74] Mohagheghian, E., Luo, J.Y., Chen, J.J., Chaudhary, G., Chen, J.W., Sun, J., Ewoldt, R.H., Wang, N., Quantifying compressive forces between living cell layers and within tissues using elastic round microgels. *Nat. Commun.*, 9, 1878, 2018.
[8.75] Mullins, R.D., Cytoskeletal mechanisms for breaking cellular symmetry. *Cold Spring Harbor Perspect. Biol.*, 2, 1, a003392, 2010.
[8.76] Murrell, M.P. and Gardel, M.L., F-actin buckling coordinates contractility and severing in a biomimetic actomyosin cortex. *Proc. Natl. Acad. Sci. U. S. A.*, 109, 51, 20820–20825, 2012.
[8.77] Nechaev, S., Non-Euclidean geometry in nature. https://arxiv.org/abs/1705.08013, 2017.
[8.78] Nechaev, S., Non-Euclidean geometry in nature, in: *Order, Disorder and Criticality: Advanced Problems of Phase Transition Theory*, vol. 5, Y. Holovatch (Ed.), pp. 61–112, World Scientific Publishing Company, Singapore, 2018.
[8.79] Pappas, J.L., Geometry and topology of diatom shape and surface morphogenesis for use in applications of nanotechnology. *J. Nanosci. Nanotechnol.*, 5, 1, 120–130, 2005.
[8.80] Pappas, J.L., Theoretical morphospace and its relation to freshwater gomphonemoid-cymbelloid diatom (Bacillariophyta) lineages. *J. Biol. Syst.*, 13, 4, 385–398, 2005.
[8.81] Pappas, J.L., More on theoretical morphospace and its relation to freshwater gomphonemoid-cymbelloid diatom (Bacillariophyta) lineages. *J. Biol. Syst.*, 16, 1, 119–137, 2008.
[8.82] Pappas, J.L., Multivariate complexity analysis of 3D surface form and function of centric diatoms at the Eocene-Oligocene transition. *Mar. Micropaleontol.*, 122, 67–86, 2016.
[8.83] Pappas, J.L., Quantified ensemble 3D surface features modeled as a window on centric diatom valve morphogenesis, in: *Diatom Morphogenesis [DIMO, Volume in the series: Diatoms: Biology & Applications, series editors: Richard Gordon & Joseph Seckbach]*, V. Annenkov, J. Seckbach, R. Gordon (Eds.), Wiley-Scrivener, Beverly, MA, USA, 2021.
[8.84] Pappas, J.L. and Miller, D.J., A generalized approach to the modeling and analysis of 3D surface morphology in organisms. *PLoS One*, 8, 10, e77551, 2013.
[8.85] Pappas, J.L., Tiffany, M.A., Gordon, R., The uncanny symmetry of some diatoms and not of others: A multi-scale morphological characteristic and a puzzle for morphogenesis [DUNC], in: *Diatom Morphogenesis [DIMO, Volume in the series: Diatoms: Biology & Applications, series editors: Richard Gordon & Joseph Seckbach]*, V. Annenkov, J. Seckbach, R. Gordon (Eds.), Wiley-Scrivener, Beverly, MA, USA, 2021.
[8.86] Parkinson, J., Brechet, Y., Gordon, R., Centric diatom morphogenesis: A model based on a DLA algorithm investigating the potential role of microtubules. *Biochim. Biophys. Acta - Mol. Cell Res.*, 1452, 1, 89–102, 1999.

[8.87] Pickett-Heaps, J.D., Schmid, A.M.M., Edgar, L.A., The cell biology of diatom valve formation. *Prog. Phycol. Res.*, 7, 1–168, 1990.
[8.88] Pickett-Heaps, J.D., Tippit, D.H., Andreozzi, J.A., Cell division in the pennate diatom *Pinnularia.* III - The valve and associated cytoplasmic organelles. *Biol. Cell.*, 35, 2, 195–198, 1979.
[8.89] Pickett-Heaps, J.D., Tippit, D.H., Andreozzi, J.A., Cell division in the pennate diatom *Pinnularia.* IV. - Valve morphogenesis. *Biol. Cell.*, 35, 2, 199–203, 1979.
[8.90] Rammerstorfer, F.G., Buckling of elastic structures under tensile loads. *Acta Mech.*, 229, 2, 881–900, 2018.
[8.91] Rayleigh, L., On the instability of a cylinder of viscous liquid under capillary force. *Philos. Mag.*, 34, 145–154, 1892.
[8.92] Robinson, D.H. and Sullivan, C.W., How do diatoms make silicon biominerals? *Trends Biochem. Sci.*, 12, 4, 151–154, 1987.
[8.93] Rogerson, A., Defreitas, A.S.W., McInnes, A.G., Observations on wall morphogenesis in *Coscinodiscus asteromphalus* (Bacillariophyceae). *Trans. Am. Microsc. Soc.*, 105, 1, 59–67, 1986.
[8.94] Round, F.E., Crawford, R.M., Mann, D.G., *The Diatoms, Biology & Morphology of the Genera*, Cambridge University Press, Cambridge, 1990.
[8.95] Saad, Y. and Schultz, M.H., GMRES: A generalized minimal residual algorithm for solving nonsymmetric linear systems. *SIAM J. Sci. Stat. Comput.*, 7, 3, 856–869, 1986.
[8.96] Sato, S., Valve and girdle band morphogenesis in a bipolar centric diatom *Plagiogrammopsis vanheurckii* (Cymatosiraceae, Bacillariophyta). *Eur. J. Phycol.*, 45, 2, 167–176, 2010, Pii 922978394.
[8.97] Scheffel, A., Poulsen, N., Shian, S., Kröger, N., Nanopatterned protein microrings from a diatom that direct silica morphogenesis. *Proc. Natl. Acad. Sci. U. S. A.*, 108, 8, 3175–3180, 2011.
[8.98] Schmid, A.M.M., The development of structure in the shells of diatoms. *Nova Hedwig.*, 64, Suppl., 219–230 + Figures 211–222, 1979.
[8.99] Schmid, A.M.M., Morphogenetic forces in diatom cell wall formation, in: *Cytomechanics: The Mechanical Basis of Cell Form and Structure*, J. Bereiter-Hahn, O.R. Anderson, W.E. Reif (Eds.), pp. 183–199, Springer-Verlag, Berlin, 1987.
[8.100] Schmid, A.M.M., Aspects of morphogenesis and function of diatom cell walls with implications for taxonomy. *Protoplasma*, 181, 1–4, 43–60, 1994.
[8.101] Schmid, A.M.M. and Schulz, D., Wall morphogenesis in diatoms: Deposition of silica by cytoplasmic vesicles. *Protoplasma*, 100, 3–4, 267–288, 1979.
[8.102] Schmid, A.M.M. and Volcani, B.E., Wall morphogenesis in *Coscinodiscus wailesii* Gran and Angst. I. Valve morphology and development of its architecture. *J. Phycol.*, 19, 4, 387–402, 1983.
[8.103] Siddiqi, K., Tannenbaum, A., Zucker, S.W., A Hamiltonian approach to the eikonal equation. *Lect. Notes Comput. Sci.*, 1654, 1–13, 1999.
[8.104] Silva, E.C.C.M., Tong, L.M., Yip, S., Van Vliet, K.J., Size effects on the stiffness of silica nanowires. *Small*, 2, 2, 239–243, 2006.
[8.105] Simoncini, V. and Szyld, D.B., Recent computational developments in Krylov subspace methods for linear systems. *Numer. Linear Algebra Appl.*, 14, 1, 1–59, 2007.
[8.106] Sodt, A.J. and Pastor, R.W., Bending free energy from simulation: Correspondence of planar and inverse hexagonal lipid phases. *Biophys. J.*, 104, 10, 2202–2211, 2013.
[8.107] Soheilypour, M., Peyro, M., Peter, S.J., Mofrad, M.R.K., Buckling behavior of individual and bundled microtubules. *Biophys. J.*, 108, 7, 1718–1726, 2015.

[8.108] Sumper, M. and Kröger, N., Silica formation in diatoms: The function of long-chain polyamines and silaffins. *J. Mater. Chem.*, 14, 14, 2059–2065, 2004.
[8.109] Tamstorf, R. and Grinspun, E., Discrete bending forces and their Jacobians. *Graph. Models*, 75, 362–370, 2013.
[8.110] Tesson, B. and Hildebrand, M., Extensive and intimate association of the cytoskeleton with forming silica in diatoms: Control over patterning on the meso- and micro-scale. *PLoS One*, 5, 12, e14300, 2010.
[8.111] Tesson, B. and Hildebrand, M., Characterization and localization of insoluble organic matrices associated with diatom cell walls: Insight into their roles during cell wall formation. *PLoS One*, 8, 4, e61675, 2013.
[8.112] Tesson, B., Lerch, S.J.L., Hildebrand, M., Characterization of a new protein family associated with the silica deposition vesicle membrane enables genetic manipulation of diatom silica. *Sci. Rep.*, 7, 13457, 2017.
[8.113] Treister, E. and Haber, E., A fast marching algorithm for the factored eikonal equation. *J. Comput. Phys.*, 324, 210–225, 2016.
[8.114] van de Meene, A.M.L. and Pickett-Heaps, J.D., Valve morphogenesis in the centric diatom *Proboscia alata* Sundstrom. *J. Phycol.*, 38, 2, 351–363, 2002.
[8.115] van de Meene, A.M.L. and Pickett-Heaps, J.D., Valve morphogenesis in the centric diatom *Rhizosolenia setigera* (Bacillariophyceae, Centrales) and its taxonomic implications. *Eur. J. Phycol.*, 39, 1, 93–104, 2004.
[8.116] van der Vorst, H.A., BI-CGSTAB: A fast and smoothly converging variant of BI-CG for the solution of nonsymmetric linear-systems. *SIAM J. Sci. Stat. Comput.*, 13, 2, 631–644, 1992.
[8.117] Vartanian, M., Desclés, J., Quinet, M., Douady, S., Lopez, P.J., Plasticity and robustness of pattern formation in the model diatom *Phaeodactylum tricornutum*. *New Phytol.*, 182, 2, 429–442, 2009.
[8.118] Vavryčuk, V., Real ray tracing in anisotropic viscoelastic media. *Geophys. J. Int.*, 175, 2, 617–626, 2008.
[8.119] Vavryčuk, V., On numerically solving the complex eikonal equation using real ray-tracing methods: A comparison with the exact analytical solution. *Geophysics*, 77, 4, T109–T116, 2012.
[8.120] Volcani, B.E., Cell wall formation in diatoms: Morphogenesis and biochemistry, in: *Silicon and Siliceous Structures in Biological Systems*, T.L. Simpson and B.E. Volcani (Eds.), pp. 157–200, Springer-Verlag, New York, NY, USA, 1981.
[8.121] von Stosch, H.A., Observations on *Bellerochea* and *Streptotheca*, including descriptions of three new planctonic diatom species. *Nova Hedwig. Beih.*, 54, Suppl., 113–166, 1977.
[8.122] Vrieling, E.G., Gieskes, W.W.C., Beelen, T.P.M., Silicon deposition in diatoms: control by the pH inside the silicon deposition vesicle. *J. Phycol.*, 35, 3, 548–559, 1999.
[8.123] Wang, J. and Pelling, A.E., An approach to visualize the deformation of the intermediate filament cytoskeleton in response to locally applied forces. *ISRN Cell Biol.*, 2012, 513546, 2011.
[8.124] Wenzl, S., Hett, R., Richthammer, P., Sumper, M., Silacidins: Highly acidic phosphopeptides from diatom shells assist in silica precipitation in vitro. *Angew. Chem.-Int. Edit.*, 47, 9, 1729–1732, 2008.
[8.125] Wilken, S., Hoffmann, B., Hersch, N., Kirchgessner, N., Dieluweit, S., Rubner, W., Hoffmann, L.J., Merkel, R., Peeken, I., Diatom frustules show increased mechanical strength and altered valve morphology under iron limitation. *Limnol. Oceanogr.*, 56, 4, 1399–1410, 2011.
[8.126] Willis, L., Page, K.M., Broomhead, D.S., Cox, E.J., Computational models of the formation of diatom pore occlusions. *Plant Ecol. Evol.*, 143, 3, 297–306, 2010.

[8.127] Wu, J., Misra, G., Russell, R.J., Ladd, A.J.C., Lele, T.P., Dickinson, R.B., Effects of dynein on microtubule mechanics and centrosome positioning. *Mol. Biol. Cell*, 22, 24, 4834–4841, 2011.

[8.128] Xiao, S.P. and Belytschko, T., A bridging domain method for coupling continua with molecular dynamics. *Comput. Methods Appl. Mech. Eng.*, 193, 17–20, 1645–1669, 2004.

[8.129] Xu, X.S., Ma, Y., Lim, C.W., Chu, H.J., Dynamic buckling of cylindrical shells subject to an axial impact in a symplectic system. *Int. J. Solids Struct.*, 43, 13, 3905–3919, 2006.

[8.130] Yang, M.-H., EECS 275 Matrix Computation, Lecture 18, University of California at Merced [slides], Merced, California, 2013, http://faculty.ucmerced.edu/mhyang/course/eecs275/lectures/lecture18.pdf.

[8.131] Zenker, J., White, M.D., Gasnier, M., Alvarez, Y.D., Lim, H.Y.G., Bissiere, S., Biro, M., Plachta, N., Expanding actin rings zipper the mouse embryo for blastocyst formation. *Cell*, 173, 1–16, 2018.

[8.132] Zumdieck, A., Kruse, K., Bringmann, H., Hyman, A.A., Jülicher, F., Stress generation and filament turnover during actin ring constriction. *PLoS One*, 2, 8, e696, 2007.

9

Are Mantle Profiles of Circular Centric Diatoms a Measure of Buckling Forces During Valve Morphogenesis?

Janice L. Pappas[1]* and Richard Gordon[2,3]

[1]*Department of Mathematics, Michigan Math and Science Scholars, University of Michigan, Ann Arbor, MI, United States*
[2]*Gulf Specimen Marine Laboratory & Aquarium, Panacea, FL, United States*
[3]*C.S. Mott Center for Human Growth & Development, Department of Obstetrics & Gynecology, Wayne State University, Detroit, MI, United States*

Abstract
In diatom valve morphogenesis, the silicalemma is important in silica precipitation and surface topography. General characteristic undulations on the valve surface are a result of cytological processes, including biochemical and biophysical processes. The resultant valve surface topography is biophysically limited because of the boundary induced by the valve mantle. Surface topography may be characterized as buckling, and because of boundary limits, the relation between a geometric characterization of surface topography and buckling may be determined as buckling forces at the valve mantle. We conducted tests and determined that the profile of valve surface topography is arbitrarily related to buckling at the valve mantle and is not a measure of buckling forces during valve morphogenesis. We also discuss complementarity and heterovalvy in diatoms during valve morphogenesis.

Keywords: Buckling, valve mantle, silicalemma, Euler's critical load, diatoms, complementarity, heterovalvy, Janus cell

9.1 Introduction

Vegetative reproduction involves morphogenesis and cell division during the diatom life cycle. Changes in the three-dimensional (3D) valve surface from an annulus to the mantle in centric diatoms and from the sternum to mantle in pennate diatoms occurs during valve morphogenesis. Size reduction during diatom cell division occurs as the daughter cells produced are smaller than mother cells [9.31]. In both cases, silica precipitation occurs, and the timing and rate of silica precipitation during valve morphogenetic stages or size reduction steps is variable.

**Corresponding author*: jlpappas@umich.edu

Vadim Annenkov, Joseph Seckback and Richard Gordon (eds.) Diatom Morphogenesis, (231–250) © 2022 Scrivener Publishing LLC

Cytologically, valve morphogenesis and size reduction involve the functioning of cell constituents (e.g., [9.31]). Within a flat, membranous silica deposition vesicle (SDV), silica condensation occurs, a the SDV is attached to the inside of the plasmalemma. Surrounding the SDV is the silicalemma membrane. Silica precipitates and is deposited such that the siliceous nanopatterning of the diatom valve occurs during valve morphogenesis. Various forces affect the diatom cell as morphogenesis proceeds, and subsequently when daughter cells are produced from mother cells as size reduction occurs during the vegetative part of the life cycle. Such forces affect siliceous structures, impacting the external features and shape of the valve face.

As division of a diatom cell occurs, the plasmalemma and SDV of the daughter cell exerts pressure on the mother cell as well as the forming of the second daughter cell. This pressure extends to the cingula. Multiple cingula are produced and added to newly formed valves throughout the life cycle [9.31]. Morphologically, cingula may be split, closed or bands of scales [9.40].

As silica deposition continues and the cell silica thickens, the valve mantle may become deformed at the cingula [9.13]. Because the diatom cell in its simplest geometric state is a cylinder, the walls of the cylinder play a role as a stopping point of construction of valve surface features. The valve mantle is that stopping point where forces may be instrumental in dictating the degree of deformation at the valve mantle and the degree to which the shape of the siliceous valve surface is affected. The valve mantle at the edge of the valve face is positioned just above the cingula when the diatom cell is viewed in girdle view [9.36].

Buckling is a prevalent characteristic in diatom morphogenesis [9.17]. The height of siliceous material on the valve surface is indicative of a kind of buckling [9.29]. Changes in the valve surface from flat to buckled occurs during valve morphogenesis. At each of the three stages of valve morphogenesis as determined by Schmid and Volcani [9.35], varying degrees of buckling may occur. Horizontal expansion of the valve surface may produce buckling near the annulus and valve central area as exhibited by *Cyclotella meneghiniana* (e.g., [9.18]). During the second and third stages of valve morphogenesis, vertical expansion of the valve surface produces varying degrees of buckling, such as that exhibited by some species of *Stephanodiscus* (e.g., [9.30, 9.31]) and *Actinoptychus senarius* (e.g., [9.19, 9.31]). As morphogenesis proceeds, valve surfaces exhibit some degree of buckling in the overall valve shape topography.

As morphogenesis is completed, the valve mantle may be formed as a stepped structure, as the SDV and silicalemma of one daughter cell presses against the other daughter cell forming and the plasmalemma of the mother cell. As the pressure occurs, silica deposition may be heavy, producing the stepped valve mantle [9.31]. In *Aulacoseira islandica*, relief valves may form with a stepped mantle, while intaglio valves form with a plain mantle [9.21]. A stepped valve mantle may form distant to the valve face [9.21] as in *Ellerbeckia arenaria* [9.34] but is evident as the result of valve formation pressure and heavy silica deposition via environmental influences ([9.31, 9.34]).

One of us (R.G.) posed the question, "Are mantle profiles of circular centric diatoms a measure of buckling forces during valve morphogenesis? We provide a simple model to indirectly study the role that the silicalemma may have in determining valve surface via buckling forces as they are applied to the mantle during valve morphogenesis. We use Euler's critical load as the vehicle to study the relation between the change in length of the silicalemma represented by the valve surface profile and the buckling forces generated at the valve mantle. Different buckling patterns that represent different diatom valve surface topographies may have implications for discerning complementarity or heterovalvy during valve formation as vegetative size reduction occurs. Buckling forces for complimentary valves in contrast to Janus cells may be one way to study the formation of anomalous valves as one form of heterovalvy.

9.2 Methods

Analyses are performed on devised valve profiles of circular centric diatoms as examples. Because of their simplified form, centric diatom valves enable the designation of a single point as the valve mantle on which to conduct analyses. Geometrically, a centric diatom is treated as a capped 3D cylinder, with the cap representing the valve face surface, and the valve mantle as the point of contact at the diameter edge.

We treat the centric diatom valve profile as a cylindrical column with an infinitesimal diameter in the z plane where it follows a fixed straight-line radius in the x, y plane, spanning the fixed annulus to the fixed valve mantle (Figures 9.1 and 9.3) or from mantle to mantle across a diameter (Figures 9.2 and 9.3). An axial load is applied at the point where the profile ends at the valve mantle causing buckling. The buckling force at the valve mantle is determined for a number of profiles using Euler's critical load formula [9.16]. Each profile represents the silicalemma growth as silica deposition occurs in order to complete valve division and formation.

From small deformation (also called infinitesimal strain) theory [9.7], Euler's critical load, $P_{cr} = \dfrac{\pi^2 EI}{(KL_{cr})^2}$, is calculated as the first mode of buckling, where P_{cr} is the critical load buckling force, E is the Young's (= elastic) modulus for amorphous silicon dioxide, I is the minimum area moment of inertia (= second moment of area), L_{cr} is the critical buckling length of the diatom profile (= single layer column), and K is the effective length factor [9.2, 9.3]. Values of the variables for diatom valve profiles are: E is 71.2 (minimum) to 74.8 (maximum) for amorphous silicon dioxide [9.5, 9.28], with the average used in calculations is 73; I is set to an arbitrary value of 1 to represent an infinitesimal cross-sectional profile diameter; L_{cr} is calculated to be the length of a polynomial curve fit, so that while the path of the profile is along a straight line in the x, y plane, the length of the

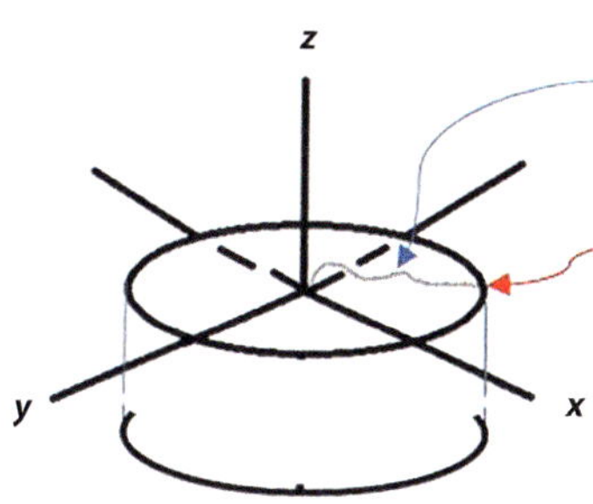

Figure 9.1 Centric diatom schematic with valve profile. Axial load at the point of contact at the valve mantle constrains the profile, inducing buckling.

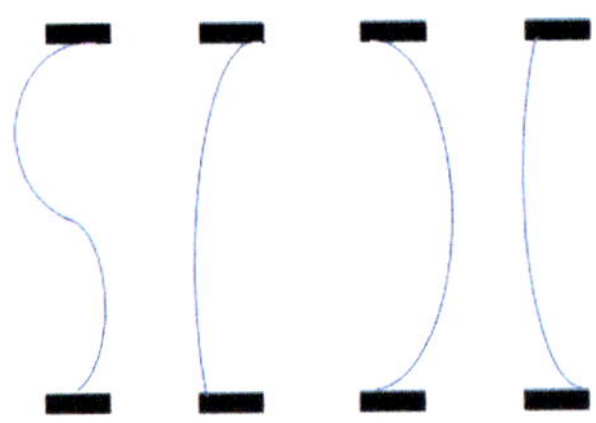

Figure 9.2 Various potential valve profiles undergoing buckling.

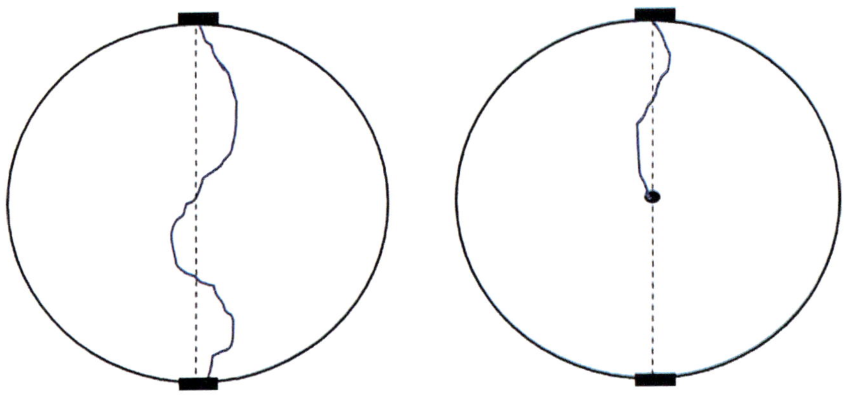

Figure 9.3 Measurement of buckling forces with respect to two fixed points along a diameter (left) or two fixed points along a radius (right) of a centric diatom valve face.

profile records the non-linear aspects of the curve in the z plane and as is based on arc length $ds = \sqrt{\left(\frac{dy}{dx}\right)^2 + 1}\, dx$; K is 0.5 representing fixed rotation and fixed translation so that the polynomial is not rotated (= no torque) and translation does not stray from the path of the polynomial (= fixed linear profile path) (Figures 9.2 and 9.3). E, I, and K are held constant for all profiles.

Assumptions for application of Euler's critical load are that silica material precipitated is isotropic, the length of the column (= profile) is large compared to its cross-sectional area of the valve, the valve cross-section is uniform, direct stress to the column is small, critical stress to the column is within the elastic strain range, and the column starts out straight so that the axial load at the valve mantle is applied in a straight line [9.16].

Three hypotheses are tested via the relation between mantle profile and buckling force. Different buckling forces should indicate different profiles and therefore enable distinguishing different valve morphogenetic paths. That is, the Null hypothesis in each of the first two tests is randomness in patterns of buckling and that there is no pattern to the relation between buckling forces and profiles. The first test involves holding the profile length constant for different profiles and calculating the buckling forces. The second test involves different profiles having different buckling forces with respect to size reduction during vegetative reproduction. Profile length is decreased at each step of size reduction so that different paths should be evident for different profiles. A question related to valve morphogenesis involves whether buckling forces with respect to profiles may be used to differentiate between complementarity and heterovalvy in diatom cells. Tracings as valve profiles from Figures 26a–e, page 34 in Round *et al.* [9.31] are a second set used to conduct the second hypothesis test. The third test is a generalized test to answer the question, "Are mantle

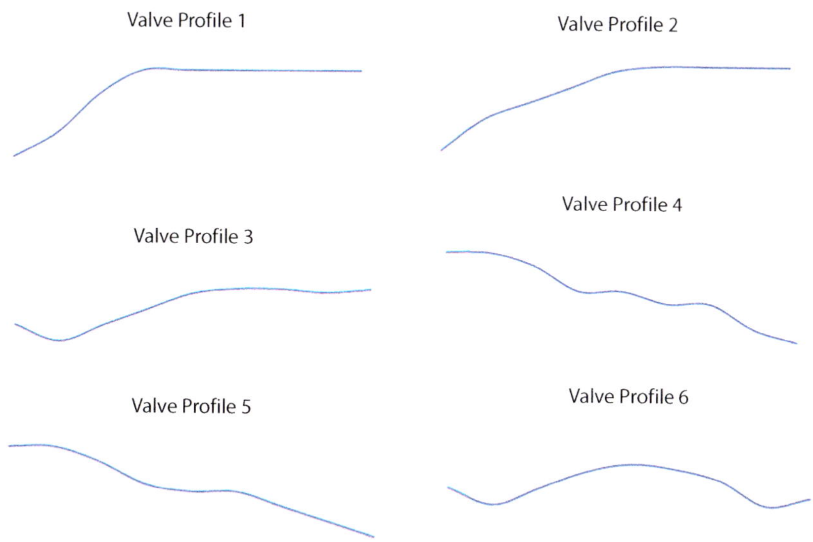

Figure 9.4 Least-squares polynomial curve fits of six profiles. In the x-direction, the annulus is represented by the lowest valued point, and the valve mantle is represented by the highest valued point.

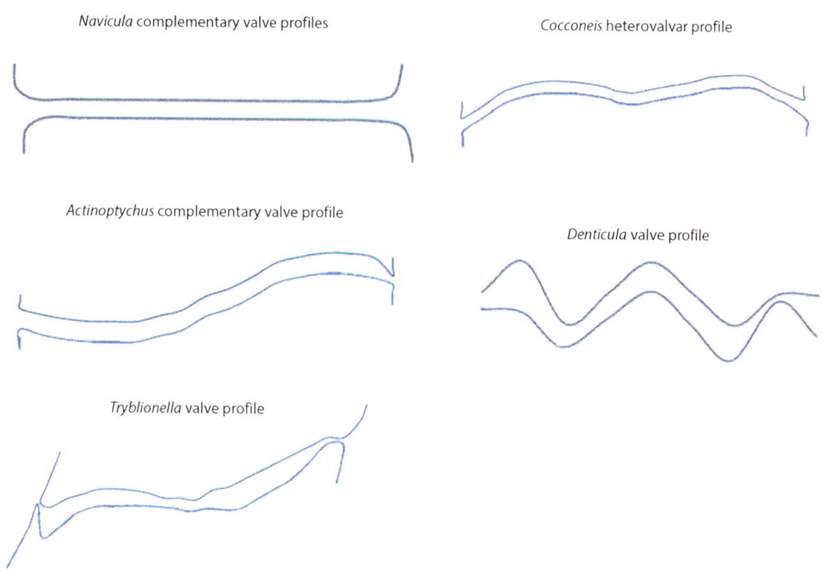

Figure 9.5 Tracings of diatom profiles after those depicted in Figures 26a–e from Round *et al.* (1990). Complementarity is illustrated in the *Navicula* and *Actinoptychus* valve profiles. Heterovalvy is illustrated in the *Cocconeis* valve profile. Valve profiles are more complex for *Denticula* and *Tryblionella*.

profiles of circular centric diatoms a measure of buckling forces during valve morphogenesis?" This test has the Null hypothesis as a profile with no buckling, that is, a straight-line profile, while the alternate hypothesis is any buckled profile, that is, any non-straight-line profile.

Profiles are created by devising polynomial least-squares curve fits representing two separate sets of potential taxa. Polynomials include straight lines which are a special case of nth-order polynomials with $n = 1$. One set consists of six potential centric valve profiles (Figure 9.4). The second set consists of the outline shape of each cell valve face in girdle view (as given in Figures 26a–e, page 34 in Round *et al.* [9.31]) (Figure 9.5). Polynomial curve fitting enables tracings from which arc length calculations are made to determine buckling forces. Valve profiles from the first set (Figure 9.4) are used to conduct the first and second tests. Size reduction is represented by decreases in arc length at discrete steps as values of x in the polynomials are changed to produce decreases in profile length. That is, successive arc lengths for each profile are used to represent each valve formation stage during size reduction. The second set (Figure 9.5) is used in analyses about complementarity and heterovalvy in diatom cells to determine whether buckling with respect to these particular valve characteristics may be informative in morphogenetic and vegetative reproduction diatom studies. The third test will produce a generalization of all results.

9.2.1 Background: Circular Centric 2D Profiles and 3D Surfaces of Revolution

Tracings of cross-sections of circular centric diatom valves are used, and each curve is fitted by an nth-order polynomial to each tracing that becomes the profile (Table 9.1, column 1). To illustrate the method, two different scanning electron micrographic digital images of *Stephanodiscus* sp. are used. Curve fittings are accomplished using trigonometric functions with boundary conditions (Table 9.1, columns 3 and 4). Surfaces of revolution for one valve are devised based on the trigonometric functions for each profile (Table 9.1, column 2).

Two surfaces of revolution for each image in Table 9.1 were used to illustrate that the resulting equations are not unique for a given diatom valve surface. Slight changes in coordinate selection on the surface, coefficients, or functions can produce multiple equations representing the same diatom valve surface. That is, the polynomials for each diatom profile on the same surface are arbitrary. Because surfaces of revolution are symmetrical about the z-axis, the equations will be polynomials of trigonometric functions, $y = f(x)$, where sines, cosines, and hyperbolic tangents were used, although hyperbolic secants could have been used as well. All of these functions are continuous. All other trigonometric functions have discontinuities and are unnecessary and therefore avoided. Most importantly, the boundary interval needs to be used with each equation; otherwise, the equations are infinite in both directions and are not useful in generating specific surfaces of revolution. The same boundary interval was used throughout, although this is not necessary as illustrated below.

Table 9.1 Circular centric diatom *Stephanodiscus* sp. and surfaces of revolution.

Original image	Surface of revolution (around the *z*-axis for one valve)*	Trigonometric equation (curve fitting the profile of the valve face)	Boundary interval
 Miscellaneous photo		$-x\ Sin\ 6x - x^3\ Cos\ 2x - x^2 Cos\ 2x - 2x^2 Sin\ 2x$	$x \in \{0, 3\}$
		$-2x^2\ Sin\ 2x + x\ Sin\ 2x^3 - 6x^4\ Cos\ 2x$	$x \in \{0, 3\}$
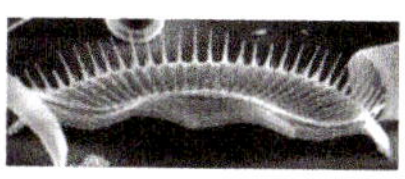 Photo: E.F. Stoermer		$-0.12x^2\ Tanh\ x^3 + 0.6x\ Sin\ 2x\ Cos\ x - 0.12x^2\ Cos\ x\ Cos\ 3x$	$x \in \{0, 3\}$
		$-2\ Tanh\ x^3 + 0.5\ Sin\ 1.5x\ Cos\ 1.5x -10x^2\ Cos\ x\ Cos\ 3x$	$x \in \{0, 3\}$

*3D Surface models created by Pappas (2000, personal data).

Circular centric diatom valve surfaces are a series of concentric "bumps" spaced variously apart so that just about any combination of polynomial terms of degree 2 or greater will model all potential buckled morphological variants. For example, is the radius curve fit of image 1, equation 2 (Table 9.1, top image in column 2). One can imagine adding "bumps" with regular or irregular spacing and potentially creating a morphological variant of a symmetric circular centric diatom and getting , which is a radius curve fit of image 2 (Table 9.1, bottom image in column 2), and an equation given as 5 $Tanh\ 3x^3 + 0.5\ Sin\ 1.5x - (0.5\ Cos 1.5x)^2$, boundary interval $x \in \{0, 7\}$ is yet another representation of image 2. Again, this illustration is important in understanding that circular centric diatom valve profiles are arbitrary representations.

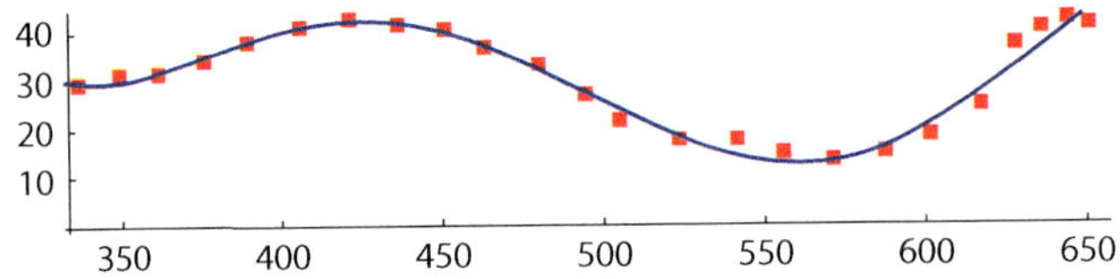

Figure 9.6 Curve fit of a sixth-order polynomial in Cartesian coordinates to the profile of *Stephanodiscus* sp. from Table 9.1, bottom image in column 2. See text for equation.

Image 1 and image 2 of *Stephanodiscus* sp. curve fittings were rotated about the *z*-axis to produce surfaces of revolution (Table 9.1, column 2). Circular centric diatom taxa such as additional *Stephanodiscus*, *Cyclotella*, *Cyclostephanos*, *Aulacoseira*, *Ellerbeckia*, *Actinocyclus*, and many others may be modeled in the same way. Because surfaces of revolution are idealized circular centric diatoms generated from trigonometric polynomials, they do not represent actual morphogenetic surfaces (Pappas, personal data and analysis). The point about this exercise is not that circular centric diatom valves are surfaces of revolution— they are not. The point is that diatom profiles are arbitrary polynomial functions and there are *n*-numbers of profiles for each diatom valve face, all of which are arbitrary polynomial functions.

Trigonometric polynomials may be converted into Cartesian coordinates to further make the point about the arbitrariness of using a profile to represent a diatom valve surface. The more "bumps" on the diatom valve surface, the more complicated the polynomial becomes. For *Stephanodiscus* sp. (Table 9.1, bottom image in column 2), a sixth-order polynomial is fitted to selected *x*, *y* coordinates along the profile of the valve (Figure 9.6) resulting in the equation, $-6.64436 \times 10^{-13}\,x^6 + 1.42583 \times 10^{-9}x^5 - 1.0464 \times 10^{-6}x^4 + 0.000219174\,x^3 + 0.0816907\,x^2 - 44.2038\,x + 5532.09$. By using Cartesian coordinates, *n*th-order polynomials may be generated as profiles and tested in terms of their relation to buckling forces and whether the outcome enables distinguishing patterns in valve morphogenesis or whether profiles as measures of buckling forces says anything about valve morphogenesis.

9.3 Results

Arbitrary-valued *x*, *y* points tracing a given profile in the *z*-direction were used to obtain fifth- and sixth-order polynomial least-squares curve fits. These profiles represent various tracings from annuli to valve mantles as would be expected on a centric diatom valve during morphogenesis. Polynomials and their coefficients of determination for those profiles are listed in Table 9.2. These same profiles are used in modeling size reduction during valve formation by varying the length of the profiles. For heterovalvar and complementary profiles given in Figures 26a–e from Round *et al.* [9.31], fifth- and sixth-order polynomial least-squares curve fits and their coefficients of determination are listed in Table 9.3.

The relation between profile and buckling forces was analyzed using the three hypothesis tests. The relation between profile and buckling forces is used to inform the role of valve profiles and buckling forces at the valve mantle in valve formation and size reduction.

Table 9.2 Polynomials and coefficients of determination for centric valve profiles depicted in Figure 9.4.

Valve profile	Polynomial	Coefficient of determination
1	$-0.0005x^5 + 0.0135x^4 - 0.1289x^3 + 0.5187x^2 - 0.6801x + 0.975$	0.9968
2	$0.0002x^5 - 0.0046x^4 + 0.0415x^3 - 0.1877x^2 + 0.5243x + 0.3283$	0.9977
3	$-0.0001x^5 + 0.0046x^4 - 0.0609x^3 + 0.3481x^2 - 0.7763x + 1.4833$	0.9977
4	$0.0002x^6 - 0.0042x^5 + 0.0406x^4 - 0.1735x^3 + 0.2912x^2 - 0.1437x + 1.5889$	0.989
5	$0.000005x^6 - 0.0011x^5 + 0.0055x^4 + 0.0194x^3 - 0.2383x^2 + 0.5355x + 1.1778$	0.9959
6	$0.000005x^5 + 0.0008x^4 - 0.0325x^3 + 0.2512x^2 - 0.6256x + 1.4033$	0.9792

Table 9.3 Polynomials and coefficients of determination for one valve profile for each cell profile depicted in Figure 9.5.

Valve profile	Polynomial	Coefficient of determination
Navicula	$0.000006x^6 - 0.0005x^5 + 0.0005x^4 - 0.003x^3 + 0.01x^2 - 0.0166x + 1.0107$	0.8828
Cocconeis	$0.0017x^6 - 0.0516x^5 + 0.6183x^4 - 3.5537x^3 + 9.9405x^2 - 11.795x + 12.805$	0.9137
Actinoptychus	$0.0004x^5 - 0.0134x^4 + 0.1154x^3 - 0.0874x^2 - 1.1844x + 9.8098$	0.979
Denticula	$-0.0012x^6 + 0.0398x^5 - 0.4964x^4 + 2.8913x^3 - 7.8387x^2 + 8.4796x + 1.3066$	0.876
Tryblionella	$-0.0006x^6 + 0.0181x^5 - 0.2249x^4 + 1.3803x^3 - 4.4559x^2 + 7.5159x - 1.6674$	0.9379

9.3.1 Approximate Constant Profile Length Representing Approximate Same Sized Valves

The first hypothesis test consists of the following:

H_0: Buckling forces are indistinguishable with respect to different diatom valve profiles. Flat and bumpy valves will show no difference in buckling forces when length is held approximately constant.

H_A: Buckling forces are distinguishable with respect to different diatom valve profiles. The expectation is that different taxa with have different buckling forces despite profiles having approximately the same length.

Six profiles constructed from least-squares polynomial curve fitting have different slopes, different number of buckles per profile, and different distribution of buckles along the profile (Figure 9.4).

In the test, L_{cr} is held approximately constant with resultant values of ~1.06 to 1.12 for the six valve profiles that are comparably the same state of cell size during vegetative reproduction. With a target length of one, the measurement error is $\mp$ 0.06. The profile length versus Euler buckling force for the six profiles indicates a narrow range with respect to a narrow range of profile lengths when including measurement error for the sake of illustration (Figure 9.7); otherwise, all the points would plot one on top of the other. When including measurement error, the range of buckling forces for all profiles is small, and even with this, two of the values plot almost one on top of the other, being almost indistinguishable.

The order of the profiles in the plot is depicted from shortest to longest length, including measurement error, reflecting the change in buckling force (Figure 9.8). A pattern from least to highest number of buckles is not established in this analysis. That is, profiles do not measure buckling forces concerning morphogenesis of *n*-taxa; the outcome is arbitrary. Therefore, the Null hypothesis holds, and the alternate hypothesis is rejected.

9.3.2 Change in Profile Length Representing Size Reduction During Valve Morphogenesis

The second hypothesis test consists of the following:

H_0: Buckling forces are indistinguishable with respect to different diatom valve profiles. Flat and bumpy valves will show no difference in buckling forces with varying length.

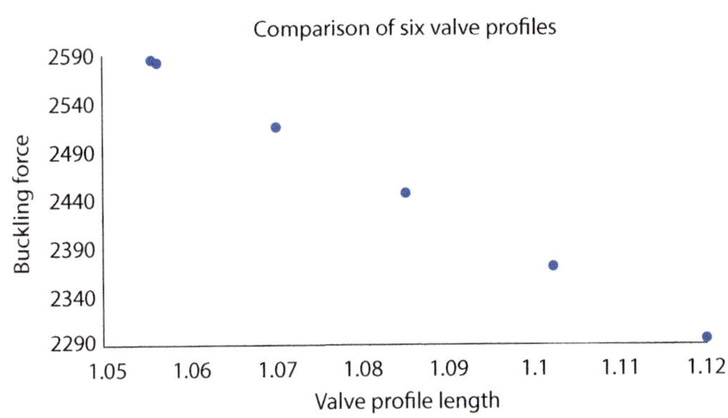

Figure 9.7 Plot of approximately constant profile length from six polynomial least-squares fits depicted in Figure 9.5 versus buckling forces. These profiles have different slopes, different number of buckles per profile, and different distributions of buckles along the profile.

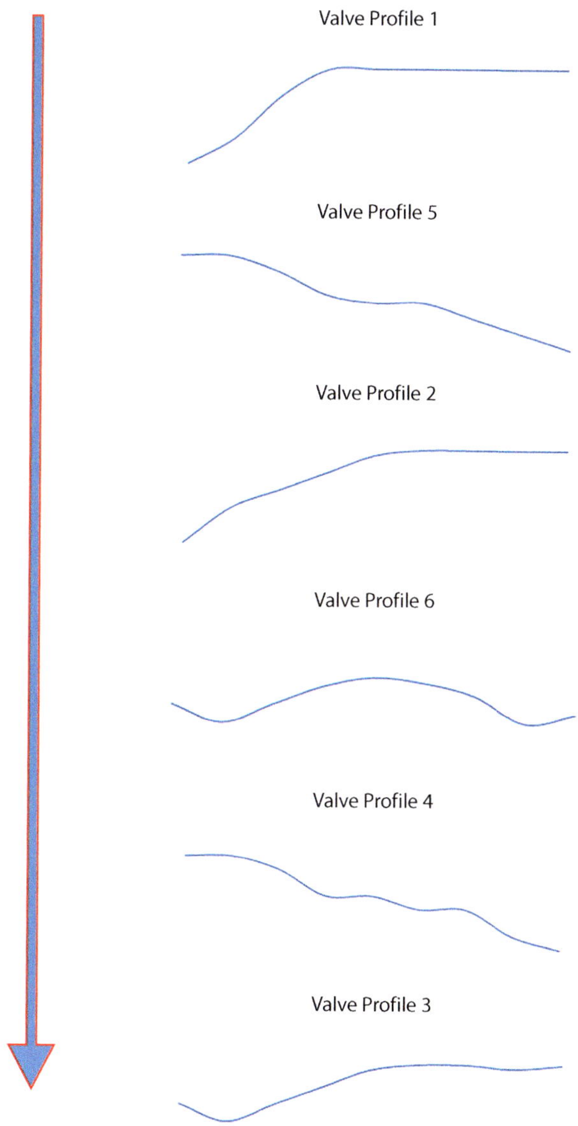

Figure 9.8 Valve profile buckling forces at the valve mantle ordered from lowest (top) to highest value (bottom).

H_A: Buckling forces are distinguishable with respect to different diatom valve profiles. The expectation is that different taxa will have different size reduction series (via varying lengths) with respect to the amount of buckling on the valve surface. Flatter valves will have smaller buckling forces at the mantle compared to bumpy valves as size reduction proceeds during valve formation. In the plane of the girdle (cingula), buckling of the valve mantle occurs as a result of a flexible girdle (cingula), inducing deformation as change in valve shape during size reduction [9.23].

The same six profiles constructed from least-squares polynomial curve fitting used in the first hypothesis test are used in the second test. These profiles have different slopes, different number of buckles per profile, and different distribution of buckles along the profile (Figure 9.4). In this test, L_{cr} varies for different valve profiles to represent size reduction during vegetative reproduction in which daughter cells are formed from division of mother cells. As L_{cr} decreases for each profile, Euler's critical buckling forces are calculated for four arbitrary steps in size reduction. As size decreases, buckling force decreases (Figure 9.9). During size reduction, the diatom valve has less buckling, and therefore lower buckling forces at the valve mantle.

Different life cycle stages have different buckling forces, but different diatom valve profiles are indistinguishable from each other with respect to buckling forces over the course of size reduction. Separate plots for each profile should be evident in order to distinguish different patterns of buckling as size reduction occurs. That is, the expectation is that size reduction stages should have different buckling forces, depending on the taxon. Test results show that buckling profiles are not measures of buckling forces during valve size reduction during morphogenesis. The results are arbitrary. Therefore, the Null hypothesis holds, and the alternate hypothesis is rejected.

9.3.2.1 *Inferences About Complementarity and Heterovalvy*

Using the results from the test of profiles from Figures 26a–e from Round *et al.* (1990) (Figure 9.5), comparison of complementarity and heterovalvy is made regarding buckling forces with the same valve profile length of 1.000. All taxa profiles had approximately the same buckling force within a narrow range of 2879.14 to 2881.33 (Figure 9.10). The *Navicula* profile exhibiting complementarity of planar valves had a buckling force that was only slightly different from the *Cocconeis* profile exhibiting heterovalvy (Figures 9.5 and 9.10). Not all concave-convex cells are heterovalvar, so additional criteria must be added to enable comparison of complementarity and heterovalvy in all diatom cells. The test is inconclusive.

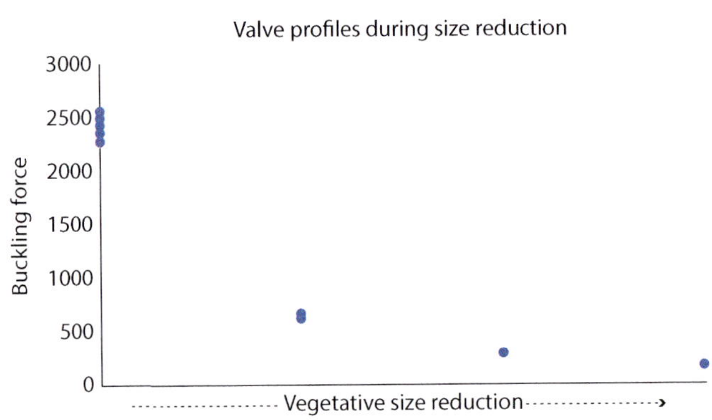

Figure 9.9 Plot of size reduction step during diatom vegetative reproduction using all Figure 9.4 tracings of valve profiles. The overall trend is a negative exponential. Buckling forces decrease with a reduction in valve profile length that represents size diminution of the diatom cell.

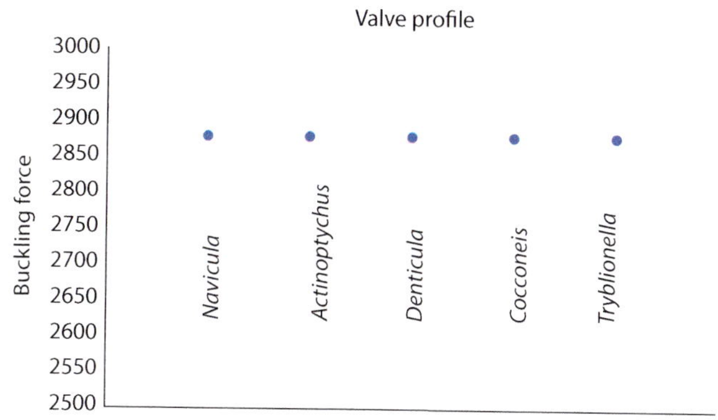

Figure 9.10 Plot of valve profile versus buckling force representing Figure 9.4 tracings. Complementarity is exhibited in profiles of *Navicula* and *Actinoptychus*. Heterovalvy is exhibited in the profile of *Cocconeis*.

9.3.3 Are Profiles Measures of Buckling Forces During Valve Morphogenesis?

The third hypothesis test consists of the following:

> H_0: Diatom valve profiles are straight lines (that is, no buckling) and of the same length.
> H_A: Diatom valve profiles are non-linear, meaning that they exhibit buckling and are of the same length.

In Table 9.4, all valve profiles used thus far are included in analysis. A straight-line profile is shown with no buckling (*Navicula*) as are the remaining buckled profiles with fifth- and sixth-order polynomial curve fits. Non-buckled and buckled profile lengths are equal by normalization to 1 to eliminate measurement error. Because, $P_{cr} = \frac{\pi^2 EI}{(KL_{cr})^2}$, Euler's critical load is inversely proportional to length. With all lengths equal, and all other values held

Table 9.4 Length profiles normalized to 1 for *n*th-order polynomial curve fits in Figures 9.4 and 9.5.

Profile	Length	Euler's critical load	Buckling?
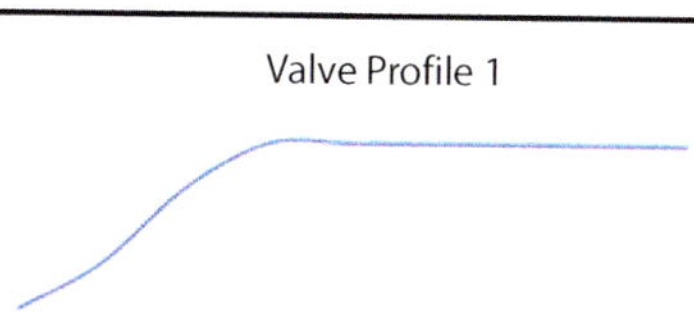	1	2881.92	YES
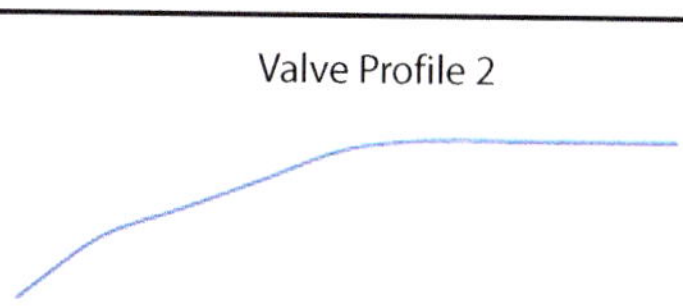	1	2881.92	YES

(*Continued*)

Table 9.4 Length profiles normalized to 1 for *n*th-order polynomial curve fits in Figures 9.4 and 9.5. (*Continued*)

Profile	Length	Euler's critical load	Buckling?
Valve Profile 3 	1	2881.92	YES
Valve Profile 4 	1	2881.92	YES
Valve Profile 5 	1	2881.92	YES
Valve Profile 6 	1	2881.92	YES
Navicula 	1	2881.92	NO
Cocconeis 	1	2881.92	YES
Actinoptychus 	1	2881.92	YES
Denticula 	1	2881.92	YES
Tryblionella 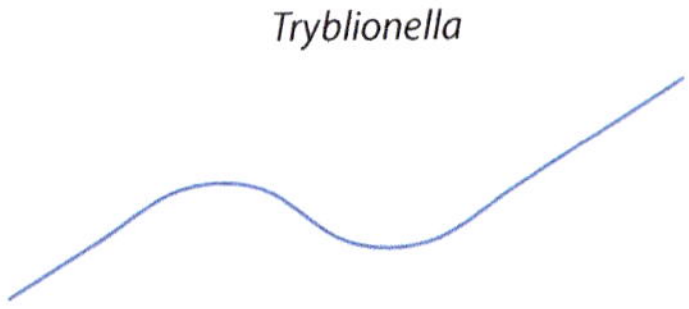	1	2881.92	YES

constant, all values of Euler's critical load are identical (Table 9.4). The result is that all buckling patterns cannot be distinguished from a straight-line pattern. Even one flat diatom valve such as some *Thalassiosira* [9.31] cannot be distinguished from another flat diatom valve such as *Melosira* to see the difference in valve morphogenesis. Both these non-buckled taxon profiles would have the same buckling force as any buckled profile such as *Cyclotella*, *Actinoptychus*, or any pennate diatom for that matter. Therefore, profiles are not measures of buckling forces during valve morphogenesis. The Null hypothesis holds, and the alternate hypothesis is rejected.

9.4 Discussion

Are mantle profiles of circular centric diatoms a measure of buckling forces during valve morphogenesis? Buckling forces are related to length of valve profiles, according to small deformation theory. However, this relation does not necessarily inform us about valve formation during morphogenesis whether looking at specific profiles such as those given Figure 9.5, or arbitrary profiles of approximately the same length that may represent specific diatom taxa undergoing size reduction (Figure 9.4). Buckling load is dependent on the shape of the function used and the distribution of buckles along the length of the profile. We do not know what the governing equations are for silica deposition during valve formation; therefore, exact solutions for buckling are not possible. Buckling occurring equally along the length of the profile will produce a higher buckling force value; uneven distribution of buckles will produce a lower value [9.16]. Stiffness versus elasticity is variable on different parts of the same cell and must be determined as well [9.1]. Multiple profiles may vary in length with respect to the number of buckles on the same valve for a given specimen, and profile length may vary on each valve of the same cell. There are many caveats to overcome in the determination of the relation between buckling force at the valve mantle and profile and how this relates to valve formation or size reduction.

Buckling forces at the mantle may be measured as a result of valve profiles at the completion of valve formation. However, such profiles were determined to be indistinguishable in terms of their particular characteristics, namely, number and distribution of buckles. In fact, flat valves with straight-line profiles are indistinguishable from buckled profiles. Looking at valve profiles alone is insufficient to determine whether buckling forces have specific outcomes in diatom valve formation during morphogenesis. However, valve profiles and their consequent buckling force measurements at the valve mantle may be useful in discerning differences between complementarity and heterovalvy in valve formation.

As new valves are formed to complete each daughter and mother cell during division, the new valves may exhibit complementarity with the corresponding original valve or heterovalvy [9.31]. In complementarity, the original valve may act as a molding surface in which the new valve shape is identical to the original, or the new valve may form to fill in peaks and valleys in the original, producing a mirror image [9.22, 9.31]. Commonly, diatom genera exhibit complementarity, with planar valves, as one example.

In heterovalvy, one valve is concave and the other valve is convex so that a tracing of each valve reveals the same shape. Heterovalvar genera are commonly seen with taxa having raphid and araphid valves in a single cell. Genera such as *Achnanthes* [9.9], *Cocconeis* [9.38], and *Gephyria* [9.39] exhibit such heterovalvy. Another mode of heterovalvy occurs in taxa that have an attached existence or form filamentous colonies such as *Paralia*, *Ellerbeckia*, *Trochosira*, and *Aulacoseira* where cells are held together with interlocking structures [9.15, 9.31, 9.34]. In *Aulacoseira*, *A. islandica* cells are linked together in chains with spatulate spines, while lanceolate spines are present on separation valves [9.21], and this phenomenon is seen in *A. granulata* and *Skeletonema costatum* [9.31] as well as in *Paralia sulcata* [9.13]. The presence of a pilus or rimoportula on one valve and the absence on the other valve is another example of heterovalvy as exhibited in *Plagiogrammopsis vanheurckii* [9.32].

Heterovalvy may suggest non-synchronicity in daughter cell morphogenesis because two different valves are forming [9.32]. *Craspedostauros australis* may have a very subtle heterovalvy as non-synchronicity was detected in formation of daughter cells [9.10]. Synchronicity is important in valve morphogenesis and warrants further study when considering heterovalvy as well [9.32].

Complementarity may have an important function with respect to valve morphogenesis and adaptation in the environment. Benthic diatoms that crawl may need to have complementary valves to enable this function. In contrast, heterovalvar diatoms live via attached raphe valves and have a need to conform to substrate. *Cocconeis* have an epilithic and epiphytic existence, while other monoraphid taxa such as *Anorthoneis* live in the psammon [9.31]. Habitat preference may be a result of the heterovalvar form of each taxon.

Functionality may play a role in heterovalvar forms as well. Colonial forms may appear to be complementary forms. However, *Aulacoseira* colonies with interlocking spatulate spines also have separation cells with sharp spines, indicating that this taxon is heterovalvar [9.31]. Heterovalvy is present in *Aulacoseira distans*, with spineless separation valves [9.14]. Having a colony that stays intact as vegetative reproduction occurs enables control of buoyancy and sinking in the water column for such planktonic forms [9.8]. Separation valves, when found, are present at the end of an *Aulacoseira* colony [9.31]. Valves exemplifying another form of heterovalvy include the presence of external structures such as portulae [9.12] on one valve with absence of the structure on the other valve as is the case in *Plagiogrammopsis* [9.32].

Buckling force measurements may serve to show the differences between complementarity and heterovalvy. For planar complimentary valve profiles such as those found in *Navicula*, buckling forces were larger than those for heterovalvar taxa such as *Cocconeis* (Table 9.3). Stiffness and buckling are two properties of importance during valve morphogenesis. Structural components such as the raphe may reflect the ability of complimentary valves to mitigate and resist cell breakage in spite of high buckling forces; both daughter and mother cell exhibit complementarity as a result of the capacity of stiffness [9.1]. In heterovalvy, flexibility rather than stiffness enables the diatom cell to undergo buckling without failure [9.16]. Concavity in concert with convexity in daughter and mother cells, respectively, may reflect the ability of such heterovalvar taxa to accommodate the valve shape necessary to suit their feeding habit [9.31].

Heterovalvy may also take the form of Janus cells. The number or density of striae differing from one valve to the other has been found in some *Gomphonema* as an example of Janus cells [9.4, 9.24]. The production of Janus cells in *Gomphonema* has been found to be a result of environmental influences producing an ecophenotypic response during morphogenesis. During valve morphogenesis, there may be taxa that produce separate valves that have an extreme level of morphological variation that may cover different orders in diatom classification [9.20]. Janus cells may be studied to understand phenotypic plasticity as diatom development occurs [9.4, 9.11].

Environmental conditions have been found to determine the occurrence of Janus cells during vegetative reproduction in taxa such as *Cyclotella* [9.25], which is unlike heterovalvar diatoms such as *Gomphonema* that have valve structural differences [9.23]. Polymorphisms (e.g., Stoermer, 1967), phenotypic plasticity, and identifications of conspecific diatoms (e.g., [9.18]) are all important in consideration of identifying particular cells as having one valve with one morphology and the other valve having another morphology (e.g., [9.24, 9.25, 9.37]). Janus cells occur intermittently as a result of environmental influences (e.g., [9.4]).

9.4.1 Laminated Structures and Mantle Buckling Forces Affecting the Valve Profile

For laminated structures, buckling force calculations follow two applicable formulae with regard to glass [9.26, 9.27]. One is Sattler and Stein's [9.33] formula for critical load for laminated glass via Euler's buckling force, $P_{cr} = \frac{\pi^2 \cdot (1+\alpha+\pi^2 \cdot \alpha \cdot \beta)}{1+\pi^2 \cdot \beta} \cdot \frac{D_0}{L_{cr}^2}$, where P_{cr} is the critical load buckling force, D_0 is the bending stiffness of laminated glass "lower limit," α is the moment of inertia coefficient, β is the stiffness coefficient, and L_{cr} is the critical buckling length.

The other is Zenkert's [9.41] formula for critical load for laminated glass via Euler's buckling force, $P_{cr} = \frac{\frac{\pi^4 \cdot D_L D_0}{D_f \cdot L_{cr}^4} + \frac{\pi^2 \cdot D_0}{L_{cr}^2}}{1+\frac{\pi^2 \cdot D_0}{D_f \cdot L_{cr}^2}}$, where P_{cr} is the critical load buckling force, D_L is the bending stiffness of single glass layers in laminated glass, D_0 is the bending stiffness of non-bonded glass layers, D_f is the bending stiffness of foil (= interlayered material used in laminated glass), and L_{cr} is the critical buckling length. In both methods, the critical buckling length is inversely proportional to critical load buckling, just as it is in Euler's critical load for a single layer, if all other variables are held constant. Therefore, the same result as that for a single layer in the previous analyses is obtained. However, buckling of laminated glass may be dependent on stiffness (e.g., [9.6]), and this factor must be considered with respect to the relation between the valve profile and mantle as well as the relation between the valve mantle and cingula of a diatom cell.

Buckling forces at the valve mantle may be a useful measure with regard to the compilation of data as a support for a role that the mantle plays in valve face formation. Characteristics of valve face shape using buckling force measures at the valve mantle may be an adjunct to studies involving complementarity and heterovalvy, including the formation of Janus cells. In order for measurement of buckling forces at the valve mantle to have utility in diatom morphogenesis and vegetative reproduction studies, additional information based on experimental evidence needs to be included such as follows: interaction effects of the valve mantle with cingula during valve formation; measurement of different thicknesses of amorphous silicon dioxide being present on different parts of a given valve as morphogenesis or size reduction occurs; silica availability and uptake dynamics of diatom cells during valve formation and division; the relation of internal versus external forces on the diatom cell; and radial forces perpendicular to axial forces on the diatom cell at the valve mantle.

9.5 Conclusion

Are mantle profiles of circular centric diatoms a measure of buckling forces during valve morphogenesis? No. Profiles are arbitrary *n*th-order polynomials of equal length so that buckling forces are arbitrary values as well. It is impossible to distinguish between no buckling and buckling using valve profiles regardless of the diatom valve being considered. These negative results are instructive in that buckling during valve morphogenesis requires a more sophisticated model that has a basis in biological evidence, as mentioned in the previous paragraph, rather than using a purely physical model.

Acknowledgement

We would like to thank Dorothea Blostein and her students, Niyousha Saeidi and Nuwan Perera, from Queens University for many discussions in 2018.

References

[9.1] Almqvist, N., Delamo, Y., Smith, B.L., Thomson, N.H., Bartholdson, Å., Lal, R., Brzezinski, M., Hansma, P.K., Micromechanical and structural properties of a pennate diatom investigated by atomic force microscopy. *J. Microsc.*, 202, Pt 3, 518-532, 2001.

[9.2] American Institute of Steel Construction (AISC), *Manual of Steel Construction*, 8th edition, American Institute of Steel Construction, Inc., Chicago, Illinois, 1980, Table C1.8.1, Effective Length Factor (*K*).

[9.3] American Concrete Institute Committee, *Building Code Requirements for Structural Concrete (ACI 318-05)*, American Concrete Institute, Farmington Hills, Michigan, 2005, Section R10.12.1, Effective Length Factor (*K*).

[9.4] Andrejić, J.Z., Spaulding, S.A., Manoylov, K.M., Edlund, M.B., Phenotypic plasticity in diatoms: Janus cells in four *Gomphonema* taxa. *Diatom Res.*, 33, 4, 453–470, 2019.

[9.5] AZo Materials, *Silica – Silicon Dioxide (SiO2)*, https://www.azom.com/properties.aspx?ArticleID=1114, AZoNetwork UK ltd., Manchester, UK, Accessed on November 1, 2019.

[9.6] Blaauwendraad, J., Buckling of laminated glass columns. *Herons*, 52, 1/2, 147–164, 2007.

[9.7] Boresi, A.P. and Schmidt, R.J., *Advanced Mechanics of Materials*, 6th edition, John Wiley & Sons, Inc., Hoboken, New Jersey, 2003.

[9.8] Bormans, M. and Webster, I.T., Modelling the spatial and temporal variability of diatoms in the River Murray. *J. Plankton Res.*, 21, 3, 581–598, 1999.

[9.9] Boyle, J.A., Pickett-Heaps, D.J., Czarnecki, D.B., Valve morphogenesis in the pennate diatom *Achnanthes coarctata*. *J. Phycol.*, 20, 563–76, 1984.

[9.10] Cox, E.J., Variation in patterns of valve morphogenesis between representatives of six biraphid diatom genera. *J. Phycol.*, 35, 1297–1312, 1999.

[9.11] Cox, E.J., Diatom identification in the face of changing species concepts and evidence of phenotypic plasticity. *J. Micropalaeontol.*, 33, 111–120, 2014.
[9.12] Crawford, R.M., Taxonomy and frustular structure of the marine centric diatom *Paralia sulcata*. *J. Phycol.*, 15, 2, 200–210, 1979.
[9.13] Crawford, R.M., Valve formation in diatoms and the fate of the silicalemma and plasmalemma. *Protoplasma*, 106, 157–166, 1981.
[9.14] Crawford, R.M. and Likhoshway, Y., The frustule structure of original material of *Aulacoseira distans* (Ehrenberg) Simonsen. *Diatom Res.*, 14, 2, 239–250, 1999.
[9.15] Davey, M.C. and Crawford, R.M., Filament formation in the diatom *Melosira granulata*. *J. Phycol.*, 22, 2, 144–150, 1986.
[9.16] Fenster, S.K. and Ugural, A.C., *Advanced Mechanics of Materials and Applied Elasticity*, 5th edition, Prentice Hall, Upper Saddle River, New Jersey, 2011.
[9.17] Gordon, R. and Tiffany, M.A., Possible buckling phenomena in diatom morphogenesis, in: *The Diatom World*, J. Seckbach and J.P. Kociolek (Eds.), pp. 245–272, Springer, Dordrecht, The Netherlands, 2011.
[9.18] Håkansson, H. and Chepurnov, V., A study of variation in valve morphology of the diatom *Cyclotella meneghiniana* in monoclonal cultures: effect of auxospore formation and different salinity conditions. *Diatom Res.*, 12, 2, 251–272, 1999.
[9.19] Kiss, K.T., Genkal, S.I., Ector, L., Molnár, L., Duleba, M., Bíró, P., Ács, E., Morphology, taxonomy and distribution of *Stephanodiscus triporus* (Bacillariophyceae) and related taxa. *Eur. J. Phycol.*, 48, 4, 363–379, 2013.
[9.20] Kociolek, J.P. and Spaulding, S.A., Morphological variation, species concepts, and classification of an unusual fossil centric diatom (Bacillariophyta) from western North America. *J. Phycol.*, 38, 4, 821–833, 2002.
[9.21] Le Cohu, R., Further observations and some comments on the fine structure of the centric diatom *Aulacoseira islandica* (Bacillariophyceae). *J. Phycol.*, 32, 333–338, 1996.
[9.22] Mann, D.G., An ontogenetic approach to diatom systematics, in: *Proceedings of the 7th International Diatom Symposium*, O. Koeltz, Koenigstein, Germany, pp. 113–141, 1984.
[9.23] Mann, D.G., The species concept in diatoms. *Phycologia*, 38, 437–495, 1999.
[9.24] McBride, S.A. and Edgar, R.K., Janus cells unveiled: Frustular morphometric variability in *Gomphonema angustatum*. *Diatom Res.*, 13, 293–310, 1998.
[9.25] Meyer, B. and Håkansson, H., Morphological variation of *Cyclotella polymorph asp.* Nov. (Bacillariophyceae). *Phycologia*, 45, 64–69, 1996.
[9.26] Nehme, K., Jakab, A., Nehme, S.G., Experiments on the buckling behaviour of glass columns. Part 1. *Mater. Sci.*, Epitoanyag-Journal of Silicate Based and Composite Materials, 65, 62-66, 2013.
[9.27] Nehme, K., Jakab, A., Nehme, S.G., Experiments on the buckling behaviour of glass columns. Part 2. *Mater. Sci.*, Epitoanyag-Journal of Silicate Based and Composite Materials, 65, 112-117, 2013b.
[9.28] Ni, H., Li, X., Gao, H., Elastic modulus of amorphous SiO_2 nanowires. *Appl. Phys. Lett.*, 88, 4, 043108, 2006.
[9.29] Pappas, J.L. and Gordon, R., Buckling: a geometric and biophysical multiscale feature of centric diatom valve morphogenesis, in: *Diatom Morphogenesis [DIMO, Volume 2 in the series: Diatoms: Biology & Applications, series editors: Richard Gordon & Joseph Seckbach]*, V. Annenkov, J. Seckbach, R. Gordon (Eds.), Wiley-Scrivener, Beverly, MA, USA, 2021.
[9.30] Round, F.E., Auxospore structure, initial valves and development of populations of *Stephanodiscus* in Farmoor Reservoir. *Ann. Bot.*, 53, 457–468, 1982.
[9.31] Round, F.E., Crawford, R.M., Mann, D.G., *The Diatoms, Biology & Morphology of the Genera*, Cambridge University Press, Cambridge, 1990.

[9.32] Sato, S., Valve and girdle band morphogenesis in a bipolar centric diatom *Plagiogrammopsis vanheurchii* (Cymatosiraceae, Bacillariophyta). *Eur. J. Phycol.*, 45, 2, 167–176, 2010.
[9.33] Sattler, K. and Stein, P., *Ingenieubauten 3, Theorie und Praxis*, Springer-Verlag, Wien, 1974.
[9.34] Schmid, A.M.M. and Crawford, R.M., *Ellerbeckia arenaria* (Bacillariophyceae): formation of auxospores and initial cells. *Eur. J. Phycol.*, 36, 4, 307–320, 2001.
[9.35] Schmid, A.M.M. and Volcani, B.E., Wall morphogenesis in *Coscinodiscus wailesii* Gran and Angst. I. Valve morphology and development of its architecture. *J. Phycol.*, 19, 4, 387–402, 1983.
[9.36] Spaulding, S.A., Bishop, I.W., Edlund, M.B., Lee, S., Furey, P., Jovanovska, E., Potapova, M., *Diatoms of North America*, 2019, https://diatoms.org/glossary/mantle/.
[9.37] Stoermer, E.F., Polymorphism in *Mastogloia. J. Phycol.*, 3, 73–77, 1967.
[9.38] Suzuki, H., Osada, K., Nagumo, T., Valve development in Cocconeis, in: *17th International Diatom Symposium*, Abstracts Book, M. Poulin, (Ed.), Ottawa, Canada, p. 140, 2002.
[9.39] Tiffany, M.A., Valve morphogenesis in the marine araphid diatom *Gephyria media* (Bacillariophyceae). *Diatom Res.*, 17, 391–400, 2002.
[9.40] von Stosch, H.A., An amended terminology of the diatom girdle. *Nova Hedwig. Beih.*, 53, 1–35, 1975.
[9.41] Zenkert, D. (Ed.), *The Handbook of Sandwich Construction*, Engineering Materials Advisory Service Ltd., Cradley Heath, West-Midland, U.K., 1997.

Part 3

PHYSIOLOGY, BIOCHEMISTRY AND APPLICATIONS

10

The Effect of the Silica Cell Wall on Diatom Transport and Metabolism*

Mark Hildebrand

Marine Biology Research Division, Scripps Institution of Oceanography University of California, San Diego, CA, United States

Abstract

The diatom silica cell wall is intrinsically interesting, but it is not commonly considered in terms of its effect on cellular transport and metabolic processes. There are several factors related to this to consider. One is that the rigid cell wall enables a variety of cell shapes that are substantially different from spherical. This has implications in terms of surface area: volume ratio, which will affect transport processes. Another factor is that the number of transport entry and exit points in the cell is limited by the characteristics of pores (number, size, and potentially shape) in the silica. Further, although pores are generally considered to be open structures, atomic force microscopy imaging has revealed that they are occluded with organic material. This could impart a level of selectivity over transport processes. Silica in the cell wall is impermeable on short time scales to gasses (CO_2 and O_2 specifically), which may affect carbon concentrating and photorespiratory mechanisms and metabolism. This chapter will explore possible ramifications to transport and metabolic processes in silicified diatoms.

*Mark Hildebrand intended to publish a chapter in our book, but he passed away after he submitted this abstract. In memorial, we list his works and some publications about him (prepared by Richard Gordon).

Vadim Annenkov, Joseph Seckback and Richard Gordon (eds.) Diatom Morphogenesis, (253–260) © 2022 Scrivener Publishing LLC

Publications by and about Mark Hildebrand

[10.1] Abbriano, R., Vardar, N., Yee, D., Hildebrand, M., Manipulation of a glycolytic regulator alters growth and carbon partitioning in the marine diatom *Thalassiosira pseudonana*. *Algal Res.-Biomass Biofuels Bioprod.*, 32, 250–258, 2018.

[10.2] https://algaeindustrymagazine.com/mark-hildebrand-leading-algal-biofuel-scientist-passes-away/

[10.3] Allison, D.P., Dufrêne, Y.F., Doktycz, M.J., Hildebrand, M., Biomineralization at the nanoscale: Learning from diatoms. *Methods Cell Biol.*, 90, 61–86, 2008.

[10.4] Armbrust, E.V., Berges, J.A., Bowler, C., Green, B.R., Martinez, D., Putnam, N.H., Zhou, S., Allen, A.E., Apt, K.E., Bechner, M., Brzezinski, M.A., Chaal, B.K., Chiovitti, A., Davis, A.K., Demarest, M.S., Detter, J.C., Glavina, T., Goodstein, D., Hadi, M.Z., Hellsten, U., Hildebrand, M., Jenkins, B.D., Jurka, J., Kapitonov, V.V., Kröger, N., Lau, W.W., Lane, T.W., Larimer, F.W., Lippmeier, J.C., Lucas, S., Medina, M., Montsant, A., Obornik, M., Parker, M.S., Palenik, B., Pazour, G.J., Richardson, P.M., Rynearson, T.A., Saito, M.A., Schwartz, D.C., Thamatrakoln, K., Valentin, K., Vardi, A., Wilkerson, F.P., Rokhsar, D.S., The genome of the diatom *Thalassiosira pseudonana*: ecology, evolution, and metabolism. *Science*, 306, 5693, 79–86, 2004.

[10.5] Beld, J., Abbriano, R., Finzel, K., Hildebrand, M., Burkart, M.D., Probing fatty acid metabolism in bacteria, cyanobacteria, green microalgae and diatoms with natural and unnatural fatty acids. *Mol. Biosyst.*, 12, 4, 1299–1312, 2016.

[10.6] Bernabeu, E., Sanchez-Brea, L.M., Siegmann, P., Martinez-Anton, J.C., Gomez-Pedrero, J.A., Wilkening, G., Koenders, L., Muller, F., Hildebrand, M., Hermann, H., Classification of surface structures on fine metallic wires. *Appl. Surf. Sci.*, 180, 3–4, 191–199, 2001.

[10.7] Cohoon, G.A., Alvarez, C.E., Meyers, K., Deheyn, D.D., Hildebrand, M., Kieu, K., Norwood, R.A., Analysis of quasi-periodic pore-network structure of centric marine diatom frustules. *Proc. SPIE*, 9341, 93410g, 2015.

[10.8] Cook, O. and Hildebrand, M., Enhancing LC-PUFA production in *Thalassiosira pseudonana* by overexpressing the endogenous fatty acid elongase genes. *J. Appl. Phycol.*, 28, 2, 897–905, 2016.

[10.9] Corbeil, L.B., Hildebrand, M., Shrestha, R., Davis, A., Schrier, R., Oyler, G.A., Rosenberg, J.N., *Diatom-based vaccines*, Patent US 09358283, The Regents of the University of California; Synaptic Research LLC, 2016.

[10.10] Curnow, P., Senior, L., Knight, M.J., Thamatrakoln, K., Hildebrand, M., Booth, P.J., Expression, purification, and reconstitution of a diatom silicon transporter. *Biochemistry*, 51, 18, 3776–3785, 2012.

[10.11] Davis, A., Abbriano, R., Smith, S.R., Hildebrand, M., Clarification of Photorespiratory Processes and the Role of Malic Enzyme in Diatoms. *Protist*, 168, 1, 134–153, 2017.

[10.12] Davis, A., Crum, L.T., Corbeil, L.B., Hildebrand, M., Expression of Histophilus somni IbpA DR2 protective antigen in the diatom *Thalassiosira pseudonana*. *Appl. Microbiol. Biotechnol.*, 101, 13, 5313–5324, 2017.

[10.13] Davis, A., Hildebrand, M., Palenik, B., Novel genes induced by short term Cu exposure in the diatom *Thalassiosira pseudonana*. *J. Phycol.*, 42, Suppl. 1, 20, 2006.

[10.14] Davis, A.K. and Hildebrand, M., A self-propagating system for Ge incorporation into nanostructured silica. *Chem. Commun.*, 37, 4495–4497, 2008.

[10.15] Davis, A.K., Hildebrand, M., Palenik, B., A stress-induced protein associated with the girdle band region of the diatom Thalassiosira pseudonana (Bacillariophyta). *J. Phycol.*, 41, 3, 577–589, 2005.

[10.16] Davis, A.K., Hildebrand, M., Palenik, B., Gene expression induced by copper stress in the diatom Thalassiosira pseudonana. *Eukaryot. Cell*, 5, 7, 1157–1168, 2006.
[10.17] Davis, A.K., Hildebrand, M., Sigel, H., Sigel, A., Molecular processes of biosilicification in diatoms, in: *Metal Ions in Life Sciences Vol 4, Biomineralization: From Nature to Application*, A. Sigel, H. Sigel, R.K.O. Sigel (Eds.), pp. 255–294, John Wiley & Sons Ltd., Chichester, UK, 2008.
[10.18] Doktycz, M.J., Hildebrand, M., Retterer, S.T., Allison, D.P., POLY 264-Understanding silica nanostructure formation using natural structures and bio-inspired techniques. *Abstracts of Papers of the American Chemical Society*, p. 234, 2007.
[10.19] Frigeri, L.G., Radabaugh, T.R., Haynes, P.A., Hildebrand, M., Identification of proteins from a cell wall fraction of the diatom *Thalassiosira pseudonana*: Insights into silica structure formation. *Mol. Cell. Proteomics*, 5, 1, 182–193, 2006.
[10.20] Grillet, N., Schwander, M., Hildebrand, M.S., Sczaniecka, A., Kolatkar, A., Velasco, J., Webster, J.A., Kahrizi, K., Najmabadi, H., Kimberling, W.J., Stephan, D., Bahlo, M., Wiltshire, T., Tarantino, L.M., Kuhn, P., Smith, R.J., Muller, U., Mutations in LOXHD1, an evolutionarily conserved stereociliary protein, disrupt hair cell function in mice and cause progressive hearing loss in humans. *Am. J. Hum. Genet.*, 85, 3, 328–337, 2009.
[10.21] Hearst Elementary School, *Voyager: Where did you get the idea to use algae as biofuel?*, Submitted by 5th graders at Hearst Elementary School, San Diego, Calif., 2008, https://scripps.ucsd.edu/news/voyager-where-did-you-get-idea-use-algae-biofuel.
[10.22] Hildebrand, M., Arabidopsis cell cultures, 2000, http://iubio.bio.indiana.edu/R27766-29127-/Network-News/bionet/genome/arabidopsis/10005.newsm.
[10.23] Hildebrand, M., Silicic acid transport and its control during cell wall silicification in diatoms, in: *Biomineralization: from Biology to Biotechnology and Medical Applications*, E. Bäuerlein (Ed.), pp. 171–188, Wiley-VCH, Weinheim, Germany, 2000.
10.24. Hildebrand, M., Lack of coupling between silicon and other elemental metabolisms in diatoms. *J. Phycol.*, 38, 5, 841–843, 2002.
[10.25] Hildebrand, M., Biological processing of nanostructured silica in diatoms. *Prog. Org. Coat.*, 47, 3–4, 256–266, 2003.
[10.26] Hildebrand, M., Biogenic nanostructured silica formation in diatoms: Proteins, genes, and structure. *Abstracts of Papers of the American Chemical Society*, p. 229, 2005, 093-POLY.
[10.27] Hildebrand, M., Cloning and functional characterization of ammonium transporters from the marine diatom Cylindrotheca fusiformis (Bacillariophyceae). *J. Phycol.*, 41, 1, 105–113, 2005.
[10.28] Hildebrand, M., The prospects of manipulating diatom silica nanostructure. *J. Nanosci. Nanotechnol.*, 5, 1, 146–157, 2005.
[10.29] Hildebrand, M., Application of molecular and genomics tools to diatoms. *J. Phycol.*, 42, Suppl. 1, 6, 2006.
[10.30] Hildebrand, M., Diatoms, biomineralization processes, and genomics. *Chem. Rev.*, 108, 11, 4855–4874, 2008.
[10.31] Hildebrand, M., Foreword, in: *The Diatom World*, J. Seckbach and J.P. Kociolek (Eds.), pp. vii–ix, Springer, Dordrecht, The Netherlands, 2011.
[10.32] Hildebrand, M., Lipid induction and gene expression of Thalassiosira pseudonana during silicon starvation, in: *1st International Conference, The Molecular Life of Diatoms*, N. Kröger, T. Mock, P. Kroth, M. Hildebrand, G. Armbrust and N. Poulsen, (Eds.), June 5th-9th 2011, Georgia Institute of Technology, Atlanta, GA USA, 2011.
[10.33] Hildebrand, M., The effect of the silica cell wall on diatom transport and metabolism [abstract only], with a list of publications by and about Mark Hildebrand, in: *Diatom Morphogenesis [DIMO, Volume 2 in the series: Diatoms: Biology & Applications, series editors: Richard*

Gordon & Joseph Seckbach], V. Annenkov, J. Seckbach, R. Gordon (Eds.), Wiley-Scrivener, Beverly, MA, USA, 2019.

[10.34] Hildebrand, M., Abbriano, R.M., Polle, J.E., Traller, J.C., Trentacoste, E.M., Smith, S.R., Davis, A.K., Metabolic and cellular organization in evolutionarily diverse microalgae as related to biofuels production. *Curr. Opin. Chem. Biol.*, 17, 3, 506–514, 2013.

[10.35] Hildebrand, M., Corey, D.K., Ludwig, J.R., Kukel, A., Feng, T.Y., Volcani, B.E., Plasmids in diatom species. *J. Bacteriol.*, 173, 18, 5924–5927, 1991.

[10.36] Hildebrand, M. and Dahlin, K., Cloning and characterization of nitrate transporter genes from the diatom *Cylindrotheca fusiformis*. *J. Phycol.*, 36, 30, 2000.

[10.37] Hildebrand, M. and Dahlin, K., Nitrate transporter genes from the diatom *Cylindrotheca fusiformis* (Bacillariophyceae): mRNA levels controlled by nitrogen source and by the cell cycle. *J. Phycol.*, 36, 4, 702–713, 2000.

[10.38] Hildebrand, M., Dahlin, K., Volcani, B.E., Characterization of a silicon transporter gene family in *Cylindrotheca fusiformis*: sequences, expression analysis, and identification of homologs in other diatoms. *Mol. Gen. Genet.*, 260, 5, 480–486, 1998.

[10.39] Hildebrand, M., Davis, A., Abbriano, R., Pugsley, H.R., Traller, J.C., Smith, S.R., Shrestha, R.P., Cook, O., Sanchez-Alvarez, E.L., Manandhar-Shrestha, K., Alderete, B., Applications of imaging flow cytometry for microalgae, in: *Imaging Flow Cytometry: Methods and Protocols*, Barteneva, N.S. and Vorobjev, I.A. (Eds.), pp. 47–67, 2016.

[10.40] Hildebrand, M., Davis, A.K., Smith, S.R., Traller, J.C., Abbriano, R., The place of diatoms in the biofuels industry. *Biofuels*, 3, 2, 221–240, 2012.

[10.41] Hildebrand, M., Doktycz, M.J., Allison, D.P., Application of AFM in understanding biomineral formation in diatoms. *Pflugers Arch.-Eur. J. Physiol.*, 456, 1, 127–137, 2008.

[10.42] Hildebrand, M., Frigeri, L.G., Davis, A.K., Synchronized growth of *Thalassiosira pseudonana* (Bacillariophyceae) provides novel insights into cell-wall synthesis processes in relation to the cell cycle. *J. Phycol.*, 43, 4, 730–740, 2007.

[10.43] Hildebrand, M., Frigeri, L.G., Thamatrakoln, K., Davis, A.K., The value of monitoring the cell cycle in relation to physiological and molecular changes in phytoplankton. *J. Phycol.*, 45, 5, 2009.

[10.44] Hildebrand, M., Hasegawa, P., Ord, R.W., Thorpe, V.S., Glass, C.A., Volcani, B.E., Nucleotide sequence of diatom plasmids: identification of open reading frames with similarity to site-specific recombinases. *Plant Mol. Biol.*, 19, 5, 759–770, 1992.

[10.45] Hildebrand, M., Higgins, D.R., Busser, K., Volcani, B.E., Silicon-responsive cDNA clones isolated from the marine diatom Cylindrotheca fusiformis. *Gene*, 132, 2, 213–218, 1993.

[10.46] Hildebrand, M., Holton, G., Joy, D.C., Doktycz, M.J., Allison, D.P., Diverse and conserved nano- and mesoscale structures of diatom silica revealed by atomic force microscopy. *J. Microsc.-Oxford*, 235, 2, 172–187, 2009.

[10.47] Hildebrand, M., Kim, S., Shi, D., Scott, K., Subramaniam, S., 3D imaging of diatoms with ion-abrasion scanning electron microscopy. *J. Struct. Biol.*, 166, 3, 316–328, 2009.

[10.48] Hildebrand, M. and Lerch, S.J.L., Diatom silica biomineralization: Parallel development of approaches and understanding. *Semin. Cell Dev. Biol.*, 46, 27–35, 2015.

[10.49] Hildebrand, M., Lerch, S.J.L., Shrestha, R.P., Understanding diatom cell wall silicification—moving forward. *Front. Mar. Sci.*, 5, 125, 2018.

[10.50] Hildebrand, M., Manandhar-Shrestha, K., Abbriano, R., Effects of chrysolaminarin synthase knockdown in the diatom *Thalassiosira pseudonana*: Implications of reduced carbohydrate storage relative to green algae. *Algal Res.-Biomass Biofuels Bioprod.*, 23, 66–77, 2017.

[10.51] Hildebrand, M. and Palenik, B., Investigation into the optical properties of nanostructured silica from diatoms, 2000, https://www.researchgate.net/publication/235192483_Investigation_into_the_Optical_Properties_of_Nanostructured_Silica_from_Diatoms.

[10.52] Hildebrand, M. and Palenik, B., *Investigation into the optical properties of nanostructured silica from diatoms [Final Performance Report, Grant F49620-01-1-0165] [https://apps.dtic.mil/dtic/tr/fulltext/u2/a413662.pdf]*, US Air Force Office Scientific Research, Department of Defense, Washington, DC, USA, 2003.

[10.53] Hildebrand, M., Volcani, B.E., Gassmann, W., Schroeder, J.I., A gene family of silicon transporters. *Nature*, 385, 6618, 688–689, 1997.

[10.54] Hildebrand, M., Waggoner, L.E., Lim, G.E., Sharp, K.H., Ridley, C.P., Haygood, M.G., Approaches to identify, clone, and express symbiont bioactive metabolite genes. *Nat. Prod. Rep.*, 21, 1, 122–142, 2004.

[10.55] Hildebrand, M., Waggoner, L.E., Liu, H.B., Sudek, S., Allen, S., Anderson, C., Sherman, D.H., Haygood, M., *bryA*: An unusual modular polyketide synthase gene from the uncultivated bacterial symbiont of the marine bryozoan *Bugula neritina*. *Chem. Biol.*, 11, 11, 1543–1552, 2004.

[10.56] Hildebrand, M. and Wetherbee, R., Components and control of silicification in diatoms, in: *Silicon Biomineralization: Biology, Biochemistry, Molecular Biology, Biotechnology*, vol. 33, W.E.G. Mueller (Ed.), pp. 11–57, Springer, Berlin, Germany, 2003.

[10.57] Hildebrand, M., York, E., Kelz, J.I., Davis, A.K., Frigeri, L.G., Allison, D.P., Doktycz, M.J., Nanoscale control of silica morphology and three-dimensional structure during diatom cell wall formation. *J. Mater. Res.*, 21, 10, 2689–2698, 2006.

[10.58] Jacobs, J.D., Ludwig, J.R., Hildebrand, M., Kukel, A., Feng, T.Y., Ord, R.W., Volcani, B.E., Characterization of two circular plasmids from the marine diatom *Cylindrotheca fusiformis*: plasmids hybridize to chloroplast and nuclear DNA. *Mol. Gen. Genet.*, 233, 1–2, 302–310, 1992.

[10.59] Kieu, K., Li, C., Fang, Y., Cohoon, G., Herrera, O.D., Hildebrand, M., Sandhage, K.H., Norwood, R.A., Structure-based optical filtering by the silica microshell of the centric marine diatom *Coscinodiscus wailesii*. *Opt. Express*, 22, 13, 15992–15999, 2014.

[10.60] Kopanska, K.S., Tesson, B., Lin, H., Meredith, J.C., Hildebrand, M., Davis, A., Morphological factors involved in adhesion of acid-cleaned diatom silica. *Silicon*, 6, 2, 95–107, 2014.

[10.61] Kröger, N., Mock, T., Kroth, P., Hildebrand, M., Armbrust, G., Poulsen, N., *1st International Conference, The Molecular Life of Diatoms*, June 5th-9th 2011, Georgia Institute of Technology, Atlanta, GA USA. http://www.lifeofdiatoms.org/, 2011.

[10.62] Manandhar-Shrestha, K. and Hildebrand, M., Development of flow cytometric procedures for the efficient isolation of improved lipid accumulation mutants in a *Chlorella* sp. microalga. *J. Appl. Phycol.*, 25, 6, 1643–1651, 2013.

[10.63] Manandhar-Shrestha, K. and Hildebrand, M., Characterization and manipulation of a DGAT2 from the diatom Thalassiosira pseudonana: Improved TAG accumulation without detriment to growth, and implications for chloroplast TAG accumulation. *Algal Res.-Biomass Biofuels Bioprod.*, 12, 239–248, 2015.

[10.64] Martin-Jézéquel, V., Hildebrand, M., Brzezinski, M.A., Silicon metabolism in diatoms: Implications for growth. *J. Phycol.*, 36, 5, 821–840, 2000.

[10.65] Norden-Krichmar, T.M., Allen, A.E., Gaasterland, T., Hildebrand, M., Characterization of the small RNA transcriptome of the diatom, *Thalassiosira pseudonana*. *PLoS One*, 6, 8, e22870, 2011.

[10.66] Pan, Z., Lerch, S.J.L., Xu, L., Li, X., Chuang, Y.-J., Howe, J.Y., Mahurin, S.M., Dai, S., Hildebrand, M., Electronically transparent graphene replicas of diatoms: A new technique for the investigation of frustule morphology. *Sci. Rep.*, 4, 6117, 2014.

[10.67] Sachs, C., Hildebrand, M., Volkening, S., Wintterlin, J., Ertl, G., Spatiotemporal self-organization in a surface reaction: From the atomic to the mesoscopic scale. *Science*, 293, 5535, 1635–1638, 2001.

[10.68] Sanchez-Alvarez, E.L., Gonzalez-Ledezma, G., Bolanos Prats, J.A., Stephano-Hornedo, J.L., Hildebrand, M., Evaluating *Marinichlorella kaistiae* KAS603 cell size variation, growth and TAG accumulation resulting from rapid adaptation to highly diverse trophic and salinity cultivation regimes. *Algal Res.-Biomass Biofuels Bioprod.*, 25, 12–24, 2017.

[10.69] Sandhage, K.H., Allan, S.M., Dickerson, M.B., Gaddis, C.S., Shian, S., Weatherspoon, M.R., Cai, Y., Ahmad, G., Haluska, M.S., Snyder, R.L., Unocic, R.R., Zalar, F.M., Zhang, Y.S., Rapp, R.A., Hildebrand, M., Palenik, B.P., Merging biological self-assembly with synthetic chemical tailoring: The potential for 3-D genetically engineered micro/nano-devices (3-D GEMS). *Int. J. Appl. Ceram. Technol.*, 2, 4, 317–326, 2005.

[10.70] Sandhage, K.H., Shian, S., Gaddis, C.S., Weatherspoon, M.R., Cai, Y., Yoo, S., Haluska, M.S., Snyder, R.L., Liu, Y., Liu, M.L., Ferrell, N., Hansford, D.J., Hildebrand, M., Palenik, B., Biologically enabled syntheses of nanostructured 3-D sensor materials: the potential for 3-D genetically engineered, microdevices (3-D GEMs). *Rare Met. Mater. Eng.*, 35, Suppl. 3, 13–14, 2006.

[10.71] Schwartz, D., AIM Interview: Scripps' Dr. Mark Hildebrand, http://www.algaeindustrymagazine.com/aim-interview-scripps-mark-hildebrand/, 2014.

[10.72] Scripps Institution of Oceanography, *Transcript level coordination of carbon pathways during silicon starvation induced lipid accumulation in the diatom Thalassiosira pseudonana [http://www.ebi.ac.uk/ena/data/view/PRJNA304974]*, European Nucleotide Archive, 2015.

[10.73] Scripps Institution of Oceanography, *Transcript level coordination of carbon pathways during silicon-starvation induced lipid accumulation in the diatom Thalassiosira pseudonana [http://www.ebi.ac.uk/ena/data/view/PRJNA304348]*, European Nucleotide Archive, 2015.

[10.74] Scripps Institution of Oceanography, Obituary Notice: Mark Hildebrand, Leading Algal Biofuel Scientist, https://scripps.ucsd.edu/news/obituary-notice-mark-hildebrand-leading-algal-biofuel-scientist, 2018.

[10.75] Shrestha, R.P. and Hildebrand, M., Evidence for a regulatory role of diatom silicon transporters in cellular silicon responses. *Eukaryot. Cell*, 14, 1, 29–40, 2015.

[10.76] Shrestha, R.P. and Hildebrand, M., Development of a silicon limitation inducible expression system for recombinant protein production in the centric diatoms *Thalassiosira pseudonana* and *Cyclotella cryptica*. *Microb. Cell Fact.*, 16, 145, 2017.

[10.77] Shrestha, R.P., Tesson, B., Norden-Krichmar, T., Federowicz, S., Hildebrand, M., Allen, A.E., Whole transcriptome analysis of the silicon response of the diatom *Thalassiosira pseudonana*. *BMC Genomics*, 13, 499, 2012.

[10.78] Smith, S.R., Abbriano, R.M., Hildebrand, M., Comparative analysis of diatom genomes reveals substantial differences in the organization of carbon partitioning pathways. *Algal Res.-Biomass Biofuels Bioprod.*, 1, 1, 2–16, 2012.

[10.79] Smith, S.R., Gle, C., Abbriano, R.M., Traller, J.C., Davis, A., Trentacoste, E., Vernet, M., Allen, A.E., Hildebrand, M., Transcript level coordination of carbon pathways during silicon starvation-induced lipid accumulation in the diatom *Thalassiosira pseudonana*. *New Phytol.*, 210, 3, 890–904, 2016.

[10.80] Smith, S.R., Sanka, R., Gle, C., Abbriano, R.M., Traller, J.C., Davis, A.K., Trentacoste, E.M., Vernet, M., Allen, A.E., Hildebrand, M., GSE75651: Transcript level coordination of carbon pathways during silicon starvation induced lipid accumulation in the diatom Thalassiosira pseudonana [http://www.ncbi.nlm.nih.gov/geo/query/acc.cgi?acc=GSE75651], 2015.

[10.81] Sudek, S., Lopanik, N.B., Waggoner, L.E., Hildebrand, M., Anderson, C., Liu, H.B., Patel, A., Sherman, D.H., Haygood, M.G., Identification of the putative bryostatin polyketide synthase gene cluster from "Candidatus endobugula sertula", the uncultivated microbial symbiont of the marine bryozoan Bugula neritina. *J. Nat. Prod.*, 70, 1, 67–74, 2007.

[10.82] Tesson, B. and Hildebrand, M., Dynamics of silica cell wall morphogenesis in the diatom *Cyclotella cryptica*: substructure formation and the role of microfilaments. *J. Struct. Biol.*, 169, 1, 62–74, 2010.
[10.83] Tesson, B. and Hildebrand, M., Extensive and intimate association of the cytoskeleton with forming silica in diatoms: Control over patterning on the meso- and micro-scale. *PLoS One*, 5, 12, e14300, 2010.
[10.84] Tesson, B. and Hildebrand, M., Characterization and localization of insoluble organic matrices associated with diatom cell walls: Insight into their roles during cell wall formation. *PLoS One*, 8, 4, e61675, 2013.
[10.85] Tesson, B., Lerch, S.J.L., Hildebrand, M., Characterization of a new protein family associated with the silica deposition vesicle membrane enables genetic manipulation of diatom silica. *Sci. Rep.*, 7, 13457, 2017.
[10.86] Thamatrakoln, K., Alverson, A.J., Hildebrand, M., Comparative sequence analysis of diatom silicon transporters: Toward a mechanistic model of silicon transport. *J. Phycol.*, 42, 4, 822–834, 2006.
[10.87] Thamatrakoln, K. and Hildebrand, M., Comparative sequence analysis of newly isolated diatom silicic acid transporters, in: *17th North American Diatom Symposium*, E. Gaiser, (Ed.), 2003.
[10.88] Thamatrakoln, K. and Hildebrand, M., Approaches for functional characterization of diatom silicic acid transporters. *J. Nanosci. Nanotechnol.*, 5, 1, 158–166, 2005.
[10.89] Thamatrakoln, K. and Hildebrand, M., Analysis of *Thalassiosira pseudonana* silicon transporters indicates distinct regulatory levels and transport activity through the cell cycle. *Eukaryot. Cell*, 6, 2, 271–279, 2007.
[10.90] Thamatrakoln, K. and Hildebrand, M., Silicon uptake in diatoms revisited: A model for saturable and nonsaturable uptake kinetics and the role of silicon transporters. *Plant Physiol.*, 146, 3, 1397–1407, 2008.
[10.91] Traller, J., Cokus, S., Lopez, D., Gaidarenko, O., Smith, S., McCrow, J., Gallaher, S., Podell, S., Thompson, M., Cook, O., Morselli, M., Jaroszewicz, A., Allen, E., Allen, A., Merchant, S., Pellegrini, M., Hildebrand, M., MOESM2 of Genome and methylome of the oleaginous diatom *Cyclotella cryptica* reveal genetic flexibility toward a high lipid phenotype, 2016, 2017-04-13. http://dx.doi.org/10.6084/m9.figshare.c.3597308_D1.
[10.92] Traller, J.C., Cokus, S.J., Lopez, D.A., Gaidarenko, O., Smith, S.R., McCrow, J.P., Gallaher, S.D., Podell, S., Thompson, M., Cook, O., Morselli, M., Jaroszewicz, A., Allen, E.E., Allen, A.E., Merchant, S.S., Pellegrini, M., Hildebrand, M., Genome and methylome of the oleaginous diatom *Cyclotella cryptica* reveal genetic flexibility toward a high lipid phenotype. *Biotechnol. Biofuels*, 9, 258, 2016.
[10.93] Traller, J.C. and Hildebrand, M., High throughput imaging to the diatom *Cyclotella cryptica* demonstrates substantial cell-to-cell variability in the rate and extent of triacylglycerol accumulation. *Algal Res.-Biomass Biofuels Bioprod.*, 2, 3, 244–252, 2013.
[10.94] Trentacoste, E.M., Shrestha, R.P., Smith, S.R., Gle, C., Hartmann, A.C., Hildebrand, M., Gerwick, W.H., Metabolic engineering of lipid catabolism increases microalgal lipid accumulation without compromising growth. *Proc. Natl. Acad. Sci. U. S. A.*, 110, 49, 19748–19753, 2013.
[10.95] Trofimov, A.A., Pawlicki, A.A., Borodinov, N., Mandal, S., Mathews, T.J., Hildebrand, M., Ziatdinov, M.A., Hausladen, K.A., Urbanowicz, P.K., Steed, C.A., Ievlev, A.V., Belianinov, A., Michener, J.K., Vasudevan, R., Ovchinnikova, O.S., Deep data analytics for genetic engineering of diatoms linking genotype to phenotype via machine learning. *NPJ Comput. Mater.*, 5, 4, 2019.

[10.96] vanWaasbergen, L.G., Hildebrand, M., Tebo, B.M., Identification and characterization of a gene cluster involved in manganese oxidation by spores of the marine *Bacillus* sp. strain SG-1. *J. Bacteriol.*, 178, 12, 3517–3530, 1996.

11

Diatom Plasticity: Trends, Issues, and Applications on Modern and Classical Taxonomy, Eco-Evolutionary Dynamics, and Climate Change

Lawrence Victor D. Vitug

Biology Department, College of Science, Pamatasan ng Lungsod ng Manila (University City of Manila) General Luna, Manila, Philippines

Abstract

Diatoms exhibit phenotypic plasticity because of environmental changes. It is known that external and internal factors play an important role in changes in the morphology of diatoms. Several changes in morphology in diatoms may result in several morphotypes that have the same genotype. This paper aims to discuss the trends and issues covering diatom phenotypes that go beyond external and internal factors that affect morphological changes in varying/specific environmental conditions; it will also further explain the influence of biogeography, adaptive and evolutionary behavior, macromolecular composition, and climate change, challenging the traditional species concept and molecular approaches that we might gain perspective and insights on diatom phenotypic plasticity. In the review, *Phaeodactylum tricornutum* and other diatoms species were chosen as diatom model systems that exhibits phenotypic plasticity. Finally, I discuss the application of phenotypic plasticity in some diatoms that undergo plasticity and an emerging field on eco-evolutionary dynamics that can be linked to phenotypic plasticity in diatoms.

Keywords: Diatom plasticity, eco-evolutionary dynamics, adaptive behavior, phenotypic plasticity

11.1 Introduction

Morphological changes in certain diatom species depend on environmental conditions. Such an example is that of two valves of a single frustules that differ morphologically. Such inherently heterovalveate or Janus Cells reveal changes with a single genotype [11.1], hence exhibiting two different morphological forms [11.2]. Janus cells are described as a kind of heterovalvy in diatom species that typically generate in homomorphic frustules [11.3].

Email: lvdvitug@plm.edu.ph
https://www.linkedin.com/in/ldvitug/;
https://www.researchgate.net/profile/Lawrence_Vitug;
https://twitter.com/DiatomProtege;
https://scholar.google.com/citations?user=29xsatMAAAAJ&hl=en;
https://orcid.org/0000-0001-8996-5914

Vadim Annenkov, Joseph Seckback and Richard Gordon (eds.) Diatom Morphogenesis, (261–268) © 2022 Scrivener Publishing LLC

Mechanisms of phenotypic plasticity involve both physiological and morphological changes in reaction to shifting environmental variations. Examples of these morphological changes are variations due to changes in temperature [11.4], interaction between temperature and CO_2 concentration [11.5], and temperature and salinity changes [11.6]. Various diatom taxa have undergone ecophenotypic plasticity in response environmental changes. These variations are in Janus cells including *Mastogloia* [11.7], *Cyclotella* [11.8], and *Proboscia Sundström*. It was reported that heterovalvy is not related to the environmental fluctuation for *Thalassiosira pseudonana* (Hustedt) Hasle [11.9]. Generally, phenotypic plasticity is a simple character state that may have an adaptive behavior differing specific environment [11.10]. Furthermore, caution is needed in naming certain diatoms as separate species that may have more than one morphological expression/variation but belong to a single genotype. Wide phenotypic plasticity influences valves in chain forming diatoms such as *Skelotonema* [11.11], with *Raphoneis amphiceros*, whose morphological plasticity is related to various life cycle phases [11.12], or toxins produced by *Psuedo-nitzchia* [11.13]. Changes in gross morphology may be an indication of responses to environmental changes such as possibly giving an ecological benefit to variation in cell morphotype. Indication of teratological forms in diatoms [11.14] may be in synonymous to phenotypic plasticity as it also provides rapid response that is sensitive to environmental contaminants and may change its morphology in certain diatoms species.

This chapter addresses certain issues and provides new pieces of evidence that challenge the traditional definition of species concept. Other aspects may need to be considered, such as biogeography, climate change, ecology, adaptive and evolutionary behavior, and macromolecular compositions in certain species of diatoms.

11.2 Model Species: *Phaeodactylum tricornutum*

A good model species is *Phaeodactylum tricornutum*, a diatom that has become heavily investigated in the field of biotechnology for various applications, such as biofuel and the feeds for the aquaculture industry, due to its fast growth rate. This diatom has several morphotypes which are oval, fusiform, and triradiate [11.15, 11.16] and, sometimes, cruciform [11.17]. Information on the plesiomorphy of *Phaeodactylum* is relatively unknown. Having several morphotypes is a signal of cell's biological reaction to the natural environment. This plasticity can be correlated to a distinct kind of the cell wall, which is only weakly silicified in contrasted to most other diatom taxa [11.18, 11.15]. This species seems to be distinctive as it is not obligate for the uptake of silicic acid. Moreover, despite a large number of genome data, the basic biological characteristics of *P. tricornutum* are somewhat inadequately explored, specifically its reproductive biology, life histories, life cycle, and its ecological importance, although *P. tricornutum* has wide usage for its physiological and biochemical exploration, for food resources for aquaculture, because of its ease in cultivation and its high lipid productivity and content. Tripolarity in other diatoms was observed [11.19] and similar morphological changes detected in *Phaeodactylum* are not due to mutation, but are perhaps an expression of morphogenetic adaptability [11.20].

11.3 Transformation Mechanisms of *P. tricornutum*

The three main morphotypes (oval, triadiate, and fusiform) have shown that varying degrees of morphology change are dependent on growth conditions. Few studies have examined the endogenous alterations of morphotype formations or conversions [11.18, 11.20, 11.21]. The polymorphism of *P. tricornutum* may represent distinctive ecophenotypes. Each may be exclusively designed for growth under specific conditions. Differences between the three morphotypes and their morphological changes have been explained by characterizing the cell surface using AFM [11.22]. We need to understand its adaption to environmental fluctuations. Conversion to oval morphotype can be caused from all strains independent of genotype; the induction can be made by triradiate and fusiform morphotypes in nutrient-depleted media [11.20].

11.4 Future Advances in the Phenotypic Plasticity on *P. tricornutum*

Future work was summarized by [11.23] using several environmental growth parameters [11.24–11.26]. Attempts to understand the roles in genes have considered the carbon pathway [11.27], pigments [11.28], nitrogen and iron metabolisms [11.29], stress acclimatization [11.30], silicification activities in cells [11.31, 11.32], and the control of cell cycles [11.33]. In some instances, an oval morphotype cell can correspond to over-expressed genes of lipid metabolism while the triradiate morphotype library showed overexpression of genes for active transport processes.

11.4.1 Genomic and Molecular Mechanisms in Diatom Phenotypic Plasticity

Genetic markers have been recognized and examined in a variety of planktonic diatoms, revealing unusually elevated genetic diversity even at the population level [11.34]. Morphological characters used to contrast species with parallel morphology of Thalassiosiroid diatoms were detected. The great deal of phenotypic plasticity prevented precise identification of species using morphological methods alone [11.36]. Molecular methods and approaches have a vital role in the discovery and delineation of diatom species, demonstrating an unseen diversity and demonstating a taxon having cryptic species. Conversely, different cell morphotypes are to be found within a single species with the same genotype but because of the presence of phenotypic plasticity [11.19, 11.35–11.37].

11.4.2 Biogeography of Diatoms

The global distribution *P. tricornutum* has a restricted relationship with morphology, biogeography, and phylogeny. Various strains has been isolated from different geographical origins [11.20]. Limited relationships were found between morphotype and genotype. There is no specific genotype that could be classified as a variation in cell morphotypes, indicating that morphological changes are very dependent relative to environmental conditions and not on the genotype [11.23].

Ocean diatoms such as *Fragilariopsis cylindrus*, *Chaetoceros simplex*, and *Pseudo-nitzschia subcurvata* have displayed vast phenotypic plasticity in response to various salinity and temperature fluctuations experienced seasonally [11.6]. It has been suggested that phenotypic plasticity is designated by the extreme variations of the polar marine environment. A marine clonal isolate of *Skeletonema* spp. was found to have one single geographical location but, interestingly from varying seasons, has a unique proportion of the photosynthetic pigments [11.38].

11.4.3 Eco-Evolutionary Dynamics Approach on Diatoms Phenotypic Plasticity

Experiments on eco-evolutionary dynamics have restricted their model systems in two interacting species (host-virus) [11.39] and (prey-predator) [11.39]. These studies give only a snapshot of the ecological and evolutionary dynamics of the focal species population involved. However, it showed that by manipulating the environment by varying nutrient medium, a distinct effect was found between prey and predator in two model system interaction (Bacteria-Protist) [11.40]. Using a multispecies model system the eco-evolutionary dynamics was significantly altered compared to two species system model system but in this case, it was done in a constant environment [11.39]. Host-parasite and host-ecosystem interactions may evolve in tandem, functionally connecting evolutionary and ecological processes [11.39, 11.42].

These approaches can be applied to diatoms. A good example of a model species that can be used is that the chain length plasticity in *Skeletonema marinoi* is beneficial in adapting to life in the pelagic environment. Chains are created by daughter cells that stay affixed after cell division, leading to grazer-stimulated chain length plasticity [11.41]. This chain-forming diatom thus exhibits flexibility in size due to widespread plasticity in the chain length.

The development of varying morphologies in the chain of cells that shows phenotypic plasticity of *S. marinoi* has various degrees of flexibility [11.42]. *S. marinoi* cells have a defense mechanism which is linked to modification in the chain length. The marine diatom *Thalassiosira weissflogii* enhances the cell wall silicification in reaction to cues from grazing copepod [11.43]. Size-selective predation plays a vital role in the evolution of chain development and length basing it on-chain cell plasticity in *T. weissflogii* colonies.

These model species such as *Skelotenema* and *T. weissflogii* colonies can be used to understand the eco-evolutionary dynamics that can be used to linked ecology and evolution in order to understand phenotypic plasticity and its role in a community's function and structure. It is important not only to measure plasticity but also to assess its adaptive importance. Plastic reactions are induced and changes in fitness are monitored. For instance, defensive responses to a predator often reduce its susceptibility to that predator.

Moreover, this approach has not yet been tested on several diatoms species, but only on chlorophytes. As many species of diatoms have certain defensive mechanisms to grazers, it is possible to test a model system for diatoms and to see it is eco-evolutionary dynamics in mesocosm experiments to replicate the natural setting and see how it will respond to a given environment, interacting with other species.

11.4.4 Adaptive Behavior and Evolutionary Changes in Diatoms Linking to Diatom Plasticity

Individual isolated cells of *Navicula* exhibit a "circular run-and-reversal" pattern behavior with variation in concentration of dissolved silicic acid (dSi), which might be linked to phenotypic plasticity [11.44]. *Pseudo-Nitszscha* plasticity may be linked to the toxin production in which various biochemical processes in nature [11.6, 11.45].

11.4.5 Climate Change and Phenotypic Plasticity

Diatoms respond to climate change by phenotypic plasticity and evolutionary adaptation of organisms within its populations. Diatoms are indicators for gaging the effects of climate change because of their role in global biogeochemical cycling and because of the high abundance of microfossils archived in the sediment.

11.5 Conclusion

In exploring phenotypic plasticity, classical phycological on morphological, physiological, ecological, and evolutionary studies are intertwined with cutting edge molecular studies on omics (genomics, transcriptomics, and metabolomics). This will facilitate learning how these individuals can withstand a continually shifting environment. We should perhaps take a deeper look at diatoms species as morphologically distinct individual species, since all genotypes may possibly be pleiomorphic in nature.

References

[11.1] Cox, E.J., Diatom identification in the face of changing species concepts and evidence of phenotypic plasticity, *J. Micropalaeontol.*, 33, 111–120, 2014, https://doi.org/10.1144/jmpaleo2014-014.

[11.2] Mann, D.G., The species concept in diatoms. *Phycologia*, *38*, 6, 437–495, 1999, https://doi.org/10.2216/i0031-8884-38-6-437.1.

[11.3] Andrejić, J.Z., Spaulding, S.A., Manoylov, K.M., Edlund, M.B., Phenotypic plasticity in diatoms: Janus cells in four Gomphonema taxa. *Diatom Res.*, *33*, 4, 453–470, 2018, https://doi.org/10.1080/0269249X.2019.1572652.

[11.4] Gsell, A.S., Domis, L.N., de, S., Przytulska-Bartosiewicz, A., Mooij, W.M., van Donk, E., Ibelings, B.W., Genotype-by-Temperature Interactions May Help to Maintain Clonal Diversity in Asterionella Formosa (bacillariophyceae). *J. Phycol.*, *48*, 5, 1197–1208, 2012, https://doi.org/10.1111/j.1529-8817.2012.01205.x.

[11.5] Zhu, Z., Qu, P., Fu, F., Tennenbaum, N., Tatters, A.O., Hutchins, D.A., Understanding the blob bloom: Warming increases toxicity and abundance of the harmful bloom diatom Pseudo-nitzschia in California coastal waters. *Harmful Algae*, *67*, 36–43, 2017, https://doi.org/10.1016/j.hal.2017.06.004.

[11.6] Sackett, O., Armand, L., Beardall, J., Hill, R., Doblin, M., Connelly, C., Heraud, P., Taxon-specific responses of Southern Ocean diatoms to Fe enrichment revealed by synchrotron radiation FTIR microspectroscopy. *Biogeosciences*, *11*, 20, 5795–5808, 2014, https://doi.org/10.5194/bg-11-5795-2014.

[11.7] Stoermer, E.F., Polymorphism in Mastogloia1. *J. Phycol.*, *3*, 2, 73–77, 1967, https://doi.org/10.1111/j.1529-8817.1967.tb04633.x.

[11.8] Schultz, M.E., Salinity-related polymorphism in the brackish-water diatom Cyclotella cryptica. *Can. J. Bot.*, *49*, 8, 1285–1289, 1971, https://doi.org/10.1139/b71-182.

[11.9] Jordan, R.W., Ligowski, L., Nöthig, E.-M., Priddle, J., The Diatom Genus Proboscia in Antarctic Waters. *Diatom Res.*, *6*, 1, 63–78, 1991, https://doi.org/10.1080/0269249X.1991.9705148.

[11.10] Nijhout, H.F., Development and evolution of adaptive polyphenisms. *Evol. Dev.*, *5*, 1, 9–18, 2003, https://doi.org/10.1046/j.1525-142X.2003.03003.x.

[11.11] Medlin, L.K., Elwood, H.J., Stickel, S., Sogin, M.L., Morphological and Genetic Variation Within the Diatom Skeletonema Costatum (bacillariophyta): Evidence for a New Species, Skeletonema Pseudocostatum1. *J. Phycol.*, *27*, 4, 514–524, 1991, https://doi.org/10.1111/j.0022-3646.1991.00514.x.

[11.12] Falasco, E., Bona, F., Badino, G., Hoffmann, L., Ector, L., Diatom teratological forms and environmental alterations: a review, *Hydrobiologia*, 623, 1–35, 2009, https://doi.org/10.1007/s10750-008-9687-3.

[11.13] Marchetti, A., Maldonado, M.T., Lane, E.S., Harrison, P.J., Iron requirements of the pennate diatom Pseudo-nitzschia: Comparison of oceanic (high-nitrate, low-chlorophyll waters) and coastal species. *Limnol. Oceanogr.*, 51, 2092–2101, 2006.

[11.14] Falasco, E., Ector, L., Wetzel, C.E., Badino, G., Bona, F., Looking back, looking forward: a review of the new literature on diatom teratological forms (2010–2020). *Hydrobiologia*, 1675–1753, 2021, https://doi.org/10.1007/s10750-021-04540-x.

[11.15] De Martino, A., Bartual, A., Willis, A., Meichenin, A., Villazán, B., Maheswari, U., Bowler, C., Physiological and Molecular Evidence that Environmental Changes Elicit Morphological Interconversion in the Model Diatom Phaeodactylum tricornutum. *Protist*, *162*, 3, 462–481, 2011, https://doi.org/10.1016/j.protis.2011.02.002.

[11.16] Borowitzka, M.A. and Volcani, B.E., The Polymorphic Diatom Phaeodactylum Tricornutum: Ultrastructure of Its Morphotypes1,2. *J. Phycol.*, *14*, 1, 10–21, 1978, https://doi.org/10.1111/j.1529-8817.1978.tb00625.x.

[11.17] He, L., Han, X., Yu, Z., A Rare Phaeodactylum tricornutum Cruciform Morphotype: Culture Conditions, Transformation and Unique Fatty Acid Characteristics. *PLoS One*, *9*, 4, 1–8, 2014, https://doi.org/10.1371/journal.pone.0093922.

[11.18] Lewin, J.C., The Taxonomic Position of Phaeodactylum tricornutum. *Microbiology*, *18*, 2, 427–432, 1958, https://doi.org/10.1099/00221287-18-2-427.

[11.19] Schmid, A.M.M., Intraclonal variation of the tripolar pennate diatom. *Nova Hedwig.*, 65, 27–45, 1997, https://doi.org/10.1127/nova.hedwigia/65/1997/27.

[11.20] Martino, A.D., Meichenin, A., Shi, J., Pan, K., Bowler, C., Genetic and phenotypic characterization of Phaeodactylum tricornutum (Bacillariophyceae) accessions1. *J. Phycol.*, *43*, 5, 992–1009, 2007, https://doi.org/10.1111/j.1529-8817.2007.00384.x.

[11.21] Bartual, A., Gálvez, J.A., Ojeda, F., Phenotypic response of the diatom Phaeodactylum tricornutum Bohlin to experimental changes in the inorganic carbon system. *Bot. Mar.*, *51*, 5, 350–359, 2008, https://doi.org/10.1515/BOT.2008.047.

[11.22] Francius, G., Tesson, B., Dague, E., Martin-Jézéquel, V., Dufrêne, Y.F., Nanostructure and nanomechanics of live Phaeodactylum tricornutum morphotypes. *Environ. Microbiol.*, *10*, 5, 1344–1356, 2008, https://doi.org/10.1111/j.1462-2920.2007.01551.x.

[11.23] Martin-Jézéquel, V. and Tesson, B., 3 Phaeodactylum tricornutum polymorphism: an overview, *Advances in Cell Biology*, pp. 43–80, 2012, https://doi.org/10.1515/9783110229615.43.

[11.24] Maheswari, U., Jabbari, K., Petit, J.-L., Porcel, B.M., Allen, A.E., Cadoret, J.-P., Bowler, C., Digital expression profiling of novel diatom transcripts provides insight into their biological functions. *Genome Biol.*, *11*, 8, R85, 2010, https://doi.org/10.1186/gb-2010-11-8-r85.

[11.25] Maheswari, U., Montsant, A., Goll, J., Krishnasamy, S., Rajyashri, K.R., Patell, V.M., Bowler, C., The Diatom EST Database. *Nucleic Acids Res.*, *33*, Database Issue, D344–D347, 2005, https://doi.org/10.1093/nar/gki121.

[11.26] Maheswari, U., Mock, T., Armbrust, E.V., Bowler, C., Update of the Diatom EST Database: a new tool for digital transcriptomics. *Nucleic Acids Res.*, *37*, suppl_1, D1001–D1005, 2009, https://doi.org/10.1093/nar/gkn905.

[11.27] Kroth, P.G., Chiovitti, A., Gruber, A., Martin-Jezequel, V., Mock, T., Parker, M.S., Bowler, C., A Model for Carbohydrate Metabolism in the Diatom Phaeodactylum tricornutum Deduced from Comparative Whole Genome Analysis. *PLoS One*, *3*, 1, 1–14, 2008, https://doi.org/10.1371/journal.pone.0001426.

[11.28] Coesel, S., Oborník, M., Varela, J., Falciatore, A., Bowler, C., Evolutionary Origins and Functions of the Carotenoid Biosynthetic Pathway in Marine Diatoms. *PLoS One*, *3*, 8, 1–16, 2008, https://doi.org/10.1371/journal.pone.0002896.

[11.29] Allen, A.E., Vardi, A., Bowler, C., An ecological and evolutionary context for integrated nitrogen metabolism and related signaling pathways in marine diatoms. *Curr. Opin. Plant Biol.*, *9*, 3, 264–273, 2006, https://doi.org/10.1016/j.pbi.2006.03.013.

[11.30] Vardi, A., Bidle, K.D., Kwityn, C., Hirsh, D.J., Thompson, S.M., Callow, J.A., Bowler, C., A Diatom Gene Regulating Nitric-Oxide Signaling and Susceptibility to Diatom-Derived Aldehydes. *Curr. Biol.*, *18*, 12, 895–899, 2008, https://doi.org/10.1016/j.cub.2008.05.037.

[11.31] Vartanian, M., Desclés, J., Quinet, M., Douady, S., Lopez, P.J., Plasticity and robustness of pattern formation in the model diatom Phaeodactylum tricornutum. *New Phytol.*, *182*, 2, 429–442, 2009, https://doi.org/10.1111/j.1469-8137.2009.02769.x.

[11.32] Sapriel, G., Quinet, M., Heijde, M., Jourdren, L., Tanty, V., Luo, G., Lopez, P.J., Genome-Wide Transcriptome Analyses of Silicon Metabolism in Phaeodactylum tricornutum Reveal the Multilevel Regulation of Silicic Acid Transporters. *PLoS One*, *4*, 10, e7458, 2009, https://doi.org/10.1371/journal.pone.0007458.

[11.33] Huysman, M.J., Martens, C., Vandepoele, K., Gillard, J., Rayko, E., Heijde, M., Vyverman, W., Genome-wide analysis of the diatom cell cycle unveils a novel type of cyclins involved in environmental signaling. *Genome Biol.*, *11*, 2, R17, 2010, https://doi.org/10.1186/gb-2010-11-2-r17.

[11.34] Rynearson, T.A. and Armbrust, E.V., DNA fingerprinting reveals extensive genetic diversity in a field population of the centric diatom Ditylum brightwellii. *Limnol. Oceanogr.*, *45*, 6, 1329–1340, 2000, https://doi.org/10.4319/lo.2000.45.6.1329.

[11.35] Cherepanova, M.V., Usol'tseva, M.V., Pushkar, V.S., Dubrovina, Yu. F., Morphogenesis in Cyclotella ocellata — complex from Lake El'gygytgyn (Chukchi Peninsula) during the Pleistocene-Holocene. *Paleontolog. J.*, *44*, 10, 1252–1261, 2010, https://doi.org/10.1134/S0031030110100035.

[11.36] Li, Y., Kociolek, J.P., Metzeltin, D., Gomphonema Sichuanensis Li & Kociolek Sp. Nov. and Gomphonema Heilongtanensis Li, Kociolek Et Metzeltin Sp. Nov. from Two High Mountain Lakes, China. *Diatom Res.*, *25*, 1, 87–98, 2010, https://doi.org/10.1080/0269249X.2010.9705831.

[11.37] Shirokawa, Y., Karino, K., Mayama, S., Developmental plasticity and genotype-environment interactions influence valve morphology in the Cyclotella meneghiniana species complex (Bacillariophyceae). *Eur. J. Phycol.*, *47*, 3, 245–253, 2012, https://doi.org/10.1080/09670262.2012.697580.

[11.38] Schmidt, K., van Oosterhout, C., Collins, S., Mock, T., The role of phenotypic plasticity and epigenetics in experimental evolution with phytoplankton. *Perspect. Phycol.*, 3, 29–36, 2016, https://doi.org/10.1127/pip/2016/0036.

[11.39] Frickel, J., Sieber, M., Becks, L., Eco-evolutionary dynamics in a coevolving host–virus system. *Ecol. Lett.*, *19*, 4, 450–459, 2016, https://doi.org/10.1111/ele.12580.
[11.40] Hiltunen, T. and Becks, L., Consumer co-evolution as an important component of the eco-evolutionary feedback. *Nat. Commun.*, *5*, 5226, 2014, https://doi.org/10.1038/ncomms6226.
[11.41] Bergkvist, J., Thor, P., Jakobsen, H.H., Wängberg, S.-Å., Selander, E., Grazer-induced chain length plasticity reduces grazing risk in a marine diatom. *Limnol. Oceanogr.*, *57*, 1, 318–324, 2012, https://doi.org/10.4319/lo.2012.57.1.0318.
[11.42] Meyer, B. and Håkansson, H., Morphological variation of Cyclotella polymorpha sp. nov. (Bacillariophyceae). *Phycologia*, *35*, 1, 64–69, 1996, https://doi.org/10.2216/i0031-8884-35-1-64.1.
[11.43] Pančić, M., Torres, R.R., Almeda, R., Kiørboe, T., Silicified cell walls as a defensive trait in diatoms. *Proc. R. Soc. B: Biol. Sci.*, *286*, 1901, 20190184, 2019, https://doi.org/10.1098/rspb.2019.0184.
[11.44] Hu, W.-S., Huang, M., Zhang, H.P., Zhang, F., Vyverman, W., Liu, Q.-X., Movement behavioral plasticity of benthic diatoms driven by optimal foraging. *bioRxiv*, 682153, 2020, https://doi.org/10.1101/682153.
[11.45] Sison-Mangus, M.P., Jiang, S., Tran, K.N., Kudela, R.M., Host-specific adaptation governs the interaction of the marine diatom, Pseudo-nitzschia and their microbiota. *ISME J.*, *8*, 1, 63–76, 2014. https://doi.org/10.1038/ismej.2013.138.

12

Frustule Photonics and Light Harvesting Strategies in Diatoms

Johannes W. Goessling[1]*, Yanyan Su[2], Michael Kühl[3,4] and Marianne Ellegaard[5]

[1]*International Iberian Nanotechnology Laboratory, Braga, Portugal*
[2]*Carlsberg Research Laboratory, Copenhagen, Denmark*
[3]*Marine Biology Section, Department of Biology, University of Copenhagen, Helsingør, Denmark*
[4]*University of Technology Sydney, Ultimo, Australia*
[5]*University College Copenhagen, Campus Carlsberg, København V., Denmark*

Abstract
The diatoms, a unique class of microalgae, are enclosed in an extracellular, silica-based porous structure called the frustule. The frustule is denser than water and reduces cell buoyancy, thus increasing the risk of sinking out of the sunlight illuminated zone. Yet, diatoms are an outstanding ecological group with respect to global abundance, species diversification, and photosynthetic efficiency. The potential adaptive role of the diatom frustule can be linked to its light modulating properties, which have been explored as templates or raw materials for potential industrial, mainly photonic application. The optical properties of diatom frustules have also led to speculations about potential photo-biological relevance. Two main functions have been proposed: the frustule 1) modulates photosynthetic light harvesting by wave-diffraction or forward scattering via its nano-scale lattice structure and 2) reduces exposure to harmful radiation, i.e., by attenuation of ultraviolet radiation or mitigation of high light intensities. This chapter explores the unique photobiology of diatoms summarizing pigmentation and biochemical and structural adaptation of photosynthesis, including motility of chloroplasts and entire cells. We also discuss frustule photonics in relation to cellular light modulation for efficient light harvesting and highlight their potential implications for niche differentiation under various light regimes in the different habitats of diatoms.

Keywords: Light energy harvesting, photopigments, light perception, photosynthesis, chloroplast movement, phototaxis, frustule optics and photonics, niche differentiation

**Corresponding author*: Johannesgoessling@gmail.com

Vadim Annenkov, Joseph Seckback and Richard Gordon (eds.) Diatom Morphogenesis, (269–300) © 2022 Scrivener Publishing LLC

12.1 Introduction

Diatoms are a major group of phototrophic microalgae that thrive in the sunlit (photic) zone of nearly all aquatic and humid terrestrial habitats on Earth. Diatoms are responsible for approximately 40% of marine primary production and account for about 20% of global carbon sequestration by photosynthesis [12.1–12.3]. The evolutionary success of diatoms seems to be closely related to high metabolic flexibility, such as the ability to duplicate biochemical pathways between cell compartments [12.4] and to molecular response mechanisms facilitating efficient photosynthesis under various light regimes [12.5, 12.6]. Efficient diatom photosynthesis can rely on the extended photosynthetic utilization of sunlight in the cyan-green spectral range [12.7], on effective dissipation of excessive, potentially harmful energy [12.8], and on dynamic structural reorganization at the cellular level such as chloroplast movement for optimal light exposure [12.9].

The first documented observation of a diatom was communicated to the Philosophical Transactions of the Royal Society of London in 1703. With a simple compound microscope, an unknown English author observed what he described as minute plants "made up of two parallelograms joyn'd longwise" [12.10]. Remarkably, the author correctly identified a phototrophic organism, although the materials for comparison were only macroscopic plants and algae [12.11]. The bipolar appearance of virtually all diatoms, defined by the frustule—an extracellular structure made of silica which encases the diatom cell, was also described for the first time. The frustule is formed by two valves that fit into one another due to their slightly different size. So-called girdle bands are coiled around the overlapping valves, thus keeping the frustule together [12.11]. Valves, and often also the girdle bands, are perforated with highly ordered miniscule structures. While these structures help in the identification of single species due to their large variety, diatoms can be roughly divided into two main groups according to their valve symmetry [12.12]. Valves of centric diatoms (the Centrics) are commonly formed like Petri-dishes and have radial symmetry. Pennate diatom species (the Pennates) have elongated valves with bilaterally symmetrical patterns (Figure 12.1).

The often periodic nano-/microscale structures of the frustule resemble artificial optical structures and photonic crystals and could interact with light in a similar way [12.13]. That these structures affect light propagation is immediately obvious when studying empty valves of diatoms under a light microscope (Figure 12.2). These optical properties of frustules are potentially of interest for a suite of industrial applications, including the design of photonic crystals [12.13], micro-lenses [12.14], solar cells [12.15], light focusing and optical switches [12.16–12.20], and finally as attenuators of ultraviolet (UV) light [12.21]. As most diatoms are mainly photo-autotrophic and therefore dependent on light for covering cellular energy, carbon and nutrient demands, it is obvious to also want to speculate about possible benefits of these optical properties for diatom photosynthesis. However, surprisingly little is known about the light modulating properties of intact diatom frustules under *in vivo*-like conditions in intact living cells, and their potential implications for diatom photosynthesis [12.22, 12.23], and/or photo-protection toward harmful light, e.g., attenuation of UV radiation [12.24]. Even less is known about possible connections between light and morphogenesis of frustule nano-structures, e.g., whether different patterns might have evolved at least partially in response to different light regimes and *vice versa* whether the diatom nanostructure might respond to different light regimes [12.25]. Most published experimental data on frustule

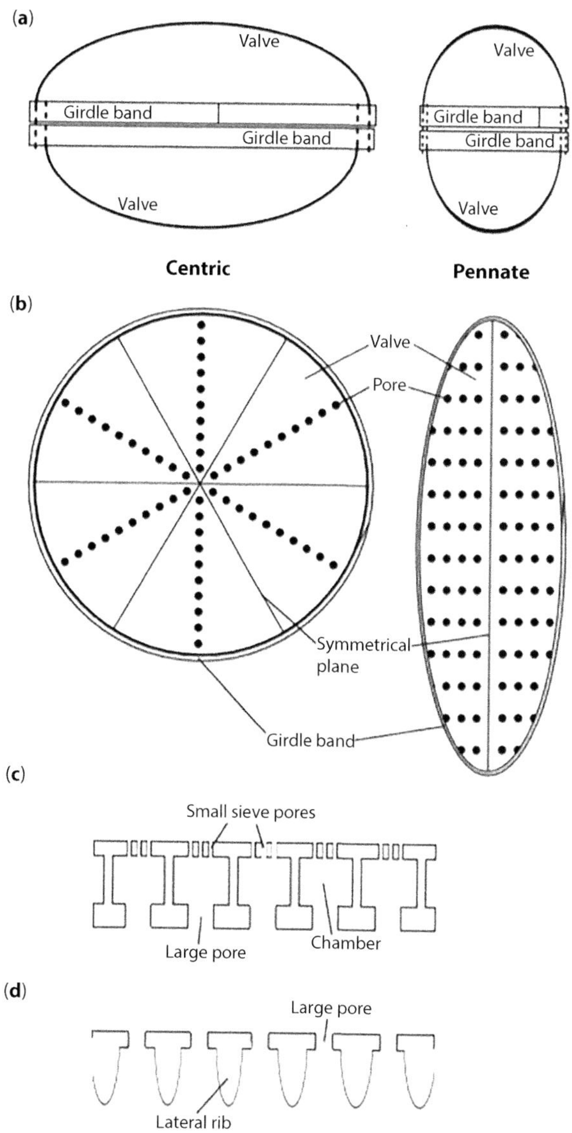

Figure 12.1 Generalized concept of centric and pennate frustule morphotypes among diatoms. (a) Side view of frustules with two valves and associated girdle bands. (b) Top view on the valves of Centrics with radial pore symmetry, and Pennates with lateral pore symmetry. (c) Cross-section through the valve with intricate periodic porous structures in the nano- to micrometer range, here inspired by the centric diatom *Coscindiscus granii* [12.23]. (d) The porous structure can be significantly different between species, such as in the pennate diatom *Achnanthes* spp., employing lateral rib structures in their frustules [12.142].

optical properties have been measured *in vitro* under artificial conditions in air. In contrast, photo-biological research has focussed mainly on the biochemical and structural regulation of the photosynthetic apparatus of diatoms and has largely not taken into account any possible role of frustule optics for diatom photosynthesis (e.g., [12.26–12.28]). Diatoms occur in a diversity of natural habitats that exhibit significant spatio-temporal variation in availability and spectral quality of solar radiation. The huge variety of periodic frustule

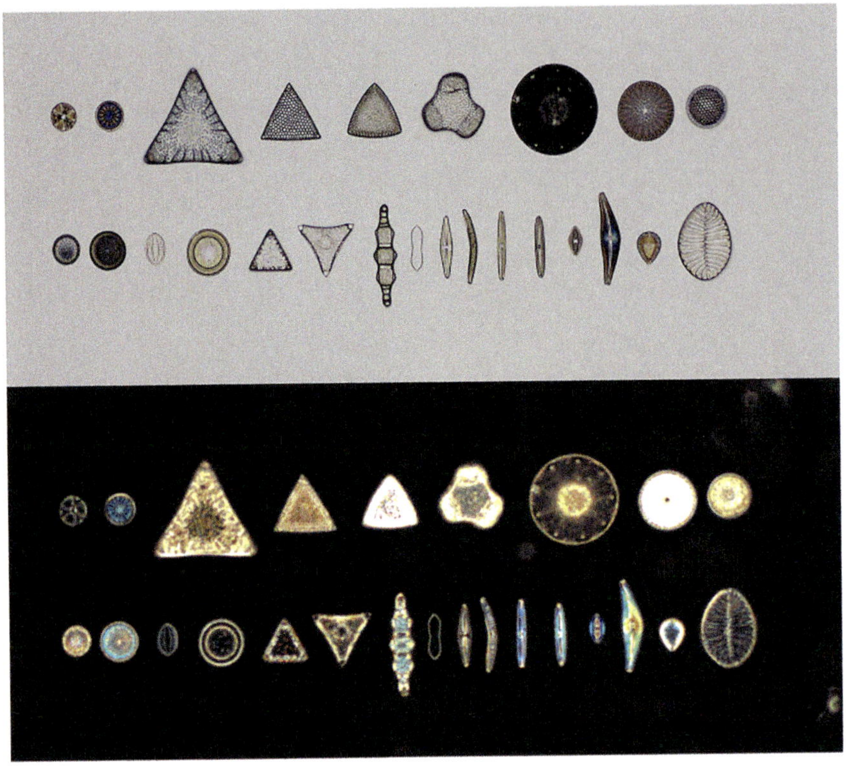

Figure 12.2 Silica valves of different diatoms in light microscopy. Micrographs were recorded at transmitted white light (top) and in the dark field at an incident angle of $\theta = 25°$ (bottom). Valves are fixed on a permanent microscope slide in resin ("Diatom Cubed®"; RI>1.70).

structures and associated optical properties could thereby play a role in niche differentiation to various light regimes.

Recent review articles suggested reciprocities between the light modulating properties of frustules and microalgal physiology [12.29] and further discussed potential functions in light of optical regulation for phototaxis and karyostrophy or signaling between single diatom cells [12.30]. This review focuses on diatom light harvesting strategies and their photosynthetic regulation related to the various light climates of diatom-inhabited environments. Furthermore, we discuss whether variation in frustule optical properties between species could be linked to distinct habitats or to the various life styles of diatoms (e.g., planktonic and benthic, motile and non-motile, and colony- and chain-forming species) (Figures 12.3a and b). This chapter thus explores the possibility that differences in light regimes could have shaped frustule optical structures for optimized light harvesting or the attenuation of harmful radiation.

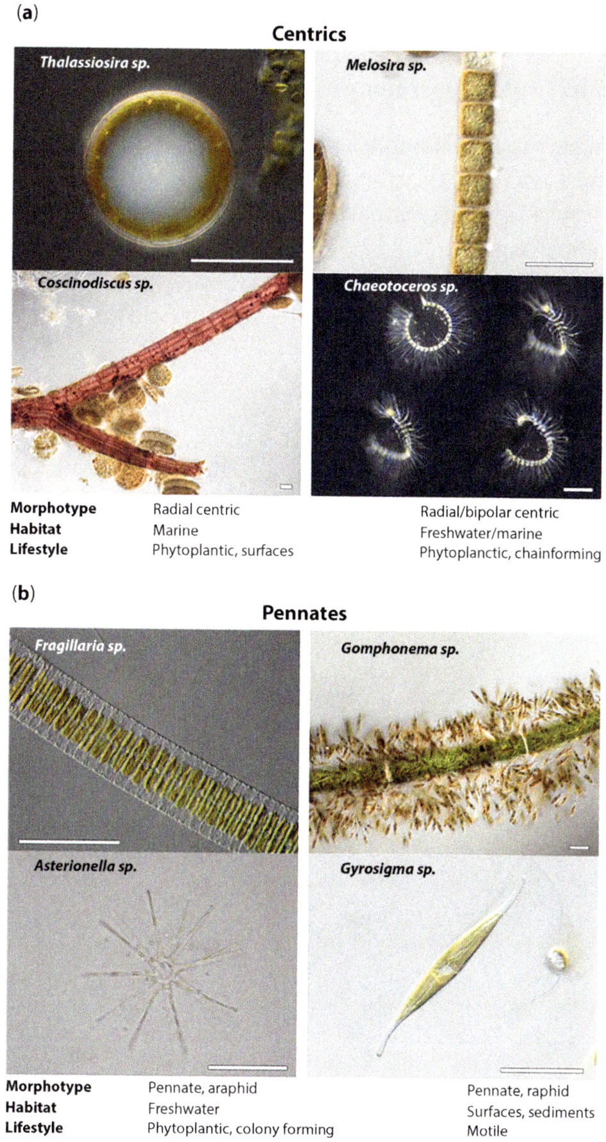

Figure 12.3 (a) Morphotypes, habitats and lifestyles of centric diatoms. Centric diatoms arose before the pennate frustule shape evolved. Scale bars: 50 μm. Micrograph copyrights and permission was granted by Wolfgang Bettighofer (http://www.protisten.de). (b) Morphotypes, habitats and lifestyles of some raphid diatoms. Raphid pennate species are the evolutionary youngest group of diatoms (Armbrust *et al.*, 2004) [12.70], exhibiting a raphe slit in their frustule facilitating locomotion by excretion of extracellular polysaccharides. Some raphid Pennates (such as *Gyrosigma* sp.) can move into the sediment to protect photosynthesis of high light intensities at the surface [12.97]. Scale bars: 50 μm. Micrograph copyrights and permission was granted by Wolfgang Bettighofer (http://www.protisten.de).

12.2 Light Spectral Characteristics and Signaling

12.2.1 Variation of Light Regimes

Diatoms populate a wide range of habitats in the plankton and benthos of marine and freshwater as well as in moist and temporarily immersed terrestrial environments, including the open ocean, coastal waters, glaciers, estuaries, lakes, rivers, marshlands, forests, grasslands, and agricultural systems (e.g., [12.31, 12.32]). These habitats are characterized by salient differences in the availability of light (Figure 12.4). Some species can live partly buried in sedimentary environments, where light attenuates rapidly, and harmful radiation is reduced

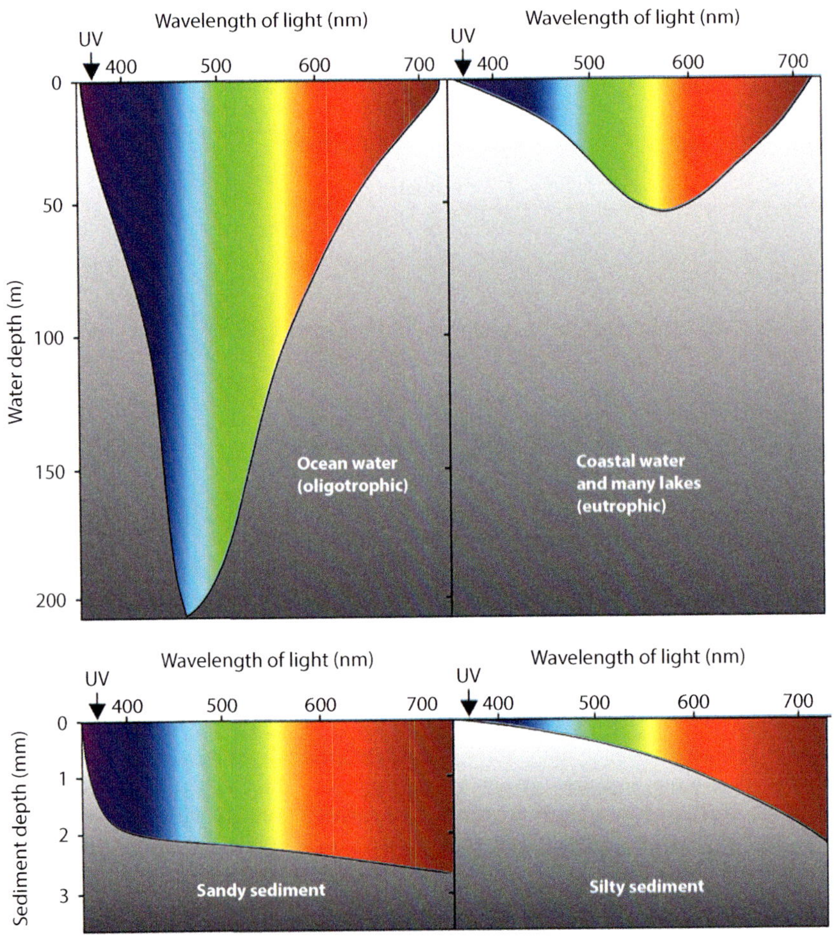

Figure 12.4 Variation of PAR attenuation in diatom dominated environments. In the open ocean, longer wavelengths are rapidly absorbed while blue light penetrates deep into the water column in the absence of large amounts of dissolved organic matter. Coastal waters and meso-eutrophic lakes exhibit stronger light attenuation due to a combination of higher absorption and scattering [12.143]. Light attenuation in sediment is dependent on the particle size. Shorter wavelengths of the PAR spectrum rapidly attenuate in silty sediments due to scattering on clay minerals [12.144]. Top panels are adapted from Kyle Carothers (NOAA-OE 52). Bottom panels are based on studies [12.135, 12.145].

(see Section 12.3.4). The availability of light can vary over time scales ranging from sub-seconds to diel variation to seasonal changes and is affected by the perpendicular rotational axis of Earth to its orbital plane, by mixing events such as weather events and cloud cover [12.33], and by wave-focusing [12.34], ocean currents, and tidal rhythms [12.35]. The light regime in aquatic environments is very different from that of terrestrial habitats, because water absorbs light more efficiently than air. Light attenuation in aquatic environments is thus heavily affected by absorption of longer (red and near infrared radiation, NIR) wavelengths of light by water itself, by the specific light absorption by dissolved organic matter, as well as by scattering due to suspended particulate matter [12.36, 12.37]. In the open ocean, the light regime is often regarded as more homogenous compared to coastal regions, where tides and currents can cause sudden changes in light availability and spectral characteristics. This might explain the observed significant differences in the photosynthetic machinery among diatoms from these habitats. For example, oceanic diatom species exhibit reduced protection of photosynthesis against sudden changes of light intensity [12.38].

The dominant spectral color of water varies between habitats and changes with depth in the water column [12.37]. Generally speaking, the dominant color in the deep and open ocean is blue, while humic lakes are yellow, and coastal waters appear rather brown [12.39, 12.40]. Light attenuation varies with the wavelength of penetrating light, i.e., longer wavelengths of visible light (red light) are rapidly absorbed by the water, while shorter wavelengths (e.g., UV or blue light) penetrate deeper into oligotrophic water, causing predominance of UV-blue-green spectral components with increasing water depth [12.6]. The presence of humic substances, organic acids, and other dissolved organic material (collectively termed "Gelbstoff") leads to strong absorption of UV and blue light leading to the yellow coloration of humic lakes or the brownish coloration of coastal waters [12.39]. As a rule of thumb, warm, tropical waters are often more oligotrophic, with deeper light penetration, compared to waters in subtropical and temperate regions [12.41]. Phototrophs absorb particular parts of light for their photosynthesis and therefore also affect the availability of light [12.42, 12.43].

12.2.2 Light Perception and Signaling

Whether light is too high or limiting is partially sensed by biochemical feedback regulation during photosynthesis [12.44]. It is also perceived by specialized photoreceptor proteins located inside the cell, which change their conformation in response to particular wavelengths of light, inducing cellular signaling cascades that trigger physiological responses [12.45]. Besides being the driving force for photosynthesis, light is also an important signal carrier, e.g., of diurnal and seasonal information. Photoreceptors thus regulate kernels of development and photo-morphogenesis [12.46] and chlorophyll biosynthesis [12.47] and mediate structural adaptation and light-induced movement, like phototropism [12.48] and chloroplast re-allocation movement [12.49]. Like chlorophytes, diatoms employ red light photoreceptors from the phytochrome family, and cryptochrome photoreceptors for UV and blue light perception, some of which are involved in repair of UV-induced DNA damage [12.50, 12.51]. Other important blue light photoreceptor proteins are the phototropins and the aureochromes, of which the latter seem unique to the Heterokonta [12.48]. Aureochromes thus seem to be absent in red algae, although these are the ancestral origin of heterokont chloroplasts [12.52]. Aureochromes contain a Light-Oxygen-Voltage (LOV) sensitive domain and a bZIP motif, indicating a regulatory function as a light induced transcription factor [12.53]. Although aureochromes were identified about a decade ago, their effect on cellular processes in diatoms remains largely unknown [12.54]. The different blue light photoreceptors present in diatoms may be part of more complex regulatory networks [12.51]. However, blue light seems to be essential in the perception of circadian and annual rhythms and might control kernels of the diatom life cell cycle, such as cell proliferation and reproduction [12.55].

12.3 Photosynthesis and Photo-Protection in Diatoms

12.3.1 Pigment-Based Light Absorption

Diatoms mainly utilize visible light (λ = 400 to 700 nm) for photosynthesis, often denoted PAR staining for photosynthetically active radiation. Only a few algae expanded light absorption for photosynthesis into the near-UV [12.56], while bacteria with anoxygenic and some cyanobacteria with oxygenic photosynthesis are capable of utilizing longer wavelengths in the far red (> λ = 710 nm) and NIR spectral range [12.57]. The pigment composition is responsible for the typical brownish coloration of diatoms [12.7]. The quantitatively dominant photo-pigments in diatoms comprise three types of chlorophyll (Chl a, c1, and c2), and four carotenoids, i.e., fucoxanthin, β-carotene, diadinoxanthin, and diatoxanthin. However, some diatom species have Chl c3 instead of Chl c1 [12.58]. Under specific light conditions such as high irradiance, minor amounts of other carotenoids (violaxanthin, antheraxanthin, and zeaxanthin) have been detected in some diatoms [12.59]. Fucoxanthin, Chl a, and Chl c are the main pigments of diatoms and are *in vivo* associated to protein complexes named fucoxanthin-Chl-c binding proteins (FCP). They form the main light harvesting antenna complexes of diatoms [12.60], where fucoxanthin collects photon energy by absorption of mainly blue and green light [12.61]. The gross amount of light harvesting pigments usually varies with light exposure and typically increases under low irradiance to support light harvesting for optimum photosynthesis [12.62]. However, in diatoms, the stoichiometry of the main photo-pigments is rather constant and apparently not significantly affected by differences in available PAR [12.26, 12.28].

The carotenoids diadinoxanthin, diatoxanthin, β-carotene, violaxanthin, antheraxanthin, and zeaxanthin are part of the xanthophyll cycle, which is activated under high irradiance [12.7, 12.63]. The xanthophyll cycle plays a central role in diatom photo-protection by dissipation of excessive energy via non-photochemical quenching [12.64]. The xanthophyll cycle involves the de-epoxidation of diadinoxanthin to diatoxanthin when the chloroplast luminal pH decreases due to high electron flux rates along the photosynthetic electron transport chain. Therefore, the ratio of diatoxanthin to diadinoxanthin generally significantly increases under high irradiance [12.65].

12.3.2 Molecular Photo-Protection Mechanisms

Photosynthesis is the central metabolic pathway for the energy supply of photo-autotrophic diatoms. During the light reaction of photosynthesis, energy from the photons of sunlight is harvested and converted into biochemical energy (ATP) and reduction equivalents

(NADPH) by charge separation and a subsequent combination of linear and cyclic electron transport processes between and around the photosynthetic reaction centers. During the dark reaction of photosynthesis, ATP and NADPH are consumed for carbon and nutrient acquisition leading to macromolecular synthesis for maintenance and formation of new biomass. While this basic principle is conserved in all oxygenic phototrophs, there are several peculiarities in the metabolic regulation of diatoms. Most research on photosynthesis involves green algae and higher plants which acquired their chloroplast by a primary endosymbiosis with a cyanobacterium-like organism [12.66]. In contrast, diatoms (and other Heterokonta) acquired their chloroplasts by a secondary endosymbiosis of a red algal-like ancestral cell [12.67]. Consequently, diatom chloroplasts possess four membranes instead of two [12.68]. Other differences in the photosynthetic apparatus and regulation of diatoms relative to plants include their special pigmentation and spectral absorption of light for photosynthesis [12.58], evidence for a photosynthetic machinery adapted to iron limited conditions in the open ocean [12.38], different redox-regulation during high-light compared to chlorophytes and the absence of the oxidative pentose phosphate pathway [12.69], as well as an altered plastidal nitrogen acquisition pathway [12.70].

As the driving force for photosynthesis, light is the key environmental factor limiting cellular productivity, but changing light regimes can also imbalance biochemical regulation. Chlorophytes and cyanobacteria can overcome short-term imbalances in available PAR by modulating their antenna complexes between the two photosystems with slightly different absorption maxima of photosystem (PS) II and PSI [12.71, 12.72]. In contrast, the light harvesting complexes associated with the photosystems of diatoms do not differ in spectral absorption, and hence, the stoichiometry of PSII and PSI seems to be constant in diatoms regardless of light conditions [12.28]. Under high irradiance, light energy can be transferred to alternative electron acceptors, and if these processes are not controlled, reactive oxygen species may form, causing damage to the photosynthetic machinery, which can ultimately induce cell death [12.73]. There are in principle two main strategies to balance over-reduction of photosynthesis: excessive energy can be remove and high photon flux can be avoided [12.74]. The above-mentioned xanthophyll cycle is particularly effective in diatoms, i.e., in a comparative study it was demonstrated that the capacity of diatoms to release absorbed energy by such non-photochemical quenching mechanisms was three- to fivefold higher compared to that of vascular plants under intermittent light regimes [12.75].

12.3.3 Intracellular Structural Adaptation in Response to Light

In addition to biochemical surveillance of redox states during the photosynthetic electron transport, many phototrophs can reduce light absorption by intracellular structural rearrangement for optimized photosynthesis [12.76]. Under supra-optimal irradiance, the chloroplasts can accumulate in the cell periphery to reduce light absorption via self-shading, thus avoiding the formation of reactive oxygen species, while chloroplasts are more evenly distributed in the cell under low light conditions [12.77, 12.78]. Such chloroplast re-allocation movement has been observed in both pennate and centric diatom species both in pelagic and benthic species inhabiting different marine and freshwater habitats [12.9, 12.79–12.81]. High-light avoidance movement of chloroplasts is mediated by blue light perception via photoreceptors upon light saturation of photosynthesis. In this case, chloroplasts accumulate to reduce the illuminated light field. In contrast, under low irradiance, chloroplasts are evenly distributed over the cell. Such low light-induced chloroplast movement is stimulated by far red light illumination and under low irradiance [12.82, 12.83]. Chloroplast movement could constitute an important mechanism for regulation of the intracellular light exposure; however, some centric diatom species do not employ such high-light avoidance movement of chloroplasts in their photo-protection. It was, e.g., observed that an alteration of the chloroplast distribution in the centric diatom *Coscinodiscus granii* after high doses of blue light was irreversible even after a diel relaxation period in the dark [12.84]. Follow-up experiments showed that cells with such an altered chloroplast distribution were dead. In another study, Furukawa *et al.* (1998) communicated that the chloroplasts of the centric diatom *Pleurosira laevis* did not move back toward an even distribution after high blue light irradiation. Interestingly, in both studies, only marginal effects on the chloroplast distribution inside the cell were observed after application of high doses of green light [12.9] and red light [12.84], respectively.

The main components of the light reaction of photosynthesis are organized along an internal thylakoid membrane system in the chloroplasts. Thylakoids of land plants are organized in so-called grana, which are highly ordered stacks that vary between 5 and 20 layers [12.85]. In land plants, the number of thylakoid stacks seems to be generally higher under low light conditions [12.86]. While this observation could indicate a regulatory role of thylakoid organization for photosynthetic light harvesting, the function of grana remains unknown [12.87]. In diatoms, thylakoids are organized in stacks of three running in parallel through the chloroplast [12.88]. While the common observation is that this scheme is unaffected by changes in the light intensities in diatoms [12.28], it was recently shown that thylakoids can form homogenously organized stacks with 10 layers when cells of *Phaeodactylum tricornutum* are grown under low red light illumination [12.89]. This observation could be related to lateral segregation, i.e., the rearrangement of the photosynthetic protein complexes to facilitate charge separation under different light conditions. However, this is so far the only evidence of lateral segregation of the photosynthetic machinery in diatoms [12.90]. Stacking of thylakoids could also affect light harvesting and propagation via Bragg-scattering. This has, e.g., been shown in *Begonia* plants where their highly ordered thylakoid membranes were found to improve the photosynthetic efficiency under the dim, diffusive light regime of the shaded forest floor [12.91].

12.3.4 Motility as a Unique Photo-Protection Mechanism

The superphylum Heterokonta is also named stramenopiles, derived from the Latin words stramen (straw) and pilus (hair), referring to the hairs on one of the two heterodynamic flagella inherent to most members of the superphylum. In diatoms, these flagella are only preserved as one flagellum on the motile sperms of some centric diatom species. In consequence, the vast majority of centric diatoms cannot actively move toward more suitable conditions. Planktic diatoms may, however, relocate over the water column by changes in their buoyancy, controlled by lipid production as the main storage product of diatom photosynthesis and by regulation of osmolites [12.26, 12.92]. Such reallocation in the water column might also be facilitated by the frustule, which can be understood as ballast, as it is relatively heavy compared to the cell [12.93]. In fact, motility is a more recently evolved trait of diatoms, and an exclusive feature of the youngest group named raphid Pennates [12.70]. Their motility is enabled by slits along the valve axis (shown in Figure 12.5b), through which extracellular polysaccharides are excreted, facilitating gliding over surfaces, i.e., other cells, solid surfaces, or in sediments [12.94]. The ability of locomotion allowed raphid Pennates to colonize benthic habitats; they are nowadays the most diverse group of diatoms [12.95]. Raphid Pennates inhabiting muddy sediments can locate vertically at depths with optimal light exposure for photosynthesis [12.96]. This movement is considered as a unique form of photo-protection (comparable to chloroplast high-light avoidance movement) leading to reduced absorption of sunlight, which can reach extremely high intensities at the sediment surface [12.97–12.99]. Such diatom migration is a key factor regulating photosynthesis in the microphytobenthos [12.100].

12.4 Frustule Photonics Related to Diatom Photobiology

12.4.1 An Extracellular Structure With Optical Properties

The potential biological function of the diatom siliceous frustule with its intricate pore system has received much attention, and a variety of different biological implications of the frustule have been proposed and tested to varying degrees. These include prevention of desiccation in moist terrestrial areas and intertidal zones [12.101], altering the functional density of the cells [12.102], sorting of nutrients versus filtration of harmful agents including bacteria and viruses [12.103], defence against predation by mechanical strength [12.104], pH buffering [12.105], and modulating the cellular light regime (e.g., [12.22]) [12.106]. In contrast to the intra-cellular structures that may adapt to changes in light, i.e., chloroplasts and thylakoid membrane stacks, the frustule is an extracellular compartment surrounding the diatom cell; yet, it has been proposed that frustules can also adapt to light spectral properties in a similar way [12.25] (see also Section 12.4.2). The size ranges and periodicity, i.e., the two-dimensional distance between repeats and the diameter of nano-pores and other frustule structures, are in many species of the same order as the wavelengths of UV-VIS light and can therefore cause different types of frustule-light interactions, largely driven by absorption, scattering, diffraction, and refraction. The immense variability in the size and nano-structural patterns of the frustule in the >10,000 described species of diatoms [12.107] indicates that frustule-light interactions may vary significantly among species (Figure 12.5). Furthermore, the optical effects of the frustule appear to be strongly dependent on the angle at which light arrives on a diatom, and there is also some dependency on which side of the valve is facing toward the light source due to variation of porous structures on the internal and external valve side [12.23]. For the species *Coscinodiscus granii*, Maibohm *et al.* (2016) showed that a single valve in air did not redirect light of wavelengths above 700 nm toward normal (i.e., the interior of the cell) at large angles of incident light, in contrast to light arriving at normal incidence [12.108].

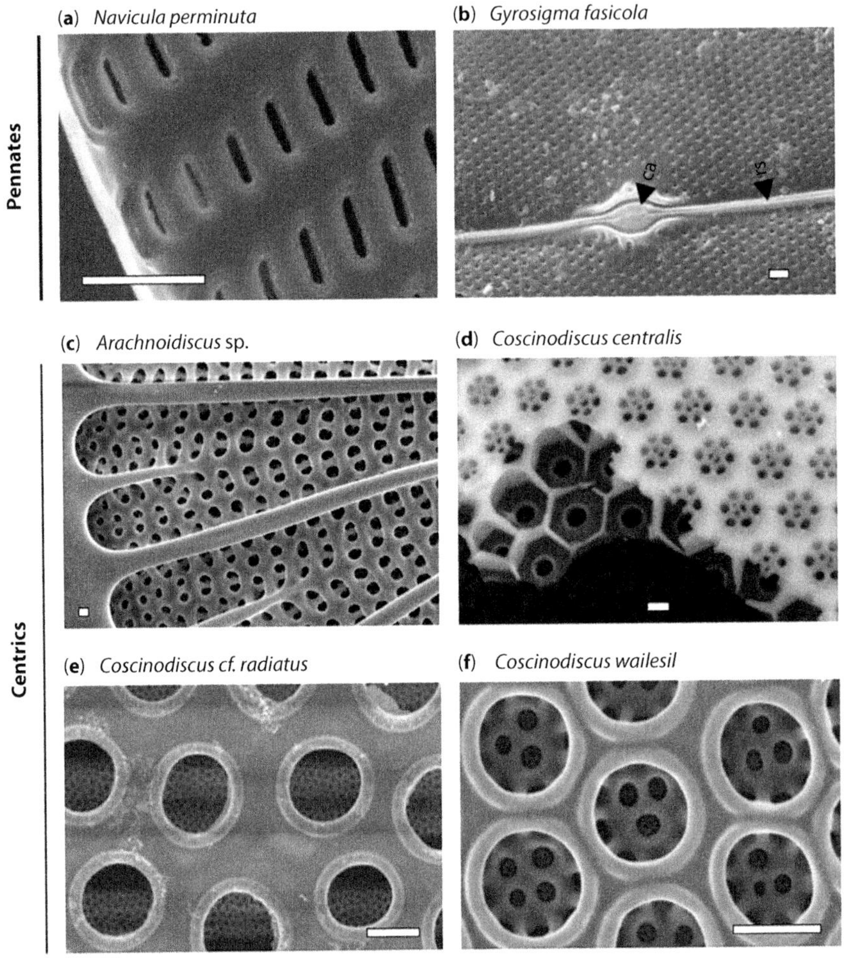

Figure 12.5 Example of frustule structural variation in centric and pennate diatoms. Frustules of raphid pennate diatoms bear a raphe slit (rs) along the symmetrical axis [shown in panel (b)], emerging from a central area (ca). Scale bars: 1 μm. Image source and courtesy: (a, e, f) Aguirre *et al.* (2018) [12.24]; (b) João Serôdio (CESAM, University of Aveiro), (c, d) reproduced from Ødegård *et al.* (2014) [12.146] with permission from the Centre National de la Recherche Scientifique (CNRS) and The Royal Society of Chemistry.

They further found strong interactions with light at 60° and 40° incidence for some wavelengths in the PAR range.

So far, most studies on optical properties of diatom frustules have been carried out on single isolated valves measured in air. This may impede the interpretation of observed phenomena with regard to their biological significance, as the behavior of light differs significantly by refractive contrast to the surrounding medium. Therefore, some of the conclusions may not be directly relatable to intact diatoms cells, with watery cell content and in a water environment.

12.4.2 Intraspecific and Intra-Individual Variation of Frustule Periodicity

If the precise frustule nanostructure is decisive for its optical properties, and if these are central for light harvesting and optimization of photosynthesis in the diatom cell, the frustule architecture could be affected by the spectral characteristics of the ambient light during growth. This might facilitate both short-term phenotypic plasticity and long-term adaptation during the evolution of diatoms. Su *et al.* (2015) tested whether the spectral light properties affect the frustule nanostructure by cultivating the large centric diatom *Coscinodiscus granii* at five different narrow spectral wavebands [12.25]. Except red-orange light, all tested light qualities (blue, green, yellow, and red light) led to significant reduction in the density of large pores in the frustule compared to when incubated under a broad spectrum of white light with similar irradiance (300 μmol photons m^{-2} s^{-1}). The same overall pattern was observed in a second experiment where *C. granii* was cultivated for several months under different narrow spectral ranges of light [12.109]. Thus, the diatom frustule nanostructure in this diatom species can apparently respond to the spectral light composition. However, a range of other environmental parameters can also affect frustule morphogenesis and nano-structuring. For example, pH and salinity have significant effects on the frustule nano-structure in different species [12.110, 12.111]. The range of variability within each single light treatment was relatively large, and the optical properties potentially also varied even within specimens from the same species. Such an observation was also communicated by Kieu *et al.* (2014) and Cohoon *et al.* (2015) who emphasized the quasi-periodic nature of the frustule nanostructure in Coscinodiscophyceae, exhibiting slightly different pore sizes in different parts of the frustule [12.112, 12.113]. Kieu *et al.* (2014) further showed how different parts of the valve face of *Coscinodiscus wailesii* varied by pore spacing (4.3 μm at the centre versus 4.7 μm near the edge), and that these differences caused minima in the transmitted light spectrum of dry valves, i.e., at 950 nm near the edge and at 650 nm close to the centre of the valve. Due to the regular array of frustule structures in the order of wavelengths of light, Fuhrmann *et al.* (2004) described the frustule as a "photonic box", where the its walls could function as photonic crystal slabs [12.13].

12.4.3 Photonic Crystal Properties

Photonic crystal properties are introduced by periodic mismatches of refractive index relative to a bulk, dielectric material (such as silica). The unit cell period must match an integer of the wavelength of light, facilitating guided and prohibited modes causing photonic band gaps. Fuhrmann *et al.* (2004) used numerical techniques to characterize photonic frustule properties in the species *Coscinodiscus granii*, as these fulfill the above-mentioned structural and material properties. Their results suggested that both valves and girdles behave like photonic crystals [12.13]. Several studies aimed at measuring photonic crystal properties using valves of different diatom species. Photonic crystal functioning has since been inferred in the centric species *Coscinodiscus centralis* [12.114], where hyper-spectral mapping of the 3D structure of a single valve in air indicated an enhancement of transmitted light around 636 and 663 nm (close to absorption maxima of Chl *a* and *c*). Using entire frustules of the freshwater centric diatom *Melosira varians* that were cleaned of cell contents, Yoneda *et al.* (2016) found that when illuminated in parallel (from one valve end of the cell) and measured at the other valve end, a dip in light transmittance was seen with a minimum at $\lambda \approx 400$–450 nm [12.115]. It has also been proposed that light could be enhanced by Fabry-Perot–like interference of radiation on the chambered walls inside the valve wall [12.116]. Different studies suggested that the asymmetry of the structure in the two sides of the valve of centric diatom species captures light effectively [12.117]. A recent study testing light transmittance on wet valves of the centric planktonic diatom *Coscinodiscus granii* indicated that valves captured incident light effectively by their asymmetric design, when light entered through the exterior diatom valve side [12.23]. Large chambered structures inside the valve wall could then function as sub-wavelength diffraction gratings, thus redirecting incident light for an even cellular exposure. In further experiments, the authors focused a tuneable laser spot of PAR onto a small area of a single live cell and observed by variable chlorophyll fluorescence imaging that chloroplasts initiated photosynthesis more or less evenly all over the cell. This was explained by redistribution of incident light through wave-guiding in the plane of the valve and optical coupling of chloroplasts to the evanescent field, thus giving first experimental proof for earlier speculations based on theoretical arguments [12.13]. Kieu *et al.* (2014) also showed that the valve of the diatom *C. wailesii* exhibited well-defined resonant bands suggesting photonic crystal function of the frustule, but they could not identify complete band gaps. Part of the reason for the apparent difficulties in measuring photonic band gaps in the diatom frustule could be related to the low refractive index contrast between the silica skeleton and surrounding medium [12.21], as high contrast facilitates resolution [12.118]. However, another explanation could be related to the fact that the above-mentioned studies were focused on the valves, where unit cell periods often exceed the wavelength range of visible light ($\lambda > 800$ nm). Recently, clear experimental proof of photonic band gaps was demonstrated for the girdle band of the diatom *C. granii* [12.119]. This structure is a ≈800-nm-thick silica slab perforated by a pore system in three dimensions with a unit cell period of approximately 280 nm. Despite low refractive index contrast when immersed in water, the girdle showed a well-defined pseudo-gap (a band gap that exists only in some directions) in the NIR. This property was highly reproducible between individual girdles and between different diatom cell lines. However, despite the apparent conservation of the feature under more natural conditions in water, no conclusions related to potential biological functionality could be derived.

12.4.4 Light Confinement and Focusing

Several studies have looked into the diffraction characteristics of diatom valves for different wavelengths in the PAR range, as focusing due to such confinement could enhance particular wavelengths inside the organism and affect light harvesting for photosynthesis. Similar light focusing has been demonstrated in plants living in the shade of the understory in tropical forests, which have modified epidermal cells functioning as lenses [12.120]. Tommasi *et al.* (2010) observed focussing of light by diatom valves at PAR wavelengths and in the NIR spectral range (λ = 532, 633, and 1,000 nm) [12.16]. Diffraction on the grating of large pores was also shown in valves of *Coscinodiscus wailesii*, *C. coccinus*, and *C. centralis*, where light was focused at a distance of ~100 µm behind the valve [12.14, 12.20, 12.117]. Maibohm *et al.* (2015a) also showed that green and red wavelengths interfered constructively only at distances of ~120 µm (thus outside the live diatom) when passing the porous network of *C. granii* valves in air. Such focussing would not fall within live cells and may thus not be relevant for photosynthesis. However, as all mentioned experiments were performed in air, different results may be expected under natural conditions, i.e., when the frustule has not been cleaned of its organic coating and is surrounded by water and the cell is filled with cytoplasm.

12.4.5 Scattering and Dispersion of Light

Maibohm *et al.* (2015) also found that the light field was diffusive at a close distance behind the valve [12.19], where the chloroplasts in live cells of centric diatoms are located [12.121]. Valve porous structures might thereby scatter PAR for more homogenous illumination of the chloroplasts. In another study where transmittance of light on wet valves of the same species *Coscinodiscus granii* was measured, Goessling *et al.* (2018) observed that blue light was forward scattered by the valve pores when light was incident at specific angles [12.23]. This was not the same when valves were measured in air. Similar forward scattering of blue light was also observed on wet valves of three raphid pennate diatoms from an intertidal mudflat, i.e., *Gyrosigma fasciola*, *G. balticum*, and *Pleurosigma angulatum* [12.122]. The angle of incident white light at which the phenomenon was visible differed between the tested species.

Other studies also observed conspicuous interactions of blue light with valves of the centric diatom *Coscinodiscus centralis* and the freshwater centric diatom *Meliosira varians* [12.117, 12.123]. In these studies, light transmittance through valves was measured with different techniques, and the observed phenomena were interpreted as absorption/attenuation of blue light. However, due to diffraction on the grating of the porous structure, the distance at which a light collecting device is positioned is crucial to differentiate superposition versus destructive interference [12.23]. Another study on *Meliosira varians* observed quasi-stop bands in the blue spectral range and they suggested that such a characteristic could assist light absorption in the living cell around $\lambda = 420$ nm [12.115]. In Figure 12.6 forward scattering of blue light is visible on the valves of multiple species fixed in resin on a permanent microscope slide. In fact, blue light seems to play a special role in diatom photosynthesis. Comparative studies of diatom growth under different light colors thus demonstrated higher growth rates under blue light [12.124]; albeit other studies suggest that this is not the case for some species [12.125]. Some studies concluded that blue light is also more efficiently transmitted from the photo-pigments toward the reaction centers of the photosystems in diatoms [12.44, 12.84, 12.126]. It has also been demonstrated that diatoms compensate for a higher absorption of blue light by reducing the number of antenna pigments associated with the photosystems after long-term incubation under equal amounts of photon irradiance for different spectral wavebands [12.127].

12.4.6 Attenuation of UV Light for Photo-Protection

As red, orange, and yellow wavelengths are most strongly absorbed by water in oligotrophic water, almost half of the integrated light energy at depth will be in the UV spectral

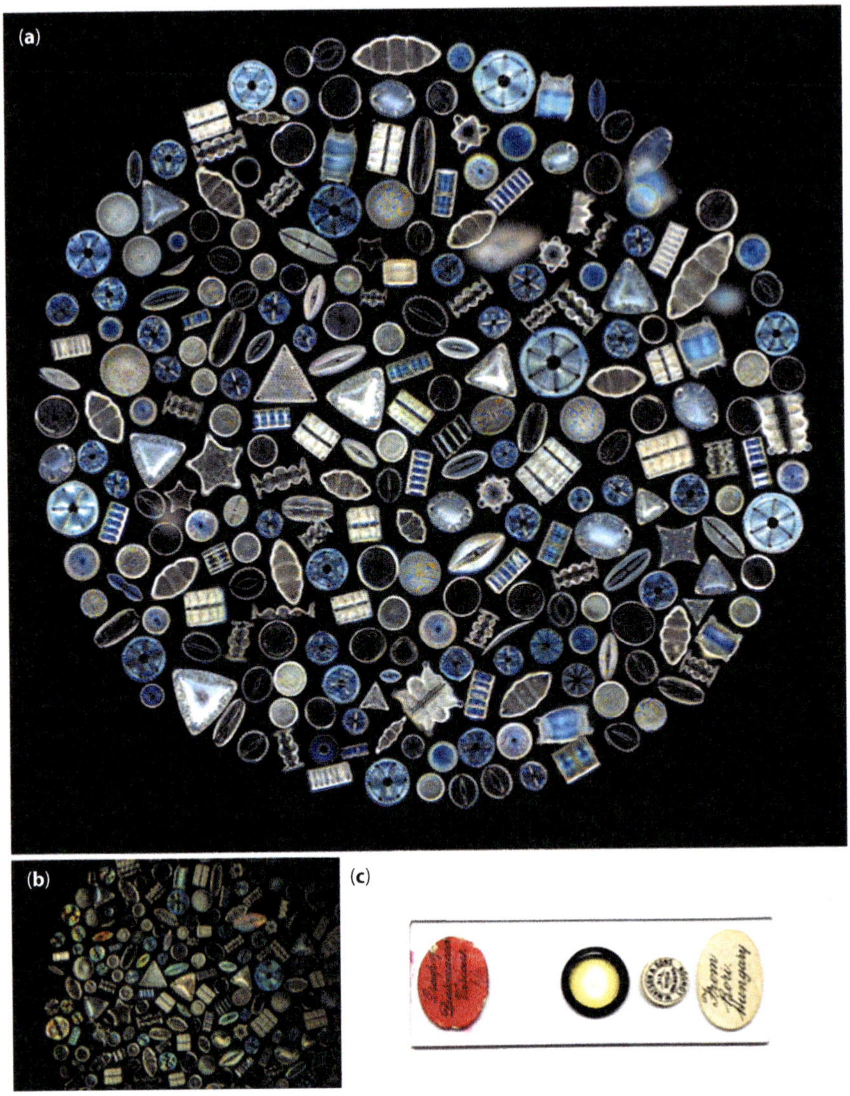

Figure 12.6 Permanent preparation diatom slide in dark field microscopy. (a) Mosaic of micrographs recorded in the dark field. Under certain incidence of transmitted white light, frustules scatter shorter wavelengths of the visible spectrum and appear blue [12.23, 12.122]. (b) The same permanent slide illustrating that those frustules appear blue only in the center of field of view. (c) Photography of the permanent slide. Images and information were provided by Martin Mach, who also permitted republication.

region compared to only about 3% at the surface [12.128]. Diatoms and other phytoplankton that spend much time below a few meters under the ocean surface, thus live in a UV and blue light dominated environment [12.129, 12.130]. It has been hypothesized that the nano-structured frustule may protect the cellular contents from exposure to harmful UV radiation. This assumption might be supported by the fact that diatoms appear to respond more variably between species to UV light in experimental settings than microalgae from other systematic groups, suggesting that a species specific mechanism might be involved in UV protection [12.21]. UV protection could, in live cells, also be mediated by mycosporine-like amino acids affiliated with the frustule [12.131].

Some studies have indeed demonstrated that the frustule itself strongly interacts with UV light. A recent study tested the effect of frustules of three morphologically very different diatoms, *Coscinodiscus wailesii*, *C. cf. radiatus*, and *Navicula perminuta* on the passage of UV light (300 nm) in air, as measured by staining on a UV sensitive material placed below the frustules [12.24]. The study concluded that the frustules shielded UV radiation to reduce damage to the DNA located in the nucleus. Several studies have shown that dry frustules reflect more (or transmit less) light at low wavelengths (in the UV range); both single frustules (simulated by [12.14]) and drop-casted frustule layers [12.132]. Other studies showed that the focus point for transmitted UV radiation is more diffuse and/or further away from the frustule as compared to the focus point of transmitted light in the visible range (PAR) [12.16, 12.18]. In some studies, this was mainly attributed to the absorption characteristics of silica itself [12.133]. However, other studies have shown that the UV reflecting properties of rinsed diatom valves in air vary between species, although the valve material properties are the same [12.132]. In a recent study, it was proposed that high UV absorption by the frustule could have shaped diatom evolution by reducing damage to the cellular DNA, though the authors measured these properties in dry frustules, where attenuation of UV light might be higher as compared to lower refractive contrast in water [12.24].

12.5 Frustule Photonics in Light of Niche Differentiation

So far, links between frustule form on the one hand, and photo-physiology, chloroplast structure, and distribution, on the other hand, have only been tested for relatively few species (Table 12.1 and references therein). To the best of our knowledge, 13 species have been investigated, with species of the centric genus Coscinodiscophyceae (*Coscinodiscus granii*, *C. wailesii*, *C. centralis*, *C. concinnus*, and *C. cf. radiatus*) accounting for most studies by far; these species have large and flat frustules and might therefore be more suitable for optical studies. In addition, the centric diatom species *Arachnoidiscus sp.*, *Melosira varians*, *Thalassiosira pseudonana*, and *T. punctifera* have been investigated, where *M. varians* is the only chain-forming (colony-forming) species. Two studies included a total of four pennate species from the genera *Gyrosigma*, *Pleurosigma*, and *Navicula*. The centric diatom *Melosira varians* is the only exclusively freshwater species tested (the pennate diatom *N. perminuta* has been reported from both marine and freshwater [12.134]). While the studies mentioned clearly hint at different adaptive values of the optical properties of the nanostructured diatom frustule, there is too little data to formulate firm conclusions, whether these different functions are linked to specific lineages or functional groups of diatoms. Although several authors suggested that the ability to reflect harmful UV radiation might be an evolutionary driver for the frustule nanostructure [12.24] and/or that the valve has been optimized for photosynthesis through millions of years of evolution (e.g., 12.113, 12.22, 12.80]), there is little experimental or phylogenetic evidence that this is indeed the case. However, we agree that the results point out potential roles of the frustule optical properties in niche adaptation, as discussed in the following.

Table 12.1 Summary of studies on the optical properties of different diatoms and proposed photo-biological function. Abbreviations in the experimental conditions column are E = Experimental and N = Numerical analysis.

	Species name	Habitat	Lifestyle	Main optical properties	Proposed biological function	Experiment conditions	Type of analysis	Ref.
Pennates (raphid)	*Gyrosigma balticum*	Intertidal sediments	Benthic; motile	• Forward scattering of blue light	• Enhanced blue light for photosynthesis	Water	E	[12.122]
	Gyrosigma fasciola	Intertidal sediments	Benthic; motile	• Forward scattering of blue light	• Enhanced blue light for photosynthesis	Water	E	[12.122]
	Pleurosigma angulatum	Intertidal sediments	Benthic; motile	• Forward scattering of blue light	• Enhanced blue light for photosynthesis	Water	E	[12.122]
	Navicula perminuta	Marine/ Freshwater	Benthic; motile	• Blocking of UV radiation	• Protecting DNA against UVR	Air	E	[12.24]
Centrics	*Arachnoidiscus* sp.	Marine	Phytoplanktic	• Spatial separation of focused light	• Optimization of PAR collection and minimized photo damage	Air	N	[12.18]
	Coscinodiscus sp.	Marine	Phytoplanktic	• Light trapping in the PAR range	• Efficient PAR capture for photosynthesis	Air	E	[12.138]

(*Continued*)

Table 12.1 Summary of studies on the optical properties of different diatoms and proposed photo-biological function. Abbreviations in the experimental conditions column are E = Experimental and N = Numerical analysis. (*Continued*)

	Species name	Habitat	Lifestyle	Main optical properties	Proposed biological function	Experiment conditions	Type of analysis	Ref.
	Coscinodiscus centralis	Marine	Phytoplanktic	• Photonic crystal function • Light trapping due to asymmetry • Attenuation of blue light	• May prevent backscattering and enhance light absorption	Air	E	[12.114]
	Coscinodiscus concinnus	Marine	Phytoplanktic	• Wave focusing by diffraction related to differences in the lattice constant	• Increasing PAR inside the cell	Air	N/E	[12.20]
	Coscinodiscus granii	Marine	Phytoplanktic	• Photonic crystal slab waveguiding • Coupling of chloroplasts to the evanescent field	• Even exposure of chloroplasts	Air/water	N	[12.13]

(*Continued*)

Table 12.1 Summary of studies on the optical properties of different diatoms and proposed photo-biological function. Abbreviations in the experimental conditions column are E = Experimental and N = Numerical analysis. (*Continued*)

	Species name	Habitat	Lifestyle	Main optical properties	Proposed biological function	Experiment conditions	Type of analysis	Ref.
				• Wave focusing by diffraction related to differences in the lattice constant	• Increasing PAR inside the cell	Air	N/E	[12.19]
				• Light enhancement by Fabry-Perot interference	• Improved light capture for photosynthesis	Air	N	[12.116]
				• Photonic crystal slab waveguiding	• Even exposure of chloroplasts	Water	E	[12.23]
				• Forward scattering of blue light	• Cellular enhancement of more productive light			
				• Shielding of UV radiation	• Protecting DNA against UVR	Air	E	[12.132]

(*Continued*)

Table 12.1 Summary of studies on the optical properties of different diatoms and proposed photo-biological function. Abbreviations in the experimental conditions column are E = Experimental and N = Numerical analysis. (*Continued*)

	Species name	Habitat	Lifestyle	Main optical properties	Proposed biological function	Experiment conditions	Type of analysis	Ref.
	Coscinodiscus cf.radiatus	Marine	Phytoplanktic	• Blocking of UV radiation	• Protecting DNA against UVR	Air	E	[12.24]
	Coscinodiscus wailesii	Marine	Phytoplanktic	• Light focusing at defined distances	• Increasing PAR inside the cell	Air	E	[12.14]
				• Higher transmittance of red compared to blue light	• Matching chl-a absorption peak	Air	E	[12.139]
				• No focusing of UVR	• UVR protection of the cell	Air	N	[12.16]
				• Wavelength dependent light concentration	• Redistribution of light and concentration in certain locations	Air	E	[12.140]
				• Position dependent diffraction gratings		Air	E	[12.112]

(*Continued*)

Table 12.1 Summary of studies on the optical properties of different diatoms and proposed photo-biological function. Abbreviations in the experimental conditions column are E = Experimental and N = Numerical analysis. (*Continued*)

	Species name	Habitat	Lifestyle	Main optical properties	Proposed biological function	Experiment conditions	Type of analysis	Ref.
				• Wave focusing by diffraction related to differences in the lattice constant	• Increasing PAR inside the cell	Air	N/E	[12.20]
				• Blocking of UV radiation	• Protecting DNA against UVR	Air	E	[12.24]
	Melosira varians	Freshwater	Chain forming Phytoplanktic	• Absorption of blue light	• Reduction of more energetic blue light	Air	E	[12.115]
				• Quasi-stop bands in the range λ = 400–500 nm	• Assisted light absorption around λ=420 nm	Air	E	[12.115]
	Thalassiosira pseudonana	Marine	Phytoplanktic	• Light trapping in the PAR range	• Efficient PAR capture for photosynthesis	Air	E	[12.141]
				• Shielding of UV radiation	• Protecting DNA against UVR	Air	E	[12.132]
	Thalassiosira punctifera	Marine	Phytoplanktic	• Shielding of UV radiation	• Protecting DNA against UVR	Air	E	[12.132]

One example is the difference between benthic diatom species in their response to blue light. Differences in the growth wresponse to blue light were wobserved in pennate species and could be an adaptive strategy toward proliferation in particular light environments [12.124]. A slurry of frustules of a community of benthic pennate diatom species living in an intertidal mudflat enhanced scalar irradiance in the blue spectral range by up to 120% relative to the incident irradiance, while these wavelengths were strongly attenuated inside the muddy sediment [12.122]. In this study, live diatoms from the sediment, i.e., exclusively raphid pennate species, also showed higher rates of photosynthesis under blue light. This blue light enhancement was explainable by scattering of these wavelengths, as demonstrated for frustules of three species collected from the same habitat. The observed interaction of blue light with valves differed between the species for different angles of light incidence. It has been speculated that such differences can be linked to micro-niche adaptation inside the sediment, where some diatoms are located at more diffusive light fields at subsurface position, while other species may prefer the surface where sunlight is more collimated [12.135, 12.136]. The planktonic, centric diatom *Coscinodiscus granii* also showed scattering of blue light by its valves at more acute angles of incident light [12.23]. The authors speculated that this could enhance more productive PAR inside the cell, when the organism is close to the water surface. Whether such an effect can indeed enhance photosynthesis remains unknown, but could be further investigated with numerical analysis of frustule-light interactions under different external light fields.

As a central cue for photosynthetic regulation, high versus low light habitats could also play a role in adaptation of the frustule nanostructure to incident sunlight. Differences in the photosynthetic architecture between oceanic and coastal diatoms have, e.g., been related to the differences in light regimes in these habitats, where oceanic species are subjected to stronger fluctuation of available PAR [12.38]. A photo-protective function of the frustule could consequently be more relevant in strongly light exposed habitats with high UV radiation, such as on sediment surfaces in intertidal mudflats or in oligotrophic water where these wavelengths are less attenuated. Optical structures could also relate to more diffused versus more coherent light fields, i.e., in the open ocean/inside sediments or sunlight exposed habitats at the surface of the water body and sediments, respectively.

12.6 Conclusion

Figure 12.7 illustrates links between characteristics of the tested species (listed in Table 12.1) and the proposed/substantiated biological function of optical and photonic phenomena. So far, no clear patterns emerge, but this overview can be used to point out directions for future research. The links between the different light spectral properties in specific habitats could be explored by targeting species with well-defined requirements/preferences to test specific hypotheses. For example, species from high light exposed sites could be targeted to test for UV light protection capabilities, compared to species from low light environments with dominance of different parts of the spectrum, e.g., red light for coastal diatoms and blue light for deep-living oceanic diatoms. Species with different frustule shape and size could be selected to broaden the dataset from the large, flat, single-celled species that dominate the literature so far. Lastly, more studies on intact, live diatom cells, with cell content, are needed, including studies on intact, live chains of diatoms. More realistic bio-optics studies

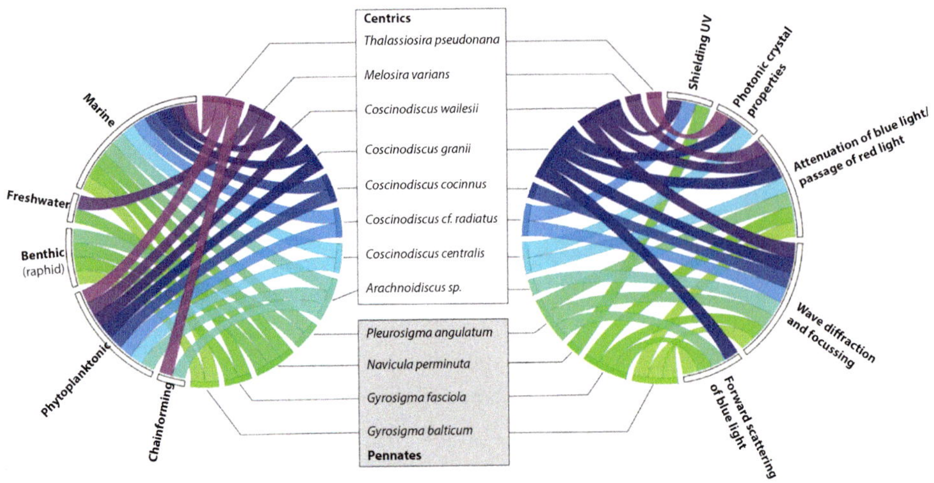

Figure 12.7 Chord diagram of studied optical properties of diatom frustules. Left hand: Species related to habitat and lifestyle. Right hand: Observed optical phenomena as indicated in the literature concluded in Table 12.1.

on intact, live diatoms may also benefit from numerical analysis, e.g., where the refractive index contrast could be adjusted to water and cytoplasm instead, to better understand the implications of light propagation behind diffractive layers inside the frustule. Media that cloak diatom frustules [12.30] might remove some of their optical effects. A first attempt considering the refractive indices of cytoplasm as pore filling fraction were performed by Tomassi *et al.* (2018) based on numerical methods, demonstrating that light confinement is well located inside the cell [12.137]. However, more research on the filling fraction *in vivo* and estimation of effective refractive indices is needed. The living diatom cell may be understood as a complex optical system, where the cellular light field is affected by light modulation through photopigments and by the special optical properties of the frustule. Light, as an important environmental factor, has shaped the photosynthetic architecture of diatoms; it may have also influenced optical structures of the frustule in response to different light regimes.

References

[12.1] Nelson, D.M. and Brzezinski, M.A., Diatom growth and productivity in an oligo-trophic midocean gyre: A 3-yr record from the Sargasso Sea near Bermuda. *Limnol. Oceanogr.*, 42, 473–486, 1997.

[12.2] Geider, R.J. *et al.*, Primary productivity of planet earth: biological determinants and physical constraints in terrestrial and aquatic habitats. *Glob. Change Biol.*, 7, 849–882, 2001.

[12.3] Yool, A. and Tyrrell, T., Role of diatoms in regulating the ocean's silicon cycle. *Global Biogeochem. Cycles*, 17, 1–21, 2003.

[12.4] Heydarizadeh, P. *et al.*, Response of CO2-starved diatom *Phaeodactylum tricornutum* to light intensity transition. *Philos. Trans. R. Soc.*, 5, 1–9, 2017.

[12.5] Kroth, P.G. *et al.*, A model for carbohydrate metabolism in the diatom *Phaeodactylum tricornutum* deduced from comparative whole genome analysis. *PLoS One*, 3, e1426, 2008.
[12.6] Depauw, F.A., Rogato, A., D'Alcalá, M.R., Falciatore, A., Exploring the molecular basis of responses to light in marine diatoms. *J. Exp. Bot.*, 63, 1575–1591, 2012.
[12.7] Kuczynska, P., Jemiola-Rzeminska, M., Strzalka, K., Photosynthetic pigments in diatoms. *Mar. Drugs*, 13, 5847–5881, 2015.
[12.8] Bertrand, M., Schoefs, B., Siffel, P., Rohacek, K., Molnar, I., Cadmium inhibits epoxidation of diatoxanthin to diadinoxanthin in the xanthophyll cycle of the marine diatom *Phaeodactylum tricornutum*. *FEBS Lett.*, 508, 153–156, 2001.
[12.9] Furukawa, T., Watanabe, M., Shihira-Ishikawa, I., Green- and blue-light-mediated chloroplast migration in the centric diatom *Pleurosira laevis*. *Protoplasma*, 203, 214–220, 1998.
[12.10] Anonymous author, Two letters from a Gentleman in the Country, relating to Mr. Leeuwenhoek's Letter in Transaction, no. 283. *Philos. Trans. R. Soc. Lond. B*, 23, 1494–1498 and 1498–1501, 1702.
[12.11] Stosch, H., An amended terminology of the diatom girdle. *Nova Hedwig. Beih.*, 54–55, 113–166, 1975.
[12.12] Round, F.E., Crawford, R.M., Mann, D.G., *The Diatoms: Biology and Morphology of the Genera*, Cambridge Univ. Press, Cambridge, UK, 19901990.
[12.13] Fuhrmann, T., Landwehr, S., El Rharbi-Kucki, M., Sumper, M., Diatoms as living photonic crystals. *Appl. Phys. B Lasers Opt.*, 78, 257–260, 2004.
[12.14] De Stefano, L., Rea, I., Rendina, I., De Stefano, M., Moretti, L., Lensless light focusing with the centric marine diatom Coscinodiscus walesii. *Opt. Express*, 15, 18082–18088, 2007.
[12.15] Ottesen, P., *Processing and Characterisation of Diatoms for Light Harvesting Materials in Solar Cells*, Norwegian University of Science and Technology, Trondheim, Norway, 2011.
[12.16] De Tommasi, E., Rea, I., Mocella, V., Moretti, L., De Stefano, M., Rendina, I., De Stefano, L., Multi-wavelength study of light transmitted through a single marine centric diatom. *Opt. Express*, 18, 12203–12212, 2010.
[12.17] Jeffryes, C., Agathos, S.N., Rorrer, G., Biogenic nanomaterials from photosynthetic microorganisms. *Curr. Opin. Biotechnol.*, 33, 23–31, 2015.
[12.18] Ferrara, M.A. *et al.*, Optical properties of diatom nanostructured biosilica in *Arachnoidiscus sp:* Micro-optics from mother nature. *PLoS One*, 9, 3–10, 2014.
[12.19] Maibohm, C., Friis, S.M.M., Ellegaard, M., Rottwitt, K., Interference patterns and extinction ratio of the diatom *Coscinodiscus granii*. *Opt. Express*, 23, 9543, 2015.
[12.20] Maibohm, C., Friis, S.M.M., Su, Y., Rottwitt, K., Comparing optical properties of different species of diatoms. *SPIE Proceedings*, 9360, 93600B, 2015.
[12.21] Ellegaard, M., Lenau, T., Lundholm, N., Maibohm, C., Michael, S., Friis, M., Rottwitt, K., Su, Y., The fascinating diatom frustule – can it play a role for attenuation of UV radiation? *J. Appl. Phycol.*, 28, 3295–3306, 2016.
[12.22] De Tommasi, E., Light manipulation by single cells: the case of diatoms. *J. Spectrosc.*, 2016, 1–13, 2016.
[12.23] Goessling, J.W. *et al.*, Structure-based optics of centric diatom frustules: Modulation of the *in vivo* light field for efficient diatom photosynthesis. *New Phytol.*, 219, 122–134, 2018.
[12.24] Aguirre, L.E., Ouyang, L., Elfwing, A., Hedblom, M., Wulff, A., Inganäs, O., Diatom frustules protect DNA from ultraviolet light. *Sci. Rep.*, 8, 1–6, 2018.
[12.25] Su, Y., Lundholm, N., Friis, S.M.M., Ellegaard, M., Implications for photonic applications of diatom growth and frustule nanostructure changes in response to different light wavelengths. *Nano Res.*, 8, 2363–2372, 2015.
[12.26] Wilhelm, C. *et al.*, The Regulation of Carbon and Nutrient Assimilation in Diatoms is Significantly Different from Green Algae. *Protist*, 157, 91–124, 2006.

[12.27] Lavaud, J., Fast regulation of photosynthesis in diatoms: mechanisms, evolution and eco-physiology. *Funct. Plant Sci. Biotechonol.*, 1, 267–287, 2007.
[12.28] Lepetit, B., Goss, R., Jakob, T., Wilhelm, C., Molecular dynamics of the diatom thylakoid membrane under different light conditions. *Photosynth. Res.*, 111, 245–257, 2012.
[12.29] Lehmuskero, A., Skogen Chauton, M., Boström, T., Light and photosynthetic microalgae: A review of cellular- and molecular-scale optical processes. *Prog. Oceanogr.*, 168, 43–56, 2018.
[12.30] Ghobara, M.M., Mazumder, N., Vinayak, V., Reissig, L., Gebeshuber, I.C., Tiffany, M.A., Gordon, R., On Light and Diatoms: A Photonics and Photobiology Review Mohamed. in: *Diatoms: Fundamentals and Applications,* Eds: Joseph Seckbach and Richard Gordon, Scrivener Publishing 2019, Beverly, USA, 1, 2018.
[12.31] Malviya, S. *et al.*, Insights into global diatom distribution and diversity in the world's ocean. *Proc. Natl. Acad. Sci.*, 113, E1516–E1525, 2016.
[12.32] Antonelli, M., Wetzel, C.E., Ector, L., Teuling, A.J., Pfister, L., On the potential for terrestrial diatom communities and diatom indices to identify anthropic disturbance in soils. *Ecol. Indic.*, 75, 73–81, 2017.
[12.33] Burgess, P., Variation in light intensity at different latitudes and seasons, effects of cloud cover, and the amounts of direct and diffused light. *Contin. Cover For. Gr. Sci. Meet.*, pp. 1–10, 2009.
[12.34] Schenk, H.J., On the Focusing of Sunlight by Ocean Waves. *J. Opt. Soc. Am.*, 47, 653–657, 1957.
[12.35] Koch, E.W. and Beer, S., Tides, light and the distribution of Zostera marina in Long Island Sound, USA. *Aquat. Bot.*, 53, 97–107, 1996.
[12.36] Markager, S. and Vincent, W., Spectral light attenuation and the absoption of UV and blue light in natural waters. *Limnol. Oceanogr.*, 45, 642–650, 2000.
[12.37] Kirk, J.T.O., *Light and Photosynthesis in Aquatic Ecosystems*, 3rd edn, Cambridge University Press, Cambridge, United Kingdom, 2010.
[12.38] Strzepek, R.F. and Harrison, P.J., Photosynthetic architecture differs in coastal and oceanic diatoms. *Nature*, 431, 689–692, 2004.
[12.39] Estlander, S., Nurminen, L., Olin, M., Vinni, M., Immonen, S., Rask, M., Ruuhijärvi, J., Horppila, J., Lehtonen, H., Diet shifts and food selection of perch Perca fluviatilis and roach Rutilus rutilus in humic lakes of varying water colour. *J. Fish Biol.*, 77, 241–256, 2010.
[12.40] Zhao, J., Temimi, M., Ghedira, H., Hu, C., Exploring the potential of optical remote sensing for oil spill detection in shallow coastal waters-a case study in the Arabian Gulf. *Opt. Express*, 22, 13755, 2014.
[12.41] Cunha, D.G.F., Calijuri M do, C., Lamparelli, M.C., A trophic state index for tropical/subtropical reservoirs (TSItsr). *Ecol. Eng.*, 60, 126–134, 2013.
[12.42] Millie, D.F., Schofield, O., Kirkpatrick, G.J., Johnsen, G., Evens, T.J., Using absorbance and fluorescence spectra to discriminate microalgae. *Eur. J. Phycol.*, 37, 313–322, 2002.
[12.43] Renema, W., Terrestrial influence as a key driver of spatial variability in large benthic foraminiferal assemblage composition in the Central Indo-Pacific. *Earth-Sci. Rev.*, 177, 514–544, 2018.
[12.44] Schellenberger Costa, B., Junandreas, A., Jakob, T., Weisheit, W., M.M., Wilhelm, C., Blue light is essential for high light acclimation and photoprotection in the diatom *Phaeodactylum tricornutum*. *J. Exp. Bot.*, 64, 483–493, 2013.
[12.45] Wolken, J.J., Light Detectors, Photoreceptors, and Imaging Systems in Nature. Oxford Univ. Press, Oxford, United Kingdom, 1995.
[12.46] Strasser, B., Sánchez-Lamas, M., Yanovsky, M.J., Casal, J.J., Cerdán, P.D., *Arabidopsis thaliana* life without phytochromes. *Proc. Natl. Acad. Sci. U. S. A.*, 107, 4776–4781, 2010.

[12.47] Stephenson, P.G. and Terry, M.J., Light signalling pathways regulating the Mg-chelatase branchpoint of chlorophyll synthesis during de-etiolation in *Arabidopsis thaliana*. *Photochem. Photobiol. Sci.*, 7, 1243–1252, 2008.
[12.48] Ishikawa, M., Takahashi, F., Nozaki, H., Nagasato, C., Motomura, T., Kataoka, H., Distribution and phylogeny of the blue light receptors aureochromes in eukaryotes. *Planta*, 230, 543–552, 2009.
[12.49] Kasahara, M., Kagawa, T., Sato, Y., Kiyosue, T., Wada, M., Phototropins mediate blue and red light-induced chloroplast movements in Physcomitrella patens. *Plant Physiol.*, 135, 1388–1397, 2004.
[12.50] König, S., Juhas, M., Jäger, S., Kottke, T., Büchel, C., The cryptochrome - photolyase protein family in diatoms. *J. Plant Physiol.*, 217, 15–19, 2017.
[12.51] Mittag, M. and Wilhelm, C., Light driven reactions in model algae. *J. Plant Physiol.*, 217, 1–3, 2017.
[12.52] Archibald, J.M., Review Endosymbiosis and Eukaryotic Cell Evolution. *Curr. Biol.*, 25, R911–R921, 2015.
[12.53] Takahashi, F. *et al.*, Areochrome, a photoreceptor required for photomorphogenesis in stramenopiles. *Proc. Natl. Acad. Sci. U. S. A.*, 104, 19625–19630, 2007.
[12.54] Kroth, P.G., Wilhelm, C., Kottke, T., An update on aureochromes: Phylogeny – mechanism – function. *J. Plant Physiol.*, 217, 20–26, 2017.
[12.55] Huysman, M.J.J. *et al.*, AUREOCHROME1a-mediated induction of the diatom-specific cyclin dsCYC2 controls the onset of cell division in diatoms (*Phaeodactylum tricornutum*). *Plant Cell*, 25, 215–28, 2013.
[12.56] Xu, J. and Gao, K., Use of UV-A energy for photosynthesis in the red macroalga gracilaria lemaneiformis. *Photochem. Photobiol.*, 86, 580–585, 2010.
[12.57] Nürnberg, D.J. *et al.*, Photochemistry beyond the red limit in chlorophyll f–containing photosystems. *Sci. (80-.)*, 360, 1210–1213, 2018.
[12.58] Stauber, J.L. and Jeffrey, S.W., Photosynthetic pigments in fifty-one species of marine diatoms. *J. Phycol.*, 24, 158–172, 1988.
[12.59] Garrido, J.L., Airs, R.L., Rodríguez, F., van Heukelem, L., Zapata, M., *New HPLC separation techniques*, Phytoplankton Pigments, Cambridge University Press, Cambridge, United Kingdom, 2011.
[12.60] Büchel, C., Wilhelm, C., Wagner, V., Mittag, M., Functional proteomics of light-harvesting complex proteins under varying light-conditions in diatoms. *J. Plant Physiol.*, 217, 38–43, 2017.
[12.61] Gelzinis, A. *et al.*, Mapping energy transfer channels in fucoxanthin-chlorophyll protein complex. *Biochim. Biophys. Acta - Bioenerg.*, 1847, 241–247, 2015.
[12.62] Nakajima, Y. and Ueda, R., Improvement of photosynthesis in dense microalgal suspension by reduction of light harvesting pigments. *J. Appl. Phycol.*, 9, 503–510, 1997.
[12.63] Schumann, A., Goss, R., Jakob, T., Wilhelm, C., Investigation of the quenching efficiency of diatoxanthin in cells of *Phaeodactylum tricornutum* (Bacillariophyceae) with different pool sizes of xanthophyll cycle pigments. *Phycologia*, 46, 113–117, 2007.
[12.64] Bilger, W. and Björkman, O., Role of the xanthophyll cycle in photoprotection elucidated by measurements of light-induced absorbance changes, fluorescence and photosynthesis in leaves of Hedera canariensis. *Photosynth. Res.*, 25, 173–185, 1990.
[12.65] Brunet, C., Chandrasekaran, R., Barra, L., Giovagnetti, V., Corato, F., Ruban, A.V., Spectral radiation dependent photoprotective mechanism in the diatom Pseudo-nitzschia multistriata. *PLoS One*, 9, pp. 1–10, 2014.
[12.66] Bhattacharya, D. and Medlin, L.K., The phylogeny of plastids: a review based on comparisons of small-subunit ribosomal RNA coding regions. *J. Phycol.*, 31, 489–498, 1995.

[12.67] Cavalier-Smith, T., Chloroplast evolution: Secondary symbiogenesis and multiple losses. *Curr. Biol.*, 12, 62–64, 2002.
[12.68] Kroth, P. and Strotmann, H., Diatom plastids: Secondary endocytobiosis, plastid genome and protein import. *Physiol. Plant*, 107, 136–141, 1999.
[12.69] Michels, A.K., Wedel, N., Kroth, P.G., Diatom plastids possess a phosphoribulokinase with an altered regulation and no oxidative pentose phosphate pathway. *Plant Physiol.*, 137, 911–920, 2005.
[12.70] Armbrust, E.V. *et al.*, The Genome of the Diatom *Thalassiosira pseudonana*: Ecology, Evolution, and Metabolism. *Science*, 306, 79–86, 2004.
[12.71] Williams, W.P. and Allen, J.F., State 1/State 2 changes in higher plants and algae. *Photosynth. Res.*, 13, 19–45, 1987.
[12.72] Mullineaux, C.W. and Allen, J.F., State 1-State 2 transitions in the cyanobacterium *Synechococcus* 6301 are controlled by the redox state of electron carriers between Photosystems I and II. *Photosynth. Res.*, 23, 297–311, 1990.
[12.73] Apel, K. and Hirt, H., Reactive oxygen species: metabolism, oxidative stress, and signal transduction. *Annu. Rev. Plant Biol.*, 55, 373–399, 2004.
[12.74] Demming-Adams, B., Adams, W.W., Mattoo, A.K., *Photoprotection, photoinhibition, gene regulation and environment*, Springer, Heidelberg, Germany, 2006.
[12.75] Ruban, A., Lavaud, J., Rousseau, B., Guglielmi, G., Horton, P., Etienne, A.L., The super-excess energy dissipation in diatom algae: Comparative analysis with higher plants. *Photosynth. Res.*, 82, 165–175, 2004.
[12.76] Kasahara, M., Kagawa, T., Oikawa, K., Chloroplast avoidance movement reduces photodamage in plants. *Nature*, 420, 829–832, 2002.
[12.77] Park, Y.I., Chow, W.S., Anderson, J.M., Chloroplast Movement in the Shade Plant *Tradescantia albiflora* Helps Protect Photosystem II against Light Stress. *Plant Physiol.*, 111, 867–875, 1996.
[12.78] Wada, M., Kagawa, T., Sato, Y., Chloroplast movement. *Annu. Rev. Plant Biol.*, 54, 455–68, 2003.
[12.79] Kiefer, D. A., Chlorophyll a fluorescence in marine centric diatoms: Responses of chloroplasts to light and nutrient stress. *Mar. Biol.*, 23, 39–46, 1973.
[12.80] Chen, S.-T. and Li, C.-W., Relationships between the Movements of Chloroplasts and Cytoskeletons in Diatoms. *Bot. Mar.*, 34, 505–511, 1991.
[12.81] Gillard, J. *et al.*, Physiological and transcriptomic evidence for a close coupling between chloroplast ontogeny and cell cycle progression in the pennate diatom Seminavis robusta. *Plant Physiol.*, 148, 1394–1411, 2008.
[12.82] DeBlasio, S.L., Mullen, J.L., Luesse, D.R., Hangarter, R.P., Phytochrome modulation of blue light-induced chloroplast movements in *Arabidopsis*. *Plant Physiol.*, 133, 1471–1479, 2003.
[12.83] Gabryś, H., Blue light-in duced orientation movements of chloroplasts in higher plants: Recent progress in the study of their mechanisms. *Acta Physiol. Plant*, 26, 473–478, 2004.
[12.84] Goessling, J.W., Cartaxana, P., Kühl, M., Photo-protection in the centric diatom *Coscinodiscus granii* is not controlled by chloroplast high-light avoidance movement. *Front. Mar. Sci.*, 2, 1–10, 2016.
[12.85] Mullineaux, C.W., Function and evolution of grana. *Trends Plant Sci.*, 10, 521–525, 200.
[12.86] Lichtenthaler, H.K., Buschmann, C., Doll, M., Bach, T., Kozel, U., Meier, D., Rahmsdorf, U., Photosynthetic activity, chloroplast ultrastructure, and leaf characteristics of high-light and low-light plants and of sun and shade leaves. *Photosynth. Res.*, 2, 115–141, 1981.
[12.87] Pribil, M., Labs, M., Leister, D., Structure and dynamics of thylakoids in land plants. *J. Exp. Bot.*, 65, 1955–1972, 2018.

[12.88] Gibbs, S., The comparative ultrastructure of the algal chloroplast. *Ann. N. Y. Acad. Sci.*, 175, 454–473, 1970.
[12.89] Bína, D., Herbstová, M., Gardian, Z., Vácha, F., Litvín, R., Novel structural aspect of the diatom thylakoid membrane: lateral segregation of photosystem I under red-enhanced illumination. *Sci. Rep.*, 6, 25583, 2016.
[12.90] Grouneva, I., Rokka, A., Aro, E., The thylakoid membrane proteome of two marine diatoms outlines both diatom-specific and species-specific features of the photosynthetic machinery. *J. Proteome Res.*, 10, 5338–5353, 2011.
[12.91] Jacobs, M., Lopez-garcia, M., Phrathep, O.-P., Lawson, T., Photonic crystal structure of *Begonia* chloroplasts enhances photosynthetic efficiency. *Nat. Plants*, 2, 1–16, 2016.
[12.92] Boyd, C.M. and Gradmann, D., Impact of osmolytes on buoyancy of marine phytoplankton. *Mar. Biol.*, 141, 605–618, 2002.
[12.93] Smetacek, V.S., Role of sinking in diatom life-hystory: ecological, evolutionary and geological significance. *Mar. Biol.*, 84, 239–251, 1985.
[12.94] Edgar, L a. and Pickett-Heaps, J.D., The Mechanism of Diatom Locomotion. I. An Ultrastructural Study of the Motility Apparatus. *Proc. R. Soc. B Biol. Sci.*, 218, 331–343, 1983.
[12.95] Kooistra, H., Gersonde, R., Medlin, L., Mann, D., The origin and evolution of the diatoms: their adaptation to a planktonic existence, in: *Evolution of Primary Producers in the Sea*, P. Falkowsk and A. Knoll (Eds.), Elsevier Academic Press, Cambridge, Massachusetts, USA 2007.
[12.96] Kromkamp, J., Barranguet, C., Peene, J., Determination of microphytobenthos PSII quantum efficiency and photosynthetic activity by means of variable chlorophyll fluorescence. *Mar. Ecol. Prog. Ser.*, 162, 45–55, 1998.
[12.97] Lassen, C., Ploug, H., Barker Jmgensen, B., Microalgal photosynthesis and spectral scalar irradiance in coastal marine sediments of Limfjorden, Denmark. *Limnol. Oceanogr.*, 37, 760–772, 1992.
[12.98] Kühl, M., Lassen, C., Jørgensen, B.B., Light penetration and light intensity in sandy marine sediments measured with irradiance and scalar irradiance fiber-optic microprobes. *Mar. Ecol. Prog. Ser.*, 105, 139–148, 1994.
[12.99] Serôdio, J., Vieira, S., Cruz, S., Coelho, H., Rapid light-response curves of chlorophyll fluorescence in microalgae: Relationship to steady-state light curves and non-photochemical quenching in benthic diatom-dominated assemblages. *Photosynth. Res.*, 90, 29–43, 2006.
[12.100] Cartaxana, P., Cruz, S., Gameiro, C., Kühl, M., Regulation of intertidal microphytobenthos photosynthesis over a diel emersion period is strongly affected by diatom migration patterns. *Front. Microbiol.*, 7, 1–11, 2016.
[12.101] Sims, P.A., Mann, D.G., Medlin, L.K., Evolution of the diatoms: insights from fossil, biological and molecular data. *Phycologia*, 45, 361–402, 2006.
[12.102] Raven, J.A. and Waite, A.M., The evolution of silicification in diatoms: Inescapable sinking and sinking as escape? *New Phytol.*, 162, 45–61, 2004.
[12.103] Losic, D., Rosengarten, G., Mitchell, J.G., Voelcker, N.H., Pore architecture of diatom frustules: potential nanostructured membranes for molecular and particle separations. *J. Nanosci. Nanotechnol.*, 6, 982–989, 2006.
[12.104] Hamm, C.E., Merkel, R., Springer, O., Jurkojc, P., Maier, C., Prechtel, K., Smetacek, V., Architecture and material properties of diatom shells provide effective mechanical protection. *Nature*, 421, 841–843, 2003.
[12.105] Milligan, A.J., A Proton Buffering Role for Silica in Diatoms. *Science*, 297, 1848–1850, 2002.
[12.106] De Tommasi, E., Light manipulation by single cells : the case of diatoms. *J. Spectro.*, 2016, 1–24, 2016.
[12.107] Guiry, M.D., How many species of algae are there? *J. Phycol.*, 48, 1057–1063, 2012.

[12.108] Maibohm, C., Nielsen, J.H., Rottwitt, K., Light interaction with nano-structured diatom frustule, from UV-A to NIR. *MRS Adv.*, **3**, 350, 2016.

[12.109] Su, Y., Lundholm, N., Ellegaard, M., Long-term cultivation of the diatom *Coscinodiscus granii* at different light spectra: effects on frustule morphology. *J. Appl. Phycol.*, 29, 1–5, 2017.

[12.110] Vrieling, E.G., Sun, Q., Tian, M., Kooyman, P.J., Gieskes, W.W.C., van Santen, R.A., Sommerdijk, N.A.J.M., Salinity-dependent diatom biosilicification implies an important role of external ionic strength. *Proc. Natl. Acad. Sci. U. S. A.*, 104, 10441–10446, 2007.

[12.111] Townley, H.E., Woon, K.L., Payne, F.P., White-Cooper, H., Parker, A.R., Modification of the physical and optical properties of the frustule of the diatom *Coscinodiscus wailesii* by nickel sulfate. *Nanotechnology*, 18, 295101, 2007.

[12.112] Kieu, K., Li, C., Fang, Y., Cohoon, G., Herrera, O.D., Hildebrand, M., Sandhage, K.H., Norwood, R.A., Structure-based optical filtering by the silica microshell of the centric marine diatom *Coscinodiscus wailesii*. *Opt. Express*, 22, 15992–9, 2014.

[12.113] Cohoon, G.A., Alvarez, C.E., Meyers, K., Deheyn, D.D., Hildebrand, M., Kieu, K., Norwood, R.A., Analysis of quasi-periodic pore-network structure of centric marine diatom frustules. *Proc. SPIE*, 9341, 2015.

[12.114] Valmalette, J., Romann, J., Røyset, A., Einarsrud, M., 3D-Hyperspectral mapping of light propagation through diatom frustule silica nanostructures. *Opt. Lett.*, 40, 3–5, 2015.

[12.115] Yoneda, S., Ito, F., Yamanaka, S., Usami, H., Optical properties of nanoporous silica frustules of a diatom determined using a 10 μm microfiber probe. *Jpn. J. Appl. Phys.*, 55, pp. 1–5, 2016.

[12.116] Dossou, K.B., Large field enhancement obtained by combining Fabry-Perot resonance and Rayleigh anomaly in photonic crystal slabs. *J. Opt.*, 19, 1–12, 2017.

[12.117] Romann, J., Valmalette, J.-C., Chauton, M.S., Tranell, G., Einarsrud, M.-A., Vadstein, O., Wavelength and orientation dependent capture of light by diatom frustule nanostructures. *Sci. Rep.*, 5, 17403, 2015.

[12.118] Yablonovitch, E., Photonic band-gap structures. *J. Opt. Soc. Am.*, 10, 283–2959, 1993.

[12.119] Goessling, J.W., Wardley, W.P., Lopez-Garcia, M., Highly Reproducible, Bio-Based Slab Photonic Crystals Grown by Diatoms. *Adv. Sci.*, 7, 1903726, 2020.

[12.120] Vogelmann, T.C., Bornman, J.F., Yates, D.J., Focusing of light by leaf epidermal cells. *Physiol. Plant*, 98, 43–56, 1996.

[12.121] Armbrecht, L.H., Smetacek, V., Assmy, P., Klaas, C., Cell death and aggregate formation in the giant diatom *Coscinodiscus wailesii* (Gran & Angst, 1931). *J. Exp. Mar. Bio. Ecol.*, 452, 31–39, 2014.

[12.122] Goessling, J.W., Frankenbach, S., Ribeiro, L., Serôdio, J., Kühl, M., Modulation of the light field related to valve optical properties of raphid diatoms: Implications for niche differentiation in the microphytobenthos. *Mar. Ecol. Prog. Ser.*, 588, 29–42, 2018.

[12.123] Yamanaka, S., Yano, R., Usami, H., Hayashida, N., Ohguchi, M., Takeda, H., Yoshino, K., Optical properties of diatom silica frustule with special reference to blue light. *J. Appl. Phys.*, 103, 1–6, 2008.

[12.124] Mercado, J.M., Sánchez-Saavedra, M.D.P., Correa-Reyes, G., Lubián, L., Montero, O., Figueroa, F.L., Blue light effect on growth, light absorption characteristics and photosynthesis of five benthic diatom strains. *Aquat. Bot.*, 78, 265–277, 2004.

[12.125] Correa-Reyes, J.G., Sanchez-saavedra, MDELP, Siqueiros-beltrones, D.A., Flores-acevedo, N., Growth of Eight Strains of Benthic Under Two Light Conditions. *J. Shellfish Res.*, 20, 603–610, 2001.

[12.126] Sanchez Saavedra, M.P. and Voltolina, D., The effect of different light quality on the food value of the diatom Chaetoceros sp. for Artemia franciscana. *Riv. Ital. Acquacolt.*, 30, 135–138, 1995.

[12.127] Su, Y., The effect of different light regimes on pigments in *Coscinodiscus granii*. *Photosynth. Res.*, 140, 301–310, 2019.

[12.128] Johnsen, S. and Sosik, H., Shedding light on light in the ocean. *Oceanogr. Mag.*, 43, 1–5, 2004.

[12.129] Karentz, D., Cleaver, J.E., Mitchell, D.L., Cell survival characteristics and molecular responses of Antarctic phytoplancton to ultraviolet-B radiation. *J. Phycol.*, 27, 326–341, 1991.

[12.130] Davidson, A.T., Bramich, D., Marchant, H.J., McMinn, A., Effects of UV-B irradiation on growth and survival of Antarctic marine diatoms. *Mar. Biol.*, 119, 507–515, 1994.

[12.131] Ingalls, A.E., Whitehead, K., Bridoux, M.C., Tinted windows: The presence of the UV absorbing compounds called mycosporine-like amino acids embedded in the frustules of marine diatoms. *Geochim. Cosmochim. Acta*, 74, 104–115, 2010.

[12.132] Su, Y., Lenau, T.A., Gundersen, E., Kirkensgaard, J.J.K., Maibohm, C., Pinti, J., Ellegaard, M., The UV filtering potential of drop-casted layers of frustules of three diatom species. *Sci. Rep.*, 8, 1–10, 2018.

[12.133] Davidson, A.T. and Marchant, H.J., The impact of ultraviolet radiation on *Phaeocystis* and selected species of Antarctic diatoms, in: *Ultrav. Radiat. Antarct. Meas. Biol. Eff*, vol. 62, pp. 187–205, 1994.

[12.134] Wasell, A. and Håkansson, H., Diatom stratigraphy in a lake on horseshoe island, antarctica: A marine-brackish-fresh water transition with comments on the systematics and ecology of the most common diatoms. *Diatom Res.*, 7, 157–194, 1992.

[12.135] Cartaxana, P., Ribeiro, L., Goessling, J., Cruz, S., Kühl, M., Light and O2 microenvironments in two contrasting diatom-dominated coastal sediments. *Mar. Ecol. Prog. Ser.*, 545, 35–47, 2016.

[12.136] Baker, E.T. and Lavelle, J.W., The Effect of Particle Size on the Light Attenuation Coefficient of Natural Suspensions. *J. Geophys. Res.*, 89, 8197–8203, 1984.

[12.137] De Tommasi, E., Congestri, R., Dardano, P., De Luca, A.C., Managò, S., Rea, I., De Stefano, M., UV-shielding and wavelength conversion by centric diatom nanopatterned frustules. *Sci. Rep.*, 8, 1–14, 2018.

[12.138] Chen, X., Wang, C., Baker, E., Sun, C., Numerical and experimental investigation of light trapping effect of nanostructured diatom frustules. *Sci. Rep.*, 5, 11977, 2015.

[12.139] Noyes, J., Sumper, M., Vukusic, P., Light manipulation in a marine diatom. *J. Mater. Res.*, 23, 3229–3235, 2008.

[12.140] Hsu, S.-H., Paoletti, C., Torres, M., Ritchie, R.J., Larkum, A.W.D., Grillet, C., Light transmission of the marine diatom *Coscinodiscus wailesii*. *Proc. SPIE*, 8339, 83390F-83390F–8, 2012.

[12.141] Stramski, D. and Sciandra, A., Effects of temperature, nitrogen, and light limitation on the optical properties of the marine diatom *Thalassiosira pseudonana*. *Limnol. Oceanogr.*, 47, 392–403, 2002.

[12.142] Deuk Lee, S., Park, J.S., Lee, J.H., Taxonomic study of the genus Achnanthes (Bacillariophyta) in Korean coastal waters. *J. Ecol. Environ.*, 36, 391–406, 2013.

[12.143] Cunningham, A., Ramage, L., McKee, D., Relationships between inherent optical properties and the depth of penetration of solar radiation in optically complex coastal waters. *J. Geophys. Res. Oceans*, 118, 2310–2317, 2013.

[12.144] Bricaud, A., Morel, A., Babin, M., Allali, K., Claustre, H., Variations of light absorption by suspended particles with chlorophyll a concentrations for bio-optical models. *J. Geophys. Res.*, 103, 31033–31044, 1998.

[12.145] Kühl, M. and Jørgensen, B.B., The light-field of microbenthic communities- radiance distribution and microscale optics of sandy coastal sediments. *Limnol. Oceanogr.*, 39, 1368–1398, 1994.

[12.146] Ødegård, I.A., Romann, J., Fossdal, A., Røyset, A., Synthesis and properties of silicon/magnesium silicon nitride diatom frustule replicas†. *J. Mater. Chem. A*, 2, 16410–16415, 2014.

13

Steps of Silicic Acid Transformation to Siliceous Frustules: Main Hypotheses and Discoveries

Vadim V. Annenkov[1]*, Elena N. Danilovtseva[1] and Richard Gordon[2,3†]

[1]*Limnological Institute of the Siberian Branch of the Russian Academy of Sciences, Irkutsk, Russia*
[2]*Gulf Specimen Marine Laboratory & Aquarium, Panacea, FL, United States*
[3]*C.S. Mott Center for Human Growth & Development, Department of Obstetrics & Gynecology Wayne State University, Detroit, MI, United States*

Abstract
Diatoms are unicellular photosynthetic organisms that live inside a silica-based glass exoskeleton (frustule). This exoskeleton consists of two valves, similar to parts of a petri dish, and girdle bands between the valves. Diatom valves are species-specific structures with a large number of submicrometric and nano-level elements: pores, spines, rimoportules, etc. Cellular mitosis takes place inside the frustule, after which a new valve is built in each daughter cell in addition to the parent valve. The formation of new valves takes several hours and requires the assimilation of large amounts of silicon from the environment in the form of silicic acid. Silicic acid is transported through the cytoplasm to the silica deposition vesicle where a new valve is synthesized. The mechanisms of these key stages of frustule morphogenesis are not sufficiently clear. We have tried to summarize the main hypotheses in this area and discuss them in the light of the available knowledge of diatom physiology, biochemistry, and chemistry of the substances involved.

***Keywords*:** Diatoms, silica, silica transport, morphogenesis

13.1 Introduction

Multiscalar siliceous frustules of diatoms are their distinctive feature [13.121] and have been an intriguing challenge for generations of biologists, biochemists, material science specialists, and everybody who is interested in the mystery of morphogenesis. Diatoms and their frustules where among the first organisms observed after invention of the microscope in the 17th century [13.65] [13.357]. Starting from the middle of the last century, intensive study of diatoms was carried out with transmission electron microscopy

**Corresponding author*: annenkov@yahoo.com
†*Corresponding author*: dickgordoncan@xplornet.com

Vadim Annenkov, Joseph Seckback and Richard Gordon (eds.) Diatom Morphogenesis, (301–348) © 2022 Scrivener Publishing LLC

(TEM), scanning electron microscopy (SEM), and biochemical approaches. Silicon was found to be a key element in diatom physiology whose depletion resulted in halting of most processes in diatom culture [13.64]. In 1988, these investigations got an impulse from nanotechnology—synthesis of silicon dioxide in highly ordered form at moderate conditions was very attractive [13.129]. Over 25 years ago, we published a detailed review on the study of diatom morphogenesis at the molecular level [13.132] and here we try to describe briefly the current status of research in this area. Nowadays, diatom physiology and formation of siliceous frustules are studied with full-genomic and transcriptomic approaches, Western blotting, solid state nuclear magnetic resonance (NMR), green fluorescent protein (GFP), and other sophisticated methods. Most of these works have been reviewed [13.96] [13.165]. Nevertheless, we have to acknowledge that our understanding of formation of diatom frustules is still a set of hypotheses which sometimes contradict each other, do not have direct experimental confirmation, or do not agree with known data from chemistry. Our article is aimed at summarizing these hypotheses and experimental facts. As a result, we will try to understand where we are and in which directions we should go.

There are 11 interrelated stages on the way from silicon in the environment to the siliceous diatom frustule:

1. Penetration of the boundary layer: the diatom as an antenna for silica.
2. Getting past the cloud of extracellular material.
3. Adsorption of silica onto the organic coat of the diatom.
4. Getting past the silica frustule or through its pores.
5. Getting past the inner organic coat, the diatotepum.
6. Transport of silica across the cell membrane.
7. Cytoplasm storage and trafficking of silica to the places of synthesis of the frustule parts.
8. Transport of silica across each silicalemma.
9. Precipitation and morphogenesis of the nascent valve within the silicalemma.
10. Thickening of the valve within the silicalemma.
11. Exteriorization of the valve.

In stages 6 through 8, we will focus on valve morphogenesis, leaving out girdle band and spines, for instance. Steps 3 and 5 may be interrelated, as in higher plants, the formation of silica phytoliths (cf. [13.259]) involves silica transport into and out of sometimes specialized cells, in some cases in a polar, i.e., localized fashion [13.29] [13.258]. This is one indication that the study of diatom morphogenesis would be enhanced if considered in the context of all silica biomineralization (Figure 13.1).

Evidently, any theory of valve synthesis has to encompass all these stages as well as to agree with chemical properties of the substances involved. Furthermore, diatoms are unique in exhibiting 8 orders of magnitude of structure in a single substance, silica [13.121], unsurpassed by any known living or artificial object. To what extent there are commonalities of mechanism across all of these levels is presently unknown. We will keep in mind these ideas as we discuss the known data and hypotheses. Diatoms are the culmination of the widespread evolution of silica biomineralization (Figure 13.1).

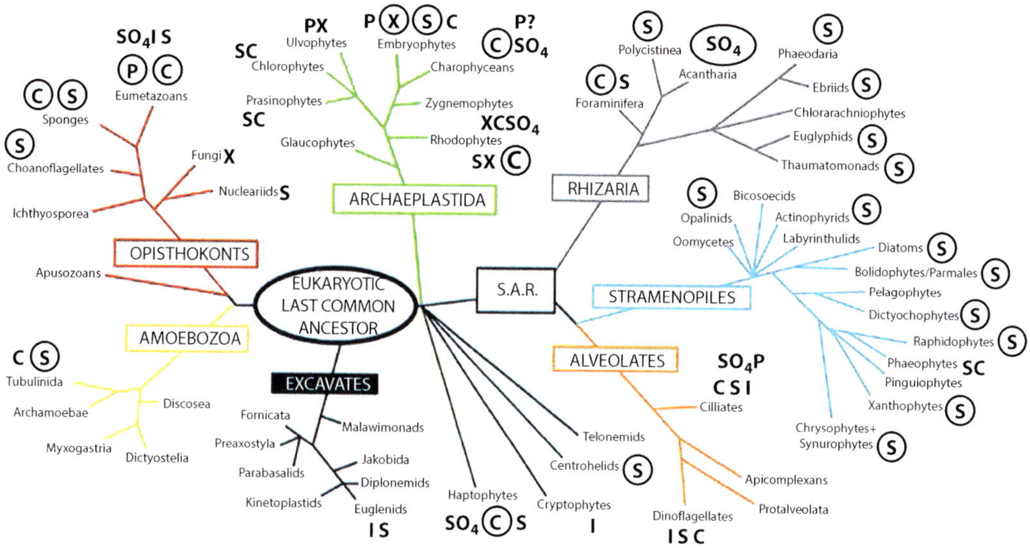

Figure 13.1 The diversity of biomineralization across the eukaryotes .... with circled letters indicating prominent and widespread use of that biomineral. S, silica; C, calcium carbonate; P, calcium phosphate; I, iron (magnetite/goethite); X, calcium oxalate; SO_4, sulfates (calcium/barium/strontium), ? denotes uncertainty in report [13.155] with permission under a Creative Commons Attribution License (CC BY). The metabolism of prokaryote communities can lead to silica precipitation [13.72]. Cf. [13.267].

13.2 Penetration of the Boundary Layer: The Diatom as an Antenna for Silica

Silicon is present in natural waters in the form of silicic acid at a concentration in most below 100 µM. $Si(OH)_4$ is a very weak acid (pK = 9.7–9.9 [13.319] [13.320]) and it is un-ionized at pH < 8. Condensation of silicic acid in aqueous medium starts at a concentration of more than 2–3 mM and gives rise to oligomers (oligosilicates) which give stable colloid solutions, gel, or solid material depending on the starting concentration, pH, temperature, ionic strength, and other factors [13.56] [13.173]. Sometimes, natural waters contain colloidal silica [13.79] [13.119] [13.261]. Some organic substances, e.g., polymeric amines, can catalyze condensation [13.173] and stabilize colloidal siliceous particles ([13.13] and references therein). Diffusion coefficients of silica depend on particle size, charge, and concentration [13.174].

The distribution of oligosilicate sizes depends on the source of the silica, which affects diatom growth [13.66] as determined by dissolving crystalline quartz, diatomaceous earth, and amorphous synthetic silica in water followed by filtration with 200-nm pores. Evidently, siliceous nanoparticles penetrated this filter and this is the reason for silicon bioavailability in [13.66]. These nanoparticles were not taken into account because silicon concentration was measured with a method which accounts for monomers and dimers of silicic acid only. While these authors found no direct interaction of centric diatoms with the surface of the silica source, as they previously hypothesized [13.282], repetition of the experiment with adherent and motile pennate diatoms might do so. While the motile pennate *Navicula* sp. was used in the earlier study by the same group [13.282], it was not considered in their subsequent study [13.66]. In more recent work, colloidal silica in the medium of different sizes and surface moieties enter diatoms at different rates [13.68].

This first stage of silica reaching the diatom has only been speculated on:

> "Perhaps the external surface of the shell is not only an 'antenna' for silica ('The Diatom as a Silica Antenna' in [13.132]) but also a catalytic surface for condensation. This might be another role of the silica embedded proteins.... Therefore, more studies are required to understand silica trafficking during the process of morphogenesis. It is not known: (i) whether colloidal size particles are ever taken in, which would seem to be more efficient when they are available in the environment; (ii) whether the valves and silica spines have roles in providing adsorption 'antennae' (another DLA [diffusion limited aggregation] problem) and surface migration 'funnels' to bring the external silica to the cell membrane, a form of dimensionality reduction [13.3] analogous to the cell nucleus and its pores [13.283]..." [13.137].

Thus, spines may be more than impediments to sinking [13.202], predation [13.122], or light pipes [13.137]. The shape of spines of, for instance, the centric diatom *Chaetoceros*, suggests that they reach out into the silica gradient around the diatom during their morphogenesis. This could be tested by altering that gradient, for example, via flow of the water under a light microscope in a microchannel experiment [13.308] or after the fact in a larger channel [13.111] [13.130]. Furthermore, the huge porous structure of diatoms [13.362] [13.363], as exemplified by their pores [13.120], may act as an antenna for silica, whether or not a given species has spines. However, see [13.118] that functional groups may not correlate with surface area. Diatoms that hoist themselves on stalks might be using those stalks in part as antennae to collect silica [13.1], as they do for metals [13.376]. Indeed, there is a gradient of silicon that is highest near the cell [13.265], suggesting the possibility of transport of silica along the stalk.

Diatoms can also actively participate in stage #1, via silica chemotaxis of motile diatoms [13.49] [13.50] [13.51], or via changes in buoyancy of "nonmotile" diatoms [13.45] [13.107] [13.203] [13.333] that might move them vertically into silica rich water layers [13.255], or by detaching from their substrate and drifting horizontally [13.19], or even by swimming [13.128]. Diatoms on stalks may be launching themselves away from the boundary layer, without detachment, and thereby access higher silica concentrations. Any movement relative to the immediately surrounding water will add convection to what would otherwise be strictly diffusion-limited access of silica in the water to the diatom's surface, reducing the boundary layer, and generally increasing the rate of silica adsorption. This effect could be enhanced if diatoms exhibit acoustic streaming [13.188], which might in part account for the high accelerations observed in motile diatoms [13.304]. Diatoms that hitch a ride on other diatoms [13.302], animals [13.235], or moving sand grains [13.176] should also have better access to silica.

13.3 Getting Past the Cloud of Extracellular Material

The presence of some substances which facilitate silicon capture on the diatom surface was mentioned in [13.214]: washing of the diatom biomass decreased silicon assimilation. Washing may have removed adherent polysaccharide exudates, such as the diatom trail material of motile diatoms. This material has yet to be investigated for its ability to absorb silica from water. It is found around *Bacillaria paradoxa* colonies [13.299], but it is not yet known if this is the analog of diatom trail material of single cell motile pennate diatoms [13.71] [13.160] [13.171] [13.367] or acquired coincidentally from other organisms (Jan Rines, personal communication, 2019). Bacteria can induce diatoms to secrete proteins [13.63] and, in turn, somehow prevent abnormal frustule morphogenesis [13.371]. Some bacteria may hydrolytically attack the organic coating on live diatoms, releasing some of their silica [13.42] [13.44] [13.278]. The bacteria-diatom association was described in [13.247] and a conclusion about mutual advantage of this coexistence was drawn. Thus, bacteria form part of the extracellular material that may not be passive in regards to access of diatoms to silica. On the other hand, axenic cultures of these diatoms were developed [13.323] which shows their self-sufficiency.

Some diatoms construct capsules [13.211] [13.215] [13.372] and others live as colonies of unattached individuals inside common tubes [13.6] [13.28] [13.58] [13.67] [13.83] [13.84] [13.85] [13.98] [13.99] [13.115] [13.150] [13.172] [13.221] [13.222] [13.223] [13.224] [13.225] [13.226] [13.227] [13.251] [13.256] [13.330] [13.369]. These too would represent barriers and/or absorption/transmission devices for the diatoms inside.

13.4 Adsorption of Silica Onto the Outer Organic Coat of the Diatom

Movement of silica the other way, namely, dissolution of the valve back into the surrounding water, is inhibited by an organic coat over the silica [13.43] [13.180] and accelerated by the presence of bacteria that attack that coating [13.255]. However, any such barrier [13.361] would be expected to work both ways, inhibiting adsorption of silica to the cell. Therefore, there may be a silica transport step, across the organic layer, that has yet to be investigated. For example, most of the influx of silica occurs during cell division [13.132] [13.335] ("surge uptake" [13.8] [13.354]). Perhaps this is accompanied by temporary reduction of the external organic layer, or alteration, such as the increase in chitin in the cell wall after silica starvation or limitation ends [13.103] or even normal cell division [13.76]. The organic layer has an electric potential and an order of magnitude more proton/hydroxyl adsorption sites than the cleaned frustule [13.118]. These sites may well be involved in the capture of silica from water and present a barrier and/or pathway for transport of silica to the cell membrane. Silica efflux does not occur in silica-free water [13.334]. While this has been attributed to binding of silica inside the cytoplasm [13.354], such a mechanism would not protect the mature frustule, again implying that the organic coat is a silica barrier, and perhaps a dynamic one at that.

Condensation of silicic acid on the outer membrane surface could proceed under the action of polysaccharides and other biopolymers which cover diatom frustules [13.112] [13.153] [13.297]. Hydroxyl enriched proteins were found around siliceous frustules [13.153] and an ability of this protein to bind silicic acid was supposed. We have studied the influence of cross-linked poly(vinyl amine) doped with Zn^{2+} on $Si(OH)_4$ condensation at undersaturated concentrations (<2 mM) and found condensation of silicic acid and coordination of the oligosilicates with the polymer gel [13.91]. Metals adsorb to the organic coat [13.117] [13.124] [13.287], and thus, its impact on silica transport across it needs to be ascertained.

13.5 Getting Past the Silica Frustule or Through Its Pores

The diatom frustule silica is probably impermeable to silica, but this may depend on the degree of silicification and sintering (Section 13.12). As the material is amorphous, it may to some extent be like a glass, which is essentially a "frozen" liquid. Thus, it behooves us to measure the diffusion of silica inside the "solid" frustule, which has not yet been done. A lower density of silanol groups occurs on diatoms subject to silica limitation [13.234], which might alter surface or 3D diffusion of silica on/within the frustule. Diffusion coefficients of substances through silica depends on its water content [13.346]. Silica can be characterized by its tortuosity for diffusing ions, molecules, and particles [13.310] [13.346] [13.347], which does not seem to have been investigated yet for diatom frustules, isolated or in intact living cells. Similarly, the mesoporosity of silica can affect diffusion coefficients [13.30] and alter the structure of molecules diffusing through it [13.249].

The self-diffusion of silica over the surface of silica was hypothesized to be the mechanism of silica sintering in diatoms [13.132]. Molecular dynamics simulations of bulk and surface self-diffusion for high temperature vitreous silica have been carried out [13.116] and should be extendable to ambient temperature diatom frustules.

Diatom pores may be open or filled with organic matter. The wide range of proposed functions for open pores has been reviewed in [13.158], including diatom viruses, with the emphasis being on their functions as filters whose structure can be bottle-shaped [13.229]. However, some specialized pores secrete extracellular polymers and most pores appear to be filled with exuding organic matter [13.356]. Either silica moves against this flow to the exterior or does not enter diatoms via their pores.

The raphe of pennate diatoms may be thought of as an extended pore. Whether it also represents a pathway for silica entry, against the flow of the raphe fluid [13.128], has yet to be investigated.

13.6 Getting Past the Inner Organic Coat, the Diatotepum

Between the silica frustule and the cell membrane is another layer, the diatotepum or diatotepic layer [13.361]. In some species this "diatopectin" layer (a name that has been discontinued [13.170] [13.361]) is retracted from portions of the valve [13.360]. It is "an insoluble organic matrix enriched in mannose" [13.350], several layers thick under the valves and girdle bands [13.87]. One species has a "reduced ocellus" in a "horn region [with]... a thin diatotepum with a hexagonal meshwork" [13.318]. The polysaccharide [13.315] plus proteoglycan [13.243] diatotepum blocks the pores in the silica frustule, except for rimoportules [13.87], "raphe openings and where girdle bands are secreted onto the surface in the region of epivalve and hypovalve overlap (i.e., none at the 'thecal junction' [13.314])" [13.243]. The diatotepum can form a lining in the space between costae occupied by mitochondria [13.252] and project into puncta [13.186]. It would appear to be a barrier that the incoming silica must pass through, unless it somehow is funneled to those few areas or through "minute pores" that have been observed in the diatotepum [13.252]. Similar "poroids" in the diatotepum were observed by [13.361] but not by [13.186].

13.7 Transport of Silica Across the Cell Membrane

Diatoms capture silicon from the environment during building of the frustule components (valves, girdle bands, spines, and other protuberances). Kinetic experiments with diatom cells after silicon starvation allowed identification of two stages of the silicon consumption: fast ("surge" [13.8] [13.354]) uptake during formation of the immature 2D valve and relatively slow assimilation as the valve thickens to its mature 3D form. The first stage proceeds from 10 minutes [13.132] up to 1 to 2 hours and the second from 1 to 3 hours [13.106] [13.132] [13.338] to 10 and more hours [13.23] [13.151]. These experiments were carried out by measuring silicon concentration in the cultural medium and with radioactive ^{68}Ge isotope as a tracker of silicon consumption. Various kinetics of silicon uptake were observed with low and high external $Si(OH)_4$ concentration too [13.23] [13.354].

Two principal mechanisms of the silicon capture with diatom cells have been discussed for decades: specific transport by a special transporter [13.41] and non-specific transport through a pore in the cell membrane which is driven by a concentration gradient [13.166] [13.354]. In 2007, Vrieling *et al.* [13.363] proposed the pinocytosis (endocytosis) hypothesis based on dependence of the structure of the siliceous frustules on salinity of the cultural medium (cf. [13.322]). This idea was verified and developed by kinetics experiments and mathematical simulation [13.53] but was criticized [13.355]. The main objection to pinocytosis was the huge amount of water which the cell would have to process if we assume that the cell takes a water bubble into the cytoplasm and concentrates silicic acid from $<100\ \mu M$ to solid silica (≈ 33 M). Of course, if antennae processes are operative (stage 1), then less water would be needed.

The specific transporter was the more attractive idea, especially starting from 1997 when the gene of a silicon transporter (SIT) was discovered [13.167]. This gene was found by studying cDNAs formed after silicon replenishment of a silicon starved culture. The sequence of the corresponding protein was typical for a transmembrane protein and injection of the corresponding RNA into *Xenopus laevis* oocytes resulted in an 84 times increase of $^{68}Ge(OH)_4$ uptake, which was applied as a silicon analogue tracer. Homologous genes occur in cyanobacteria [13.62] [13.267], loricate choanoflagellates [13.239], haptophytes [13.101], sponges, and mammals [13.102] [13.142] [13.291]. Recently, two decades were devoted to attempts to find the SIT protein itself and to localize it in diatom cells. A family of five SIT genes was identified [13.163], with some localized to the cell membrane [13.102] [13.307] [13.324], and found in all diatoms studied and also in chrysophycean algae [13.220]. Study of several dozens of diatom species allowed the discovery of several conservative motifs in the gene sequence (CMLD, DXDID, EGXQ, xQxxxQx) [13.145] [13.238] [13.352] but there is as yet no experimental evidence nor theoretical simulations describing an ability of these sequences to transport silicic acid from the environment through the cell membrane.

Several works were devoted to measurement of SIT si-RNA levels at various stages of diatom growth [13.93] [13.325] [13.353] and to synthesis of the SIT protein by genetic engineering followed by introduction of the protein into model liposomes or by visualization of the SIT protein with GFP technology [13.90] [13.185] [13.325]. None of these investigations clarified the physiological role of SIT proteins and a suggestion was proposed that SIT is a molecular signal of low silicon level [13.325]. Kinetics study of silicon assimilation with the centric diatom *Thalassiosira pseudonana* at various $Si(OH)_4$ concentrations [13.354] showed a change of the uptake mechanism at relatively high (>30 μM) silica concentration from transporter mediated to nonsaturable (diffusion driven). A similar observation for the pennate diatom *Phaeodactylum tricornutum* was mentioned in [13.97] and preferable consumption of $Si(OH)_3O^-$ ion was found based on experiments at pH = 9.5.

In 2016, two teams [13.102] [13.240] pointed out that SIT genes were found in various non-diatom species, including protist ancestors and bacteria. SIT presence in non-silicifying organisms confuses understanding of the role of SIT. Explanations of this phenomenon included a hypothesis about SIT action in detoxification of organisms from silicon in Precambrian times when silicon content in oceans was extremely high. According to the hypothesis, silicon at high concentration (>100 μM and close to condensation level at 2 mM) allows penetration into cells and the cells elaborated a special system (SIT proteins) to remove silicon [13.240]. In later times, the silicon content in oceans dropped and many organisms lost or inactivated the SIT systems. Diatoms, radiolarians and chrysophytes changed over to SIT use for silicon capture from the environment. Comprehension of the SIT role in silicifying and non-silicifying organisms requires answers to two questions: where the protein is positioned in the cell and what is the chemical mechanism of SIT silicon transport? How an ancient SIT presumably membrane protein would flip during evolution from silica secretion to uptake also deserves investigation.

Another silicon transporter, termed as NBCSA ($Na^+/HCO_3^-[Si(OH)_4]$) co-transporter, was proposed for siliceous sponges [13.316]. Searching for similar genes and proteins in diatoms would seem to be promising. An understanding of the mechanism of transmembrane silicon transport must take into account the state of the silicon after passing the cell membrane into the cytoplasm. The most probable form of it is as oligosilicates, described above. The presence of condensed silicic acid is necessary for any diffusion-like mechanism because the concentration of monomeric $Si(OH)_4$ near the outer surface of the cell membrane must be decreased to provide gradient caused diffusion. We have done several experiments with amine containing fluorescent dyes which are able to penetrate into growing siliceous structures during diatom cultivation [13.8] [13.14] [13.18]. Submicrometer fluorescent vesicles were found in diatom cytoplasm several seconds after adding silicic acid to a culture synchronized by silicon depletion. The most convincing data were obtained with Q-N2 dye which changes its fluorescence color from blue to cyan after interaction with condensed silicic acid. As shown in Figure 13.2, against a blue fluorescence background of the dye in cytoplasm, cyan particles appear several seconds after silicon addition.

Thus, silicic acid just after passing the cell membrane is found in the condensed form. Here, we can remember the pinocytosis hypothesis [13.53] [13.363]. Certainly, uptake of environmental water with the aim to extract silicic acid is not a good strategy but if we assume condensation of $Si(OH)_4$ on the outer surface of the cell membrane (stage 1 antenna hypothesis above), then pinocytosis becomes not so impossible. The typical size of oligosilicate particles is tens of nanometers and this is comparable with the size of Percoll particles (30-nm silica coated with polyvinylpyrrolidone) which are captured into a visceral yolk sac by pinocytosis [13.290]. Thus, the question "How diatom cells capture silicic acid from the environment" is still unanswered. A progress in this area can come from two directions:

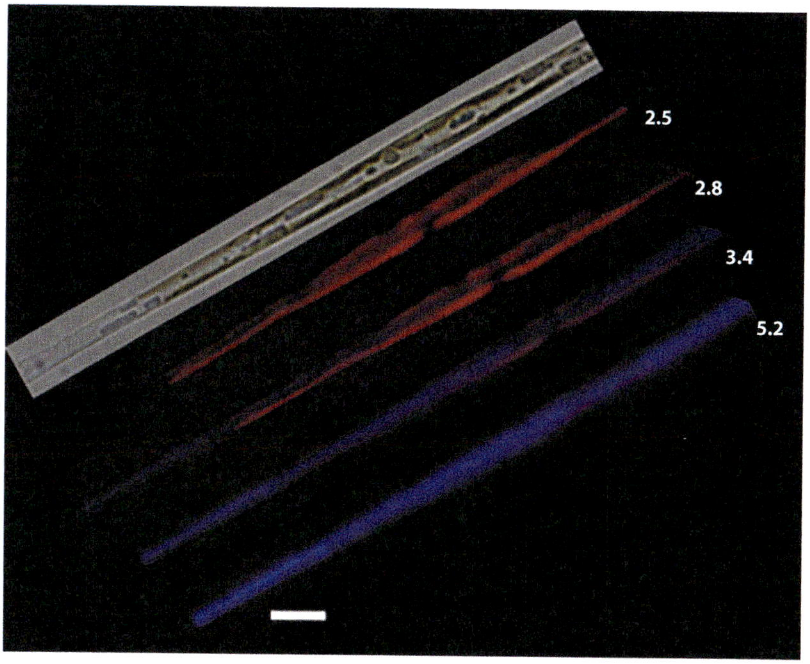

Figure 13.2 Light and fluorescence microscopy images of *Ulnaria ferefusiformis* cells growing after the addition of silicic acid (100 μM) and Q-N2 (0.5 μM) to the synchronized culture. Time (s) after the replenishment of silicic acid is presented near images. Red fluorescence comes from chloroplasts, blue comes from Q-N2 in cytoplasm, and cyan comes from condensed siliceous particles stained with Q-N2. 365 nm long-pass excitation filter was applied. Scale bar represents 10 μm.

- study of the membrane proteins, including physicochemical, structural analysis, and quantum-chemical simulations;
- monitoring of silicic acid from the cell vicinity to cytoplasm applying special markers, tracers, and fluorescent tagged $Si(OH)_4$.

13.8 Cytoplasm Storage and Trafficking of Silica to the Places of Synthesis of the Frustule Parts

There are several hypotheses about the cytoplasmic form of silicon:

- Free silicic acid. Experiments with silicon extraction from diatom biomass [13.241] [13.336] [13.337] (and references therein) showed the presence of easily extractable silicic acid (e.g., with hot water) whose concentration in cytoplasm is estimated as 19–680 mM. These works resulted in the term "free silicic acid" which some investigators interpreted as monomeric $Si(OH)_4$. The presence of silicic acid in such huge concentrations contradicts the expected fast condensation of $Si(OH)_4$ at concentrations above 2–3 mM. Moreover, at concentrations of more than 50 mM, we should obtain a silica gel which would then immobilize biopolymers and vesicles in the cytoplasm.

This could be checked via changes in cytoplasmic streaming [13.285] with changes in silica availability.
- Organic-silicon compounds. This hypothesis originates from silicon containing compounds found in some plants and products of their microbial degradation [13.245]. These are coordinating compounds based on dihydroxybenzene. Organic-silicon compounds were not reliably detected in diatoms and taking into account that at least one dihydroxybenzene group is needed for stabilization of one silicon atom, it is evidently that diatoms would have to contain unrealistic amounts of these silicon stabilizers.
- Solid silica nanoparticles of several tens of nanometers in diameter. This idea originates from observations of silicon transport vesicles (STVs) [13.313] as electron dense formations in cytoplasm. Silica in STVs was treated as building blocks for construction of the diatom frustules. This agrees with structure of the diatom silica which consists of nanoparticles of the same size range [13.88] [13.89] [13.132] [13.164] [13.230] [13.231] [13.313]. STVs as the cytoplasmic silicon pool were criticized [13.168] because the observed number of STVs was very small compared to the amount of extractable silicic acid. STVs were treated as a source of membrane material for the silicalemma [13.218], or both [13.132] [13.137]. STVs may be clathrin coated [13.266] [13.378].
- Partially condensed silicic acid (oligosilicates). Stabilized oligosilicates were proposed as a cytoplasmic form of silicon at least 40 years ago [13.21] [13.336] [13.337]. This hypothesis does not contradict the known facts: i) condensation of silicic acid decreases intracellular $Si(OH)_4$ concentration which facilitates transmembrane transport; ii) oligosilicates are easily hydrolysable providing high concentrations of the extractable silicic acid; and iii) oligosilicates are not as electron dense as silica and thus they are invisible with TEM.

Direct solid state NMR experiments [13.36] [13.60] [13.147] [13.148] [13.178] [13.198] have partially resolved this dispute: no free silicic acid or organic-silicon compounds were found in living diatoms. The number of Si-OH moieties in diatom biomass was higher than in the cleaned frustules. This points to oligosilicates as the most probable form of the cytoplasmic silicon pool. However, we also must mention that the STV hypothesis implies silica nanoparticles below 50 nm in diameter and at this scale the difference between silica and oligosilicates is very low, as <50-nm particles have large surface areas with Si-OH groups and must be easily hydrolysable (extractable). The remaining question is whether or not the cytoplasmic silica at this scale is bound inside membranous vesicles [13.137] or is free in the cytoplasm. Furthermore, there may be vesicles containing materials other than silica that fuse with the silicalemma [13.132] [13.312].

The key possibility that oligosilicates can be stabilized with organic polymers was demonstrated in works on silicic acid condensation in the presence of amine containing polymers [13.12] [13.16] [13.25] [13.61] [13.381]. The condensation reaction gives rise to nanoparticles of <100-nm diameter with core-shell or raspberry like structure.

The oligosilicate hypothesis agrees with *in vivo* and model experiments on germanium assimilation in diatoms. The germanium isotope ^{68}Ge was often applied as a tracker of silicon consumption because of its similar chemical properties to silicon [13.46]. Substitution of considerable amounts of silicon with germanium in cultural medium results in cell depression (1%–5% of Ge) or death at more than 5%–10% of Ge [13.22] [13.26] [13.73] [13.305]. Toxic action of $Ge(OH)_4$ is accompanied by formation of solid, electron dense sub-micrometer particles. Model experiments with oligosilicates stabilized with organic polymers showed [13.26] [13.305] a decrease of oligosilicate stability with formation of solid insoluble products. It was hypothesized that similar processes in diatom cytoplasm result in untimely condensation of oligosilicates before their entering into the silica deposition vesicle (SDV) [13.100] [13.217] [13.296] [13.331].

Cytoplasmic trafficking of silica transport vesicles (STVs) to the silicalemma via microtubules has been postulated [13.137] [13.276], an idea supported by [13.378] and not by [13.47], but not yet tested.

13.9 Transport and Patterning of Silica Across the Silicalemma

Silicalemmas are found in a few groups of organisms other than diatoms (radiolarians, sponges, testate amoeba [13.263] [13.264], Bolidophyceae [13.197], chrysophytes [13.59], synurophytes, dictyochophytes, and choanoflagellates [13.206]), etc. (Figure 13.1), so in piecing together a sketch of silica transport across the diatom silicalemma, a comparative anatomy approach may prove fruitful. Other features, such as lack of silica starvation synchrony in the order Parmales [13.379], may prove useful to understanding this phenomenon in diatoms [13.80] [13.81] [13.82] [13.200] [13.268] [13.269] [13.271] [13.277] [13.298].

Curiously, silicalemmas have not been reported around phytoliths, undoubtedly due to their generally being external to multicellular plant cells [13.258]. However, it is the plasmalemma of cells around a phytolith that must transport silica outside the cell. Thus, we anticipate that the plant plasmalemma shares some properties of a silicalemma.

The testate amoeba *Netzeliu tuberculata* [13.7] forms its silicalemma from "fibril-containing vesicles secreted by the Golgi body". Spherical silica particles of 10- to 20-nm diameter appear on the inner surface of the silicalemma, sinter into short chains and diffuse within the silicalemma to a "particle" that grows by precipitation and further sintering ("condensation [13.219]"). These spheres, or short chains of them, appear to cross the silicalemma. As the electron density increases, it is assumed that this represents decreasing hydration:

> "Hypothetically, fusion of GFV [Golgi-secreted fibrillar vesicle] with a localized part of the membrane transforms it into a silicalemma and initiates silica deposition at that particular site within the [digestive] vacuole.... When the GFV fuses with a vacuolar membrane, the fibrillar material is released into the vacuolar space.... The conversion of digestive vacuoles into silica-secreting vacuoles [secreting from the cytoplasm into the vacuole] raises the intriguing question of how this process arose during evolution of the species" [13.7].

Note that the spherical silica particles were not observed to be membrane bound, and that other "vesicles containing a finely fibrillar substance" [13.7] are presumed to be the source of the increasing surface of the silicalemma, contrary to our assumption that cytoplasmic vesicles are the immediate sources of both silica and membrane for the diatom silicalemma [13.132] [13.137] [13.276].

Release of the mature diatom valve (Section 13.12) occurs via fusion of the inner portion of the silicalemma with the cell membrane (plasmalemma) [13.32] [13.74]. As the silicalemma becomes part of the cell membrane, it has been hypothesized that the reverse process could initiate formation of a silicalemma [13.32], giving an alternative to the Golgi hypothesis [13.94]. This may be the case in sponge spicule formation [13.74] [13.328].

The observation that silica "fibrils" are thicker in the middle of the SDV [13.88] is consistent with a source of colloidal silica over the whole inner surface of the silicalemma rather than just its perimeter [13.132] [13.137] [13.276] and that the precipitated silica at first forms the fine branching patterns of the diffusion limited model followed by sintering [13.132]. However, silica particles external to the silicalemma are larger than the spacing of the silica "microfibrils" (Figure 1D in [13.88]). This suggests either that the external particles are the result of sintering of smaller particles during specimen preparation or that they break apart on transport across the silicalemma.

An early review allowed for two general patterning mechanisms:

> "Various mechanisms of silicification in diatoms have been proposed, and these hypotheses can be divided into two groups: 1) protein particles on the silicalemma serving as the template for silica deposition [13.153] [13.300] [13.317]; and 2) organic materials in the SDV serving as the matrix for silica deposition [13.286] [13.314] [13.359]. However, the nature of the protein particles and the organic matrix has never been identified" [13.216].

Nor has the supposed prepattern of the organic matrix been explained [13.132] [13.154]. To this has been added components of the cytoskeleton [13.33], namely, microtubules and microfilaments (actin filaments):

> "...during cell wall morphogenesis actin filaments were found interdigitated with forming silica structures or matching the detailed mesoscale patterns of the valves [13.349]. These cytoskeletal elements, located in the cytoplasm, are thought to impact the patterning of silica inside the SDV lumen, a process that would require the transmission of organizational patterns across the SDV membrane (the silicalemma) [13.348] [13.349]. A model for this process implicates putative silicalemma spanning proteins which simultaneously interact with cytoskeletal elements in the cytoplasm and silica polymerizing determinants in the SDV lumen, resulting in a replication of cytoskeletal patterns in silica structures [13.300]. These putative proteins have not yet been identified" [13.351].

In the interim since [13.216] specific molecules have been identified that might fit these roles. Long-chain polyamines (LCPAs) are capable of creating stacks of membranes on artificial lipid layers whose lipid composition matches that of diatoms. This is presumed to somehow relate to silica patterning [13.54] [13.143]. Glucans "associated with the forming valve... [may be] involved in shaping the valve at the microscale" [13.350]. In an explicit model, aquaporins, which are membrane embedded proteins (also found in plants [13.258] and animals [13.142], perhaps including humans [13.123]), are hypothesized as the source of silica patterning via dewatering of the adjacent silica gel beneath each aquaporin membrane pore [13.145]. The aquaporins may correspond to the (irregularly spaced) intramembranous particles seen in the silicalemma within which sponge spicules form [13.213]. Silaffin peptides of different sizes, inside the SDV, result in silica spheres of different diameters [13.192], hypothesized to control "frustule ornamentation" [13.209]. A few Silicalemma Associated Proteins (SAPs) have been identified [13.189] [13.351]. Silicanin-1 is a membrane bound protein [13.141], presumably in the silicalemma: "Although the [knockout] mutants grow normally, they exhibit reduced biosilica content and morphological aberrations, which drastically compromise the strength and stiffness of their cell wall" [13.141].

Less characterized are the "insoluble" organic matrices sometimes found inside the SDV or inside the precipitated silica, especially in sponge spicules [13.9]. "It has been suggested that a polysaccharide matrix is embedded inside the silica and controls its morphogenesis [13.75] [13.284] [13.309]" [13.350]. However, in diatoms: "We did not identify any organic matrix embedded within the silica" [13.350] (but cf. [13.154]). Earlier reports for and against organic matter inside diatom silica were reviewed with the suggestion that this material is trapped during sintering [13.132]. UV protection in diatoms includes organic molecules trapped in the silica [13.175]. Indeed, entrapment of various enzymes on or within silica anchors them for various nanotechnology applications [13.40] [13.77] [13.207] [13.208] [13.232] [13.250] [13.280].

Microtubules may have "...a cytoskeletal function... in controlling the shape of a specific region of the SDV" [13.52] (cf. [13.31] [13.35] [13.38] [13.39] [13.47] [13.161] [13.168] [13.183] [13.217] [13.288] [13.331] [13.349], but there is no one-to-one correlation between the location of microtubules and the pattern of silica within the SDV. The role of the cytoskeleton has been visualized as less direct, but precise:

> "During valve formation cytoskeletal elements could move the proteins around in a genetically determined choreography, condensing silica at precise locations to produce the complex valve architecture" [13.300].

In choanoflagellates the microtubules adjacent to the SDV on one side are purported to provide rigidity until the silica hardens [13.206], a function that would be enhanced by intermediate filaments attached to the microtubules [13.57] (if any). An indirect relationship has been computer simulated, in which the microtubules adjacent to the SDV, arrayed in a microtubule organizing center (MTOC), carry STVs to the periphery (perimeter) of the silicalemma in centric diatoms, with the STVs fusing with the silicalemma and releasing their silica contents into the SDV where they undergo diffusion limited aggregation [13.276]. In pennate diatoms, the MTOC may be involved in raphe formation [13.286]. Similarly, while microfilaments are "nearby" the SDV [13.52], no direct correlation with silica deposition within the SDV has been shown. In a testate amoeba a "sheet-like row of microtubules is tightly associated with the surface of the convex side of a [forming silica] scale in the SDV" [13.263]. A peripheral microfilament ring outside the growing silicalemma, forming its perimeter [13.286], has been postulated to exert an inward force on the growing, nascent diatom valve [13.137] which may be involved in buckling of the valve [13.273]. Mantle formation may involve the microfilament ring [13.132] [13.137] [13.274]. This is again an indirect relationship. Microfilaments are attached to one side of the SDV, with microtubules nearby, in a chrysomonad [13.59]. Salinity effects on morphology may involve microtubules and microfilaments [13.322].

A possible contradiction to a role for the microfilament ring in diatoms is the observation of "...extension and enlargement of the previously formed vesicles [that]... are filled with silica as they form, until the entire valve [and its silicalemma] is completed. By this time, the deposition vesicles constitute one continuous vesicle or envelope with regularly spaced constrictions; the envelope closely follows the configuration of the deposited shell [13.74]" [13.359]. Anna-Maria Schmid also concluded that the silicalemma is tightly appressed against an irregularly growing silica valve [13.311], which would seem to preclude a microfilament ring around the "perimeter". Resolution of this matter requires 3D imaging and/or fluorescence microscopy, perhaps aided by diatom cloaking [13.121], or reconstruction from serial TEM sections, applicable to microtubules [13.366] and probably microfilament bundles.

In general, the completed silicalemma appears to form before silica deposition in various groups but seems to grow as the silica precipitates in diatoms [13.244]. In the former case, the silicalemma is regarded as a mold [13.332], but this seems unlikely for diatoms. "Actinophryids and in other organisms, such as collar-flagellates, synurophytes, chrysophytes (sensu stricto), and various amoebae [13.159] [13.204] [13.205] [13.246] [13.279], in which the SDV reaches full size prior to silicification, tend to have siliceous structures with relatively simple forms" [13.260]. The formation of sometimes transient "bulbs" along the length of a sponge spicule, which can be manipulated by high or low silica or germanium [13.327], shows a way to experiment on the growth of the silicalemma relative to silica precipitation. The regular spacing of these bulbs suggests an instability phenomenon akin to drop formation from cylinders of fluids [13.133] [13.134] [13.292] [13.293] [13.294]. Active sintering of colloidal silica [13.132] within an extended SDV would indeed be analogous to fluid flow under surface tension. Sintering may thus be the actual mechanism that makes two-fluid models of diatom morphogenesis appear to have some success [13.212]. An alternative to instabilities due to sintering is the movement of membrane silica transporters to regions of the silicalemma that have specific curvatures [13.370]. Such a two-dimensional instability could involve capping of membrane proteins. Note that some membrane proteins that cap interact with microtubules [13.5] [13.199] and microfilaments [13.104] [13.177], which in diatoms could provide a way of bridging structure across different spatial levels, i.e., provide an avenue for exploring the multilevel structure of diatom valves [13.121], as, for instance, single microtubules can bridge across a whole cell (up to 50 μm [13.368]). As the outer surface of part of the silicalemma becomes the inner surface of part of the cell membrane (Section 13.12), such capping protein/cytoskeletal interactions in the silicalemma would be consistent with cytoplasmically internal location of the cytoskeleton.

All told, despite the advances in finding specific molecules that bind to the silicalemma and/or silica, the assumption that these are arrayed in specific patterns that then dictate the valve pattern is unproven. Furthermore, even if that were so, the origin of these prepatterns [13.132] would still require explanation, lest we have an infinite regress. In a way, we now know some of the bricks and mortar involved in diatom morphogenesis, but are far from knowing how they are assembled, let alone in so many different ways across the up to 200,000 species of diatoms [13.236]. The cellular analogs of architecture and building are still unsolved, not explaining the 8 orders of magnitude of diatom silica [13.121]. A pile of bricks does not self-assemble into a Taj Mahal.

13.10 Precipitation and Morphogenesis of the Nascent Valve Within the Silicalemma

The main question is—how the cell guides formation of the fine structured silica inside SDV? There are several levels of organization of the diatom silica: i) micro level—shape and size of the valve (girdle band) which are measured in μm; ii) meso level—ribs, portulae, areolae rows, and other elements of submicrometer scale; iii) nano level—structure of primary siliceous particles below 100 nm. This classification was put forward in [13.169].

Many more levels, spanning eight orders of magnitude, were defined in [13.121]. Some agents, such as cadmium, may alter only specific levels [13.156].

Diatom silica consists of spherical particles of several tens of nanometers according to SEM, AFM, and Xray diffraction analysis [13.88] [13.132] [13.164] [13.230] [13.231] [13.313]. This silica is produced at ambient conditions. Diatoms live in aqueous environments and we could expect the structure and properties of the diatom silica and silica obtained by precipitation from aqueous solution, e.g., by neutralization of sodium silicate with hydrochloric acid, to be similar. But siliceous frustules are made of highly dry silica. This is clearly seen when we try to melt silica gel and siliceous frustules (Figure 13.3). Silica gel contains much adsorbed water and Si-OH groups which results in high gas release during melting and the glass is porous and light scattering. Silica frustules consist of dry SiO_2 with low surface area and melting gives rise to a clear silica glass [13.144]. Siliceous valves and girdle bands are highly strong constructions. Their thickness is several hundreds of nanometers with length and width up to hundreds of micrometers and having this large length/thickness ratio and low water content, their mechanical strength is comparable with quartz glass [13.229] [13.231].

Siliceous frustules are relatively pure material. While they contain organic compounds tightly connected to the silica surface [13.149] [13.153] [13.190] [13.350], these can be removed with strong acids and oxidizing agents. Diatom silica contains <0.04% of metal impurities and for some elements it is comparable with pure quartz (Table 13.1). Treatment of the diatom biomass with HCl/bleach/H_2O_2 resulted in a material containing 0.14% of carbon and 0.025% of nitrogen [13.175]. The C and N may represent "small

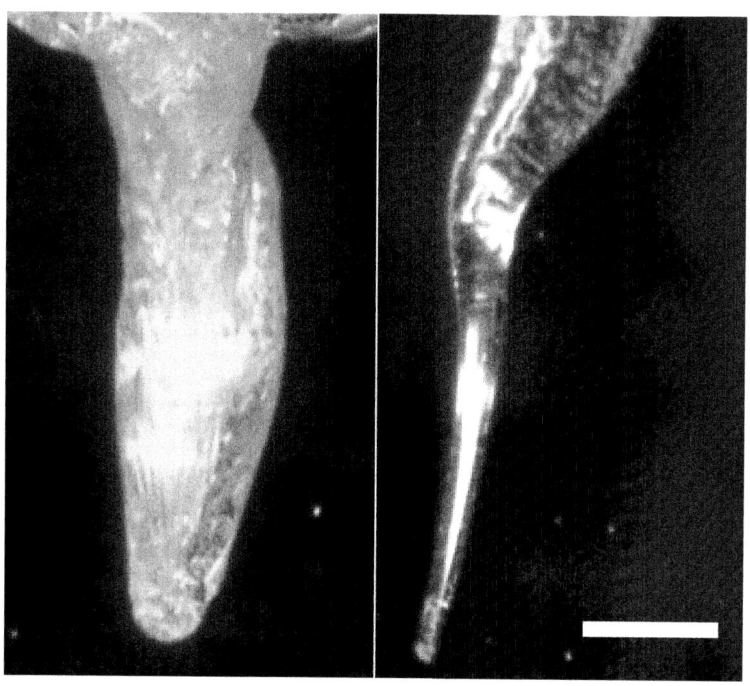

Figure 13.3 Melted silica gel (left) and siliceous diatom frustules (right). Scale bar, 10 mm.

Table 13.1 Chemical composition of *Ulnaria ferefusiformis* [13.196] frustules cleaned with HNO_3+HCl compared to quartz.

Element (ppm)	Li	B	Na	Mg	Al	P	K
U. ferefusiformis	0.045	1.3	2.1	4.4	17	42	0.68
IOTA-STD*	0.5	0.1	1.0	0.04	14	0.1	0.7
Element (ppm)	**Mn**	**Fe**	**Cr**	**Ni**	**Cu**	**Ca**	**Ti**
U. ferefusiformis	4.7	0.7	0.014	0.015	0.026	280	6.9
IOTA-STD	0.04	0.3	0.006	0.001	0.028	0.6	1.2

*Typical values for pure quartz [13.326].

polypeptides and proteins that are encapsulated inside the biosilica" [13.157]. Silica frustules can accumulate some amine containing compounds, e.g., fluorescent or spin probes [13.11] [13.18] [13.194] [13.321] and these substances remains in the silica after aggressive washing. Thus, diatom silica is closer to melted quartz glass than to silica gel obtained in water.

The common shape of the valve is correlated with cytoskeleton which is presumed to control the shape of the SDV. This was observed more than 50 years ago [13.331] and confirmed later [13.31] [13.35] [13.37] [13.39] [13.47] [13.48] [13.162] [13.168] [13.183] [13.217] [13.288] [13.348] [13.349] [13.351].

How does the cell regulate mesoscale structure of the valve? Immature valves have a "2D" structure (Figure 13.4) [13.179] [13.273] [13.284] [3.313], with clearly visible mesoscale elements, e.g., areolae rows. From general consideration, we can imagine two ways of the cell control under processes inside SDV: external control and inner self-organization. The external control idea was formulated in [13.300] as cytoskeleton "choreography" with some substances introduced through the silicalemma in the places were condensation of silica precursors is needed. How the cytoskeleton gets spatiotemporally organized to accomplish this was not specified.

A considerable achievement happened in 1999 when silaffins were discovered [13.185] [13.191]. Silaffins are proteins with posttranslational modifications: grafted polyamine chains (up to 20 and more nitrogen atoms) and phosphorylated units [13.192]. Polyamine chains are partially methylated (Figure 13.5) and can include oxygen containing groups and methylated nitrogens. Silaffins were extracted from frustules slightly washed with surfactant. The LCPAs were also obtained from frustules including acid cleaned material and ancient frustules from sediments [13.55]. Silaffins were found in marine diatoms: *Chaetoceros didymium*, *Coscinodiscus Asteromphalus*, *Coscinodiscus concinnus*, *Coscinodiscus granii*, *Coscinodiscus wailesii*, *Cylindrotheca fusiformis*, *Eucampia zodiacus*, *Nitzschia angularis*, *Stephanopyxis turris*, and *Thalassiosira pseudonana* [13.341] [13.342] [13.343], and also in the haptophyte *Prymnesium neolepis* [13.101]. Recently, we found LCPA in freshwater diatoms *Stephanodiscus meyeri* and *Ulnaria ferefusiformis* and in

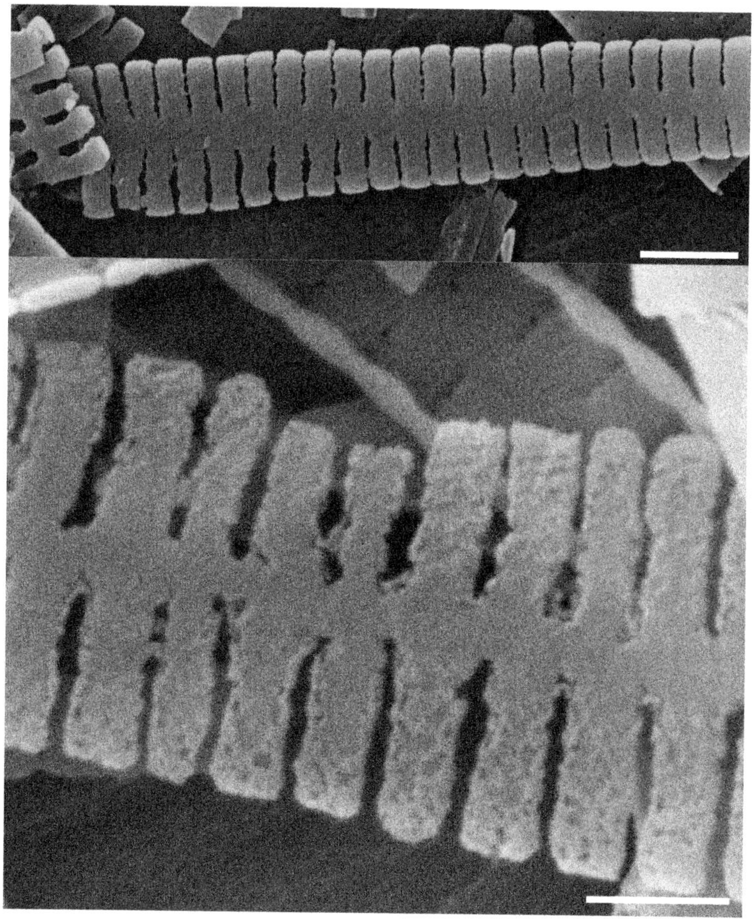

Figure 13.4 Initial 2D stage of valve morphogenesis of the diatom *Ulnaria ferefusiformis*, SEM images. Scale bars represent 2 μm (top) and 1 μm (bottom).

sediments from Lake Baikal [13.17]. In the case of *S. meyeri* and sediments, unusual polyamines containing tetramethylene fragments between nitrogens were detected.

Silaffins and/or LCPA were found in diverse diatom species but their physiological role is still something of a mystery. The first hypothesis on the role of the silaffins consisted in self-organization of silaffins and possibly some other biopolymers giving a matrix (prepattern) around which silica precursors condense and thus the fine structure of frustules is built [13.193] [13.340]. There are several contradictions to this scenario:

1. Silicic acid condensation in the presence of silaffins, LCPA or model systems can produce ordered structures, e.g., spherical or hexagonal [13.193]. Formation of immature valves with axial symmetry (Figure 13.4) seems to be impossible under self-organization of macromolecules in solution.

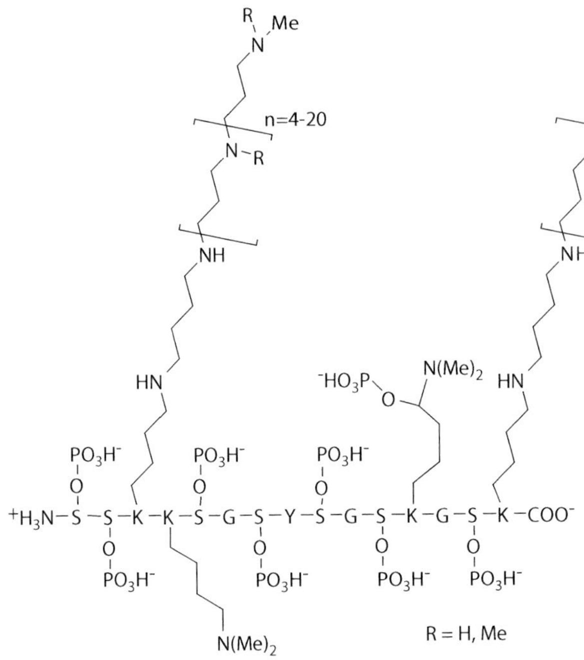

Figure 13.5 Typical structure of silaffin chain.

2. Where is the matrix? The diatom silica, as we showed above, is relatively pure silicon dioxide, 0.1%–0.2% of carbon corresponds to LCPA in low mass-spectrometry amounts and some other compounds, e.g., UV protectors [13.175]. A lot of experiments were undertaken on silicic acid condensation in the presence of various organic polymers as matrix and the materials obtained were composites containing more than 30% of organic polymers [13.10] [13.12] [13.27] [13.69] [13.181] [13.272]. Any attempts to remove these polymers (calcination, treatment with strong acids) resulted in highly porous material or fine powder.
3. Polyamine chains in silaffins and free LCPA show highly variable structure which could be treated as species specificity and the reason for high diversity of the frustule morphology. On the other hand, there are no a definite correlations between polyamine structure and frustule morphology. Moreover, an unusual tetramethylene LCPA were found [13.17] in diatoms of genus *Stephanodiscus* which is close to marine *Stephanodiscus* diatoms containing trimethylene based LCPA. The absence of a definite relation between polyamine structure and morphology of the siliceous frustules points on noninvolvement of polyamines in key stages of valve formation or to insensibility of the silica deposition in SDV to polyamine structure. Both contradict the matrix hypothesis.

An hypothesis was formulated in [13.145] proposing that silica deposition in the SDV undergoes desiccation which is caused by aquaporins—transmembrane proteins which are responsible for water removal from various organelles [13.4] [13.301]. There are several supports for this idea:

1. The SDV before silica deposition contains a large amount of water. If we assume a silicon concentration in the SDV as 1 M (an unrealistically high concentration), this means 60 g of SiO_2 per liter and so formation of the silica frustule involves removing >94% of the water from the SDV and some mechanism for doing this must exist.
2. Condensation of silica precursors accelerates with concentration and our experiments with gravity-induced concentration of model organo-siliceous nanoparticles showed [13.16] formation of a silica with a low content of organic polymer (<4%).
3. Aquaporin delivery via cytoskeleton is a possible way to control mesoscale structure of the frustule from the outside of silicalemma.

Unfortunately, there is no direct data on aquaporin involvement in frustule morphogenesis. Recently, the matrix hypothesis began to transform into the idea of cytoplasmic control with transmembrane proteins in the silicalemma which have highly phosphorylated segments inside the SDV [13.93] [13.95] [13.149] [13.189] [13.190] [13.289] [13.351]. In this connection, silaffins and LCPA work as mediators between the growing silica precipitate and negatively charged segments of the transmembrane proteins. This hypothesis explains the absence of a significant amount of organic substances in diatom silica, the presence of silaffins on the frustule surface [13.95, 13.149], enhanced phosphorylation of proteins during valve synthesis [13.295] and model experiments on silica deposition using organic insoluble components that were isolated from biosilica [13.281]. We must mention that the idea of a matrix role for the silicalemma was first formulated in [13.153].

Evidently, formation of siliceous structure inside the SDV is a dynamic process which includes not only silica deposition but also a supply of silica precursor. The place and intensity of this supply also must influence the valve morphology, as was shown in diffusion-limited aggregation (DLA) simulations comparing perimeter versus silicalemma face arrival of silica [13.132] [13.276]. Formation of immature valves of centric diatoms was successfully simulated starting from simulated silica nanoparticles which can aggregate on collision. This model is based on approximately solid silica nanoparticles whose presence in the SDV has not been proven. It is more probable that oligosilicate particles condense into solid silica via catalysts (polyamines?) and/or desiccation, but any theory of valve morphogenesis must take into account a gradual and localized SDV feeding of a silica precursor.

13.11 Thickening of the Valve Within the Silicalemma

Once the nascent valve has quickly grown out, the valve thickens over a much more extended time (Section 13.7). The essentially 2D structure becomes three-dimensional, often with two layers, each with its own pattern of pores, subtended by arched structures and chambers [13.237] which may contain mitochondria [13.132]. As buckling may have occurred in the thin, nascent valve, mechanically thickening not only adds strength, but locks in the buckled structure, in a manner similar to that of corrugated materials [13.139]. Thickening may involve a shift from silicalemma perimeter input to silicalemma surface input of silica, such as has been simulated by DLA (Section 13.10). Instability buckling phenomena in the third dimension may partially account for generation of 3D asymmetrical structure [13.132] [13.139] [13.273] [13.274]. The thickness of a diatom valve can depend on the light intensity and spectrum [13.333]. Silica starvation during the thickening phase might help us better understand this step. Staining with various vital fluorescence dyes is also a powerful tool for study the thickening phase, e.g., change of the dyes allows (Figure 13.6) to visualize formation of the initial 2D structure and further growth. In the case of

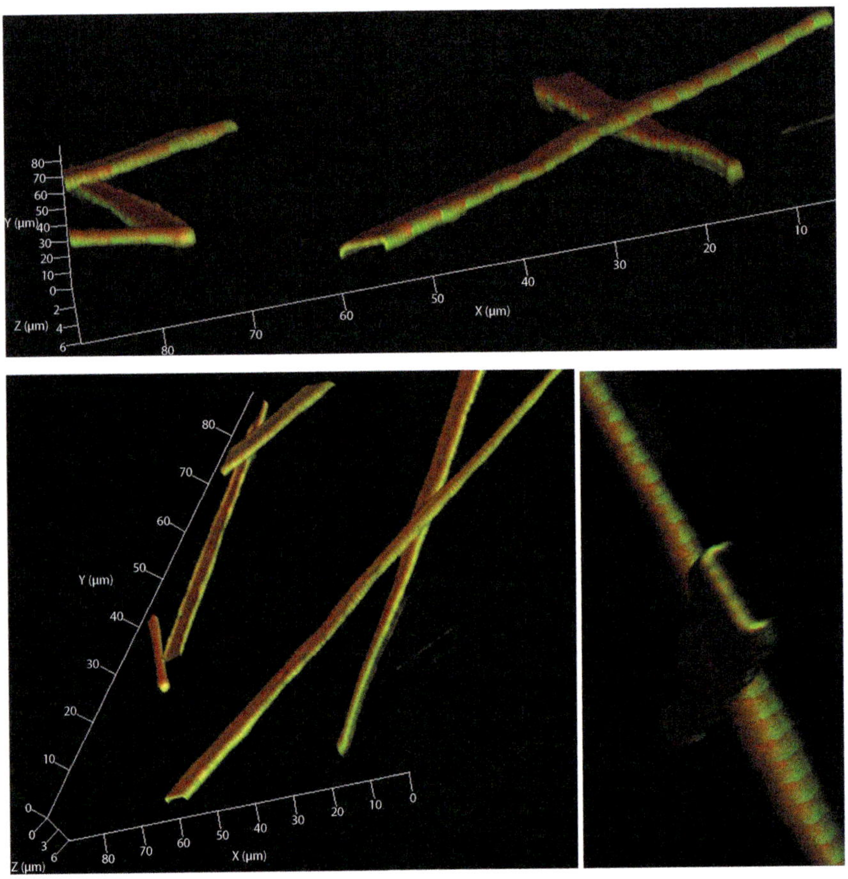

Figure 13.6 3D confocal images of siliceous valves of *Ulnaria ferefusiformis*. The diatoms were synchronized by silicon starvation and cultivated 30 min with addition of 30 μM sodium silicate and 0.5 μM of the vital red-fluorescent dye Rhod-N3 [13.92]. After that, Rhod-N3 was changed with NBD-N2 dye (green fluorescence) [13.14] and cultivation continued for 24 hours. Siliceous frustules of diatoms were isolated from the biomass by treatment with a mixture of concentrated sulfuric acid and 35% hydrogen peroxide (1:1, v/v) for 24 hours at room temperature. Excitation 488 nm, detector slit 491–520 and 619–680 nm.

U. ferefusiformis, the bottom of the valve casket is built in the first 30 min and sidewalls are formed later.

13.12 Exteriorization of the Valve

The new valve (hypotheca) on each diatom is softer than its old valve (epitheca), indicating that valve morphogenesis is not quite complete after exteriorization [13.182]. Perhaps this involves ongoing sintering of the silica. (See [13.233] for a review of spatial variations in frustule hardness.) The diatotepum (Section 13.6) is formed under the new valve and girdle bands after they are exocytosed [13.87].

13.13 Future Work Needed

Thus, we have to conclude that in spite of great efforts during recent decades, our understanding of diatom frustule morphogenesis is far from achieving a satisfactory theory. The main difficulty is the nature of the diatom: the silica frustule protects the cell from the environment threats and also from investigators. It is highly difficult to study silica penetration through the silica frustule. It is evident that the silicalemma is the key object in understanding of formation of diatom silica in the SDV, but no one has isolated an intact silicalemma from inside its diatom's frustule. Our analysis allows us to propose the following ways to understanding of the diatom morphogenesis:

1. Elaboration of new silicon trackers and tracking methods. We need direct answers on questions: Where and when monomeric silicic acid condenses into oligosilicates? Which compounds are presented near oligosilicates when they become solid silica?
2. Genomic, proteomic, and peptidomic investigations must be correlated with study of silicon containing and related compounds.
3. New methods of the cell biotomy must be developed and applied, such as:

   - Terahertz laser [13.15]
   - Soft x-ray microscopy [13.2] [13.382], with computed tomography [13.375]
   - Focused ion beam scanning electron microscope [13.373], with reconstruction from serial sections
   - AFM (atomic force microscopy) [13.233]
   - SIMS (secondary ion mass spectrometry) [13.210]
   - SEM (scanning electron microscopy) with stereo pairs [13.257]
   - HVEM (high voltage electron microscopy) [13.113] [13.306]
   - 3D light microscopy [13.201] [13.270] [13.377]
   - Optical cloaking of valves [13.121]
   - Merged imaging modalities [13.110] [13.131]

One of the major obstacles to understanding diatom morphogenesis is that we do not yet quantitatively know the full structure of any single diatom valve, let alone its 3D structure over time. However, various imaging tools are available that could be applied toward this goal [13.242]. At the light microscopy level, one approach would be to rotate a diatom, or better a colonial chain of diatoms, inside a capillary tube [13.184], to obtain multiple views for computed tomography. Large diatoms, some from Antarctica, might solve resolution limitations [13.140] [13.195] [13.254] [13.262] [13.329] [13.339] [13.358]. Optical tweezers can also be used to rotate a diatom on its long axis [13.308]. Light sheet microscopy has been applied to diatoms [13.377] and is capable of high resolution without damaging cells by high light intensities [13.70]. Any optical technique would be improved if appropriate, nontoxic agents such as methylcellulose are shown to successfully cloak or enhance the outer silica surface [13.121].

Somewhere in all of the steps toward the new valve, isotopic fractionation of silica occurs [13.345]. While there is a substantial literature on the importance of silica isotope fractionation to biogeochemistry [13.109] [13.344], dissection of when and where it occurs on or inside the living diatom may provide new insights to diatom morphogenesis.

The experimental separation of cell division from valve morphogenesis, in which bent valves that are otherwise similar to regular valves form within a cell that has been prevented from dividing [13.24], may provide insights not available from the normal tight coupling of these phenomena.

Computer simulation is a powerful method in the study of various natural processes. There are at least two aims in these simulations: to visualize and better understand some theoretical ideas and to cut off unrealistic hypotheses. The most successful simulations allow us to build a real model of a complex system starting from well-defined physical parameters. Simulation of the diatom valve growth started from [13.132] [13.276]. This work showed the possibility of explaining the morphology of diatoms at micro- and meso-levels by manipulating parameters outside the silicalemma: silicalemma shape and size, places and rate of silica entry, physicochemical parameters of movement of silica particles, association,

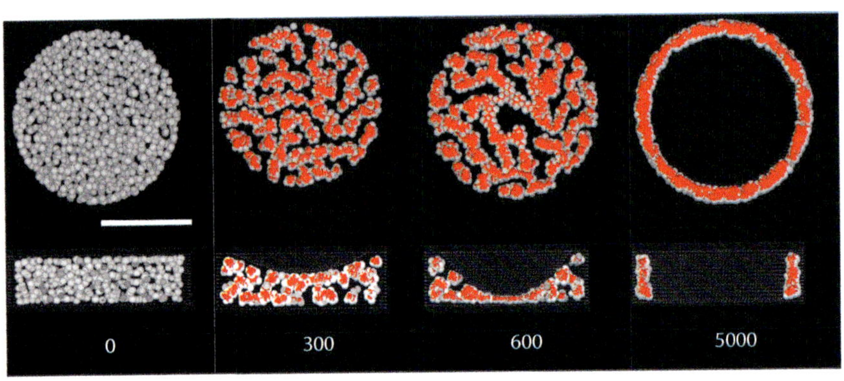

Figure 13.7 Condensation of silica nanoparticles under the action of decrease of the SDV volume (aquaporin induced desiccation). The aquaporin action is simulated as membrane invagination. Scale bar represents 500 nm. Bottom row presents 90° rotated and 200 nm central sliced image, representing the thickness of the silicalemma. The numbers show simulated time in μs. Red, areas of merging of the particle shells.

and sintering. Another model [13.35] was elaborated for pennate diatoms and includes some predetermined raphe. The aquaporin hypothesis [13.145] was supported with a simple model demonstrating formation of the 2D structure of a pennate diatom without any matrix inside SDV. We now elaborate a 3D model which describes interaction of siliceous nanoparticles (oligosilicate) during Brownian movement in a space bordered by a membrane of various shapes. Particles can bear a charge and associate by merging their surface shells. The membrane is simulated with a set of points which repulse siliceous nanoparticles. The model allows us to simulate the behavior of the system under changes of the membrane, particle concentration and addition of oligosilicate at definite locations on the membrane. Figure 13.7 shows condensation under the influence of aquaporin: local desiccation is simulated as membrane invagination and we simulate concentration induced condensation of siliceous particles into solid silica.

13.14 Conclusion

The molecular approach to understanding diatom morphogenesis has so far failed to come anywhere close to explaining the multiscalar structure of diatoms and their enormous species variations, even though there is no doubt that "biomolecules actively contribute to the silica polymerization process during valve formation" [13.157]. The alternatives are computer simulation [13.34] [13.35] [13.86] [13.132] [13.228] [13.276] [13.370] and emulation [13.126] [13.380]. For example, the fractal nature of DLA automatically generates self-similar structure at many scales [13.129] [13.132]. For emulation, we proposed the construction of an artificial silicalemma (Figure 13.8) [13.126]. A recent emulation started with arrays of silica spheres, which filled in between pores at rates different from surface precipitation onto the spheres, when placed in supersaturated solutions [13.275]. No proteins or other organic molecules were used. This emulation is more akin to the thickening phase of valve morphogenesis, whereas the approach in Figure 13.8 could address the initial two-dimensional growth of the nascent (artificial) valve, and, if constrained [13.135], buckling of the mantle emulated [13.274]. Even the simpler structure of a silica sponge spicule may be hierarchically constructed, perhaps assisted by various silica binding proteins, which by no means act as prepatterns (Figure 13.9). Branching fractal structures similar to those simulated in [13.132] have been emulated by silica precipitating in the presence of silicatein-α and silintaphin-1 [13.364]. We thus suggest that it is time to drop the prepattern approach entirely and ask what shapes and patterns can be generated by silica alone, and then by silica in combination with the various identified silica binding and membrane proteins and cytoskeleton, backed by observations of their spatiotemporal appearances during valve morphogenesis. Eventually, we should be able to construct an artificial silicalemma and understand how to manipulate it in ways the cell can, to produce various diatom valve patterns. This might be a boon to diatom nanotechnology [13.127].

The modern level of computer facilities allows us to simulate huge systems of thousands of atoms, molecules, or colloidal particles, and biosilicification is a great challenge and a favorable area for such digital models. Several steps should be considered on the way to reliable description of siliceous valves formation:

i. behavior of silicic acid in aqueous medium, including condensation of several $Si(OH)_4$ molecules, depending on pH, ionic strength, concentration, and presence of organic substances, e.g., amines;
ii. formation of primary silica nanoparticles from tens of silicic acid molecules. This stage as well as i) seems to be simple but if we remember that there are four active silanol groups in each $Si(OH)_4$ molecule and change of acidity and activity of those groups depend on the state of condensation, we will understand that these first stages require a thoroughly analysis using accurate quantum-chemical methods. As we mentioned, $Si(OH)_4$ condensation proceeds in relatively diluted aqueous solutions and water molecules should be considered explicitly;
iii. interaction between primary silica nanoparticles and organic substances (proteins, polysaccharides, polyamines, and complex compounds). Various scenarios should be simulated here: change of silica condensation state and formation of soluble composite nano(micro)particles or solid phase. The dependence of these reactions depends on concentration;
iv. interaction of $Si(OH)_4$ and particles from i) to iii) with various surfaces and porous membranes. These interactions are critical for silicon capture from the environment and for processes inside the SDV.

Simulation must include a whole set of methods, from accurate *ab initio* approaches to semi-empirical and molecular dynamics methods. Calculations should take into account current experimental data and stimulate new experiments. We believe that a complex simulation project could be the nucleus of a large biosilicification program that will solve the mysteries of diatom cells and give us new scientific methods and useful processes.

A top-down mechanism for diatom morphogenesis has been proposed [13.165] [13.348] [13.349]. This is conceptually similar to building inside a scaffold, which would then just be another form of prepattern. It would also seem to violate reductionist approaches, for

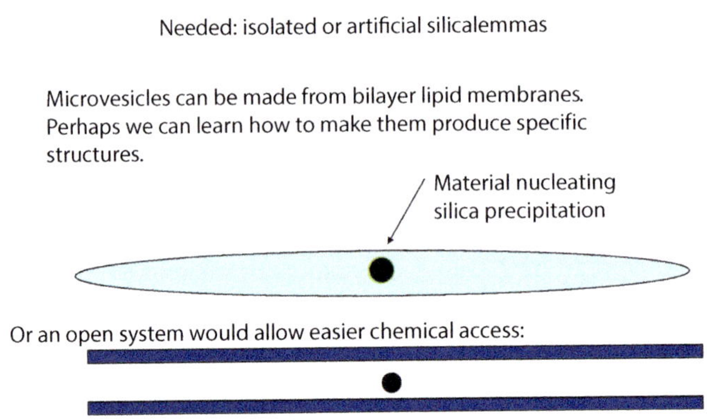

Figure 13.8 Proposal for two forms of artificial SDVs (silica deposition vesicles) with and without an artificial silicalemma, from [13.126]. An artificial silicalemma would block silica transport, allowing testing of passive diffusion of various forms of silica in the external solution. Adding membrane bound proteins such as aquaporin would permit testing their effects on the morphology of the precipitate that should form inside, pH, temperature, and gradient conditions permitting. Silica access would be over the whole artificial silicalemma surface. The bottom system restricts access only along its perimeter, such as in [13.274] [13.276]. Both surface and perimeter access were simulated by computer in [13.132].

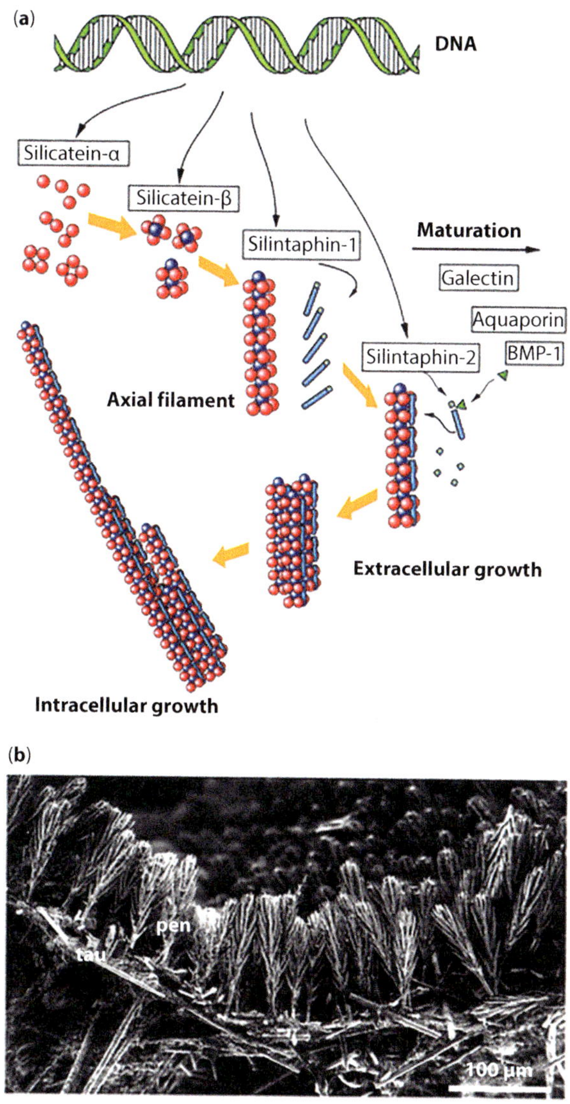

Figure 13.9 Proposed hierarchical genetic model controlling the synthesis of the siliceous spicules. (a) The spicule synthesis starts intracellularly in specialized cells, the sclerocytes, after expression of the silicateins, with their isoforms silicatein-α and silicatein-β. In a stoichiometric manner, four silicatein-α molecules form a tetramer allowing one silicatein-β molecule to associate in the middle of the tetrad. Through axial assembly, filaments are formed that are stabilized by the silicatein interactor silintaphin-1. The axial filament is harbored in the axial canal, and mediates the formation of an immature spicule rod. Subsequently, after the release into the extracellular space the spicule continues to grow through appositional layering of silica lamellae, thereby thickening of the spicule occurs, by telescoping organic cylinders formed from galectin and silicatein, in association with silintaphin-2. Silintaphin-2 becomes proteolytically activated by the bone morphogenetic protein-1 (BMP-1). Finally, the formed spicules are shaped within a collagen cast and processed and aged through a process involving syneresis (expulsion of water) from the "soft" biosilica formed by silicatein, a process facilitated by aquaporin channels inserted into the cell membranes of the adjacent cells. (b) Example of highly organized arrangements of spicules in *Monorhaphis chuni*, the dermal pinular pentactines (pen), arranged along the longer tauactines (tau)" [13.365] with permission of the Royal Society of Chemistry. Cf. [13.253].

whence the scaffold, especially after auxospore formation. An alternative is one proposed for multicellular embryogenesis, involving the bottom-up hierarchical construction of cybernetic feedback systems, where one of the "goals" at each level is construction of the next one or two cybernetic systems [13.138]. In this manner, the multiscalar structure of the frustule [13.121], and its variations between species, might be accounted for, along with the "refinement" of auxospore toward vegetative morphology over a few generations [13.303]. In general, the morphogenesis of single cell eukaryotes, such as diatoms and ciliates [13.20] [13.114] [13.146], is still an unsolved problem, which might be approachable by taking a broad overview of both multicellular and one cell morphogenetic mechanisms [13.125]. One lesson from ciliates is that their cortical structure, normal or aberrant, is passed on from at least one vegetative generation to the next without direct control by the cell nucleus, a property that may be shared with diatoms [13.108]. Many diatom characters have low heritability [13.105] [13.374], and cultured diatoms are often aberrant (reviewed in [13.187]), including clones [13.108], suggesting non-nuclear inheritance. Tests in microgravity have been proposed [13.136].

While diatoms are relatively late comers on Earth, their uncanny ability to build with silica may stem from an evolving relationship between organisms and silica minerals [13.152] [13.248] and the ocean [13.78] reaching back to the origin of life [13.123].

References

[13.1] Aboal, M., Marco, S., Chaves, E., Mulero, I., García-Ayala, A., Ultrastructure and function of stalks of the diatom *Didymosphenia geminata*. *Hydrobiologia*, 695, 1, 17–24, 2012.

[13.2] Abramson, L., Wirick, S., Lee, C., Jacobsen, C., Brandes, J.A., The use of soft X-ray spectromicroscopy to investigate the distribution and composition of organic matter in a diatom frustule and a biomimetic analog. *Deep-Sea Res. Part II-Top. Stud. Oceanogr.*, 56, 18, 1369–1380, 2009.

[13.3] Adam, G. and Delbrück, M., Reduction of dimensionality in biological diffusion processes, in: *Structural Chemistry and Molecular Biology, A Volume Dedicated to Linus Pauling by his Students, Colleagues, and Friends*, A. Rich and N. Davidson (Eds.), pp. 198–215, W.H. Freeman, New York, 1968.

[13.4] Agre, P., Preston, G.M., Smith, B.L., Jung, J.S., Raina, S., Moon, C., Guggino, W.B., Nielsen, S., Aquaporin CHIP: The archetypal molecular water channel. *Am. J. Physiol.*, 265, 4, F463–F476, 1993.

[13.5] Albertini, D.F. and Clark, J.I., Membrane-microtubule interactions: Concanavalin A capping induced redistribution of cytoplasmic microtubules and colchicine binding proteins. *Proc. Natl. Acad. Sci. U. S. A.*, 72, 12, 4976–4980, 1975.

[13.6] An, S.M., Noh, J.H., Choi, D.H., Lee, J.H., Yang, E.C., Repeat region absent in mitochondrial genome of tube-dwelling diatom Berkeleya fennica (Naviculales, Bacillariophyceae). *Mitochondrial DNA Part A*, 27, 3, 2137–2138, 2016.

[13.7] Anderson, O.R., Fine structure of silica deposition and the origin of shell components in a testate ameba *Netzelia tuberculata*. *J. Protozool.*, 35, 2, 204–211, 1988.

[13.8] Annenkov, V.V., Basharina, T.N., Danilovtseva, E.N., Grachev, M.A., Putative silicon transport vesicles in the cytoplasm of the diatom *Synedra acus* during surge uptake of silicon. *Protoplasma*, 250, 5, 1147–1155, 2013.

[13.9] Annenkov, V.V. and Danilovtseva, E.N., Spiculogenesis in the siliceous sponge *Lubomirskia baicalensis* studied with fluorescent staining. *J. Struct. Biol.*, 194, 1, 29–37, 2016.

[13.10] Annenkov, V.V., Danilovtseva, E.N., Filina, E.A., Likhoshway, Y.V., Interaction of silicic acid with poly(1-vinylimidazole). *J. Polym. Sci. Part A-Polym. Chem.*, 44, 2, 820–827, 2006.
[13.11] Annenkov, V.V., Danilovtseva, E.N., Khutsishvili, S.S., Pal'shin, V.A., Polienko, Y.F., Saraev, V.V., Vakul'skaya, T.I., Zelinskiy, S.N., Grigor'ev, I.A., Polyamine-based spin probes for the study of siliceous structures. *Microporous Mesoporous Mater.*, 242, 74–81, 2017.
[13.12] Annenkov, V.V., Danilovtseva, E.N., Pal'shin, V.A., Aseyev, V.O., Petrov, A.K., Kozlov, A.S., Patwardhan, S.V., Perry, C.C., Poly(vinyl amine)-silica composite nanoparticles: models of the silicic add cytoplasmic pool and as a silica precursor for composite materials formation. *Biomacromolecules*, 12, 5, 1772–1780, 2011.
[13.13] Annenkov, V.V., Danilovtseva, E.N., Pal'shin, V.A., Verkhozina, O.N., Zelinskiy, S.N., Krishnan, U.M., Silicic acid condensation under the influence of water-soluble polymers: from biology to new materials. *RSC Adv.*, 7, 34, 20995–21027, 2017.
[13.14] Annenkov, V.V., Danilovtseva, E.N., Zelinskiy, S.N., Basharina, T.N., Safonova, T.A., Korneva, E.S., Likhoshway, Y.V., Grachev, M.A., Novel fluorescent dyes based on oligopropylamines for the in vivo staining of eukaryotic unicellular algae. *Anal. Biochem.*, 407, 1, 44–51, 2010.
[13.15] Annenkov, V.V., Kozlov, A.S., Danilovtseva, E.N., Basharina, T.N., Petrov, A.K., Dissection of the frustules of the diatom *Synedra acus* under the action of picosecond impulses of sub-millimeter laser irradiation. *Eur. Biophys. J. Biophys. Lett.*, 42, 7, 587–590, 2013.
[13.16] Annenkov, V.V., Pal'shin, V.A., Verkhozina, O.N., Larina, L.I., Danilovtseva, E.N., Composite nanoparticles: A new way to siliceous materials and a model of biosilica synthesis. *Mater. Chem. Phys.*, 165, 227–234, 2015.
[13.17] Annenkov, V.V., Verkhozina, O.G.N., Zelinskiy, S.N., Shishlyannikova, T.A., Bridoux, M.C., Danilovtseva, E.N., Unusual polyamines from baikalian diatoms. *ChemistrySelect*, 3, 33, 9708–9713, 2018.
[13.18] Annenkov, V.V., Zelinskiy, S.N., Pal'shin, V.A., Larina, L.I., Danilovtseva, E.N., Coumarin based fluorescent dye for monitoring of siliceous structures in living organisms. *Dyes Pigm.*, 160, 336–343, 2019.
[13.19] Araújo, C.V.M., Romero-Romero, S., Lourençato, L.F., Moreno-Garrido, I., Blasco, J., Gretz, M.R., Moreira-Santos, M., Ribeiro, R., Going with the flow: Detection of drift in response to hypo-saline stress by the estuarine benthic diatom *Cylindrotheca closterium*. *PLoS One*, 8, 11, #e81073, 2013.
[13.20] Aufderheide, K.J., In memoriam: Gary W. Grimes (1946-1999). *J. Eukaryotic Microbiol.*, 47, 3, 323–324, 2000.
[13.21] Azam, F., Hemmings, B.B., Volcani, B.E., Role of silicon in diatom metabolism V. Silicic acid transport and metabolism in the heterotrophic diatom *Nitzschia alba*. *Arch. Microbiol.*, 97, 2, 103–114, 1974.
[13.22] Azam, F., Hemmingsen, B.B., Volcani, B.E., Germanium incorporation into the silica of diatom cell walls. *Arch. Mikrobiol.*, 92, 11–20, 1973.
[13.23] Azam, F. and Volcani, B.E., Role of silicon in diatom metabolism VI. Active transport of germanic acid in the heterotrophic diatom *Nitzschia alba*. *Arch. Microbiol.*, 101, 1, 1–8, 1974.
[13.24] Badour, S.S., Experimental separation of cell division and silica shell formation in *Cyclotella cryptica*. *Arch. Mikrobiol.*, 62, 17–33, 1968.
[13.25] Baranovsky, V.Y., Suchishvili, S.A., Kasaikin, V.A., Kabanov, V.A., Polycondensation of silicic-acid in the presence of poly(*N*-ethyl-4-vinylpyridinium bromide) and properties of the reaction products. *Eur. Polym. J.*, 29, 1, 111–114, 1993.
[13.26] Basharina, T.N., Danilovtseva, E.N., Zelinskiy, S.N., Klimenkov, I.V., Likhoshway, Y.V., Annenkov, V.V., The effect of titanium, zirconium and tin on the growth of diatom *Synedra acus* and morphology of its silica valves. *Silicon*, 4, 4, 239–249, 2012.

[13.27] Bassindale, A.R., Taylor, P.G., Abbate, V., Brandstadt, K.F., Simple and mild preparation of silica-enzyme composites from silicic acid solution. *J. Mater. Chem.*, 19, 41, 7606–7609, 2009.
[13.28] Basson, P.W. and Mohamed, S.A., The tube-dwelling diatom *Nitzschia martiana* (C A. Agardh) Van Heurck as recorded from Bahrain (Arabian Gulf). *Bot. Mar.*, 32, 3, 253–257, 1989.
[13.29] Bäuerlein, E., Biomineralization of unicellular organisms: An unusual membrane biochemistry for the production of inorganic nano- and microstructures. *Angew. Chem.-Int. Ed.*, 42, 6, 614–641, 2003.
[13.30] Baumert, J., Asmussen, B., Gutt, C., Kahn, R., Pore-size dependence of the self-diffusion of hexane in silica gels. *J. Chem. Phys.*, 116, 24, 10869–10876, 2002.
[13.31] Bedoshvili, Y., Gneusheva, K., Popova, M., Morozov, A., Likhoshway, Y., Anomalies in the valve morphogenesis of the centric diatom alga *Aulacoseira islandica* caused by microtubule inhibitors. *Biol. Open*, 2018, 7, 8, bio035519.
[13.32] Bedoshvili, Y.D., Gneusheva, K.V., Popova, M.S., Avezova, T.N., Arsentyev, K.Y., Likhoshway, Y.V., Frustule morphogenesis of raphid pennate diatom *Encyonema ventricosum* (Agardh) Grunow. *Protoplasma*, 255, 3, 911–921 2017. http://www.doc88.com/p-3844995788605.html.
[13.33] Bedoshvili, Y.D. and Likhoshway, Y., The effects of cytoskeletal inhibitors in diatom valve morphogenesis, in: *Diatom Morphogenesis [DIMO, Volume in the series: Diatoms: Biology & Applications, series editors: Richard Gordon & Joseph Seckbach]*, V. Annenkov, J. Seckbach, R. Gordon (Eds.), pp. 349–364, Wiley-Scrivener, Beverly, MA, USA: Submitted, 2020.
[13.34] Bentley, K., Clack, C., Cox, E.J., Diatom colony formation: A computational study predicts a single mechanism can produce both linkage and separation valves due to an environmental switch. *J. Phycol.*, 48, 3, 716–728, 2012.
[13.35] Bentley, K., Cox, E.J., Bentley, P.J., Nature's batik: A computer evolution model of diatom valve morphogenesis. *J. Nanosci. Nanotechnol.*, 5, 1, 25–34, 2005.
[13.36] Bertermann, R., Kröger, N., Tacke, R., Solid-state ^{29}Si MAS NMR studies of diatoms: Structural characterization of biosilica deposits. *Anal. Bioanal. Chem.*, 375, 5, 630–634, 2003.
[13.37] Bertrand, J., Diatom movements VIII: Synthesis and hypothesis. *Diatom Res.*, 23, 1, 19–29, 2008.
[13.38] Bertrand, J., Mouvements des diatomées VIII: synthèse et hypothèse [Diatom movements VIII: Synthesis and hypothesis, French]. *Diatom Res.*, 23, 1, 19–29, 2008.
[13.39] Bertrand, J., Diatom movements VIII: Synthesis and hypothesis. Translation of: Bertrand, J. (2008). Mouvements des diatomées VIII: synthèse et hypothèse. *Diatom Res.*, 23, 1, 19–29, 2020, by Gordon, R., Laviale, M., Serieyssol, K.K., in consultation with Jean Bertrand, in: *Diatom Gliding Motility [DIGM, Volume in the series: Diatoms: Biology & Applications, series editors: Richard Gordon & Joseph Seckbach]*, S.A. Cohn, K.M. Manoylov, R. Gordon (Eds.), pp. 283–294, Wiley-Scrivener, Beverly, MA, USA, In press.
[13.40] Betancor, L., Luckarift, H.R., Seo, J.H., Brand, O., Spain, J.C., Three-dimensional immobilization of β-galactosidase on a silicon surface. *Biotechnol. Bioeng.*, 99, 2, 261–267, 2008.
[13.41] Bhattacharyya, P. and Volcani, B.E., Sodium-dependent silicate transport in the apochlorotic marine diatom *Nitzschia alba*. *Proc. Natl. Acad. Sci. U.S.A.*, 77, 11, 6386–6390, 1980.
[13.42] Bidle, K.D. and Azam, F., Accelerated dissolution of diatom silica by marine bacterial assemblages. *Nature*, 397, 508–512, 1999.
[13.43] Bidle, K.D. and Azam, F., Bacterial control of silicon regeneration from diatom detritus: Significance of bacterial ectohydrolases and species identity. *Limnol. Oceanogr.*, 46, 7, 1606–1623, 2001.

[13.44] Bidle, K.D., Brzezinski, M.A., Long, R.A., Jones, J.L., Azam, F., Diminished efficiency in the oceanic silica pump caused by bacteria-mediated silica dissolution. *Limnol. Oceanogr.*, 48, 5, 1855–1868, 2003.

[13.45] Bienfang, P.K., Harrison, P.J., Quarmby, L.M., Sinking rate response to depletion of nitrate, phosphate and silicate in four marine diatoms. *Mar. Biol.*, 67, 3, 295–302, 1982.

[13.46] Blank, G.S. and Sullivan, C.W., Diatom mineralization of silicic acid. III. $Si(OH)_4$ binding and light dependent transport in *Nitzschia angularis*. *Arch. Microbiol.*, 123, 157–164, 1979.

[13.47] Blank, G.S. and Sullivan, C.W., Diatom mineralization of silicic acid VII. Influence of microtubule drugs on symmetry and pattern formation in valves of *Navicula saprophila* during morphogenesis. *J. Phycol.*, 19, 3, 294–301, 1983.

[13.48] Blank, G.S. and Sullivan, C.W., Diatom mineralization of silicic acid. VI. The effects of microtubule inhibitors on silicic acid metabolism in *Navicula saprophila*. *J. Phycol.*, 19, 1, 39–44, 1983.

[13.49] Bondoc, K.G., Heuschele, J., Gillard, J., Vyverman, W., Pohnert, G., Selective silicate-directed motility in diatoms. *Nat. Commun.*, 7, #10540, 2016.

[13.50] Bondoc, K.G.V., *Directed Motility of Benthic Diatoms*, Friedrich-Schiller-Universität, Jena, Germany, 2017.

[13.51] Bondoc, K.G.V., Lembke, C., Vyverman, W., Pohnert, G., Selective chemoattraction of the benthic diatom *Seminavis robusta* to phosphate but not to inorganic nitrogen sources contributes to biofilm structuring. *Microbiologyopen*, 8, 4, #e694, 2019.

[13.52] Boyle, J.A., Pickett-Heaps, J.D., Czarnecki, D.B., Valve morphogenesis in the pennate diatom *Achnanthes coarctata*. *J. Phycol.*, 20, 563–573, 1984.

[13.53] Brasser, H.J., van der Strate, H.J., Gieskes, W.W.C., Krijger, G.C., Vrieling, E.G., Wolterbeek, H.T., Compartmental analysis suggests macropinocytosis at the onset of diatom valve formation. *Silicon*, 4, 1, 39–49, 2012.

[13.54] Bridoux, M.C., Annenkov, V.V., Keil, R.G., Ingalls, A.E., Widespread distribution and molecular diversity of diatom frustule bound aliphatic long chain polyamines (LCPAs) in marine sediments. *Org. Geochem.*, 48, 9–20, 2012.

[13.55] Bridoux, M.C., Keil, R.G., Ingalls, A.E., Analysis of natural diatom communities reveals novel insights into the diversity of long chain polyamine (LCPA) structures involved in silica precipitation. *Org. Geochem.*, 47, 9–21, 2012.

[13.56] Brinker, C.J.S.G.W., *Sol-Gel Science: The Physics and Chemistry of Sol-Gel Processing*, Academic Press, Orlando, 1990.

[13.57] Brodland, G.W. and Gordon, R., Intermediate filaments may prevent buckling of compressively-loaded microtubules. *J. Biomech. Eng.*, 112, 3, 319–321, 1990.

[13.58] Brown, T.A., Belt, S.T., Cabedo-Sanz, P., Identification of a novel di-unsaturated C_{25} highly branched isoprenoid in the marine tube-dwelling diatom *Berkeleya rutilans*. *Environ. Chem. Lett.*, 12, 3, 455–460, 2014.

[13.59] Brugerolle, G. and Bricheux, G., Actin microfilaments are involved in scale formation of the chrysomonad cell *Synura*. *Protoplasma*, 123, 3, 203–212, 1984.

[13.60] Brunner, E., Gröger, C., Lutz, K., Richthammer, P., Spinde, K., Sumper, M., Analytical studies of silica biomineralization: Towards an understanding of silica processing by diatoms. *Appl. Microbiol. Biotechnol.*, 84, 4, 607–616, 2009.

[13.61] Brunner, E., Lutz, K., Sumper, M., Biomimetic synthesis of silica nanospheres depends on the aggregation and phase separation of polyamines in aqueous solution. *Phys. Chem. Chem. Phys.*, 6, 4, 854–857, 2004.

[13.62] Brzezinski, M.A., Krause, J.W., Baines, S.B., Collier, J.L., Ohnemus, D.C., Twining, B.S., Patterns and regulation of silicon accumulation in *Synechococcus* spp. *J. Phycol.*, 53, 4, 746–761, 2017.

[13.63] Buhmann, M.T., Schulze, B., Förderer, A., Schleheck, D., Kroth, P.G., Bacteria may induce the secretion of mucin-like proteins by the diatom *Phaeodactylum tricornutum*. *J. Phycol.*, 52, 3, 463–474, 2016.

[13.64] Busby, W.F. and Lewin, J., Silicate uptake and silica shell formation by synchronously dividing cells of the diatom *Navicula pelliculosa* (Bréb.) Hilse. *J. Phycol.*, 3, 127–131, 1967.

[13.65] Two letters from a gentleman in the country, relating to Mr. Leuwenhoeck's letter in Transaction, No. 283. *Philos. Trans. R. Soc. London*, 23, 288, 1494–1501, 1703.

[13.66] Capellacci, S., Battocchi, C., Casabianca, S., Giovine, M., Bavestrello, G., Penna, A., Bioavailability of different chemical forms of dissolved silica can affect marine diatom growth. *Mar. Ecol.- Evol. Persp.*, 34, 1, 103–111, 2013.

[13.67] Carr, J.M. and Hergenrader, G.L., Occurrence of three *Nitzschia* (Bacillariophyceae) taxa within colonies of tube-forming diatoms. *J. Phycol.*, 23, 1, 62–70, 1987.

[13.68] Casabianca, S., Penna, A., Capellacci, S., Cangiotti, M., Ottaviani, M.F., Silicification process in diatom algae using different silicon chemical sources: Colloidal silicic acid interactions at cell surface. *Colloids Surf. B-Biointerfaces*, 161, 620–627, 2018.

[13.69] Chaikittisilp, W., Khunsupat, R., Chen, T.T., Jones, C.W., Poly(allylamine)-mesoporous silica composite materials for CO_2 capture from simulated flue gas or ambient air. *Ind. Eng. Chem. Res.*, 50, 24, 14203–14210, 2011.

[13.70] Chatterjee, K., Pratiwi, F.W., Wu, F.C.M., Chen, P.L., Chen, B.C., Recent progress in light sheet microscopy for biological applications. *Appl. Spectrosc.*, 72, 8, 1137–1169, 2018.

[13.71] Chen, L., Weng, D., Du, C., Wang, J.D., Cao, S., Contribution of frustules and mucilage trails to the mobility of diatom *Navicula* sp. *Sci. Rep.*, 9, #7342, 2019.

[13.72] Chen, Y., Li, Y.L., Zhou, G.T., Li, H., Lin, Y.T., Xiao, X., Wang, F.P., Biomineralization mediated by anaerobic methane-consuming cell consortia. *Sci. Rep.*, 4, #5696, 2014.

[13.73] Chiappino, M.L., Azam, F., Volcani, B.E., Effect of germanic acid on developing cell walls of diatoms. *Protoplasma*, 93, 191–204, 1977.

[13.74] Chiappino, M.L. and Volcani, B.E., Studies on the biochemistry and fine structure of silica shell formation in diatoms. VII. Sequential wall development in the pennate *Navicula pelliculosa*. *Protoplasma*, 93, 2–3, 205–221, 1977.

[13.75] Chiovitti, A., Harper, R.E., Willis, A., Bacic, A., Mulvaney, P., Wetherbee, R., Variations in the substituted 3-linked mannans closely associated with the silicified walls of diatoms. *J. Phycol.*, 41, 6, 1154–1161, 2005.

[13.76] Chiriboga, O.G. and Rorrer, G.L., Control of chitin nanofiber production by the lipid-producing diatom *Cyclotella* sp. through fed-batch addition of dissolved silicon and nitrate in a bubble-column photobioreactor. *Biotechnol. Progr.*, 33, 2, 407–415, 2017.

[13.77] Choi, O., Kim, B.-C., An, J.-H., Min, K., Kim, Y.H., Um, Y., Oh, M.-K., Sang, B.-I., A biosensor based on the self-entrapment of glucose oxidase within biomimetic silica nanoparticles induced by a fusion enzyme. *Enzyme Microb. Technol.*, 49, 5, 441–445, 2011.

[13.78] Conley, D.J., Frings, P.J., Fontorbe, G., Clymans, W., Stadmark, J., Hendry, K.R., Marron, A.O., De La Rocha, C.L., Biosilicification drives a decline of dissolved Si in the oceans through geologic time. *Front. Mar. Sci.*, 4, #Unsp 397, 2017.

[13.79] Conrad, C.F., Icopini, G.A., Yasuhara, H., Bandstra, J.Z., Brantley, S.L., Heaney, P.J., Modeling the kinetics of silica nanocolloid formation and precipitation in geologically relevant aqueous solutions. *Geochim. Cosmochim. Acta*, 71, 3, 531–542, 2007.

[13.80] Coombs, J., Darley, W.M., Holm-Hansen, O., Volcani, B.E., Studies on the biochemistry and fine structure of silica shell formation in diatoms. Chemical composition of *Navicula pelliculosa* during silicon-starvation synchrony. *Plant Physiol.*, 42, 11, 1601–1606, 1967.

[13.81] Coombs, J., Halicki, P.J., Holm-Hansen, O., Volcani, B.E., Studies on the biochemistry and fine structure of silica shell formation in diatoms. II. Changes in concentration of nucleoside triphosphates in silicon-starvation synchyrony of *Navicula pelliculosa* (Bréb.) Hilse. *Exp. Cell Res.*, 47, 1, 315–328, 1967.

[13.82] Coombs, J., Spanis, C., Volcani, B.E., Studies on the biochemistry and fine structure of silica shell formation in diatoms. Photosynthesis and respiration in silicon-starvation synchrony of *Navicula pelliculosa*. *Plant Physiol.*, 42, 11, 1607–1611, 1967.

[13.83] Cox, E.J., The distribution of tube-dwelling diatom species in the Severn Estuary. *J. Mar. Biolog. Assoc. U. K.*, 57, 1, 19–27, 1977.

[13.84] Cox, E.J., Tube-dwelling diatom flora at two sites in the Severn Estuary. *Bot. Mar.*, 20, 2, 111–119, 1977.

[13.85] Cox, E.J., Mucilage tube morphology of three tube-dwelling diatoms and its diagnostic value. *J. Phycol.*, 17, 1, 72–80, 1981.

[13.86] Cox, E.J., Willis, L., Bentley, K., Integrated simulation with experimentation is a powerful tool for understanding diatom valve morphogenesis. *BioSystems*, 109, 3, Special Issue on Biological Morphogenesis, 450–459, 2012.

[13.87] Crawford, R.M., The organic component of the cell wall of the marine diatom *Melosira nummuloides* (Dillw.) C. Ag. *Br. Phycol. J.*, 8, 3, 257–266, 1973.

[13.88] Crawford, S., Chiovitti, T., Pickett-Heaps, J., Wetherbee, R., Micromorphogenesis during diatom wall formation produces siliceous nanostructures with different properties. *J. Phycol.*, 45, 6, 1353–1362, 2009.

[13.89] Crawford, S.A., Higgins, M.J., Mulvaney, P., Wetherbee, R., Nanostructure of the diatom frustule as revealed by atomic force and scanning electron microscopy. *J. Phycol.*, 37, 4, 543–554, 2001.

[13.90] Curnow, P., Senior, L., Knight, M.J., Thamatrakoln, K., Hildebrand, M., Booth, P.J., Expression, purification, and reconstitution of a diatom silicon transporter. *Biochemistry*, 51, 18, 3776–3785, 2012.

[13.91] Danilovtseva, E., Aseyev, V., Karesoja, M., Annenkov, V., Sorption of silicic acid from non-saturated aqueous solution by a complex of zinc ions with poly(vinylamine). *Eur. Polym. J.*, 45, 5, 1391–1396, 2009.

[13.92] Danilovtseva, E.N., Pal'shin, V.A., Zelinskiy, S.N., Annenkov, V.V., Fluorescent dyes for the study of siliceous sponges. *Limnol. Freshw. Biol.*, 2, 5, 302–307, 2019.

[13.93] Davis, A.K., Hildebrand, M., Sigel, H., Sigel, A., Molecular processes of biosilicification in diatoms, in: *Metal Ions in Life Sciences Vol 4, Biomineralization: From Nature to Application*, A. Sigel, H. Sigel, R.K.O. Sigel (Eds.), pp. 255–294, John Wiley & Sons Ltd., Chichester, UK, 2008.

[13.94] Dawson, P.A., Observations on the structure of some forms of *Gomphonema parvulum*Kütz II. Internal organization. *J. Phycol.*, 9, 2, 165–175, 1973.

[13.95] De Sanctis, S., Wenzler, M., Kröger, N., Malloni, W.M., Sumper, M., Deutzmann, R., Zadravec, P., Brunner, E., Kremer, W., Kalbitzer, H.R., PSCD domains of pleuralin-1 from the diatom *Cylindrotheca fusiformis*: NMR structures and interactions with other biosilica-associated proteins. *Structure*, 24, 7, 1178–1191, 2016.

[13.96] De Tommasi, E., Gielis, J., Rogato, A., Diatom frustule morphogenesis and function: A multidisciplinary survey. *Mar. Genomics*, 35, 1–18, 2017.

[13.97] Del Amo, Y. and Brzezinski, M.A., The chemical form of dissolved Si taken up by marine diatoms. *J. Phycol.*, 35, 6, 1162–1170, 1999.

[13.98] Drum, R.W., Cellular and colonial morphology of tube-dwelling diatom *Amphipleura rutilans* [abstract]. *J. Phycol.*, S 4, 6–7, 1968.

[13.99] Drum, R.W., Light and electron microscope observations on the tube-dwelling diatom *Amphipleura rutilans* (Trentepohl) Cleve. *J. Phycol.*, 5, 1, 21–26, 1969.

[13.100] Drum, R.W. and Pankratz, H.S., Post mitotic fine structure of *Gomphonema parvulum*. *J. Ultrastruct. Res.*, 10, 217–223, 1964.

[13.101] Durak, G.M., Taylor, A.R., Walker, C.E., Probert, I., de Vargas, C., Audic, S., Schroeder, D., Brownlee, C., Wheeler, G.L., A role for diatom-like silicon transporters in calcifying coccolithophores. *Nat. Commun.*, 7, #10543, 2016.

[13.102] Durkin, C.A., Koester, J.A., Bender, S.J., Armbrust, E.V., The evolution of silicon transporters in diatoms. *J. Phycol.*, 52, 5, 716–731, 2016.

[13.103] Durkin, C.A., Mock, T., Armbrust, E.V., Chitin in diatoms and its association with the cell wall. *Eukaryot. Cell*, 8, 7, 1038–1050, 2009.

[13.104] Dürre, K., Keber, F.C., Bleicher, P., Brauns, F., Cyron, C.J., Faix, J., Bausch, A.R., Capping protein-controlled actin polymerization shapes lipid membranes. *Nat. Commun.*, 9, #1630, 2018.

[13.105] Edgar, S.M. and Theriot, E.C., Heritability of mantle areolar characters in *Aulacoseira subarctica* (Bacillariophyta). *J. Phycol.*, 39, 6, 1057–1066, 2003.

[13.106] Eppley, R.W., Holmes, R.W., Paasche, E., Periodicity in cell division and physiological behavior of *Ditylum brightwellii*, a marine planktonic diatom, during growth in light-dark cycles. *Arch. Mikrobiol.*, 56, 4, 305–323, 1967.

[13.107] Erga, S.R., Lie, G.C., Aaro, L.H., Frette, O., Hamre, B., Migratory behaviour of *Skeletonema grethae* (Bacillariophyceae) in stratified waters. *Diatom Res.*, 30, 1, 13–25, 2015.

[13.108] Estes, A. and Dute, R.R., Valve abnormalities in diatom clones maintained in long-term culture. *Diatom Res.*, 9, 2, 249–258, 1994.

[13.109] Fabre, S., Jeandel, C., Zambardi, T., Roustan, M., Almar, R., An overlooked silica source of the modern oceans: Are sandy beaches the key? *Front. Earth Sci.*, 7, #Unsp 231, 2019.

[13.110] Ferrara, M.A., De Tommasi, E., Coppola, G., De Stefano, L., Rea, I., Dardano, P., Diatom valve three-dimensional representation: A new imaging method based on combined microscopies. *Int. J. Mol. Sci.*, 17, 10, 10, #1645, 2016.

[13.111] Finlay, J.A., Schultz, M.P., Cone, G., Callow, M.E., Callow, J.A., A novel biofilm channel for evaluating the adhesion of diatoms to non-biocidal coatings. *Biofouling*, 29, 4, 401–411, 2013.

[13.112] Francius, G., Tesson, B., Dague, E., Martin-Jézéquel, V., Dufrêne, Y.F., Nanostructure and nanomechanics of live *Phaeodactylum tricornutum* morphotypes. *Environ. Microbiol.*, 10, 5, 1344–1356, 2008.

[13.113] Frank, J., McEwen, B.F., Radermacher, M., Turner, J.N., Rieder, C.L., Three-dimensional tomographic reconstruction in high voltage electron microscopy. *J. Electron Microsc. Tech.*, 6, 2, 193–205, 1987.

[13.114] Frankel, J., *Pattern Formation, Ciliate Studies and Models*, Oxford University Press, New York, 1989.

[13.115] Fricke, A., Kihara, T.C., Kopprio, G.A., Hoppenrath, M., Anthropogenically driven habitat formation by a tube dwelling diatom on the Northern Patagonian Atlantic coast. *Ecol. Indic.*, 77, 8–13, 2017.

[13.116] Garofalini, S.H. and Conover, S., Comparison between bulk and surface self-diffusion constants of Si and O in vitreous silica. *J. Non-Crystall. Solids*, 74, 1, 171–176, 1985.

[13.117] Gelabert, A., Pokrovsky, O., Schott, J., Boudou, A., Metal adsorption by diatoms: A surface complexation model [abstract]. *Geochim. Cosmochim. Acta*, 70, 18, A197, 2006.

[13.118] Gélabert, A., Pokrovsky, O.S., Schott, J., Boudou, A., Feurtet-Mazel, A., Mielczarski, J., Mielczarski, E., Mesmer-Dudons, N., Spalla, O., Study of diatoms/aqueous solution interface. I. Acid-base equilibria and spectroscopic observation of freshwater and marine species. *Geochim. Cosmochim. Acta*, 68, 20, 4039–4058, 2004.

[13.119] George, S., Steinberg, S.M., Hodge, V., The concentration, apparent molecular weight and chemical reactivity of silica from groundwater in Southern Nevada. *Chemosphere*, 40, 1, 57–63, 2000.

[13.120] Ghobara, M., Gordon, R., Tiffany, M.A., Estimating adsorption areas of diatom valves from scanning electron micrographs (SEMs), in: *Diatom Morphogenesis [DIMO, Volume 2 in the series: Diatoms: Biology & Applications, series editors: Richard Gordon & Joseph Seckbach]*, V. Annenkov, J. Seckbach, R. Gordon (Eds.), Wiley-Scrivener, Beverly, MA, USA: In preparation, 2019.

[13.121] Ghobara, M.M., Mazumder, N., Vinayak, V., Reissig, L., Gebeshuber, I.C., Tiffany, M.A., Gordon, R., On light and diatoms: A photonics and photobiology review [Chapter 7], in: *Diatoms: Fundamentals & Applications [DIFA, Volume 1 in the series: Diatoms: Biology & Applications, series editors: Richard Gordon & Joseph Seckbach]*, J. Seckbach and R. Gordon (Eds.), pp. 129–190, Wiley-Scrivener, Beverly, MA, USA, 2019.

[13.122] Gifford, D.J., Bohrer, R.N., Boyd, C.M., Spines on diatoms: Do copepods care? *Limnol. Oceanogr.*, 26, 6, 1057–1061, 1981.

[13.123] Golokhvast, K., Pamirsky, I., A.M, P., Homology of bacteria proteins, diatoms and sponges participating in biomineralization, and human proteins, and other animals. *Pac. Sci. Rev.*, 13, 39–46, 2011.

[13.124] González, A.G., Shirokova, L.S., Pokrovsky, O.S., Emnova, E.E., Martínez, R.E., Santana-Casiano, J.M., González-Dávila, M., Pokrovski, G.S., Adsorption of copper on *Pseudomonas aureofaciens*: Protective role of surface exopolysaccharides. *J. Colloid Interface Sci.*, 350, 1, 305–314, 2010.

[13.125] Gordon, R., *The Hierarchical Genome and Differentiation Waves: Novel Unification of Development, Genetics and Evolution [HGDW]*, World Scientific & Imperial College Press, Singapore & London, 1999.

[13.126] Gordon, R., The fascination of diatoms: Thought experiments by an armchair diatomist [PowerPoint presentation], in: *Workshop on Diatom Nanotechnology, 17th North American Diatom Symposium*, E. Gaiser, (Ed.), Florida Sea Base, Lower Matecumbe Island, Florida Keys, October 21-26, 2003, Florida International University, Miami, 2003.

[13.127] Gordon, R., Diatoms and nanotechnology: Early history and imagined future as seen through patents, in: *The Diatoms: Applications for the Environmental and Earth Sciences*, 2nd, J.P. Smol and E.F. Stoermer (Eds.), pp. 585–602, Cambridge University Press, Cambridge, 2010.

[13.128] Gordon, R., The whimsical history of proposed motors for diatom motility, in: *Diatom Gliding Motility [DIGM, Volume in the series: Diatoms: Biology & Applications, series editors: Richard Gordon & Joseph Seckbach]*, S.A. Cohn, K.M. Manoylov, R. Gordon (Eds.), pp. 335–420, Wiley-Scrivener, Beverly, MA, USA: In press, 2020.

[13.129] Gordon, R. and Aguda, B.D., Diatom morphogenesis: Natural fractal fabrication of a complex microstructure, in: *Proceedings of the Annual International Conference of the IEEE Engineering in Medicine and Biology Society, Part 1/4: Cardiology and Imaging*, G. Harris and C. Walker, (Eds.), New Orleans, LA, USA, 4-7 Nov. 1988, Institute of Electrical and Electronics Engineers, New York, pp. 273–274, 1988.

[13.130] Gordon, R., Björklund, N.K., Robinson, G.G.C., Kling, H.J., Sheared drops and pennate diatoms. *Nova Hedwig.*, 112, Festschrift for Prof. T.V. Desikachary, 287–297, 1996.

[13.131] Gordon, R. and Coumans, J., Combining multiple imaging techniques for in vivo pathology: A quantitative method for coupling new imaging modalities. *Med. Phys.*, 11, 1, 79–80, 1984.

[13.132] Gordon, R. and Drum, R.W., The chemical basis of diatom morphogenesis. *Int. Rev. Cytol.*, 150, 243–372, 421–422, 1994.

[13.133] Gordon, R., Goel, N.S., Steinberg, M.S., Wiseman, L.L., A rheological mechanism sufficient to explain the kinetics of cell sorting. *J. Theor. Biol.*, 37, 1, 43–73, 1972.

[13.134] Gordon, R., Goel, N.S., Steinberg, M.S., Wiseman, L.L., A rheological mechanism sufficient to explain the kinetics of cell sorting, in: *Mathematical Models for Cell Rearrangement*, G.D. Mostow (Ed.), pp. 196–230, Yale University Press, New Haven, 1975.

[13.135] Gordon, R., Hanczyc, M.M., Denkov, N.D., Tiffany, M.A., Smoukov, S.K., Chapter 18: Emergence of polygonal shapes in oil droplets and living cells: The potential role of tensegrity in the origin of life, in: *Habitability of the Universe Before Earth [Volume 1 in series: Astrobiology: Exploring Life on Earth and Beyond, eds. Pabulo Henrique Rampelotto, Joseph Seckbach & Richard Gordon]*, R. Gordon and A.A. Sharov (Eds.), pp. 427–490, Elsevier B.V., Amsterdam, 2017.

[13.136] Gordon, R., Hoover, R.B., Tuszynski, J.A., de Luis, J., Camp, P.J., Tiffany, M.A., Nagy, S.S., Fayek, M., Lopez, P.J., Lerner, B.E., Diatoms in space: Testing prospects for reliable diatom nanotechnology in microgravity. *Proc. SPIE*, 6694, V1–V15, 2007.

[13.137] Gordon, R., Losic, D., Tiffany, M.A., Nagy, S.S., Sterrenburg, F.A.S., The Glass Menagerie: Diatoms for novel applications in nanotechnology. *Trends Biotechnol.*, 27, 2, 116–127, 2009.

[13.138] Gordon, R. and Stone, R., 5. Cybernetic embryo, in: *Biocommunication: Sign-Mediated Interactions between Cells and Organisms*, R. Gordon and J. Seckbach (Eds.), pp. 111–164, World Scientific Publishing, London, 2016.

[13.139] Gordon, R. and Tiffany, M.A., Possible buckling phenomena in diatom morphogenesis, in: *The Diatom World*, J. Seckbach and J.P. Kociolek (Eds.), pp. 245–272, Springer, Dordrecht, The Netherlands, 2011.

[13.140] Gordon, R., Witkowski, A., Gebeshuber, I.C., Allen, C.S., The diatoms of Antarctica and their potential roles in nanotechnology, in: *Antarctica: Time of Change*, M. Masó (Ed.), pp. 84–95, ACTAR, Barcelona, 2010.

[13.141] Görlich, S., Pawolski, D., Zlotnikov, I., Kröger, N., Control of biosilica morphology and mechanical performance by the conserved diatom gene *Silicanin-1*. *Commun. Biol.*, 2, #245, 2019.

[13.142] Götz, W., Tobiasch, E., Witzleben, S., Schulze, M., Effects of silicon compounds on biomineralization, osteogenesis, and hard tissue formation. *Pharmaceutics*, 11, 3, #117, 2019.

[13.143] Gräb, O., Abacilar, M., Daus, F., Geyer, A., Steinem, C., 3D-membrane stacks on supported membranes composed of diatom lipids induced by long-chain polyamines. *Langmuir*, 32, 39, 10144–10152, 2016.

[13.144] Grachev, M.A., Annenkov, V.V., A.L, V., *Method of production of the quartz glass [Patent RU2319672]*, Federal Service for Intellectual Property (ROSPATENT), Ministry of Economic Development of the Russian Federation, Moscow, Russia, 2008.

[13.145] Grachev, M.A., Annenkov, V.V., Likhoshway, Y.V., Silicon nanotechnologies of pigmented heterokonts. *BioEssays*, 30, 4, 328–337, 2008.

[13.146] Grimes, G.W. and Aufderheide, K.J., *Cellular Aspects of Pattern Formation: The Problem of Assembly*, Karger, Basel, 1991.

[13.147] Gröger, C., Lutz, K., Brunner, E., NMR studies of biomineralisation. *Prog. Nucl. Magn. Reson. Spectrosc.*, 54, 1, 54–68, 2009.

[13.148] Gröger, C., Sumper, M., Brunner, E., Silicon uptake and metabolism of the marine diatom *Thalassiosira pseudonana*: Solid-state ^{29}Si NMR and fluorescence microscopic studies. *J. Struct. Biol.*, 161, 1, 55–63, 2008.

[13.149] Gröger, P., Poulsen, N., Klemm, J., Kröger, N., Schlierf, M., Establishing super-resolution imaging for proteins in diatom biosilica. *Sci. Rep.*, 6, #36824, 2016.

[13.150] Hamsher, S.E. and Saunders, G.W., Marine tube-forming diatoms and their cohabitants: A floristic survey of canadian waters using DNA barcoding [abstract]. *J. Phycol.*, 43, s1, 47, 2007.
[13.151] Hazelaar, S., van der Strate, H.J., Gieskes, W.W.C., Vrieling, E.G., Monitoring rapid valve formation in the pennate diatom *Navicula salinarum* (Bacillariophyceae). *J. Phycol.*, 41, 2, 354–358, 2005.
[13.152] Hazen, R.M., Sverjensky, D.A., Azzolini, D., Bish, D.L., Elmore, S.C., Hinnov, L., Milliken, R.E., Clay mineral evolution. *Am. Mineral.*, 98, 11–12, 2007–2029, 2013.
[13.153] Hecky, R.E., Mopper, K., Kilham, P., Degens, E.T., The amino acid and sugar composition of diatom cell walls. *Mar. Biol.*, 19, 4, 323–331, 1973.
[13.154] Heintze, C., Formanek, P., Pohl, D., Hauptstein, J., Rellinghaus, B., Kroger, N., An intimate view into the silica deposition vesicles of diatoms. *BMC Mater.*, 2, 11, 2020. https://doi.org/10.1186/s42833-020-00017-8
[13.155] Hendry, K.R., Marron, A.O., Vincent, F., Conley, D.J., Gehlen, M., Ibarbalz, F.M., Quéguiner, B., Bowler, C., Competition between silicifiers and non-silicifiers in the past and present ocean and its evolutionary impacts. *Front. Mar. Sci.*, 5, #Unsp 22, 2018.
[13.156] Heredia, A., Figueira, E., Rodrigues, C.T., Rodriguez-Galvan, A., Basiuk, V.A., Vrieling, E.G., Almeida, S.F.P., Cd^{2+} affects the growth, hierarchical structure and peptide composition of the biosilica of the freshwater diatom *Nitzschia palea* (Kutzing) W. Smith. *Phycol. Res.*, 60, 3, 229–240, 2012.
[13.157] Heredia, A., van der Strate, H.J., Delgadillo, I., Basiuk, V.A., Vrieling, E.G., Analysis of organo-silica interactions during valve formation in synchronously growing cells of the diatom *Navicula pelliculosa. Chembiochem*, 9, 4, 573–584, 2008.
[13.158] Herringer, J.W., Lester, D., Dorrington, G.E., Rosengarten, G., Can diatom girdle band pores act as a hydrodynamic viral defense mechanism? *J. Biol. Phys.*, 45, 2, 213–234, 2019.
[13.159] Hibberd, D.J., Ultrastructure of cyst formation in *Ochromonas tuberculata* (Chrysophyceae). *J. Phycol.*, 13, 4, 309–320, 1977.
[13.160] Higgins, M.J., Crawford, S.A., Mulvaney, P., Wetherbee, R., The topography of soft, adhesive diatom 'trails' as observed by atomic force microscopy. *Biofouling*, 16, 2–4, 133–139, 2000.
[13.161] Hildebrand, M., Biological processing of nanostructured silica in diatoms. *Prog. Org. Coat.*, 47, 3–4, 256–266, 2003.
[13.162] Hildebrand, M., The Prospects of Manipulating Diatom Silica Nanostructure [slides], in: *Workshop on Diatom Nanotechnology, 17th North American Diatom Symposium*, R. Gordon, F.A.S. Sterrenburg and K. Sandhage, (Eds.), Florida Sea Base, Lower Matecumbe Island, Florida Keys, October 21-26, 2003, 2003.
[13.163] Hildebrand, M., Dahlin, K., Volcani, B.E., Characterization of a silicon transporter gene family in *Cylindrotheca fusiformis*: sequences, expression analysis, and identification of homologs in other diatoms. *Mol. Gen. Genet.*, 260, 5, 480–486, 1998.
[13.164] Hildebrand, M., Doktycz, M.J., Allison, D.P., Application of AFM in understanding biomineral formation in diatoms. *Pflugers Arch.-Eur. J. Physiol.*, 456, 1, 127–137, 2008.
[13.165] Hildebrand, M. and Lerch, S.J.L., Diatom silica biomineralization: Parallel development of approaches and understanding. *Semin. Cell Dev. Biol.*, 46, 27–35, 2015.
[13.166] Hildebrand, M., Lerch, S.J.L., Shrestha, R.P., Understanding diatom cell wall silicification—moving forward. *Front. Mar. Sci.*, 5, #125, 2018.
[13.167] Hildebrand, M., Volcani, B.E., Gassmann, W., Schroeder, J.I., A gene family of silicon transporters. *Nature*, 385, 6618, 688–689, 1997.
[13.168] Hildebrand, M. and Wetherbee, R., Components and control of silicification in diatoms, in: *Silicon Biomineralization: Biology, Biochemistry, Molecular Biology, Biotechnology*, vol. 33, W.E.G. Müller (Ed.), pp. 11–57, Springer, Berlin, Germany, 2003.

[13.169] Hildebrand, M., York, E., Kelz, J.I., Davis, A.K., Frigeri, L.G., Allison, D.P., Doktycz, M.J., Nanoscale control of silica morphology and three-dimensional structure during diatom cell wall formation. *J. Mater. Res.*, 21, 10, 2689–2698, 2006.

[13.170] Hoagland, K.D., Rosowski, J.R., Gretz, M.R., Roemer, S.C., Diatom extracellular polymeric substances: function, fine structure, chemistry, and physiology. *J. Phycol.*, 29, 5, 537–566, 1993.

[13.171] Hopkins, J.T., The diatom trail. *Microscopy*, 30, 209–217, 1967.

[13.172] Houpt, P.M., Marine tube-dwelling diatoms and their occurrence In the Netherlands. *Neth. J. Aquat. Ecol.*, 28, 1, 77–84, 1994.

[13.173] Iler, R.K., *The Chemistry of Silica: Solubility, Polymerization, Colloid and Surface Properties, and Biochemistry*, John Wiley & Sons, New York, 1979.

[13.174] Imhof, A., Vanblaaderen, A., Maret, G., Mellema, J., Dhont, J.K.G., A comparison between the ion-time self-diffusion and low shear viscosity of concentrated dispersions of charged colloidal silica spheres. *J. Chem. Phys.*, 100, 3, 2170–2181, 1994.

[13.175] Ingalls, A.E., Whitehead, K., Bridoux, M.C., Tinted windows: The presence of the UV absorbing compounds called mycosporine-like amino acids embedded in the frustules of marine diatoms. *Geochim. Cosmochim. Acta*, 74, 1, 104–115, 2010.

[13.176] Jewson, D.H., Lowry, S.F., Bowen, R., Co-existence and survival of diatoms on sand grains. *Eur. J. Phycol.*, 41, 2, 131–146, 2006.

[13.177] Jimenez-Lopez, J.C., Wang, X., Kotchoni, S.O., Huang, S.J., Szymanski, D.B., Staiger, C.J., Heterodimeric capping protein from *Arabidopsis* is a membrane-associated, actin-binding protein. *Plant Physiol.*, 166, 3, 1312–1328, 2014.

[13.178] Johnston, M.R., Gascooke, J.R., Ellis, A.V., Leterme, S.C., Diatoms response to salinity changes: investigations using single pulse and cross polarisation magic angle spinning ^{29}Si NMR spectra. *Analyst*, 143, 20, 4930–4935, 2018.

[13.179] Kaluzhnaya, O.V. and Likhoshway, Y.V., Valve morphogenesis in an araphid diatom *Synedra acus* subsp *radians*. *Diatom Res.*, 22, 1, 81–87, 2007.

[13.180] Kamatani, A., Dissolution rates of silica from diatoms decomposing at various temperatures. *Mar. Biol.*, 68, 1, 91–96, 1982.

[13.181] Kang, K.K., Oh, H.S., Kim, D.Y., Shim, G., Lee, C.S., Synthesis of silica nanoparticles using biomimetic mineralization with polyallylamine hydrochloride. *J. Colloid Interface Sci.*, 507, 145–153, 2017.

[13.182] Karp-Boss, L., Gueta, R., Rousso, I., Judging diatoms by their cover: Variability in local elasticity of *Lithodesmium undulatum* undergoing cell division. *PLoS One*, 9, 10, #e109089, 2014.

[13.183] Kharitonenko, K.V., Bedoshvili, Y.D., Likhoshway, Y.V., Changes in the micro- and nanostructure of siliceous valves in the diatom *Synedra acus* under the effect of colchicine treatment at different stages of the cell cycle. *J. Struct. Biol.*, 190, 1, 73–80, 2015.

[13.184] King, G.M., Gordon, R., Karmali, K., Biberman, L.J., A new method for the immobilisation of teleost embryos for time-lapse studies of development. *J. Exp. Zool.*, 220, 2, 147–151, 1982.

[13.185] Knight, M.J., Senior, L., Nancolas, B., Ratcliffe, S., Curnow, P., Direct evidence of the molecular basis for biological silicon transport. *Nat. Commun.*, 7, #11926, 2016.

[13.186] Kociolek, J.P., Sicko-Goad, L., Stoermer, E.F., Cytoplasmic fine structure of two Encyonema species, in: *Proceedings of the 11th International Diatom Symposium*, K.J. P., (Ed.), San Francisco, California, 12-17 August 1990, California Academy of Sciences, San Francisco, California, USA, pp. 235–245, 1993.

[13.187] Kociolek, J.P. and Stoermer, E.F., Variation and polymorphism in diatoms: the triple helix of development, genetics and environment. a review of the literature. *Vie Milieu - Life Environ.*, 60, 75–87, 2010.

[13.188] Koiller, J., Ehlers, K.M., Chalub, F., Acoustic streaming, the "small invention" of cyanobacteria? *Arbor-Ciencia Pensamiento y Cultura*, 186, 746, 1089–1115, 2010.
[13.189] Kotzsch, A., Gröger, P., Pawolski, D., Bomans, P.H.H., Sommerdijk, N., Schlierf, M., Kröger, N., Silicanin-1 is a conserved diatom membrane protein involved in silica biomineralization. *BMC Biol.*, 15, #65, 2017.
[13.190] Kotzsch, A., Pawolski, D., Milentyev, A., Shevchenko, A., Scheffel, A., Poulsen, N., Shevchenko, A., Kröger, N., Biochemical composition and assembly of biosilica-associated insoluble organic matrices from the diatom *Thalassiosira pseudonana*. *J. Biol. Chem.*, 291, 10, 4982–4997, 2016.
[13.191] Kröger, N., Deutzmann, R., Sumper, M., Polycationic peptides from diatom biosilica that direct silica nanosphere formation. *Science*, 286, 5442, 1129–1132, 1999.
[13.192] Kröger, N., Lorenz, S., Brunner, E., Sumper, M., Self-assembly of highly phosphorylated silaffins and their function in biosilica morphogenesis. *Science*, 298, 5593, 584–586, 2002.
[13.193] Kröger, N. and Poulsen, N., Diatoms: From cell wall biogenesis to nanotechnology. *Annu. Rev. Genet.*, 42, 83–107, 2008.
[13.194] Kucki, M. and Fuhrmann-Lieker, T., Staining diatoms with rhodamine dyes: Control of emission colour in photonic biocomposites. *J. R. Soc. Interface*, 9, 69, 727–733, 2012.
[13.195] Kulikovskiy, M., Chudaev, D., Glushchenko, A., Kuznetsova, I., Krivova, Z., Kociolek, J.P., *Navicula gogorevii*-a new, large-celled diatom species from Vietnam (Southeast Asia). *Phytotaxa*, 428, 1, 60–66, 2020.
[13.196] Kulikovskiy, M., Lange-Bertalot, H., Annenkova, N., Gusev, E., Kociolek, J.P., Morphological and molecular evidence support description of two new diatom species from the genus *Ulnaria* in Lake Baikal. *Fottea*, 16, 1, 34–42, 2016.
[13.197] Kuwata, A., Yamada, K., Ichinomiya, M., Yoshikawa, S., Tragin, M., Vaulot, D., dos Santos, A.L., Bolidophyceae, a sister picoplanktonic group of diatoms - a review. *Front. Mar. Sci.*, 5, #Unsp 370, 2018.
[13.198] La Vars, S.M., Johnston, M.R., Hayles, J., Gascooke, J.R., Brown, M.H., Leterme, S.C., Ellis, A.V., $^{29}Si\{^{1}H\}$ CP-MAS NMR comparison and ATR-FTIR spectroscopic analysis of the diatoms *Chaetoceros muelleri* and *Thalassiosira pseudonana* grown at different salinities. *Anal. Bioanal. Chem.*, 405, 10, 3359–3365, 2013.
[13.199] Lallemand-Breitenbach, V., Quesnoit, M., Braun, V., El Marjou, A., Poüs, C., Goud, B., Perez, F., CLIPR-59 is a lipid raft-associated protein containing a cytoskeleton-associated protein glycine-rich domain (CAP-Gly) that perturbs microtubule dynamics. *J. Biol. Chem.*, 279, 39, 41168–41178, 2004.
[13.200] Lane, T.W., Pawate, A., Lane, P.D., Lindquist, E.A., Kirton, E.S., Simmons, B.A., Digital transcriptomic analysis of silicate starvation induced triacylglycerol formation in the marine diatom *Thalassiosira pseudonana*. *Phycologia*, 48, Special Issue 4, 71–71, 2009.
[13.201] Langhans, M. and Meckel, T., Single-molecule detection and tracking in plants. *Protoplasma*, 251, 2, 277–291, 2014.
[13.202] Laurenceau-Cornec, E.C., Trull, T.W., Davies, D.M., De La Rocha, C.L., Blain, S., Phytoplankton morphology controls on marine snow sinking velocity. *Mar. Ecol. Prog. Ser.*, 520, 35–56, 2015.
[13.203] Lavoie, M., Raven, J.A., Levasseur, M., Energy cost and putative benefits of cellular mechanisms modulating buoyancy in aflagellate marine phytoplankton. *J. Phycol.*, 52, 2, 239–251, 2016.
[13.204] Leadbeater, B.S.C., Silicification of 'cell walls' of certain protistan flagellates. *Philos. Trans. R. Soc. B-Biol. Sci.*, 304, 1121, 529–+, 1984.
[13.205] Leadbeater, B.S.C., Silicon deposition and lorica assembly in choanoflagellates, in: *Biomineralization in Lower Plants and Animals: Proceedings of an International Symposium*

Held at the University of Birmingham, April 1985, B.S.C. Leadbeater and R. Riding, (Eds.), Clarendon Press, Oxford, UK, pp. 345–359, 1986.
[13.206] Leadbeater, B.S.C., *Choanoflagellates: Evolution, Biology and Ecology*, Cambridge University Press, Cambridge, 2015.
[13.207] Lechner, C.C. and Becker, C.F.W., Modified silaffin R5 peptides enable encapsulation and release of cargo molecules from biomimetic silica particles. *Bioorg. Med. Chem.*, 21, 12, 3533–3541, 2013.
[13.208] Lechner, C.C. and Becker, C.F.W., Immobilising proteins on silica with site-specifically attached modified silaffin peptides. *Biomater. Sci.*, 3, 2, 288–297, 2015.
[13.209] Lee, R.E., *Phycology*, 4th, Cambridge University Press, Cambridge, UK, 2008.
[13.210] Leefmann, T., Heim, C., Kryvenda, A., Siljestrom, S., Sjovall, P., Thiel, V., Biomarker imaging of single diatom cells in a microbial mat using time-of-flight secondary ion mass spectrometry (ToF-SIMS). *Org. Geochem.*, 57, 23–33, 2013.
[13.211] Leinweber, K. and Kroth, P.G., Capsules of the diatom *Achnanthidium minutissimum* arise from fibrillar precursors and foster attachment of bacteria. *PeerJ*, 3, #e858, 2015.
[13.212] Lenoci, L. and Camp, P.J., Diatom structures templated by phase-separated fluids. *Langmuir*, 24, 1, 217–223, 2008.
[13.213] Lethias, C., Garrone, R., Mazzorana, M., Fine structures of sponge cell membranes: Comparative study with freeze-fracture and conventional thin section methods. *Tissue Cell*, 15, 4, 523–535, 1983.
[13.214] Lewin, J. and Chen, C.H., Silicon metabolism in diatoms.VI. Silicic acid uptake by a colorless marine diatom *Nitzschia alba* Lewin and Lewin. *J. Phycol.*, 4, 2, 161–166, 1968.
[13.215] Lewin, J.C., The capsule of the diatom *Navicula pelliculosa*. *J. Gen. Microbiol.*, 13, 1, 162–&, 1955.
[13.216] Li, C.W., Chu, S., Lee, M., Characterizing the silica deposition vesicle of diatoms. *Protoplasma*, 151, 2–3, 158–163, 1989.
[13.217] Li, C.W. and Volcani, B.E., Aspects of silicification in wall morphogenesis of diatoms. *Philos. Trans. R. Soc. Lond. Ser. B-Biol. Sci.*, 304, 1121, 519–528, 1984.
[13.218] Li, C.W. and Volcani, B.E., Studies on the biochemistry and fine structure of silica shell formation in diatoms VIII. Morphogenesis of the cell wall in a centric diatom, *Ditylum brightwellii*. *Protoplasma*, 124, 1–2, 10–29, 1985.
[13.219] Liang, C.L., Liu, W.L., Zheng, Y.H., Ji, X.L., Li, S.S., Yin, W.J., Guo, X.H., Song, Z.T., Fractal nature of non-spherical silica particles via facile synthesis for the abrasive particles in chemical mechanical polishing. *Colloids Surf. A-Physicochem. Eng. Asp.*, 500, 146–153, 2016.
[13.220] Likhoshway, Y.V., Masyukova, Y.A., Sherbakova, T.A., Petrova, D.P., Grachev, M.A., Detection of the gene responsible for silicic acid transport in chrysophycean algae. *Dokl. Biol. Sci.: Proc. Acad. Sci. USSR, Biol. Sci. sections*, 408, 256–260, 2006.
[13.221] Lobban, C.S., Colony and frustule morphology of three tube-dwelling diatoms from eastern Canada. *J. Phycol.*, 19, 3, 281–289, 1983.
[13.222] Lobban, C.S., Marine tube-dwelling diatoms of eastern Canada: Descriptions, checklist, and illustrated key. *Can. J. Bot.-Revue Can. Botanique*, 62, 4, 778–794, 1984.
[13.223] Lobban, C.S., Marine tube-dwelling diatoms of the Pacific coast of North America. I. *Berkeleya*, *Haslea*, *Nitzschia*, and *Navicula* sect. *Microstigmaticae*. *Can. J. Bot.-Revue Can. Botanique*, 63, 10, 1779–1784, 1985.
[13.224] Lobban, C.S., Environmental factors, plant responses, and colony growth in relation to tube-dwelling diatom blooms in the Bay of Fundy, Canada, with a review of the biology of tube-dwelling diatoms. *Diatom Res.*, 4, 1, 89–109, 1989.

[13.225] Lobban, C.S., Marine tube-dwelling diatoms of the Pacific coast of North America. II. *Navicula* subg. *Navicula* and a key to the tube-dwelling diatoms of the region. *Can. J. Bot.-Revue Can. Botanique*, 68, 4, 707–712, 1990.

[13.226] Lobban, C.S. and Mann, D.G., The systematics of the tube-dwelling diatom *Nitzschia martiana* and *Nitzschia* section *Spathulatae*. *Can. J. Bot.-Revue Can. Botanique*, 65, 11, 2396–2402, 1987.

[13.227] Lobban, C.S. and Navarro, J.N., *Gato hyalinus gen. et sp. nov.*, an unusual araphid tube-dwelling diatom from Western Pacific and Caribbean islands. *Phytotaxa*, 127, 1, 22, 2013.

[13.228] Lobel, K.D., West, J.K., Hench, L.L., Computational model for protein mediated biomineralization of the diatom frustule. *Mar. Biol.*, 126, 3, 353–360, 1996.

[13.229] Losic, D., Mitchell, J.G., Voelcker, N.H., Diatomaceous lessons in nanotechnology and advanced materials. *Adv. Mater.*, 21, 29, 2947–2958, 2009.

[13.230] Losic, D., Pillar, R.J., Dilger, T., Mitchell, J.G., Voelcker, N.H., Atomic force microscopy (AFM) characterisation of the porous silica nanostructure of two centric diatoms. *J. Porous Mater.*, 14, 1, 61–69, 2007.

[13.231] Losic, D., Short, K., Mitchell, J.G., Lal, R., Voelcker, N.H., AFM nanoindentations of diatom biosilica surfaces. *Langmuir*, 23, 9, 5014–5021, 2007.

[13.232] Luckarift, H.R., Spain, J.C., Naik, R.R., Stone, M.O., Enzyme immobilization in a biomimetic silica support. *Nat. Biotechnol.*, 22, 2, 211–213, 2004.

[13.233] Luís, A.T., Hlúbikova, D., Vaché, V., Choquet, P., Hoffmann, L., Ector, L., Atomic force microscopy (AFM) application to diatom study: Review and perspectives. *J. Appl. Phycol.*, 29, 6, 2989–3001, 2017.

[13.234] Ma, J., Zhou, B.B., Duan, D.D., Wei, Y., Pan, K., Silicon limitation reduced the adsorption of cadmium in marine diatoms. *Aquat. Toxicol.*, 202, 136–144, 2018.

[13.235] Majewska, R., Kociolek, J.P., Thomas, E.W., De Stefano, M., Santoro, M., Bolanos, F., Van de Vijver, B., *Chelonicola* and *Poulinea*, two new gomphonemoid diatom genera (Bacillariophyta) living on marine turtles from Costa Rica. *Phytotaxa*, 233, 3, 236–250, 2015.

[13.236] Mann, D.G. and Droop, S.J.M., Biodiversity, biogeography and conservation of diatoms. *Hydrobiologia*, 336, 1–3, 19–32, 1996.

[13.237] Manoylov, K.M. and Ghobara, M., Tutorial on diatom morphology, in: *Diatom Morphogenesis [DIMO, Volume 2 in the series: Diatoms: Biology & Applications, series editors: Richard Gordon & Joseph Seckbach]*, V. Annenkov, J. Seckbach, R. Gordon (Eds.), Wiley-Scrivener, Beverly, MA, USA: In preparation, 2020.

[13.238] Marchenkov, A.M., Petrova, D.P., Morozov, A.A., Zakharova, Y.R., Grachev, M.A., Bondar, A.A., A family of silicon transporter structural genes in a pennate diatom *Synedra ulna* subsp. *danica* (Kütz.) Skabitsch. *PLoS One*, 13, 8, #e0203161, 2018.

[13.239] Marron, A.O., Alston, M.J., Heavens, D., Akam, M., Caccamo, M., Holland, P.W.H., Walker, G., A family of diatom-like silicon transporters in the siliceous loricate choanoflagellates. *Proc. R. Soc. B-Biol. Sci.*, 280, 1756, #20122543, 2013.

[13.240] Marron, A.O., Ratcliffe, S., Wheeler, G.L., Goldstein, R.E., King, N., Not, F., de Vargas, C., Richter, D.J., The evolution of silicon transport in eukaryotes. *Mol. Biol. Evol.*, 33, 12, 3226–3248, 2016.

[13.241] Martin-Jézéquel, V., Hildebrand, M., Brzezinski, M.A., Silicon metabolism in diatoms: Implications for growth. *J. Phycol.*, 36, 5, 821–840, 2000.

[13.242] Mazumder, N. and Gordon, R. (Eds.), *Diatom Microscopy [DIMI, Volume in the series: Diatoms: Biology & Applications, series editors: Richard Gordon & Joseph Seckbach, in preparation]*, Wiley-Scrivener, Beverly, MA, USA, 2021.

[13.243] McConville, M.J., Wetherbee, R., Bacic, A., Subcellular location and composition of the wall and secreted extracellular sulphated polysaccharides/proteoglycans of the diatom *Stauroneis amphioxys* Gregory. *Protoplasma*, 206, 1–3, 188–200, 1999.

[13.244] McGrory, C.B. and Leadbeater, B.S.C., Ultrastructure and deposition of silica in the Chrysophyceae, in: *Silicon and Siliceous Structures in Biological Systems*, T.L. Simpson and B.E. Volcani (Eds.), pp. 201–230, Springer, New York, NY, USA, 1981.

[13.245] Meyer, M.L. and Bloom, P.R., Boric and silicic acid adsorption and desorption by a humic acid. *J. Environ. Qual.*, 26, 1, 63–69, 1997.

[13.246] Mignot, J.P. and Brugerolle, G., Scale formation in Chyromonad flagellates. *J. Ultrastruct. Res.*, 81, 13–26, 1982.

[13.247] Mikhailov, I.S., Zakharova, Y.R., Volokitina, N.A., Petrova, D.P., Likhoshway, Y.V., Bacteria associated with planktonic diatoms from Lake Baikal. *Acta Biol. Sibirica*, 4, 4, 89–94, 2018.

[13.248] Mikhailovsky, G.E. and Gordon, R., Chapter 16: Symbiosis: Why was the transition from microbial prokaryotes to eukaryotic organisms a cosmic gigayear event?, in: *Habitability of the Universe Before Earth [in series: Astrobiology: Exploring Life on Earth and Beyond, eds. Pabulo Henrique Rampelotto, Joseph Seckbach & Richard Gordon]*, R. Gordon and A.A. Sharov (Eds.), pp. 355–405, Elsevier B.V., Amsterdam, 2017.

[13.249] Milischuk, A.A. and Ladanyi, B.M., Structure and dynamics of water confined in silica nanopores [Correction: 135, #174709]. *J. Chem. Phys.*, 135, 17, #174709, 2011.

[13.250] Miller, S.A., Hong, E.D., Wright, D., Rapid and efficient enzyme encapsulation in a dendrimer silica nanocomposite. *Macromol. Biosci.*, 6, 10, 839–845, 2006.

[13.251] Mizuno, M., Phenology and seasonal size change of the marine tube-dwelling diatom *Berkeleya rutilans*. *Jpn. J. Phycol.*, 33, 3, 262–268, 1984.

[13.252] Moffat, M.C., An ultrastructural study of *Didymosphenia geminata* (Bacillariophyceae). *Trans. Am. Microsc. Soc.*, 113, 1, 59–71, 1994.

[13.253] Murr, M.M. and Morse, D.E., Fractal intermediates in the self-assembly of silicatein filaments. *Proc. Natl. Acad. Sci. U. S. A.*, 102, 33, 11657–11662, 2005.

[13.254] Nagai, S. and Manabe, T., Auxospore formation of a giant diatom, *Coscinodiscus wailesii* (Bacillariophyceae), in culture. *Bull. Plankton Soc. Japan*, 40, 2, 151–167, 1994.

[13.255] Natori, Y., Haneda, A., Suzuki, Y., Vertical and seasonal differences in biogenic silica dissolution in natural seawater in Suruga Bay, Japan: Effects of temperature and organic matter. *Mar. Chem.*, 102, 3–4, 230–241, 2006.

[13.256] Navarro, J.N., Observations of a tube-dwelling diatom - *Navicula hamulifera* (Bacillariophyceae). *J. Phycol.*, 23, 1, 164–170, 1987.

[13.257] Navarro, J.N., Three-dimensional imaging of diatom ultrastructure with high resolution low-voltage SEM. *Phycologia*, 32, 2, 151–156, 1993.

[13.258] Nawaz, M.A., Zakharenko, A.M., Zemchenko, I.V., Haider, M.S., Ali, M.A., Imtiaz, M., Chung, G., Tsatsakis, A., Sun, S., Golokhvast, K.S., Phytolith formation in plants: From soil to cell. *Plants-Basel*, 8, 8, #249, 2019.

[13.259] Neethirajan, S., Gordon, R., Wang, L., Potential of silica bodies (phytoliths) for nanotechnology. *Trends Biotechnol.*, 27, 8, 461–467, 2009.

[13.260] Newman, P.J. and Patterson, D.J., The role of an organic matrix during the formation of siliceous scales in the heliozoon *Actinophrys sol* (Actinophryida, Protista). *Eur. J. Protistol.*, 29, 3, 334–343, 1993.

[13.261] Ning, R.Y., Reactive silica in natural waters - A review. *Desalin. Water Treat.*, 21, 1–3, 79–86, 2010.

[13.262] Nishikawa, T., Miyahara, K., Nagai, S., Effects of temperature and salinity on the growth of the giant diatom *Coscinodiscus wailesii* isolated from Harima-Nada, Seto Inland Sea, Japan. *Nippon Suisan Gakk.*, 66, 6, 993–998, 2000.

[13.263] Nomura, M. and Ishida, K.-i., Fine-structural observations on siliceous scale production and shell assembly in the testate amoeba *Paulinella chromatophora*. *Protist*, 167, 4, 303–318, 2016.

[13.264] Nomura, M., Nakayama, T., Ishida, K.-i., Detailed process of shell construction in the photosynthetic testate amoeba *Paulinella chromatophora* (Euglyphid, Rhizaria). *J. Eukaryotic Microbiol.*, 61, 3, 317–321, 2014.

[13.265] Novarino, G., Presence of minerals in the mucilage stalk of the diatom *Achnanthes longipes*. *Diatom Res.*, 8, 1, 199–202, 1993.

[13.266] Nunn, B.L., Aker, J.R., Shaffer, S.A., Tsai, Y.H., Strzepek, R.F., Boyd, P.W., Freeman, T.L., Brittnacher, M., Malmstrom, L., Goodlett, D.R., Deciphering diatom biochemical pathways via whole-cell proteomics. *Aquat. Microb. Ecol.*, 55, 3, 241–253, 2009.

[13.267] Ohnemus, D.C., Krause, J.W., Brzezinski, M.A., Collier, J.L., Baines, S.B., Twining, B.S., The chemical form of silicon in marine *Synechococcus*. *Mar. Chem.*, 206, 44–51, 2018.

[13.268] Okita, T.W. and Volcani, B.E., Role of silicon in diatom metabolism. IX. Differential synthesis of DNA polymerases and DNA-binding proteins during silicate starvation and recovery in *Cylindrotheca fusiformis*. *Biochim. Biophys. Acta*, 519, 1, 76–86, 1978.

[13.269] Okita, T.W. and Volcani, B.E., Role of silicon in diatom metabolism. X. Polypeptide labelling patterns during the cell cycle, silicate starvation and recovery in *Cylindrotheca fusiformis*. *Exp. Cell Res.*, 125, 2, 471–481, 1980.

[13.270] Olarte, O.E., Andilla, J., Gualda, E.J., Loza-Alvarez, P., Light-sheet microscopy: A tutorial. *Adv. Opt. Photonics*, 10, 1, 111–179, 2018.

[13.271] Orefice, I., Lauritano, C., Procaccini, G., Ianora, A., Romano, G., Insights into possible cell-death markers in the diatom *Skeletonema marinoi* in response to senescence and silica starvation. *Mar. Genomics*, 24, 1, 81–88, 2015.

[13.272] Papisov, I.M., Bolyachevskaya, K.I., Litmanovich, A.A., Matveenko, V.N., Volchkova, I.L., Structural effects in matrix polycondensation of silicic acid. *Eur. Polym. J.*, 35, 11, 2087–2094, 1999.

[13.273] Pappas, J.L. and Gordon, R., Buckling: A geometric and biophysical multiscale feature of centric diatom valve morphogenesis, in: *Diatom Morphogenesis [DIMO, Volume 2 in the series: Diatoms: Biology & Applications, series editors: Richard Gordon & Joseph Seckbach]*, V. Annenkov, J. Seckbach, R. Gordon (Eds.), pp. 195–230, Wiley-Scrivener, Beverly, MA, USA: In press, 2020.

[13.274] Pappas, J.L., Gordon, R., Tiffany, M.A., Are mantle profiles of circular centric diatoms a measure of buckling forces during valve morphogenesis? [MANT], in: *The Mathematical Biology of Diatoms [DMTH, Volume in the series: Diatoms: Biology & Applications, series editors: Richard Gordon & Joseph Seckbach]*, J.L. Pappas and R. Gordon (Eds.), Wiley-Scrivener, Beverly, MA, USA: In preparation, 2020.

[13.275] Park, Y.S., Kim, G., Lee, J.S., Anisotropic silicification of nanostructured surfaces by local liquid-phase deposition. *Langmuir*, 35, 39, 12656–12664, 2019.

[13.276] Parkinson, J., Brechet, Y., Gordon, R., Centric diatom morphogenesis: A model based on a DLA algorithm investigating the potential role of microtubules. *Biochim. Biophys. Acta-Mol. Cell Res.*, 1452, 1, 89–102, 1999.

[13.277] Parslow, J.S., Harrison, P.J., Thompson, P.A., Saturated uptake kinetics: Transient response of the marine diatom *Thalassiosira pseudonana* to ammonium, nitrate, silicate or phosphate starvation. *Mar. Biol.*, 83, 1, 51–59, 1984.

[13.278] Patrick, S. and Holding, A.J., The effect of bacteria on the solubilization of silica in diatom frustules. *J. Appl. Bacteriol.*, 59, 7–16, 1985.

[13.279] Patterson, D.J., On the organization and affinities of the amoeba, *Pompholyxophrys punicea* Archer, based on ultrastructural examination of individual cells from wild material. *J. Protozool.*, 32, 2, 241–246, 1985.

[13.280] Patwardhan, S.V., Shiba, K., Schroder, H.C., Muller, W.E.G., Clarson, S.J., Perry, C.C., The interaction of 'silicon' with proteins: Part 2. The role of bioinspired peptide and recombinant proteins in silica polymerization, in: *Science and Technology of Silicones and Silicone-Modified Materials*, vol. 964, Clarson, S.J., Fitzgerald, J.J., Owen, M.J., Smith, S.D., VanDyke, M.E. (Eds.), pp. 328–347, 2007.

[13.281] Pawolski, D., Heintze, C., Mey, I., Steinem, C., Kröger, N., Reconstituting the formation of hierarchically porous silica patterns using diatom biomolecules. *J. Struct. Biol.*, 204, 1, 64–74, 2018.

[13.282] Penna, A., Magnani, M., Fenoglio, I., Fubini, B., Cerrano, C., Giovine, M., Bavestrello, G., Marine diatom growth on different forms of particulate silica: evidence of cell/particle interaction. *Aquat. Microb. Ecol.*, 32, 3, 299–306, 2003.

[13.283] Peters, R., Translocation through the nuclear pore complex: selectivity and speed by reduction-of-dimensionality. *Traffic*, 6, 5, 421–427, 2005.

[13.284] Pickett-Heaps, J.D., Schmid, A.M.M., Edgar, L.A., The cell biology of diatom valve formation. *Prog. Phycol. Res.*, 7, 1–168, 1990.

[13.285] Pickett-Heaps, J.D. and Spurck, T.P., Studies on kinetochore function in mitosis. II. The effects of metabolic inhibitors on mitosis and cytokinesis in the diatom *Hantzschia amphioxys*. *Eur. J. Cell Biol.*, 28, 1, 83–91, 1982.

[13.286] Pickett-Heaps, J.D., Tippit, D.H., Andreozzi, J.A., Cell division in the pennate diatom *Pinnularia*. IV. - Valve morphogenesis. *Biol. Cell.*, 35, 2, 199–203, 1979.

[13.287] Pokrovsky, O.S., Pokrovski, G.S., Gélabert, A., Schott, J., Boudou, A., Speciation of Zn associated with diatoms using X-ray absorption spectroscopy. *Environ. Sci. Technol.*, 39, 12, 4490–4498, 2005.

[13.288] Pollock, F.M. and Pickett-Heaps, J.D., Spatial determinants in morphogenesis: Recovery from plasmolysis in the diatom *Ditylum*. *Cell Motil. Cytoskeleton*, 60, 2, 71–82, 2005.

[13.289] Poulsen, N., Scheffel, A., Sheppard, V.C., Chesley, P.M., Kröger, N., Pentalysine clusters mediate silica targeting of silaffins in *Thalassiosira pseudonana*. *J. Biol. Chem.*, 288, 28, 20100–20109, 2013.

[13.290] Pratten, M.K. and Lloyd, J.B., Uptake of microparticles by rat visceral yolk sac. *Placenta*, 18, 7, 547–552, 1997.

[13.291] Ratcliffe, S., Jugdaohsingh, R., Vivancos, J., Marron, A., Deshmukh, R., Ma, J.F., Mitani-Ueno, N., Robertson, J., Wills, J., Boekschoten, M.V., Muller, M., Mawhinney, R.C., Kinrade, S.D., Isenring, P., Bélanger, R.R., Powell, J.J., Identification of a mammalian silicon transporter. *Am. J. Physiol.-Cell Physiol.*, 312, 5, C550–C561, 2017.

[13.292] Rayleigh, L., On the capillary phenomena of jets. *Proc. R. Soc.*, 29, 71–97, 1879.

[13.293] Rayleigh, L., On the instability of jets. *Proc. London Math. Soc.*, 10, 4–13, 1879.

[13.294] Rayleigh, L., On the instability of a cylinder of viscous liquid under capillary force. *Philos. Mag.*, 34, 145–154, 1892.

[13.295] Reeves, C.D. and Volcani, B.E., Role of silicon in diatom metabolism: Patterns of protein phosphorylation in *Cylindrotheca fusiformis* during recovery from silicon starvation. *Arch. Microbiol.*, 137, 4, 291–294, 1984.

[13.296] Reimann, B.E.F., Deposition of silica inside a diatom cell. *Exp. Cell Res.*, 34, 605–608, 1964.

[13.297] Reimann, B.E.F., Lewin, J.C., Volcani, B.E., Studies on the biochemistry and fine structure of silica shell formation in diatoms. I. The structure of the cell wall of *Cylindrotheca fusiformis* Reimann and Lewin. *J. Cell Biol.*, 24, 1, 39–55, 1965.

[13.298] Řezanka, T., Lukavský, J., Nedbalová, L., Kolouchová, I., Sigler, K., Effect of starvation on the distribution of positional isomers and enantiomers of triacylglycerol in the diatom *Phaeodactylum tricornutum*. *Phytochemistry*, 80, 17–27, 2012.
[13.299] Rines, J., Plankton Theater, *Bacillaria paxillifera*, a motile, colonial pennate diatom, Narragansett Bay, Rhode Island, Fall [movie], 2001. http://thalassa.gso.uri.edu/plankton/theater/alphabetical.htm [cobweb].
[13.300] Robinson, D.H. and Sullivan, C.W., How do diatoms make silicon biominerals? *Trends Biochem. Sci.*, 12, 4, 151–154, 1987.
[13.301] Roche, J.V. and Törnroth-Horsefield, S., Aquaporin protein-protein interactions. *Int. J. Mol. Sci.*, 18, 11, #2255, 2017.
[13.302] Roubeix, V. and Laviale, M., A free ride: Diatoms epiphytic on motile diatoms, in: *Diatom Gliding Motility [DIGM, Volume in the series: Diatoms: Biology & Applications, series editors: Richard Gordon & Joseph Seckbach]*, S.A. Cohn, K.M. Manoylov, R. Gordon (Eds.), Wiley-Scrivener, Beverly, MA, USA: I press, 2020.
[13.303] Round, F.E., Crawford, R.M., Mann, D.G., *The Diatoms, Biology & Morphology of the Genera*, Cambridge University Press, Cambridge, 1990.
[13.304] Sabuncu, A.C., Gordon, R., Richer, E., Manoylov, K.M., Beskok, A., The kinematics of explosively jerky diatom motility: A natural example of active nanofluidics, in: *Diatom Gliding Motility [Volume 3 in the series: Diatoms: Biology & Applications, series editors: Richard Gordon & Joseph Seckbach]*, S.A. Cohn, K.M. Manoylov, R. Gordon (Eds.), pp. 33–64, Wiley-Scrivener, Beverly, MA, USA: In press, 2020.
[13.305] Safonova, T.A., Annenkov, V.V., Chebykin, E.P., Danilovtseva, E.N., Likhoshway, Y.V., Grachev, M.A., Aberration of morphogenesis of siliceous frustule elements of the diatom *Synedra acus* in the presence of germanic acid. *Biochem. (Moscow)*, 72, 11, 1261–1269, 2007.
[13.306] Sannomiya, T., Arai, Y., Nagayama, K., Nagatani, Y., Transmission electron microscope using a linear accelerator. *Phys. Rev. Lett.*, 123, 15, #150801, 2019.
[13.307] Sapriel, G., Quinet, M., Heijde, M., Jourdren, L., Tanty, V., Luo, G.Z., Le Crom, S., Lopez, P.J., Genome-wide transcriptome analyses of silicon metabolism in *Phaeodactylum tricornutum* reveal the multilevel regulation of silicic acid transporters. *PLoS One*, 95400J 4, 10, #e7458, 2009.
[13.308] Savoia, R., Memmolo, P., Merola, F., Miccio, L., D'Ippolito, G., Fontana, A., Ferraro, P., Full 3D morphology of diatoms flowing in a microfluidic channel by digital holographic microscopy. *Proc. SPIE*, 9540, 2015.
[13.309] Scheffel, A., Poulsen, N., Shian, S., Kröger, N., Nanopatterned protein microrings from a diatom that direct silica morphogenesis. *Proc. Natl. Acad. Sci. U. S. A.*, 108, 8, 3175–3180, 2011.
[13.310] Schlayer, S., Pusch, A.-K., Pielenz, F., Beckert, S., Peksa, M., Horch, C., Moschkowitz, L., Einicke, W.-D., Stallmach, F., X-nuclei NMR self-diffusion studies in mesoporous silica foam and microporous MOF CuBTC. *Materials*, 5, 4, 617–633, 2012.
[13.311] Schmid, A.M.M., Schalenmorphogenese in Diatomeen/Valve morphogenesis in diatoms, in: *Diatomeen I, Schalen in Natur und Technik/Diatoms I, Shells in Nature and Technics*, K. Bach and B. Burkhardt (Eds.), pp. 300–317, Cramer Verlag, Stuttgart, 1984.
[13.312] Schmid, A.M.M., Organization and function of cell structures in diatoms and their morphogenesis, in: *Proceedings of the 8th International Diatom Symposium*, M. Ricard, (ed.), Paris, Aug, 1986, O. Koeltz, Koenigstein, pp. 271–292, 1986.
[13.313] Schmid, A.M.M. and Schulz, D., Wall morphogenesis in diatoms: Deposition of silica by cytoplasmic vesicles. *Protoplasma*, 100, 3–4, 267–288, 1979.
[13.314] Schnepf, E., Deichgräber, G., Drebes, G., Morphogenetic processes in *Attheya decora* (*Bacillariophyceae, Biddulphiineae*). *Plant Syst. Evol.*, 135, 3–4, 265–277, 1980.

[13.315] Schnepf, E. and Drebes, G., The structure of the frustule of *Attheya decora* West (Bacillariophyceae, Biddulphiineae) with special reference to the organic compounds. *Br. Phycol. J.*, 12, 145–154, 1977.
[13.316] Schröder, H.C., Peroviae-Ottstadt, S., Rothenberger, M., Wiens, M., Schwertner, H., Batel, R., Korzhev, M., Müller, I.M., Müller, W.E.G., Silica transport in the demosponge *Suberites domuncula*: Fluorescence emission analysis using the PDMPO probe and cloning of a potential transporter. *Biochem. J.*, 381, 665–673, 2004.
[13.317] Schulz, D., Silicalemma membrane pattern and valve formation of a centric diatom, in: *9th International Congress on Electron Microscopy*, J.M. Sturgess, (ed.), Toronto, vol. 2, pp. 410–411, 1978.
[13.318] Schulz, D., Drebes, G., Lehmann, H., Jank-Ladwig, R., Ultrastructure of *Anaulus creticus* Drebes & Schulz with special reference to its reduced ocelli. *Eur. J. Cell Biol.*, 33, 1, 43–51, 1984.
[13.319] Schwarz, R. and Müller, W.D., Zur Kenntnis der Kieselsäuren. XIV. Die wasserlösliche Monokieselsäure. *Z. Anorg. Allg. Chem.*, 296, 1-6, 273–279, 1958.
[13.320] Shcherba, I.P., К вопросу о некоторых возможных формах переноса кремнияг в водных растворах и об условиях образования кремнезема [On the problem of some possible ways of silicon transfer in aqueous solutions and on conditions of silica formation] {Russian]. *Dokl. Akad. Nauk SSSR*, 177, 5, 1200–1203, 1967.
[13.321] Shimizu, K., Del Amo, Y., Brzezinski, M.A., Stucky, G.D., Morse, D.E., A novel fluorescent silica tracer for biological silicification studies. *Chem. Biol.*, 8, 11, 1051–1060, 2001.
[13.322] Shirokawa, Y., Karino, K., Mayama, S., Developmental plasticity and genotype-environment interactions influence valve morphology in the *Cyclotella meneghiniana* species complex (Bacillariophyceae). *Eur. J. Phycol.*, 47, 3, 245–253, 2012.
[13.323] Shishlyannikov, S.M., Zakharova, Y.R., Volokitina, N.A., Mikhailov, I.S., Petrova, D.P., Likhoshway, Y.V., A procedure for establishing an axenic culture of the diatom *Synedra acus* subsp. *radians* (Kutz.) Skabibitsch from Lake Baikal. *Limnol. Oceanogr.-Methods*, 9, 478–484, 2011.
[13.324] Shrestha, R.P. and Hildebrand, M., Evidence for a regulatory role of diatom silicon transporters in cellular silicon responses. *Eukaryot. Cell*, 14, 1, 29–40, 2015.
[13.325] Shrestha, R.P., Tesson, B., Norden-Krichmar, T., Federowicz, S., Hildebrand, M., Allen, A.E., Whole transcriptome analysis of the silicon response of the diatom *Thalassiosira pseudonana*. *BMC Genomics*, 13, #499, 2012.
[13.326] Sibelco, IOTA® High Purity Quartz, 2019, http://www.iotaquartz.com/.
[13.327] Simpson, T.L., Refolo, L.M., Kaby, M.E., Effects of germanium on the morphology of silica deposition in a freshwater sponge. *J. Morphol.*, 159, 3, 343–353, 1979.
[13.328] Simpson, T.L. and Vaccaro, C.A., An ultrastructural study of silica deposition in the freshwater sponge *Spongilla lacustris*. *J. Ultrastruct. Res.*, 47, 3, 296–309, 1974.
[13.329] Smetacek, V., Oceanography - The giant diatom dump. *Nature*, 406, 6796, 574–575, 2000.
[13.330] Sterrenburg, F.A.S., Studies on tube-dwelling *Gyrosigma* populations. *Diatom Res.*, 4, 1, 143–151, 1989.
[13.331] Stoermer, E.F., Pankratz, H.S., Bowen, C.C., Fine structure of the diatom *Amphipleura pellucida*. II. Cytoplasmic fine structure and frustule formation. *Am. J. Bot.*, 52, 10, 1067–1078, 1965.
[13.332] Stoermer, E.F., Pankratz, H.S., Bowen, C.C., Formation of frustule elements in the diatom *Amphipleura pellucida* [abstract]. *Am. J. Bot.*, 52, 6, 613, 1965.
[13.333] Su, Y., Lundholm, N., Ellegaard, M., The effect of different light regimes on diatom frustule silicon concentration. *Algal Res.-Biomass Biofuels Bioprod.*, 29, 36–40, 2018.

[13.334] Sullivan, C.W., Diatom mineralization of silicic acid. I. $Si(OH)_4$ transport characteristics in *Navicula pelliculosa*. *J. Phycol.*, 12, 4, 390–396, 1976.
[13.335] Sullivan, C.W., Diatom mineralization of silicic acid. II. Regulation of $Si(OH)_4$ transport rates during the cell cycle of *Navicula pelliculosa*. *J. Phycol.*, 13, 1, 86–91, 1977.
[13.336] Sullivan, C.W., Diatom mineralization of silicic acid. IV. Kinetics of soluble Si pool formation in exponentially growing and synchronized *Navicula pelliculosa*. *J. Phycol.*, 15, 210–216, 1979.
[13.337] Sullivan, C.W., Silicification by diatoms. *Ciba Found. Symp.*, 121, 59–89, 1986.
[13.338] Sullivan, C.W. and Volcani, B.E., Silicon in the cellular metabolism of diatoms, in: *Silicon and Siliceous Structures in Biological Systems*, T.L. Simpson and B.E. Volcani (Eds.), pp. 15–42, Springer-Verlag, New York, NY, USA, 1981.
[13.339] Sullivan, M.J. and Wear, D.J., A morphological study of the giant diatoms *Ardissonea formosa* and *Synedra bacillaris*. *Diatom Res.*, 10, 1, 179–190, 1995.
[13.340] Sumper, M., Biomimetic patterning of silica by long-chain polyamines. *Angew. Chem. Int. Ed.*, 43, 2251–2254, 2004.
[13.341] Sumper, M. and Brunner, E., Learning from diatoms: Nature's tools for the production of nanostructured silica. *Adv. Funct. Mater.*, 16, 1, 17–26, 2006.
[13.342] Sumper, M. and Brunner, E., Silica biomineralisation in diatoms: The model organism *Thalassiosira pseudonana*. *ChemBioChem*, 9, 8, 1187–1194, 2008.
[13.343] Sumper, M. and Lehmann, G., Silica pattern formation in diatoms: Species-specific polyamine biosynthesis. *Chembiochem*, 7, 9, 1419–1427, 2006.
[13.344] Sutton, J.N., Andre, L., Cardinal, D., Conley, D.J., de Souza, G.F., Dean, J., Dodd, J., Ehlert, C., Ellwood, M.J., Frings, P.J., Grasse, P., Hendry, K., Leng, M.J., Michalopoulos, P., Panizzo, V.N., Swann, G.E.A., A review of the stable isotope bio-geochemistry of the global silicon cycle and its associated trace elements. *Front. Earth Sci.*, 5, #Unsp 112, 2018.
[13.345] Sutton, J.N., Varela, D.E., Brzezinski, M.A., Beucher, C.P., Species-dependent silicon isotope fractionation by marine diatoms. *Geochim. Cosmochim. Acta*, 104, 300–309, 2013.
[13.346] Takahashi, R., Sato, S., Sodesawa, T., Nishida, H., Effect of Si content on diffusion coefficient of nickel nitrate in wet silica gel. *J. Ceram. Soc. Jpn.*, 109, 10, 840–845, 2001.
[13.347] Tantemsapya, N. and Meegoda, J.N., Estimation of diffusion coefficient of chromium in colloidal silica using digital photography. *Environ. Sci. Technol.*, 38, 14, 3950–3957, 2004.
[13.348] Tesson, B. and Hildebrand, M., Dynamics of silica cell wall morphogenesis in the diatom *Cyclotella cryptica*: substructure formation and the role of microfilaments. *J. Struct. Biol.*, 169, 1, 62–74, 2010.
[13.349] Tesson, B. and Hildebrand, M., Extensive and intimate association of the cytoskeleton with forming silica in diatoms: Control over patterning on the meso- and micro-scale. *PLoS One*, 5, 12, #e14300, 2010.
[13.350] Tesson, B. and Hildebrand, M., Characterization and localization of insoluble organic matrices associated with diatom cell walls: Insight into their roles during cell wall formation. *PLoS One*, 8, 4, #e61675, 2013.
[13.351] Tesson, B., Lerch, S.J.L., Hildebrand, M., Characterization of a new protein family associated with the silica deposition vesicle membrane enables genetic manipulation of diatom silica. *Sci. Rep.*, 7, #13457, 2017.
[13.352] Thamatrakoln, K., Alverson, A.J., Hildebrand, M., Comparative sequence analysis of diatom silicon transporters: Toward a mechanistic model of silicon transport. *J. Phycol.*, 42, 4, 822–834, 2006.
[13.353] Thamatrakoln, K. and Hildebrand, M., Analysis of *Thalassiosira pseudonana* silicon transporters indicates distinct regulatory levels and transport activity through the cell cycle. *Eukaryot. Cell*, 6, 2, 271–279, 2007.

[13.354] Thamatrakoln, K. and Hildebrand, M., Silicon uptake in diatoms revisited: A model for saturable and nonsaturable uptake kinetics and the role of silicon transporters. *Plant Physiol.*, 146, 3, 1397–1407, 2008.

[13.355] Thamatrakoln, K. and Kustka, A.B., When to say when: Can excessive drinking explain silicon uptake in diatoms? *BioEssays*, 31, 3, 322–327, 2009.

[13.356] Underwood, G.J.C. and Paterson, D.M., The importance of extracellular carbohydrate production by marine epipelic diatoms. *Adv. Bot. Res.*, 40, 183–240, 2003.

[13.357] Ussing, A.P., Gordon, R., Ector, L., Buczkó, K., Desnitskiy, A.G., VanLandingham, S.L., *The Colonial Diatom "Bacillaria paradoxa": Chaotic Gliding Motility, Lindenmeyer Model of Colonial Morphogenesis, and Bibliography, with Translation of O.F. Müller (1783), "About a peculiar being in the beach-water"*, p. 5, A.R.G. Ganter Verlag Kommanditgesellschaft, Ruggell, 2005.

[13.358] Villareal, T.A., McKay, R.M.L., Al-Rshaidat, M.M.D., Boyanapalli, R., Sherrell, R.M., Compositional and fluorescence characteristics of the giant diatom *Ethmodiscus* along a 3000 km transect (28 degrees N) in the central North Pacific gyre [Corrigendum: 54(10), 1852]. *Deep-Sea Res. Part I-Oceanogr. Res. Pap.*, 54, 8, 1273–1288, 2007.

[13.359] Volcani, B.E., Role of silicon in diatom metabolism and silicification, in: *Biochemistry of Silicon and Related Problems*, G. Bendz and I. Lindqvist (Eds.), pp. 177–204, Plenum Press, New York, NY, USA, 1978.

[13.360] von Stosch, H.A., Observations on *Bellerochea* and *Streptotheca*, including descriptions of three new planctonic diatom species. *Nova Hedwig. Beih.*, 54, Suppl., 113–166, 1977.

[13.361] von Stosch, H.A., Structural and histochemical observations on the organic layers of the diatom cell wall, in: *Proceedings of the 6th International Symposium on Recent and Fossil Diatoms*, R. Ross, (Ed.), O. Koeltz, Koenigstein, West Germany, pp. 231–252, 1981.

[13.362] Vrieling, E.G., Beelen, T.P.M., van Santen, R.A., Gieskes, W.W.C., Diatom silicon biomineralization as an inspirational source of new approaches to silica production. *J. Biotechnol.*, 70, 1–3, 39–51, 1999.

[13.363] Vrieling, E.G., Sun, Q.Y., Tian, M., Kooyman, P.J., Gieskes, W.W.C., van Santen, R.A., Sommerdijk, N., Salinity-dependent diatom biosilicification implies an important role of external ionic strength. *Proc. Natl. Acad. Sci. U. S. A.*, 104, 25, 10441–10446, 2007.

[13.364] Wang, X.H., Schloßmacher, U., Schroder, H.C., Müller, W.E.G., A milestone in biomineralization: From a structural assessment to a genetic concept of mineral formation. Or how the enzyme silicatein forms a solid silica rod, in: *Progress in Silicones and Silicone-Modified Materials*, vol. 1154, S.J. Clarson, M.J. Owen, S.D. Smith, M. Van Dyke, M. Brook, J. Mabry (Eds.), pp. 157–176, Amer Chemical Soc, Washington, 2013.

[13.365] Wang, X.H., Schroder, H.C., Wang, K., Kaandorp, J.A., Müller, W.E.G., Genetic, biological and structural hierarchies during sponge spicule formation: from soft sol-gels to solid 3D silica composite structures. *Soft Matter*, 8, 37, 9501–9518, 2012.

[13.366] Weber, B., Tranfield, E.M., Höög, J.L., Baum, D., Antony, C., Hyman, T., Verbavatz, J.-M., Prohaska, S., Automated stitching of microtubule centerlines across serial electron tomograms. *PLoS One*, 9, 12, #e113222, 2014.

[13.367] Wetherbee, R., Diatom Trail Formation [movie], http://www.botany.unimelb.edu.au/RW/media/trails.html, 2004.

[13.368] Wikipedia, Microtubule, 2019, https://en.wikipedia.org/wiki/Microtubule.

[13.369] Williams, R.B., Unusual motility of tube-dwelling pennate diatoms. *J. Phycol.*, 1, 4, 145–146, 1965.

[13.370] Willis, L., Page, K.M., Broomhead, D.S., Cox, E.J., Computational models of the formation of diatom pore occlusions. *Plant Ecol. Evol.*, 143, 3, 297–306, 2010.

[13.371] Windler, M., Bova, D., Kryvenda, A., Straile, D., Gruber, A., Kroth, P.G., Influence of bacteria on cell size development and morphology of cultivated diatoms. *Phycol. Res.*, 62, 4, 269–281, 2014.

[13.372] Windler, M., Leinweber, K., Bartulos, C.R., Philipp, B., Kroth, P.G., Biofilm and capsule formation of the diatom *Achnanthidium minutissimum* are affected by a bacterium. *J. Phycol.*, 51, 2, 343–355, 2015.

[13.373] Witkowski, A., Płociński, T., Grzonka, J., Zgłobicka, I., Bąk, M., Dąbek, P., Gomes, A.I., Kurzydłowski, K.J., Application of focused ion beam technique in taxonomy-oriented research on ultrastructure of diatoms [Chapter 6], in: *Diatoms: Fundamentals & Applications [DIFA, Volume 1 in the series: Diatoms: Biology & Applications, series editors: Richard Gordon & Joseph Seckbach]*, J. Seckbach and R. Gordon (Eds.), pp. 115–127, Wiley-Scrivener, Beverly, MA, USA, 2019.

[13.374] Wood, A.M., Lande, R., Fryxell, G.A., Quantitative genetic analysis of morphological variation in an Antarctic diatom grown at two light intensities. *J. Phycol.*, 23, 42–54, 1987.

[13.375] Wu, J., Lerotic, M., Collins, S., Leary, R., Saghi, Z., Midgley, P., Berejnov, S., Susac, D., Stumper, J., Singh, G., Hitchcock, A.P., Optimization of three-dimensional (3D) chemical imaging by soft x-ray spectro-tomography using a compressed sensing algorithm. *Microsc. Microanal.*, 23, 5, 951–966, 2017.

[13.376] Wysokowski, M., Bartczak, P., Żołtowska-Aksamitowska, S., Chudzińska, A., Piasecki, A., Langer, E., Bazhenov, V.V., Petrenko, I., Noga, T., Stelling, A.L., Ehrlich, H., Jesionowski, T., Adhesive stalks of diatom *Didymosphenia geminata* as a novel biological adsorbent for hazardous metals removal. *Clean-Soil Air Water*, 45, 11, #1600678, 2017.

[13.377] Xiao, F., Li, X.Y., Lam, K.M., Wang, D.S., Investigation of the hydrodynamic behavior of diatom aggregates using particle image velocimetry. *J. Environ. Sci.*, 24, 7, 1157–1164, 2012.

[13.378] Xu, B., Luo, C.-S., Liang, J.-R., Chen, D.-D., Zhuo, W.-H., Gao, Y.-H., Chen, C.-P., Song, S.-S., Cellular metabolic responses of the marine diatom *Pseudo-nitzschia* multiseries associated with cell wall formation. *Mar. Genomics*, 16, 29–38, 2014.

[13.379] Yamada, K., Yoshikawa, S., Ichinomiya, M., Kuwata, A., Kamiya, M., Ohki, K., Effects of silicon-limitation on growth and morphology of *Triparma laevis* NIES-2565 (Parmales, Heterokontophyta). *PLoS One*, 9, 7, #e103289, 2014.

[13.380] Yang, S.H., Park, J.H., Cho, W.K., Lee, H.S., Choi, I.S., Counteranion-directed, biomimetic control of silica nanostructures on surfaces inspired by biosilicification found in diatoms. *Small*, 5, 17, 1947–1951, 2009.

[13.381] Yermakova, L.N., Aleksandrova, T.A., Nuss, P.V., Vasserman, A.M., Kasaikin, V.A., Zezin, A.B., Kabanov, V.A., Conformation of ionogenic macromolecules adsorbed on the surface of particles of polysilicic acid sols. *Polym. Sci. USSR*, 27, 9, 2073–2080, 1985.

[13.382] Yun, W.B., Kirz, J., Sayre, D., Observation of the soft-x-ray diffraction pattern of a single diatom. *Acta Crystallogr. Sect. A*, 43, 131–133, 1987.

14

The Effects of Cytoskeletal Inhibitors on Diatom Valve Morphogenesis

Yekaterina D. Bedoshvili* **and Yelena V. Likhoshway**

Limnological Institute, Russian Academy of Sciences, Irkutsk, Russia

Abstract

The formation of diatom silica structures (valves and girdle bands) is controlled by the cytoskeleton. This mini-review covers the evidence for the role of microtubules and actin microfilaments in this process, as revealed by inhibiting these cytoskeletal elements in diatoms with various symmetry types including centric, as well as raphid and araphid pennates. The degree to which diatoms are affected by inhibitors varies between species; exact morphological abnormalities induced in the valve depend on the stage of valve formation during which inhibitors are administered. These abnormalities are documented and classified. Inhibitors of microtubule polymerization can significantly affect the structure of deposited silica, implying an effect on micromorphogenesis. We also discuss potential applications of cytoskeleton inhibitors in studying the interplay of microtubules and actin microfilaments during valve morphogenesis.

Keywords: Cytoskeleton, inhibitors, colchicine, paclitaxel, cytochalasin D, valve abnormalities

14.1 Introduction

The cytoskeleton, as its name implies, serves as a scaffolding of the cell [14.36] and consists of three major components: microtubules, actin microfilaments, and intermediate filaments. Although initially discovered in eukaryotes, precursor forms of cytoskeleton proteins are present in prokaryotes as well [14.20, 14.78]. Diatom algae, like virtually all eukaryotes, contain microtubules and actin microfilaments [14.54, 14.71]; intermediate filaments, generally confined to multicellular metazoans, have not yet been documented in this group. Since the shape of a diatom cell is determined mostly by its silica cell wall, the structural function of the cytoskeleton in generating cell shape may not be as important for them as it is for softer cells. However, the cytoskeleton remains involved in other important processes such as vesicular transport and morphogenesis of silica structures.

**Corresponding author*: bedoshvilied@list.ru

Vadim Annenkov, Joseph Seckback and Richard Gordon (eds.) Diatom Morphogenesis, (349–364) © 2022 Scrivener Publishing LLC

One of the approaches for studying the role of the cytoskeleton in these cellular processes is based on inhibiting cytoskeleton activity in living cells. Inhibitors of various intracellular processes have been used in studies of vesicular transport [14.41], enzymatic activity [14.26], and localization of cellular processes such as translation [14.30]. Due to the importance of cytoskeletal structures in cell division and movement, inhibition of cytoskeleton function has also been used for suppressing tumor growth [14.9]. Studies of diatom valve morphogenesis in the presence of cytoskeleton inhibitors have proven invaluable to understanding the role of the cytoskeleton in this process. Treating synchronized cultures and developing stages of diatoms with inhibitors have elucidated the role of microtubules and actin at different stages of morphogenesis [14.5, 14.16, 14.31].

Current studies are dedicated not to proving that cytoskeleton is involved in diatom valve formation but to documenting its exact functions at different morphogenesis stages and better understanding the mechanisms responsible for its regulation. Use of multiple inhibitors which target different elements of cytoskeleton makes it possible to study certain features of the cytoskeleton activity. Such understanding could potentially lead to control over diatom morphogenesis for nanotechnological applications. To best analyze cytoskeleton involvement in diatoms, it is important to investigate the effect of cytoskeleton inhibition on valve structure in a large number of species with different forms of valve morphology and symmetry since diatoms have such a wide range of shapes and sizes. Development of these methods in these specialized eukaryotes could further lead to a more general understanding of how genetically determined structures are built and controlled [14.34]. Single-cell morphogenesis is a widely neglected field, from prokaryotic Bacteria and Archaea, to eukaryotes, including ciliates and desmids, let alone diatoms. In this work, we present considerable data on the effect of cytoskeleton inhibitors on diatom morphogenesis and discuss the promise of this approach.

14.2 Cytoskeleton and Its Role in Cell Morphogenesis

Microtubules and actin microfilaments are dynamic structures which are kept in equilibrium by cytoskeleton-regulating proteins and intracellular conditions [14.3, 14.36, 14.26]. We review below some of the evidence that both of these cytoskeletal systems are important in the regulation of diatom morphogenesis.

While microtubules are ubiquitous among eukaryotes, there are several families of tubulin proteins that drive their function. In most cases, their functions are performed via polymerization and depolymerization of 25-nm hollow tubular structures formed by α/β-tubulin dimers. Besides α- and β-tubulin, there are six other tubulin families with secondary roles in microtubule functioning [14.22, 14.73]. In addition, microtubules can be regulated by posttranslational modifications and association with microtubule regulatory proteins and organizing centers [14.23, 14.56]. Polymerization/depolymerization equilibrium, microtubule branching and interactions with other structures are regulated by proteins binding to either any of various active sites in tubulin molecules [14.1, 14.63, 14.64, 14.76] or along the tubulin polymer [14.25, 14.75]. There are at least 36 families of proteins (a more exact number has not yet been calculated) associated with microtubules, one way or another, carrying out their growth-shrinkage, as well as branching, severing, and stabilization [14.25]. Three tubulin families (α-, β-, and γ-) and several tubulin-regulating proteins are known in diatoms [14.19, 14.22]. While the organized microtubule structure of the diatom spindle provided the first detailed evidence of the nature of interdigitated microtubules in the structures of cell division [14.38, 14.55, 14.72], the role of microtubules in organizing structures of silica deposition during diatom morphogenesis has been less clear.

Actin microfilaments (F-actin), another critical component of cellular organization, consist of polymers of actin molecules. As with microtubules, they are also regulated by a number of both assembly/disassembly and polymer binding proteins. Normally, the cell contains globular monomeric G-actin and its linear polymer, F-actin. Polymerization of G-actin into filaments is reversible and requires monovalent and bivalent cations [14.29, 14.50]. Polymeric F-actin is what serves as a part of cytoskeleton. Its functioning requires not only actin itself but also a number of proteins which regulate filament assembly, disassembly, and positioning. Approximately, 150 proteins are known to bind G-actin or actin filaments [14.21]. The number of actin genes in different organisms may vary from a single copy to 100 [14.4, 14.39]. Diatom genomes, as a rule, contain one or two copies of the actin gene [14.2]. Among the actin-regulating genes, several families of actin-related proteins (ARPs) have been found that significantly affect actin regulation and perform a variety of functions from membrane capping to cutting the actin filament; two gelsoline subfamilies performing both capping and dividing functions were also distinguishing characteristics for diatoms [14.42].

The involvement of cytoskeleton in morphogenesis is not unique to diatoms, so studying their role in diatoms will likely be valuable for understanding the process in other biological systems. Despite the large evolutionary distance between diatom algae and plants, mechanisms of morphogenetic control can be similar between these two groups. In plants, the cytoskeleton plays a major role in cell morphogenesis. For example, microtubules can help determine the direction of cell growth by orienting the depositing cellulose microfibrils during cell wall biogenesis [14.51] and affecting the polarization of cell elongation and cellular differentiation by affecting intracellular transport [14.27]. *Arabidopsis thaliana* trichomes, single cells with three or four branches on their apex, are used as a model for studying the role of cytoskeleton in defining cell shape [14.65, 14.66]. Cortical microtubules are known to affect the initial stages of trichome branching. Inhibition of microtubule polymerization with paclitaxel and oryzalin, as well as disruptive mutations in the genes encoding microtubule-associated proteins, causes trichomes to develop an abnormal number of branches [14.13, 14.14, 14.32, 14.33, 14.37, 14.44, 14.49]. Despite the mounting evidence for the concerted activity of microtubules and actin filaments during plant cell morphogenesis [14.53, 14.59, 14.60], the molecular mechanisms underlying and regulating their interaction are mostly unknown [14.70].

The morphogenesis of the frustule elements occurs inside the cell. Each valve and girdle band is synthesized in a separate specialized organelle—silica deposition vesicle (SDV; [14.57]). In diatoms with different symmetry types, valve morphogenesis starts from different structures (annulus for centrics or sternum for pennates) and occurs centrifugally. Silica ribs branch out from the annulus and sternum, the gaps between which are partially filled in with silica according to species-specific patterns [14.8], and after the mantle is formed and all openings and outgrowths take their place, the valve thickening or more complex vertical differentiation (for species with chamber valves) occurs [14.62]. Positions of microtubules and actin filaments during diatom valve morphogenesis were studied using fluorescent and laser microscopy with specific staining. It was shown that bundled microtubules concentrate around a forming proboscis as a supporting sleeve [14.74, 14.75] or a large structure with microtubule bundles radiating from the center of the forming valve [14.68, 14.69]. As for the actin filaments, they are typically located near an edge of the valve SDV, possibly controlling valve diameter [14.74, 14.75], are involved in the apical growth of thin siliceous outgrowths, and form a network between areolae [14.68].

14.3 Abnormalities of Diatom Valve Morphogenesis Induced by Cytoskeleton Inhibitors

Morphological effects of cytoskeleton inhibitors can be related to the exact mechanism of the inhibition. For example, colchicine (for microtubules) and cytochalasin D (for actin) can inhibit polymerization of their associated cytoskeletal filaments [14.12, 14.15, 14.18, 14.24, 14.48], thus producing structures that are forming under the shortage (or complete absence) of cytoskeleton. Paclitaxel [14.40, 14.58] and phalloidin [14.77] have an opposite effect, inhibiting depolymerization and thus creating an overabundance of microtubules and filaments.

Table 14.1 contains the results of works studying the effects of various cytoskeleton inhibitors on silica transport and morphogenesis in diatoms. Colchicine, a broad-range inhibitor, causes most and various abnormalities from lateral valve formation [14.7, 14.47] to changes in raphe morphology and disordered striae [14.5, 14.11, 14.16]. In studies by Blank and Sullivan [14.10, 14.11] most structural anomalies were related specifically to raphe structure and areolae rows. Curiously, in some cases, the effect of colchicine did not depend on its concentration. This is probably explained by its effect on cell division: during certain mitosis stages, incorrect microtubule activity has a catastrophic effect which arrests the division. Therefore, a number of undivided cells should be mostly determined by the number of cells which were at this stage during colchicine exposure. The number of cells with lateral valves exactly shows the number of mitotic cells in culture. Probably, if a high enough concentration of the colchicine stops cell division, then its increase will not affect the number of lateral valves. However, an inhibitor lethal dose can stop the polymerization of all microtubules and cause cell death.

Formation of lateral valves, i.e., a transition of valve face toward a side of the cell (or, in case of *Aulacoseira*, toward a side of a linear colony), can be explained by colchicine affecting the centrosome, which leads to chromosome nondisjunction and consequently blocks karyokinesis but valve morphogenesis still occurs. It is possible that lateral valve formation is restricted to diatoms with central symmetry, since it was never observed in pennates, perhaps due to mutations in the binding sites in pennate tubulin or alteration of the regulating machinery of microtubule dynamics. Possibly, more global differences in the location of the nucleus (and microtubule center) in centric and pennate diatoms during cell division can be significant for lateral valve formation.

It is known that there are three main classes of sites to which drugs generally bind with tubulin—the paclitaxel site, the Vinca domain, and the colchicine domain [14.28]. As microtubule regulation is complex and can be accomplished either by post-translational modification of tubulin and direct binding of regulatory proteins to microtubules or by more indirect post-translational modification of MAP [14.45], which can occur at several protein sites, decreasing or increasing the time of exposure to MAP. There are more sites for

Table 14.1 Effects of various cytoskeleton inhibitors on valve morphogenesis of several diatom species.

Species	Inhibitor	Effect	References
Microtubule inhibitors			
Aulacoseira islandica (O.Müller) Simonsen	Colchicine 5, 10, and 20 μg/ml	Formation lateral valve caused by non-separation of daughter nuclei Underdeveloped or missing spines	[14.7]
	Paclitaxel 0.1, 1, 10, and 100 μM	Spines of overcomplicated shape	
Navicula saprophila Lange-Bertalot & Bonik	Colchicine, 10^{-4}–10^{-7} M	Small decrease in silica consumption Small decrease in cell division rate Increased abundance of aberrant valves (almost unaffected by colchicine concentration)	[14.10, 14.11]
	Lumi-colchicine, 10^{-4}–10^{-7} M	Decrease in silica consumption Small decrease in cell division rate Slightly increased abundance of aberrant valves	
	Podophyllotoxin, 10^{-8}–10^{-4} M	Decrease in silica consumption Small decrease in cell division rate Increased abundance of aberrant valves	
	Picropodophyllotoxin, 10^{-8}–10^{-4} M	Decrease in silica consumption Long-term decrease in cell division rate Increased abundance of aberrant valves	
	N-isophenyl propyl carbamate, 10^{-3}–10^{-5} M	Decrease in silica consumption Decrease in cell division rate Increased abundance of aberrant valves (almost unaffected by N-IPC concentration)	

(*Continued*)

Table 14.1 Effects of various cytoskeleton inhibitors on valve morphogenesis of several diatom species. (*Continued*)

Species	Inhibitor	Effect	References
Pinnularia sp. Ehrenberg	Colchicine, 0.2%	Minimal effect on valve morphology; in some cases, intermittent raphe was observed	[14.16]
Surirella robusta Ehrenberg	Colchicine, 0.2%	Malformed valvar wings Malformed and shifted raphe	
Hantzschia amphyoxis (Ehrenberg) Grunow	Colchicine, 0.2%	Abnormal raphe structure, disordered rows of areolae	
Fragilaria radians (*Synedra acus* subsp. *radians*) (Kützing) D.M.Williams & Round	Colchicine, 0.01, 0.05, 0.1, 0.5 and 1 µg/ml	Structural anomaly abundance increases with colchicine concentrations Abnormal valve shape and displacement of rimoportula Irregular rows of areolae Grown-over areolae	[14.31]
	Paclitaxel 1, 2, 3, and 5 nM	Structural anomaly abundance increases with colchicine concentrations Abnormal valve shape and labiate process positioning Irregular rows of areolae Large holes in the valve Areolae with abnormal velum	[14.5]
Proboscia alata (Brightwell) Sundström	Oryzalin, 0.5 µM	Curved proboscis	[14.75]
Rhizosolenia setigera Brightwell	Oryzalin, 0.5 µM	Curved proboscis	[14.74]
Cyclotella cryptica Reimann, J.C.Lewin & Guillard	Colchicine 80 µg/ml	Formation of lateral valve	[14.47]
Stephanodiscus sp. Ehrenberg	Colchicine, 80 µg/ml	Formation of lateral valve	
Cyclotella cryptica	Colchicine, 80 µg/ml	Formation of lateral valve	[14.3]

(*Continued*)

Table 14.1 Effects of various cytoskeleton inhibitors on valve morphogenesis of several diatom species. (*Continued*)

Species	Inhibitor	Effect	References
Ditylum brightwelli (T.West) Grunow	Amiprophosmethyl, 3×10^{-6} M, 1.5×10^{-4} M, and 6×10^{-7} M	Formation of valves with two face parts	[14.35]
Thalassiosira eccentric (Ehrenberg) Cleve	Amiprophosmethyl, 10^{-6} and 10^{-4} %	Formation of lateral valve and multinuclear cells	[14.61]
Anomoeoneis sphaerophora Pfitzer 1871	Amiprophosmethyl, 10^{-6} and 10^{-4} %	Formation of internal valves	
Cyclotella cryptica	Colchicine, 50 μM	Disordered rows of areolae Shifted fultoportulae	[14.69]
	Oryzalin, 0.2 μM	Disordered rows of areolae Shifted fultoportulae	
Actin filament inhibitors			
Cyclotella cryptica	Cytochalasin D, 3 μM	Inhibition of fultoportulae formation Large holes in the valve	[14.69]
Hantzschia Grunow	Cytochalasin D, 25 μg/ml	Shifted raphe Increased distance between pores	[14.16]
Proboscia alata	Cytochalasin D, 0.1 μM	Proboscises very narrow or completely absent	[14.75]

binding the drugs with actin filaments—latrunculin A clamps the ATP binding site, suggesting that this compound inhibits actin polymerization by inhibiting nucleotide exchange [14.43], and phalloidin (and other actin filament stabilizers) bind in the gap between the two long-pitch F-actin strands, and it has been speculated that the connection between the strands might be the key to F-actin stabilization [14.46]. So, the resistance of some diatoms to microtubule and actin filament inhibitors has several possible causes, as protein mutations, changes in cytoskeleton dynamics resulting from tubulin and actin posttranslational modification, and altered expression or binding of cytoskeleton-regulatory proteins. However, in the absence of research of diatom cytoskeleton regulation, all of these assumptions are very speculative.

The results produced in our lab show that non-separation of daughter nuclei is not the only cause of lateral valve formation. Treating *Aulacoseira islandica* with cytochalasin D induced lateral valve formation just like colchicine treatment [14.7]. In this case, lateral valves had two centers of symmetry, which probably correspond to two unseparated valve faces (Figure 14.1). Confocal microscopy has shown that the cell had two nuclei, each under its center of

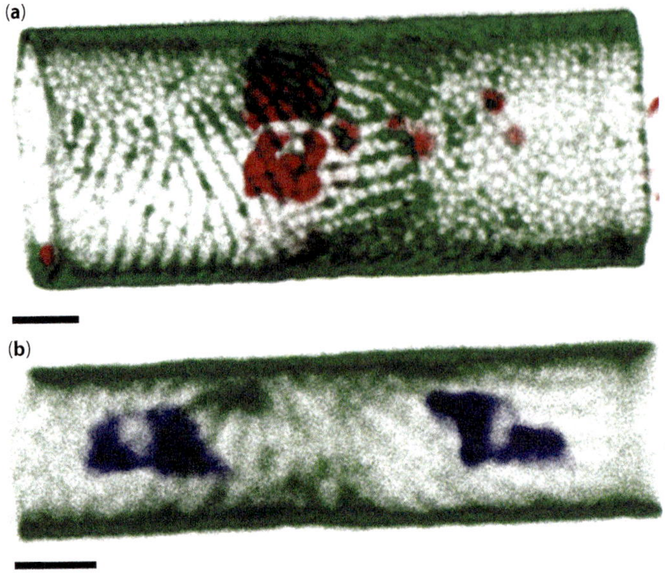

Figure 14.1 Lateral valves of *A. islandica* after cytochalasin D exposure (laser scanning microscopy). (a) Lysotracker Yellow valve staining and chloroplast autofluorescence; (b) Lysotracker Yellow valve staining and DAPI nucleus staining. Scale bar – 2 μm.

valve symmetry, although cytokinesis did not happen. It is known that cell segregation is induced by an actin ring which forms during cytokinesis and then tightens until the protoplast is separated into two daughter cells [14.67]. It appears that inhibition of actin polymerization prevents this ring from forming, thus blocking cytokinesis and leading to formation of the lateral valve. Thus, in *Aulacoseira* and hypothetically *Cyclotella* and *Stephanodiscus*, as well as *Ditylum brightwelli* [14.3, 14.7, 14.35, 14.47, 14.61], shortage or absence of correctly assembled microtubules and actin filaments induces abnormal valve production, but does not arrest morphogenesis altogether. The microtubule center determines the center of symmetry of the valve (for review, see [14.8]). Probably, the effect of colchicine stops mitosis and causes the formation of lateral valves in centric diatoms, including by blocking the separation of the microtubule center. However, as a center of symmetry, this structure continues to act, as well as when cytochalasin D is exposed to a cell.

Studies on synchronized cultures [14.5, 14.31] show that many morphological anomalies, such as displacement of the rimoportula and abnormal valve shapes, can be related to a shifted axial area in a forming valve. A short-term (30 min) colchicine exposure during areolae formation causes the areolae to be filled with silica, while paclitaxel exposure under the same conditions creates valves with large holes (some valves are even partially destroyed) and areolae with over-silicified velums.

Overall, the data suggests that microtubule inhibitors prevent microtubules from correctly performing their structural and transport functions without disabling them altogether, thus leading to finer morphological anomalies [14.7]. Even though microtubules are not completely eliminated, inhibitors often disrupt microtubule organization and structure enough to alter their effectiveness in directing cell organization. It was shown that a number of *A. islandica* valves carrying abnormal spines increases with concentrations of both colchicine and paclitaxel, although abnormalities themselves were opposite in nature since these two inhibitors have opposite effects. Colchicine treatment creates underdeveloped or altogether missing spines; paclitaxel exposure, on the other hand, leads to overcomplicated spine shapes. It is reasonable to assume that the degradation or elimination of microtubules likely leads to the inability to fully transport silicification material to where they need to be, and addition of taxol generates overabundance and disorganization of microtubule frameworks leading to too much and unorganized transport of silicification vesicles. Such results suggest that the addition of material to the enlarging SDV is an independent process from its organization of placement via the cytoskeleton.

The role of microtubules for material transport during diatom valve and girdle band morphogenesis, which was hypothesized earlier [14.52], has now been experimentally shown in two species of genus *Aulacoseira* (Figure 14.2; [14.7]). For *A. baicalensis* the experiment was performed in field during the bloom of this species in Lake Baikal in early June, 2014. The collected sample was treated with colchicine at the concentration of 20 μg/ml; after 12 hours of exposure, the material was fixed for scanning electron microscopy. In this case, just like with *A. islandica*, abnormal appearance of girdle bands on a surface of underdeveloped valves implies the shortage of silica in forming structures caused by microtubule inhibition. Paclitaxel, a microtubule depolymerization inhibitor, affects silica nanostructure as well (Figure 14.3). Normally (Figures 14.3a–c), virgae (first-order branchings) and vimini (second-order branchings) form sequentially [14.17]. As Figure 14.3c shows, vimini form along the entire length of each virga. Clearly visible nanostructure is ordered, with more density toward the center of the valve and early-forming areas [14.6]. Exposure of *Encyonema ventriculosum* to paclitaxel changes silica nanostructure

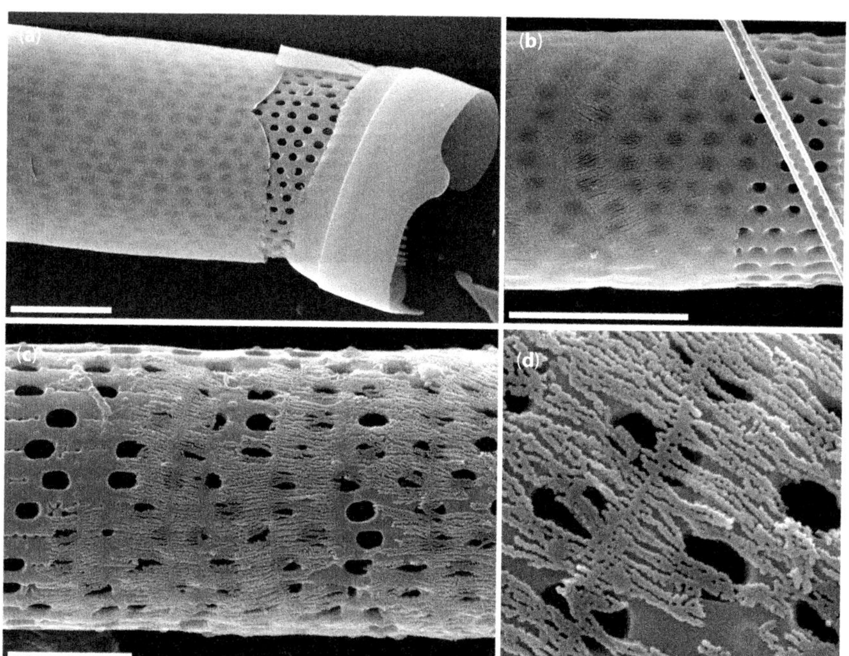

Figure 14.2 The structure of girdle bands on the valves of normal (a) and colchicine-treated (b–d) *A. baicalensis* cells (scanning electron microscopy). Scale bars: a, b – 10 μm; c – 5 μm; d – 1 μm.

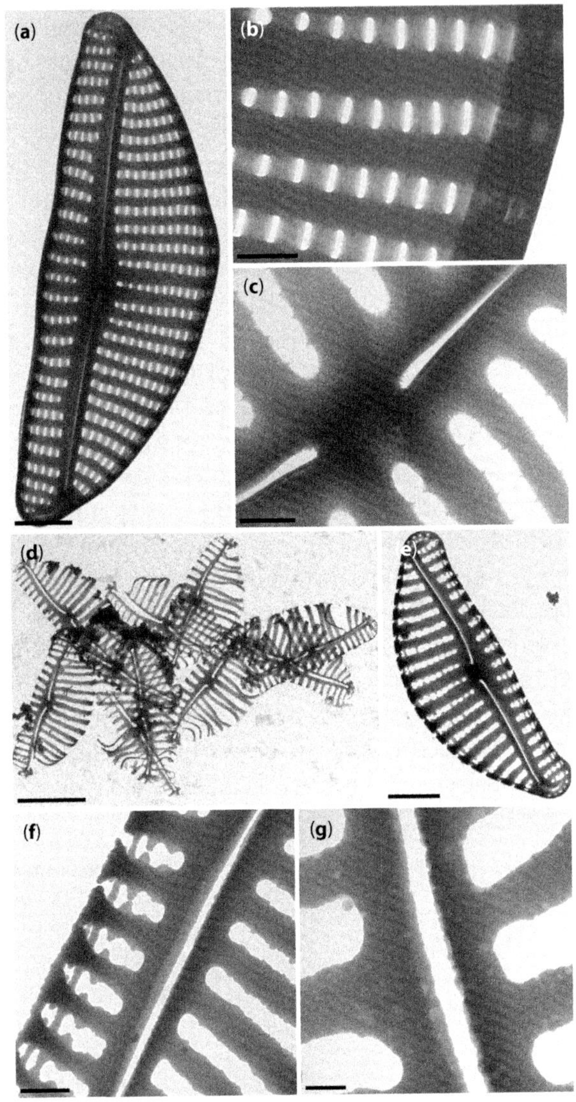

Figure 14.3 Valve ultrastructure in normal (a–c) and paclitaxel-treated (d–g) *E. ventricosum* cells. a, b – mature valve; c – valve during vimini formation; d – valves lacking virgae; e – a valve with partially formed mantle and missing virgae; f – a fragment of Figure 14.3e at greater resolution; g – a fragment of valve with missing virgae. Scale bars: a, e – 2 μm; b, c, f – 500 nm; d – 5 μm; g – 200 nm.

(Figure 14.3g) and blocks vimini formation in approximately 70% of the valves studied in transmission electron microscope (Figures 14.3d–f), causing mature valves to look like those at the vimini formation stage [14.6].

14.4 Conclusion

Based on experiments using cytoskeletal inhibitors, it is clear that the proper organization and density of cytoskeletal filaments is required for the normal morphogenesis of diatom

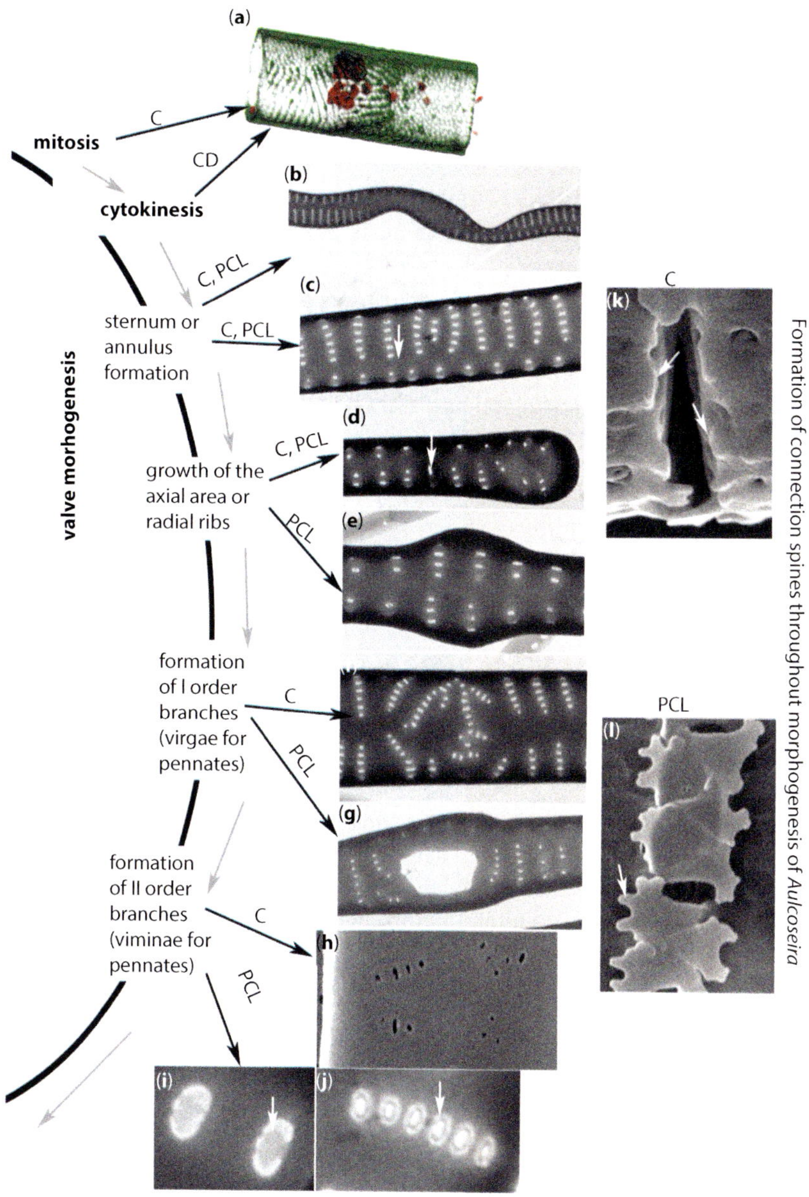

Figure 14.4 Influence of inhibitors with different mechanisms of action on valve morphogenesis by the example of *Aulacoseira* (a, k, l) and Fragilaria radians (b-j). a – lateral valve formed as a result of exposure to colchicine or cytochalasin D; b – curved valve of *F. radians*; c – Local displacement of the axial area; d – Displacement of the rimoportula; e – valve expansion; f – Misalignment of the rows of areolae; g – large opening in the valve; h – overgrown areoles; i – hyper-silicified velums on areolae; j – altered velum structure, "ring-in-ring"; k – valves with lack connecting spines; l – spines with abnormal structure. Legend: C – colchicine, CD – cytochalasin D; PCL – paclitaxel.

valves to occur (Figure 14.4). Diverse effects of these inhibitors can be explained by two major mechanisms: disruption of mitosis and cytokinesis lead to lateral valve formation and valves with two faces, while incorrect cytoskeleton activity during silica deposition leads to finer anomalies, including nanostructural ones.

Development of molecular biology produces new tools and broadens the range of possible experiments. Applications of cytoskeleton inhibitors for studying diatom morphogenesis may appear outdated, because many of them have a broad spectrum of effects and may be unspecific. In addition, their mechanisms of action are not perfectly clear. Despite this, the use of subtoxic inhibitor concentrations and studies on synchronized cultures or on cells observed to be at specific developmental stages, provides novel understanding of the role of cytoskeleton in diatom morphogenesis and paves the road to the creation of artificial siliceous nanostructures. We believe application of cytoskeletal inhibitors still has potential. We further believe that such molecular tools can be particularly useful if their use can be expanded for understanding additional diatom species with different forms of valve morphogenesis, use of synchronized cultures, and correlation of the inhibitors with the underlying modifications of cytoskeletal and ultrastructural organization.

Acknowledgment

This work was financially supported by the budget project of Limnological Institute, Siberian Branch of the Russian Academy of Sciences # 0279-2021-0008 in the Shared Research Facilities "Ultramicroanalysis".

References

[14.1] Akhmanova, A. and Steinmetz, M.O., Tracking the ends: a dynamic protein network controls the fate of microtubule tips. *Nat. Rev. Mol. Cell Biol.*, 9, 4, 309, 2008.

[14.2] Aumeier, C., *The Cytoskeleton of Diatoms Structural and Genomic Analysis*, Mathematisch-Naturwissenschaftlichen Fakultät Rheinischen Friedrich-Wilhelms-Universität Bonn, Germany (Doctoral dissertation), 2012.

[14.3] Badour, S.S., Experimental separation of cell division and silica shell formation in *Cyclotella cryptica. Arch. Mikrobiol.*, 62, 17–33, 1968.

[14.4] Baird, W. and Meagher, R., A complex gene superfamily encodes actin in petunia. *EMBO J.*, 6, 11, 3223–31, 1987.

[14.5] Bedoshvili, Y., Gneusheva, K., Likhoshway, Y., Changing of silica valves of diatom *Synedra acus* subsp. *radians* influenced by paclitaxel. *Tsitologia*, 59, 53–61, 2017, (In Russian).

[14.6] Bedoshvili, Y., Gneusheva, K., Popova, M., Avezova, T., Arsentyev, K., Likhoshway, Y., Frustule morphogenesis of raphid pennate diatom *Encyonema ventricosum* (Agardh) Grunow. *Protoplasma*, 255, 911–921, 2018b.

[14.7] Bedoshvili, Y., Gneusheva, K., Popova, M., Morozov, A., Likhoshway, Y., Anomalies in the valve morphogenesis of the centric diatom alga *Aulacoseira islandica* caused by microtubule inhibitors. *Biol. Open*, 7, bio035519, 2018a.

[14.8] Bedoshvili, Y. and Likhoshway, Y., *Cellular Mechanisms of Diatom Valve Morphogenesis. – Diatoms: Fundamentals & Applications [DIFA, Volume 1 in the series: Diatoms: Biology & Applications*, R. Gordon and J. Seckbach (Eds.), pp. 99–114, Wiley-Scrivener, Beverly, MA, USA, 2019.

[14.9] Blagg, J. and Workman, P., Choose and use your chemical probe wisely to explore cancer biology. *Cancer Cell*, 32, 9–25, 2017.

[14.10] Blank, G. and Sullivan, C., Diatom mineralization of silicon acid. VI. The effects of microtubule inhibitors on silicic acid metabolism in *Navicula saprophila*. *J. Phycol.*, 19, 39–44, 1983a.
[14.11] Blank, G. and Sullivan, C., Diatom mineralization of silicon acid. VII. Influence of microtubule drugs on symmetry and pattern formation in valves of *Navicula saprophila* during morphogenesis. *J. Phycol.*, 19, 294–301, 1983b.
[14.12] Borisy, G. and Taylor, E., The mechanism of action of colchicine. *J. Cell Biol.*, 34, 525–533, 1967.
[14.13] Burk, D.H., Liu, B., Zhong, R., Morrison, W.H., Ye, Z.H., A katanin-like protein regulates normal cell wall biosynthesis and cell elongation. *Plant Cell*, 13, 807–827, 2001.
[14.14] Buschmann, H., Hauptmann, M., Niessing, D., Lloyd, C.W., Schaffner, A.R., Helical growth of the Arabidopsis mutant tortifolia2 does not depend on cell division patterns but involves handed twisting of isolated cells. *Plant Cell*, 21, 2090–2106, 2009.
[14.15] Cassimeris, L., Plopper, G., Lingappa, V.R., *Lewin's Cells*, Jones & Bartlett Publishers, Sudbury, MA, 2011.
[14.16] Cohn, S., Nash, J., Pickett-Heaps, J., The effects of drugs on diatom valve morphogenesis. *Protoplasma*, 149, 130–143, 1989.
[14.17] Cox, E., Ontogeny, homology, and terminology — wall morphogenesis as an aid to character recognition and character state definition for pennate diatom systematics. *J. Phycol.*, 48, 1–31, 2012.
[14.18] Creighton, T., *The encyclopedia of molecular biology*, vol. 1–4, Wiley, New York, 1999.
[14.19] De Martino, A., Amato, A., Bowler, C., Mitosis in diatoms: rediscovering an old model for cell division. *Bioessays*, 31, 8, 874–84, 2009.
[14.20] Díaz-Celis, C., Risca, V.I., Hurtado, F., Polka, J.K., Hansen, S.D., Maturana, D., Lagos, R., Mullins, R.D., Monasterio, O., Bacterial tubulins A and B exhibit polarized growth, mixed-polarity bundling, and destabilization by GTP hydrolysis. *J. Bacteriol.*, 199, 19, e00211-17, 2017.
[14.21] Dos Remedios, C.G., Chhabra, D., Kekic, M., Tsubakihara, M., Berry, D.A., Nosworthy, N.J., Actin binding proteins: regulation of cytoskeletal microfilaments. *Physiol. Rev.*, 83, 2, 433–473, 2003.
[14.22] Findeisen, P., Mühlhausen, S., Dempewolf, S., Hertzog, J., Zietlow, A., Carlomagno, T., Kollmar, M., Six subgroups and extensive recent duplications characterize the evolution of the eukaryotic tubulin protein family. *Genome Biol. Evol.*, 6, 9, 2274–88, 2014.
[14.23] Garnham, C.P. and Roll-Mecak, A., The chemical complexity of cellular microtubules: tubulin post-translational modification enzymes and their roles in tuning microtubule functions. *Cytoskeleton*, 69, 7, 442–463, 2012.
[14.24] Goddette, D.W. and Frieden, C., Actin polymerization. The mechanism of action of cytochalasin D. *J. Biol. Chem.*, 261, 34, 15974–80, 1986.
[14.25] Goodson, H.V. and Jonasson, E.M., Microtubules and microtubule-associated proteins. *Cold Spring Harbor Perspect. Biol.*, 10, 6, a022608, 2018.
[14.26] Holdgate, G.A., Meek, T.D., Grimley, R.L., Mechanistic enzymology in drug discovery: a fresh perspective. *Nat. Rev.*, 17, 115–132, 2018.
[14.27] Hussey, P.J., Ketelaar, T., Deeks, M.J., Control of the actin cytoskeleton in plant cell growth. *Annu. Rev. Plant Biol.*, 57, 109–125, 2006.
[14.28] Jordan, M.A. and Wilson, L., Microtubules as a target for anticancer drugs. *Nat. Rev. Cancer*, 4, 253–265, 2004.
[14.29] Kabsch, W., Mannherz, H.G., Suck, D., Pai, E.F., Holmes, K.C., Atomic structure of the actin: DNase I complex. *Nature*, 347, 6288, 37, 1990.

[14.30] Kawashima, S.A., Chen, Z., Aoi, Y., Patgiri, A., Kobayashi, Y., Nurse, P., Kapoor, T.M., Potent, reversible, and specific chemical inhibitors of eukaryotic ribosome biogenesis. *Cell*, 167, 512–524, 2016.

[14.31] Kharitonenko, K., Bedoshvili, Y., Likhoshway, Y., Changes in the micro- and nanostructure of siliceous frustule valves in the diatom *Synedra acus* under the effect of colchicine treatment at different stages of the cell cycle. *J. Struct. Biol.*, 190, 1, 73–80, 2015.

[14.32] Kong, Z., Hotta, T., Lee, Y.R., Horio, T., Liu, B., The γ-tubulin complex protein GCP4 is required for organizing functional microtubule arrays in *Arabidopsis thaliana*. *Plant Cell*, 22, 191–204, 2010.

[14.33] Krishnakumar, S. and Oppenheimer, D.G., Extragenic suppressors of the Arabidopsis zwi-3 mutation identify new genes that function in trichome branch formation and pollen tube growth. *Development*, 126, 3079–3088, 1999.

[14.34] Kroth, P.G., Bones, A.M., Daboussi, F., Ferrante, M.I., Jaubert, M., Kolot, M., Nymark, M., Río Bártulos, C., Ritter, A., Russo, M.T., Serif, M., Winge, P., Falciatore, A., Genome editing in diatoms: achievements and goals. *Plant Cell Rep.*, 37, 10, 1401–1408, 2018.

[14.35] Li, C.-W. and Volcani, B.E., Aberrant valve formation in a centric diatom, *Ditylum brightwellii*. *Protoplasma*, 145, 1–6, 1988.

[14.36] Lodish, H., Berk, A., Kaiser, C., Krieger, M., Scott, M., Bretcher, A., Ploegh, H., Matsudaira, P., *Molecular cell biology*, Sixth ed., W.H. Freeman and Company, New York, 2007.

[14.37] Mathur, J., Spielhofer, P., Kost, B., Chua, N.-H., The actin cytoskeleton is required to elaborate and maintain spatial patterning during trichome cell morphogenesis in *Arabidopsis thaliana*. *Development*, 126, 5559–5568, 1999.

[14.38] McDonald, K.L., Pfister, K., Masuda, H., Wordeman, L., Staiger, C., Cande, W.Z., Comparison of spindle elongation *in vivo* and *in vitro* in *Stephanopyxis turris*. *J. Cell Sci.*, 5, 205–227, 1986.

[14.39] McDowell, J., Huang, S., McKinney, E., Structure and evolution of the actin gene family in *Arabidopsis thaliana*. *Genetics*, 142, 587–602, 1996.

[14.40] McGrogan, B.T., Gilmartin, B., Carney, D.N., McCann, A., Taxanes, microtubules and chemoresistant breast cancer. *Biochem. Boiphys. Acta*, 1785, 96–132, 2008.

[14.41] Mishev, K., Dejonghe, W., Russinova, E., Small molecules for dissecting endomembrane trafficking: a cross-systems view. *Chem. Biol.*, 20, 475–486, 2013.

[14.42] Morozov, A.A., Bedoshvili, Y.D., Popova, M.S., Likhoshway, Y.V., Novel subfamilies of actin-regulating proteins. *Mar. Genomics*, 37, 128–134, 2018.

[14.43] Morton, W.M., Ayscough, K.R., McLaughlin, P.J., Latrunculin alters the actin-monomer subunit interface to prevent polymerization. *Nat. Cell Biol.*, 2, 376–378, 2000.

[14.44] Nakamura, M. and Hashimoto, T., A mutation in the Arabidopsis gamma- tubulin-containing complex causes helical growth and abnormal microtubule branching. *J. Cell Sci.*, 122, 2208–217, 2009.

[14.45] Nehlig, A., Molina, A., Rodrigues-Ferreira., S., Honoré, S., Nahmias, C., Regulation of end-binding protein EB1 in the control of microtubule dynamics. *Cell. Mol. Life Sci.*, 74, 13, 2381–2393, 2017.

[14.46] Oda, T., Crane, Z.D., Dicus, C.W., Sufi, B.A., Bates, R.B., Dolastatin 11 connects two long-pitch strands in F-actin to stabilize microfilaments. *J. Mol. Biol.*, 328, 319–324, 2003.

[14.47] Oey, J.L. and Schnepf, E., Uber die AuslSsung der Valvenbildung bei der Diatomee *Cyclotella cryptica*. *Arch. Mikrobiol.*, 71, 199–213, 1970.

[14.48] Olmsted, J. and Borisy, G., Characterization of microtubule assembly in porcine brain extracts by viscometry. *Biochemistry*, 12, 4282–4289, 1973.

[14.49] Oppenheimer, D.G., Pollock, M.A., Vacik, J., Szymanski, D.B., Ericson, B., Feldmann, K., Marks, M.D., Essential role of a kinesin-like protein in Arabidopsis trichome morphogenesis. *PNAS*, 94, 6261–6266, 1997.

[14.50] Otterbein, L.R., Graceffa, P., Dominguez, R., The crystal structure of uncomplexed actin in the ADP state. *Science*, 293, 5530, 708–711, 2001.
[14.51] Paradez, A., Wright, A., Ehrhardt, D.W., Microtubule cortical array organization and plant cell morphogenesis. *Curr. Opin. Plant Biol.*, 9, 571–578, 2006.
[14.52] Parkinson, J., Brechet, Y., Gordon, R., Centric diatom morphogenesis: a model based on a DLA algorithm investigating the potential role of microtubules. *Biochim. Biophys. Acta*, 1452, 89–102, 1999.
[14.53] Petrasek, J. and Schwarzerova, K., Actin and microtubule cytoskeleton interactions. *Curr. Opin. Plant Biol.*, 12, 728–734, 2009.
[14.54] Pickett-Heaps, J.D., Cell division and morphogenesis of the centric diatom *Chaetoceros decipiens* (Bacillariophyceae) II. Electron microscopy and a new paradigm for tip growth. *J. Phycol.*, 34, 995–1004, 1998.
[14.55] Pickett-Heaps, J.D., McDonald, K.L., Tippit, D.H., Cell division in pennate diatom *Diatoma vulgare*. *Protoplasma*, 86, 205–242, 1975.
[14.56] Portran, D., Schaedel, L., Xu, Z., Théry, M., Nachury, M.V., Tubulin acetylation protects long-lived microtubules against mechanical ageing. *Nat. Cell Biol.*, 19, 4, 391–398, 2017.
[14.57] Reimann, B.E.F., Deposition of silica inside a diatom cell. *Exp. Cell Res.*, 34, 605–608, 1964.
[14.58] Rowinsky, E.K., Eisenhauer, E.A., Chaudhry, V., Arbuck, S.G., Donehower, R.C., Clinical toxicities encountered with paclitaxel (Taxol). *Semin. Oncol.*, 20, 1–15, 1993.
[14.59] Sambade, A., Findlay, K., Schaffner, A.R., Lloyd, C.W., Buschmann, H., Actin-dependent and -independent functions of cortical microtubules in the differentiation of *Arabidopsis* leaf trichomes. *Plant Cell*, 26, 1629–1644, 2014.
[14.60] Sampathkumar, A., Lindeboom, J.J., Debolt, S., Gutierrez, R., Ehrhardt, D.W., Ketelaar, T., Persson, S., Live cell imaging reveals structural associations between the actin and microtubule cytoskeleton in *Arabidopsis*. *Plant Cell*, 23, 2302–2313, 2011.
[14.61] Schmid, A.-M.M., Wall morphogenesis in diatoms: the anti-microtubule action of osmotic pressure. *Eur. J. Cell Biol.*, 20, 134, 1979.
[14.62] Schmid, A.-M. and Schulz, D., Wall morphogenesis in diatoms: deposition of silica by cytoplasmic vesicles. *Protoplasma*, 100, 267–288, 1979.
[14.63] Schroer, T.A., Microtubules don and doff their caps: dynamic attachments at plus and minus ends. *Curr. Opin. Cell Biol.*, 13, 1, 92–96, 2001.
[14.64] Schuyler, S.C. and Pellman, D., Microtubule "plus-end-tracking proteins": the end is just the beginning. *Cell*, 105, 4, 421–424, 2001.
[14.65] Smith, L.G., Cytoskeletal control of plant cell shape: getting the fine points. *Curr. Opin. Plant Biol.*, 6, 63–73, 2003.
[14.66] Szymanski, D.B., Breaking the WAVE complex: the point of *Arabidopsis* trichomes. *Curr. Opin. Plant Biol.*, 8, 103–112, 2005.
[14.67] Tanaka, A., De Martino, A., Amato, A., Montsant, A., Mathieu, B., Rostaing, P., Tirichine, L., Bowler, C., Ultrastructure and membrane traffic during cell division in the marine pennate diatom *Phaeodactylum tricornutum*. *Protist*, 166, 506–521, 2015.
[14.68] Tesson, B. and Hildebrand, M., Extensive and intimate association of the cytoskeleton with forming silica in diatoms: control over patterning on the meso- and micro-scale. *PLoS One*, 5, e14300, 2010a.
[14.69] Tesson, B. and Hildebrand, M., Dynamics of silica cell wall morphogenesis in the diatom *Cyclotella cryptica*: Substructure formation and the role of microfilaments. *J. Struct. Biol.*, 169, 62–74, 2010b.
[14.70] Tian, J., Han, L., Feng, Z., Wang, G., Liu, W., Ma, Y., Yu, Y., Kong, Z., Orchestration of microtubules and the actin cytoskeleton in trichome cell shape determination by a plant-unique kinesin. *eLife*, 4, e09351, 2015.

[14.71] Tippit, D.H., Mcdonald, K.L., Pickett-Heaps, J.D., Cell division in the centric diatom *Melosira varians*. *Cytobiologie*, 12, 1, 52–73, 1975.
[14.72] Tippit, D.H., Pickett-Heaps, J.D., Leslie, R., Cell division in two large pennate diatoms, Hantzschia and Nitzschia. III. A new proposal for kinetochore function during prometaphase. *J. Cell Biol.*, 86, 402–416, 1980.
[14.73] Turk, E., Wills, A.A., Kwon, T., Sedzinski, J., Wallingford, J.B., Stearns, T., Zeta-tubulin is a member of a conserved tubulin module and is a component of the centriolar basal foot in multiciliated cells. *Curr. Biol.*, 25, 16, 2177–2183, 2015.
[14.74] Van de Meene, A. and Pickett-Heaps, J., Valve morphogenesis in the centric diatom *Rhizosolenia setigera* (Bacillariophyceae, Centrales) and its taxonomic implications. *Eur. J. Phycol.*, 39, 93–104, 2004.
[14.75] Van de Meene, A. and Pickett-Heaps, J., Valve morphogenesis in the centric diatom *Proboscia alata* Sundström. *J. Phycol.*, 38, 351–363, 2002.
[14.76] Walczak, C.E. and Shaw, S.L., A MAP for bundling microtubules. *Cell*, 142, 364–367, 2010.
[14.77] Wehland, J., Osborn, M., Weber, K., Phalloidin-induced actin polymerization in the cytoplasm of cultured cells interferes with cell locomotion and growth. *PNAS*, 74, 12, 561–5617, 1977.
[14.78] Yutin, N. and Koonin, E.V., Archaeal origin of tubulin. *Biol. Direct*, 7, 1, 10, 2012.

15

Modeling Silicon Pools in Diatoms Using the Chemistry Toolbox

Argyro Spinthaki and Konstantinos D. Demadis*

Crystal Engineering, Growth and Design Laboratory, Department of Chemistry, University of Crete, Heraklion, Crete, Greece

Abstract

Diatoms are intriguing biomineralizing organisms, capable of transporting, processing and metabolizing various forms of "Si", thus taking part in the Earth's "Si" cycle. They admirably manage the basic (albeit complex) aspects of silica chemistry when they promptly form their cell wall using amorphous silica, starting with monosilicic acid ($Si(OH)_4$), at mild pH. In spite of extensive studies, the mechanisms for the collection, transport and stabilization of silicic acid from natural waters by diatoms, as well as the formation of their silica cell wall remain elusive. This chapter focuses on various chemical aspects of silicic acid polycondensation, particularly on its stabilization at very high concentrations, prior to biosilicification. The existence of specially designed compartments in the diatom, the so-called silicon pools has been proposed, without solid experimental proof. While the *in vitro* silicic acid stabilization at a circumneutral pH can reach up to 120 ppm (~ 2mM), various diatom species can stabilize almost x136 higher concentration (~ 340mM). We attempt to shed light on plausible stabilization mechanisms, by exploring bioinspired chemical approaches involving polyelectrolytes. These interact in various ways with silicic acid, resulting in its stabilization, and thus preventing its polycondensation.

Keywords: Diatoms, biosilicification, stabilization, hydrogen bond, electrostatic interactions, biopolymers, silica, silicic acid

15.1 Diatoms

Diatoms are widely known to the scientific community as an exceptional example of inspirational biomineralizing organisms [15.1, 15.2]. They play perhaps the most predominant role in the world's silicon cycle by metabolizing "Si" into silicon dioxide or silica (SiO_2) [15.3]. It is well-known that silicon oxides and silicates are present in argillaceous minerals, which contribute to soil fertility because they adsorb water and trace elements essential for the plants' nutrition [15.4–15.6]. Diatoms form their cell wall starting out from monosilicic acid [$Si(OH)_4$],

**Corresponding author*: demadis@uoc.gr

Vadim Annenkov, Joseph Seckback and Richard Gordon (eds.) Diatom Morphogenesis, (365–382) © 2022 Scrivener Publishing LLC

in a circumneutral pH environment, a procedure resulting in one of the most remarkable nanopatterned siliceous products, with incredible consistency and reproducibility, presumably due to genetic instructions involved [15.7, 15.8]. This inorganic construction is proposed to be surrounded by an organic coating, for prevention of SiO_2 dissolution [15.9]. It was suggested, quite early, that interaction of silicon species with the organic compounds of the coating is crucial for the biosilicification procedure and a thorough investigation of these moieties could lead to a better understanding of the entire process [15.10]. Since then, great efforts have been put forth aiming at revealing the precise role of the organic coating. An important study mentions that hydroxyl-containing biomolecules, such as modified peptides and proteins, do not play a significant role [15.11]. Diatoms utilize a series of "Si" transport proteins (SITs), for collection of silicic acid from their surrounding environment [15.12]. These proteins play an important role in "Si" collection, sequestration, and distribution because the concentration of silicic acid in natural waters is low [15.13–15.15]. After entering into the diatom, silicic acid is believed to be transported in a condensed form, within vesicular structures termed silicon transport vesicles (STVs) [15.16, 15.17].

A surprising fact regarding diatom's biosilicification process is its capability of performing controlled precipitation of silicon dioxide within minutes at ambient temperatures, in an extremely wide variety of nanostructured frameworks [15.18–15.20]. The chemical synthesis of such structures *in situ*, especially when starting from undersaturated silicic acid solutions, requires not only extreme temperatures but also high pressure conditions. The diatoms cell wall, or frustule, consists of two overlapping valves. During mitosis, the reproduction mechanism of an abundance of microorganisms including diatoms, fast formation of the second valve is required. This procedure is performed inside of a specially designed membrane vesicle called the silica deposition vesicle (SDV) [15.21].

15.2 "Silicon Pools" Biology

Due to lack of conclusive evidence, the scientific community has produced several hypotheses regarding the mechanisms of silicic acid stabilization used by diatoms. Most of them include the involvement of different types of post-translationally modified proteins, commonly known as silaffins (see Figure 15.1) [15.22–15.24], plus the key role of the so-called "silicon pool", although its existence has not been experimentally proven [15.25–15.27].

Nevertheless, within the diatom, levels of soluble silicic acid can reach up to about 20,000 ppm (340 mM) [15.28], much higher than the saturation concentration at pH-values below 9 of 60–120 ppm (1–2 mM) [15.29, 15.30]. Some studies involving nuclear magnetic resonance spectroscopy revealed that this soluble pool consists, if true, of both monosilicic acid, and various condensed silica-species [15.31–15.33].

15.3 Silica Particle Formation From Silicic Acid

Understanding the silicic acid [$Si(OH)_4$] polymerization (polycondensation) process could be proven crucial in revealing the secrets of the biosilicification mechanisms and has therefore become the subject of intense research [15.34–15.37]. Silicic acid autocondensation *in vitro*

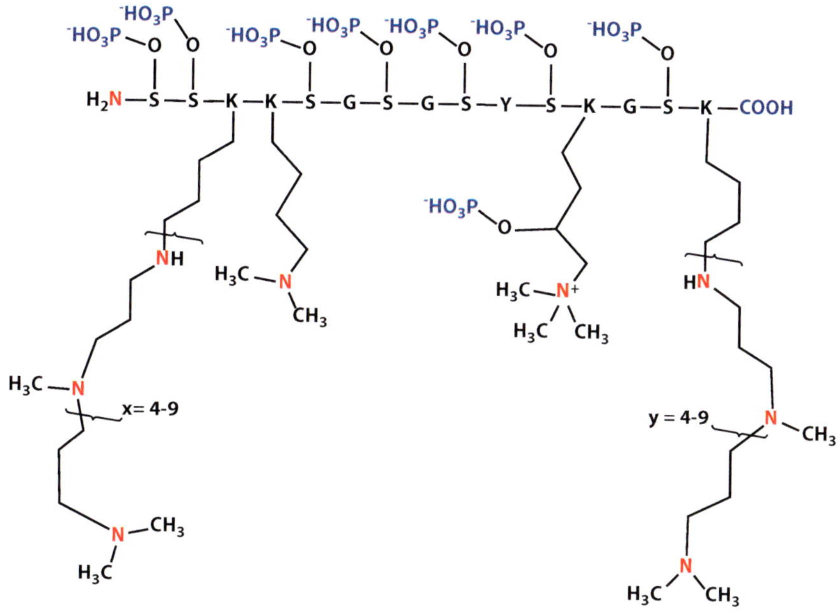

Figure 15.1 Schematic structure of native silaffin-1A1 from C. fusiformis. Cationic moieties are highlighted in red, anionic in blue.

and in the absence of externally added chemical compounds usually takes place under specific pH values (from 5 to 9) in supersaturated solutions. Its concentration must be at least 150 ppm (or ~2 mM) [15.38–15.40]. Hence, it is worthwhile to take a closer look at some significant details of its chemistry. In diluted aqueous solutions, the main soluble form of "Si" is monosilicic acid $Si(OH)_4$. This monomer exists in two forms in an about-neutral pH environment, the fully protonated form [the major form, $Si(OH)_4$)] and the monodeprotonated form [minor form, $Si(OH)_3O^-$] [15.41]. As pH increases, the concentration of the deprotonated form [$Si(OH)_3O^-$] also tends to increase. Increasing the silicate concentration and adjusting the pH value to ~7 leads to condensation of two monomers, with simultaneous loss of one water molecule. The procedure follows a S_{N2}-type mechanism, in which a deprotonated silicic acid molecule attacks a fully protonated one (see Figure 15.2) [15.42].

Disilicic acid (which is more acidic than monosilic acid) condenses with one molecule of monosilicic acid, to form a trimer. Continuation of this process will eventually lead to a silica nanoparticle of dimensions <2 nm [15.43]. Importantly, the surface of these silica nanoparticles contains silanol groups, which can be deprotonated at pH > 4, thus rendering the particle anionic.

The aforementioned step is the most crucial one as far as the kinetics of the polycondensation is concerned and is the rate-determining step in the complex silica polycondensation process [15.44].

It is worth-mentioning that the pKa of the polysilicic acids at these stages of the polymerization is about 6.5 and therefore negatively charged silicate species become predominant above pH 7–8 [15.45]. In these conditions, silicic acid on the surface of colloidal species exists in equilibrium with dissolved silicic acid molecules and, thus, further particle growth occurs via the Ostwald ripening process, leading to stable sols, see Figure 15.3.

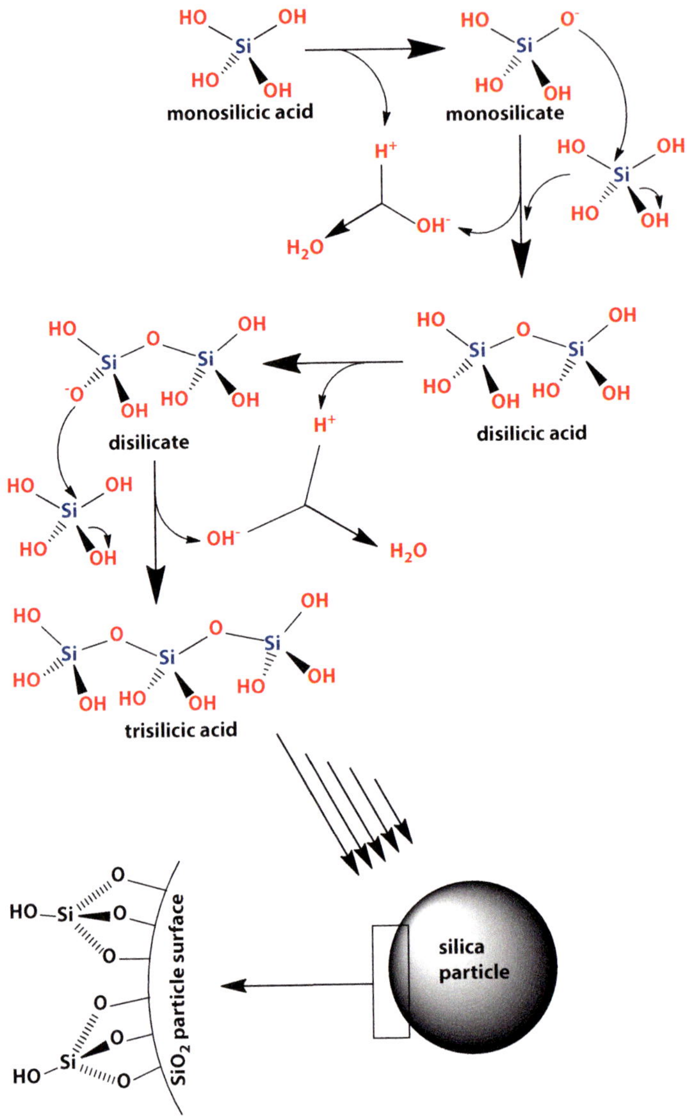

Figure 15.2 S_{N2}-like mechanism of silicic acid polycondensation.

In contrast, below pH 7, particles are only slightly charged so no electrostatic repulsion prevents them from aggregation, leading to gel formation [15.46].

The pH dependence of silicic acid polymerization was measured at 500 ppm starting concentration and the results are presented in Figure 15.4. The pH region where silicic acid polycondensation is fastest and more pronounced is 7–8. Lower or higher pH values cause slower autocondensation. This can be explained based on the S_{N2}-type mechanism, shown in Figure 15.2. At pH < 7, the neutral, protonated form of silicic acid [$Si(OH)_4$] is dominant, hence nucleophilic attack between two monomers to form the dimer is suppressed. At pH > 8, the monodeprotonated form of silicic acid [silicate, $Si(OH)_3O^-$] is the dominant species,

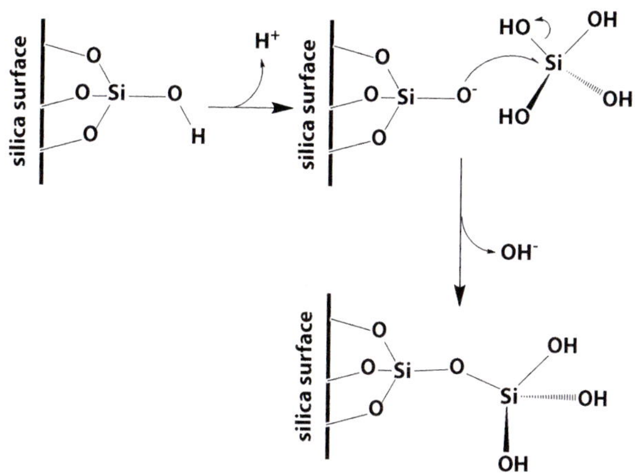

Figure 15.3 Monomeric silicic acid incorporation on the surface of a silica particle.

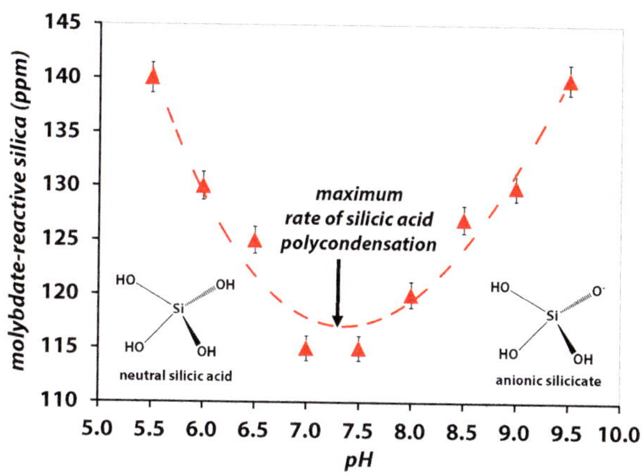

Figure 15.4 Dependence of silicic acid polymerization on solution pH.

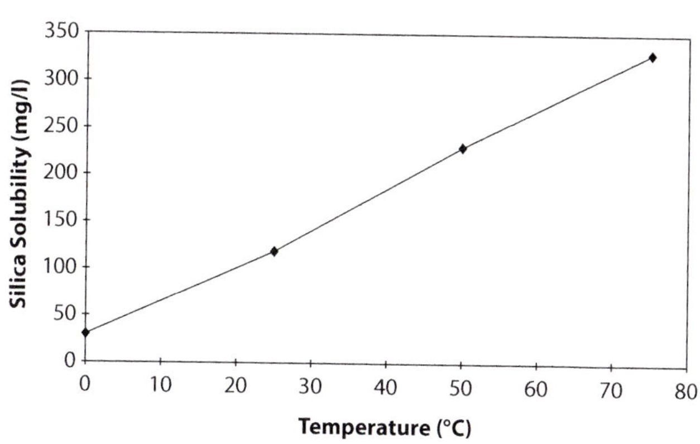

Figure 15.5 Dependence of silicic acid polymerization on temperature.

hence there is electrostatic repulsion between the anionic monomers, which, again, results in reduced tendency for nucleophilic attack.

Lastly, the dependence of silica solubility on temperature appears to be linear, ie., as temperature increases, silica solubility increases as well (see Figure 15.5). This is in contrast to other inorganic biomaterials, such as $CaCO_3$, which demonstrate inverse solubility features.

15.4 Stabilization of "Soluble" Silica Species (Monosilicic and Disilicic Acids)

Frequently, there is a misconception about the use of the terms "silica", "silicate", and "silicic acid". In this chapter the term "silica" indicates the product of silicic acid polycondensation. The term "silicate" indicates all the forms of the species $Si(OH)_4$ at various deprotonated states $[Si(OH)_{4-x}(O)_x]^{x-}$, x = 0, 1, 2, 3, 4; 4 indicates that silicic acid is fully deprotonated] which exist at pH regions > 9 [15.41]. Note that mono- and disilicic acid rapidly interconvert and they both contribute to molybdate-reactive silica. Molybdate-reactive silica can be reliably quantified using the well-established silicomolybdate spectrophotometric method with a reported precision of ± 5% [15.47–15.51]. Hence, the term "soluble silica" refers to "molybdate-reactive silica", which for all practical purposes includes mono- and disilicic acids.

Numerous studies have provided valuable information on the complex interplay between organic compounds and different silica forms, ranging from orthosilicic (monosilicic) acid to colloidal amorphous silica. To develop a deeper understanding of the complexity of the *in vivo* systems, however, an *in vitro* modeling approach involving tailored polymeric compounds is desirable. These macromolecules can be fine-tuned in terms of several variables, such as charge-distribution and structural features of the monomeric units. The polymeric molecular architecture and charge can dictate the type of interactions between polymer and silicic acid. Hence, it is reasonable to approach silicic acid stabilization based on the charge of the polymers. It is noted that there are numerous reports on polymer induced formation of silica (from appropriate "Si" precursors). These will not be presented herein, because their action as "organic catalysts" is outside the scope of this chapter. The reader is referred to an excellent recent review [15.52].

15.4.1 Cationic Polymers

Polycationic compounds were found to retard the condensation of silicic acid when added at low dosage (up to around 100 ppm) to supersaturated solutions of silicic acid (500 ppm, ~8 mm). Examples of such compounds are poly(1-vinylimidazole) [15.53], amine-terminated polyaminoamidedendrimers (PAMAMs) [15.54, 15.55], poly(acrylamide-*co*-diallyldimethylammoniumchloride) [15.56], polyallylamine hydrochloride, [15.57], polyethyleneimine (branched and linear) [15.58], cationically modified inulin [15.59], and phosphonium-grafted polyethylene bolaamphiphiles [15.60]. The schematic structures of these compounds are shown in Figure 15.6.

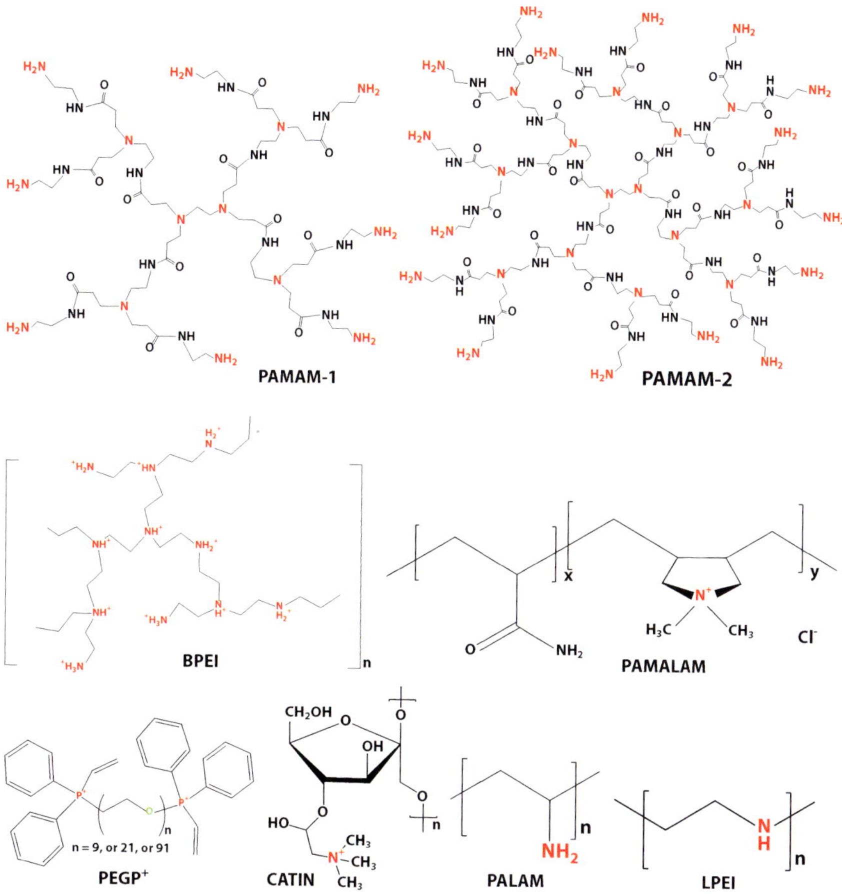

Figure 15.6 Schematic chemical structures of cationic silicic acid stabilizers. Abbreviations: PAMAM-1, amine-terminated polyaminoamide dendrimer of generation 1; PAMAM-2, amine-terminated polyaminoamide dendrimer of generation 2; BPEI, branched polyethyleneimine; PAMALAM, poly(acrylamide-co-diallyl-dimethylammonium chloride); $PEGP^+$, phosphonium-grafted polyethyleneglycol; CATIN, cationic inulin; PALAM, polyallylamine; LPEI, linear polyethyleneimine. Positively charged moieties are highlighted in red.

Data collected from silicic acid stabilization experiments are compiled in Figure 15.7 and refer to measurements of molybdate-reactive silica after 2 hours. This particular time was selected so it is relevant to diatom biosilicification chemistry, which occurs within 1–2 hours during mitosis. Of the four quadrants in the graph, the important one is the top left. This means high silicic acid stabilization efficiency, with the lowest possible stabilizer concentration.

The most efficient stabilizer is $PEGP^+$ (with phosphonium cationic groups), which quantitatively stabilizes 500 ppm (8.3 mM) molybdate-reactive silica at concentration as low as 40 ppm. Then, the two dendrimers PAMAM-1 and PAMAM-2 (with 8 and 16 peripheral cationic amine groups, respectively) can stabilize 450 ppm silica at concentration 20 ppm. Lastly, PAMALAM (with organoammonium cationic groups) can stabilize ~430 ppm silicic acid, at 60 ppm concentration.

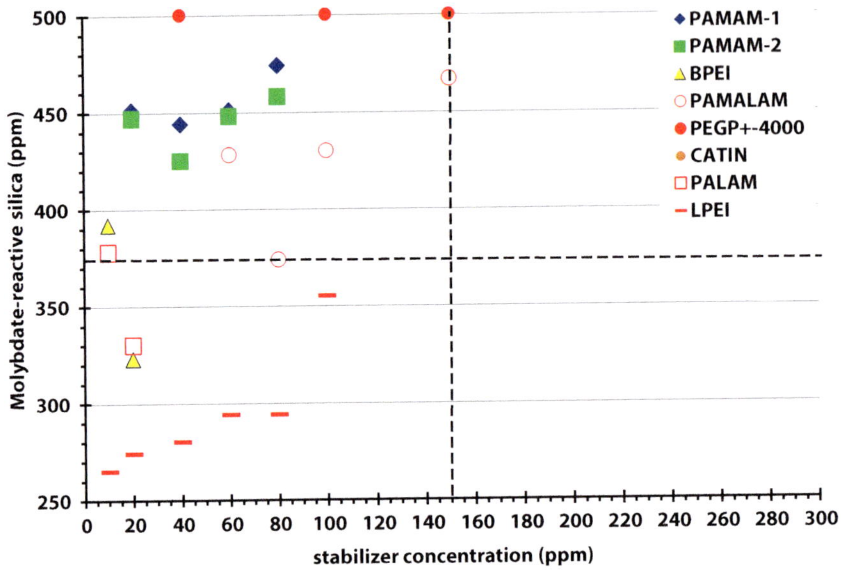

Figure 15.7 Ability of cationic polymers to stabilize silicic acid (initial concentration 500 ppm, 8.3 mM) after 2 hours of polycondensation time.

15.4.2 Neutral (Uncharged) Polymers

Examples of studied neutral polymers are polyethylene glycol (PEG) [15.61], polyethyloxazoline [15.62, 15.63], and polyvinylpyrrolidone (PVP) [15.64]; see Figure 15.8. Presumably, the stabilization of soluble mono- or di-silicic acids results from

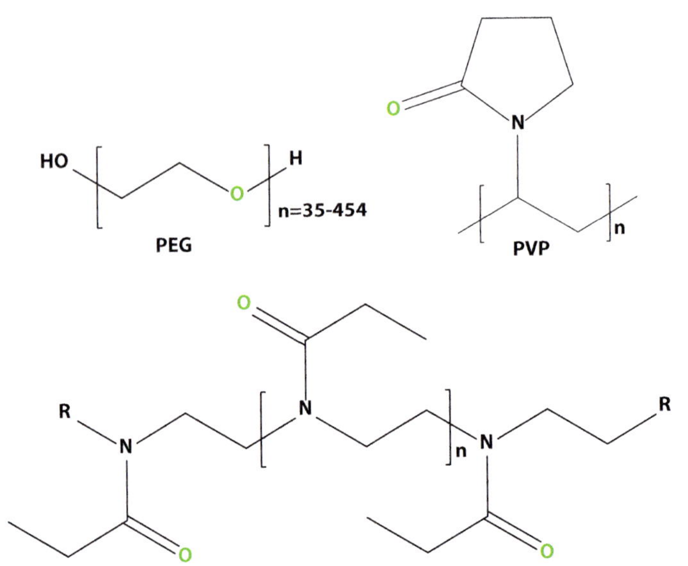

Figure 15.8 Schematic chemical structures of neutral (uncharged) silicic acid stabilizers. Abbreviations: PEG, polyethyleneglycol; PVP, polyvinylpyrrolidone; PEOX, polyethyloxazoline. Neutral moieties are highlighted in green.

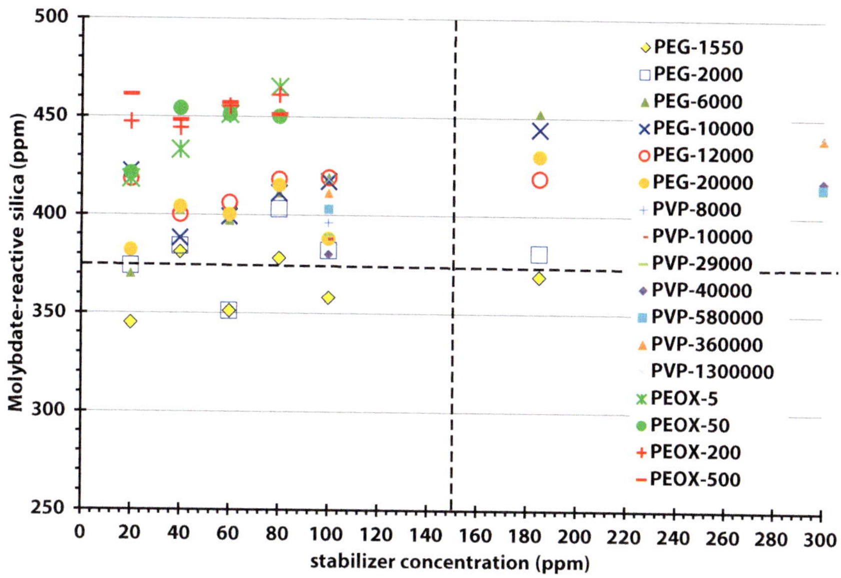

Figure 15.9 Ability of neutral (uncharged) polymers to stabilize silicic acid (initial concentration 500 ppm, 8.3 mM) after 2 hours of polycondensation time.

hydrogen-bonding or electrostatic interactions (or a combination of these), which retard the silicic acid polycondensation. This mechanism will be thoroughly explained in the following part (Section 15.5).

Of the multitude of stabilizers tested (see Figure 15.9), the most efficient are of the polyethyloxazoline family (PEOX). PEOX-500 (the number 500 indicates the monomer repeating units) stabilizes ~460 ppm silica at 20 ppm concentration. Lowering the molecular weight of the PEOX slightly reduces its efficiency, with PEOX-5 stabilizing ~420 ppm silica at 20 ppm concentration. PEGs are less effective than PEOXs. The "shorter" PEG (PEG-1550, the number indicating the polymeric chain molecular weight) appears in the lower left quadrant. Molecular weight dependence is similar to that of the PEOXs. Hence, PEG-12000 can stabilize ~420 ppm silica at 20 ppm concentration. There is a systematic increase in efficiency as PEG concentration increases. Finally, the PVP family demonstrates lower efficiency, and increased stabilizer concentrations are needed. For example, the "shorter" PVPs (PVP-8000, PVP-10000, and PVP-29000, numbers indicate polymer chain molecular weight) can stabilize a little less than 400 ppm silica, albeit at 100 ppm concentration.

15.4.3 Zwitterionic Polymers

Zwitterionic polymers tested are presented in Figure 15.10. These include phosphonomethylated polyethyleneimine [15.65, 15.66], chitosan-based macromolecules [15.67, 15.68], the polypeptide P5S3 [15.69], and histidine-containing poly-MHist polyampholyte polymer [15.70]. They all seem to have a positive effect on the stabilization of silicic acid.

Specifically, among these zwitterions, one can find some efficient silicic acid stabilizers, see Figure 15.11. For example, the histidine-containing poly-MHist polymer shows a linear dosage-dependence, with the lowest concentration of 10 ppm being capable

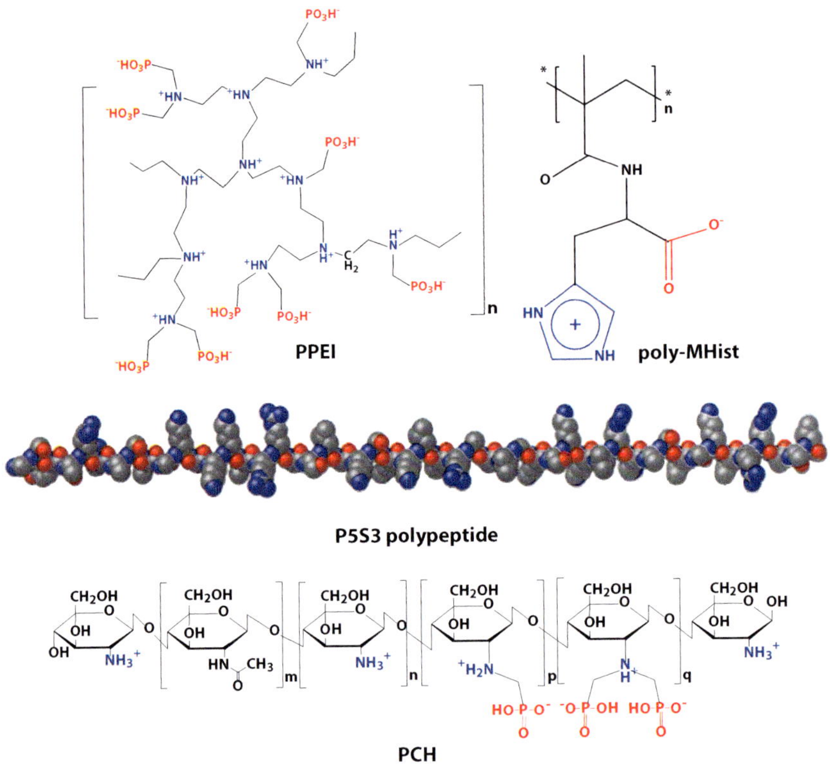

Figure 15.10 Schematic chemical structures of zwitterionic silicic acid stabilizers. Abbreviations: PPEI, phosphonomethylatedpolyethyleneimine; P5S3 polypeptide, H_2N-GSYS**RR**ASLG[**KSKK**L**RR**ASLG]$_3$ **K**L**RR**ASL-COOH; PCH, phosphonomethylated chitosan. Positively charged moieties are highlighted in blue and anionic groups in red.

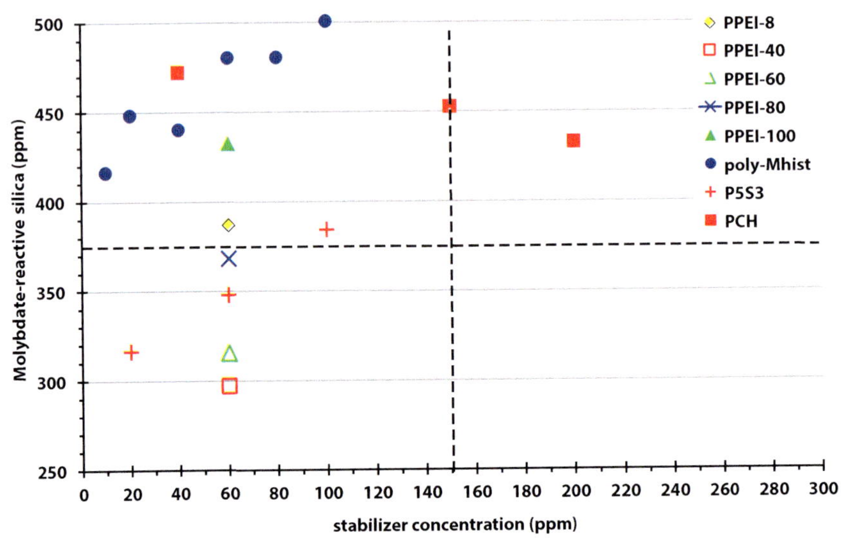

Figure 15.11 Ability of zwitterionic polymers to stabilize silicic acid (initial concentration 500 ppm, 8.3 mM) after 2 hours of polycondensation time.

of stabilizing ~420 ppm silica. If concentration increases to 100 ppm, then it stabilizes 500 ppm silica quantitatively. Phosphonated polyethylemeimines of varying phosphonomethylation degrees (8%, up to 100%) were also evaluated. PPEI-100 is the most efficient, stabilizing ~430 ppm silica, at 60 ppm concentration. Phosphonomethylated chitosan (PCH) can stabilize ~470 ppm silica at 40 ppm concentration. Finally, the polypeptide P5S3 is less effective, with only 380 ppm stabilization efficiency at 60 ppm concentration.

15.4.4 Blends of Cationic/Anionic Polymers

There are cases in which the cationic polymeric stabilizer acts as a flocculant for silica nanoparticles that form, due to incomplete stabilization. The cause of this phenomenon is the extensive cationic charge on the polycation, which interacts with the negatively charged silica nanoparticles. Hence, in order to "relieve" the positive charge combinations (blends) of cationic/anionic polymers were sought [15.71, 15.72]. Such assemblies include the polycations shown in Figure 15.6 and polyanions shown in Figure 15.12 [15.73–15.75].

Various combinations of cationic-anionic stabilizer systems yield different performances, see Figure 15.13. These depend on several factors, such as (a) the degree of cationic or anionic charge, (b) the molecular weight of the polycation or the polyanion, (c) the chemical structure of the cation (e.g., branched vs. linear), and (d) the chemical nature of the cationic or anionic groups.

The most efficient systems are the ones involving the PAMAM dendrimers and an anionic add-on, e.g., polyacrylate or CMI. Then, combinations of branched polyethyleneimine (BPEI)

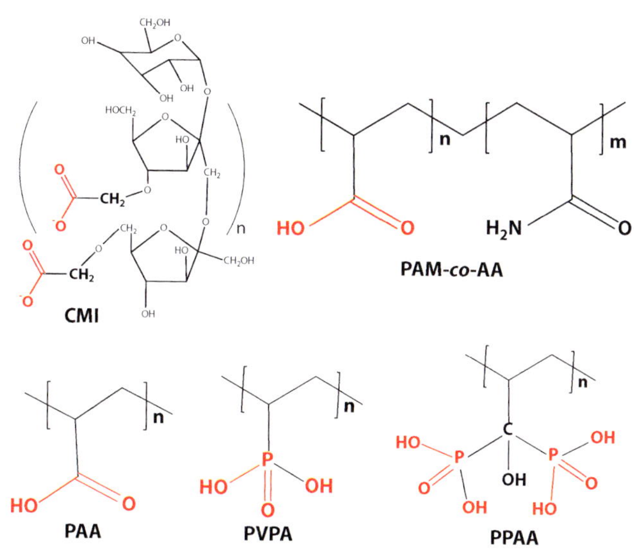

Figure 15.12 Schematic chemical structures of anionic polymers used in blends of silicic acid stabilizers. Abbreviations: CMI, carboxymethylated inulin; PAM-co-AA, copolymer of polyacrylamide and polyacrylic acid; PAA, polyacrylic acid; PVPA, poly(vinylphosphonic acid); PPAA, phosphonated polyacrylic acid. Negatively charged moieties are highlighted in red.

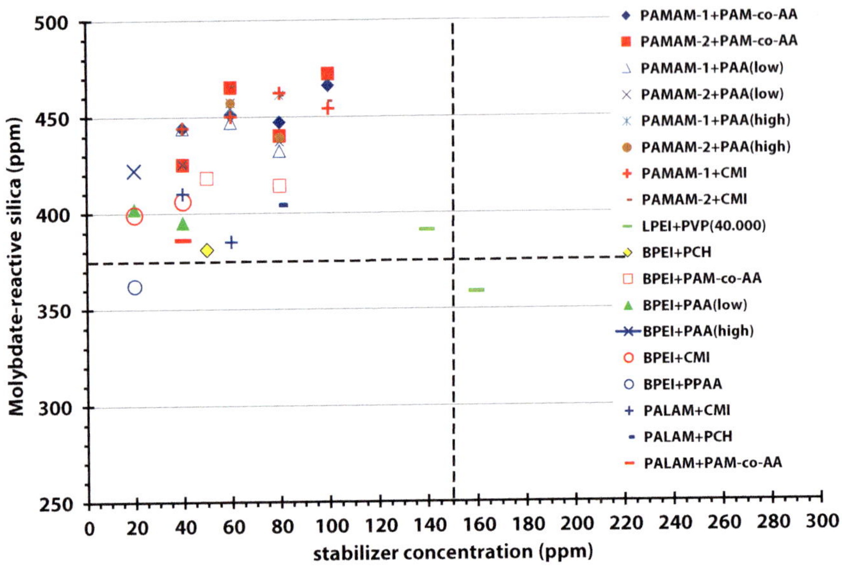

Figure 15.13 Ability of blends of cationic and anionic polymers to stabilize silicic acid (initial concentration 500 ppm, 8.3 mM) after 2 hours of polycondensation time. The stabilizer concentration indicates the total "polycation + polyanion" concentration.

with polyanions such as polyacrylate or polyacrylamide/polyacrylate co-polymer offer lower stabilizing activity.

15.5 Chemical Mechanisms

In the previous sections, we outlined the efficacy of a multitude of polymeric species (cationic, neutral, zwitterionic, and admixtures of cationic and anionic) on the stabilization of "soluble forms" of silica. As noted, these include principally monosilicic acid and, to a lesser extent, disilicic acid. Depending on the pH, water chemistry and other species present, silicic acid can interact via several interactions with the polymeric stabilizers. These are shown schematically in Figure 15.14.

The silanol group (Si-O-H) can participate in hydrogen bonding with appropriate moieties by being a hydrogen donor. Such interactions can be envisioned with chemical groups that possess one or more lone pairs of electrons. These include primary, secondary, tertiary amine nitrogens, ether oxygens, carbonyl oxygens, and N-heterocycle nitrogens. These are highlighted in red in the upper part of Figure 15.14. In addition, the monodeprotonated silanol group ($Si-O^-$) can form similar hydrogen bonds, but now acting as a proton acceptor from charged groups, such as protonated primary, secondary, and tertiary amine nitrogens. Lastly, the monodeprotonated silanol group ($Si-O^-$) can form ionic interactions (no hydrogen bonds) with quaternary orgaammonium or organophosphonium cations. These interactions are highlighted in green at the lower part of Figure 15.14.

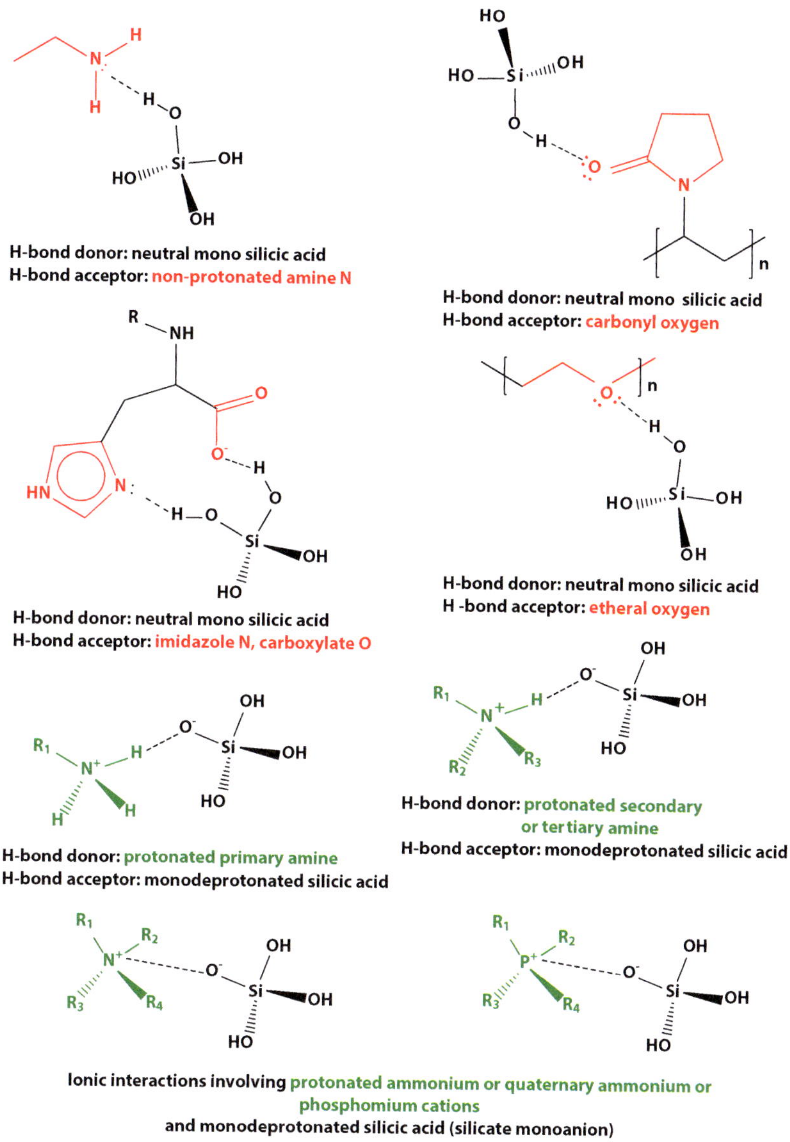

Figure 15.14 Stabilizing interactions of silicic acid with neutral moieties (in red) and charged moieties (in green).

15.6 Conclusions/Perspectives

The exciting field of biosilicification offers great opportunities for research in several related disciplines. In this chapter, we attempted to give a chemical perspective to the process that precedes biosilica formation, and this is the silicic acid stabilization in the putative "silicon pools". Although the precise chemical intricacies that take place in the diatom cell are still elusive, there is plenty of room for chemical inspiration and modeling to flourish, allowing further development of scientific disciplines such as polymer chemistry, materials science, colloidal chemistry, biochemistry, biology, and, last but not least, water chemistry [15.76].

Acknowledgments

Several past group members are acknowledged for their valuable contributions to the biosilica-related chemistry research efforts over the years. They are (alphabetically): Antonakaki, Ioanna; Aristodemou, Ioannis; Evaggelopoulou, Maria; Kamaratou, Michaela; Ketsetzi, Antonia; Mavredaki, Eleftheria; Michailidis, Dimitris; Neofotistou, Eleftheria; Pachis, Kostas; Petrou, Irena; Preari, Melina; Skordalou, Georgia; Somara, Maria; Stathoulopoulou, Aggeliki; Theodoropoulos, George; Tsiknakis, Michalis; Tsistraki, Anna. We also thank the Research Committee of the University of Crete for financial support (Projects KA 03121, and KA 10694), and the EU for funding the Research Program SILICAMPS-153, under the ERA.NET-RUS Pilot Joint Call for Collaborative S&T projects. Lastly, the following companies from the industrial sector are acknowledged for their financial support of various aspects of silica chemistry: Sychem S. A. (Greece), Dendritech Inc. (USA), Rhodia Inc. (France), and Kurita Europe APW GmbH (Germany).

References

[15.1] Demadis, K.D., Biosilica: Structure, function, science, technology, inspiration. *Am. Mineral.*, 103, 1009–1010, 2018.

[15.2] Wysokowski, M., Jesionowski, T., Ehrlich, H., Biosilica as a source for inspiration in biological materials science. *Am. Mineral.*, 103, 665–691, 2018.

[15.3] Bidle, K.D., Manganelli, M., Azam, F., Regulation of oceanic silicon and carbon preservation by temperature control on bacteria. *Science*, 298, 1980–1984, 2002.

[15.4] Brady, N.C. and Weil, R., *The Nature and Properties of Soils*, 14th ed., Prentice Hall, New Jersey, USA, 2007.

[15.5] Farooq, M.A. and Dietz, K.-J., Silicon as versatile player in plant and human biology: Overlooked and poorly understood. *Front. Plant Sci.*, 6, article # 994, 2015.

[15.6] Wiese, H., Nikolic, M., Roemheld, V., Silicon in plant nutrition. Effect of zinc, manganese and boron leaf concentrations and compartmentation, in: *The Apoplast of Higher Plants: Compartment of Storage, Transport and Reactions*, B. Sattelmacher and W.J. Horst (Eds.), pp. 33–47, Springer, Dordrecht, 2007.

[15.7] Ehrlich, H., *Biological materials of marine origin: Invertebrates*, Springer, New York, 2010.

[15.8] Perry, C.C., Silicification: The processes by which organisms capture and mineralize silica. *Rev. Mineral. Geochem.*, 54, 291–327, 2003.

[15.9] Simpson, T.L. and Volcani, B.E., *Silicon and siliceous structures in biological systems*, Springer-Verlag, New York, 1981.

[15.10] Hecky, R.E., Mopper, K., Kilham, P., Degens, E.T., The amino acid and sugar composition of diatom cell walls. *Mar. Biol.*, 19, 323–331, 1973.

[15.11] Tilburey, G.E., Patwardhan, S.V., Huang, J., Kaplan, D.L., Perry, C.C., Are hydroxyl-containing biomolecules important in biosilicification? A model study. *J. Phys. Chem. B*, 111, 4630–4638, 2007.

[15.12] Hildebrand, M., Volcani, B.E., Gassmann, W., Schroeder, J., II, A gene family of silicon transporters. *Nature*, 385, 688–689, 1997.

[15.13] Siever, R., The silica budget in the sedimentary cycle. *Am. Mineral.*, 42, 821–841, 1957.

[15.14] Truguer, P., Nelson, D.M., Van Bennekom, A.J., Demaster, D.J., Leynaert, A., Queguiner, B., The silica balance in the world ocean: a reestimate. *Science*, 268, 375–379, 1995.

[15.15] Li, S.-X., Wang, Z.-H., Stewart, B.A., Responses of crop plants to ammonium and nitrate. *Adv. Agron.*, 118, 205–397, 2013.
[15.16] Annenkov, V.V., Basharina, T.N., Danilovtseva, E.N., Grachev, M.A., Putative silicon transport vesicles in the cytoplasm of the diatom *Synedraacus* during surge uptake of silicon. *Protoplasma*, 250, 1147–1155, 2013.
[15.17] Schmid, A.-M.M. and Schulz, D., Wall morphogenesis in diatoms: Deposition of silica by cytoplasmic vesicles. *Protoplasma*, 100, 267–288, 1979.
[15.18] Mann, S., Molecular tectonics in biomineralization and biomimetic materials chemistry. *Nature*, 365, 499–505, 1993.
[15.19] Mann, S. and Ozin, G., Synthesis of inorganic materials with complex form. *Nature*, 382, 313–318, 1996.
[15.20] Oliver, S., Kuperman, A., Coombs, N., Lough, A., Ozin, G.A., Lamellar aluminophosphates with surface patterns that mimic diatom and radiolarian microskeletons. *Nature*, 378, 47–50, 1995.
[15.21] Pickett-Heaps, J., Schmid, A.-M.M., Edgar, L.A., The cell biology of diatom valve formation, in: *Progress in Phycological Research*, vol. 7, F.E. Round, and D.J. Chapman, (Eds.), pp. 1–169, Biopress, Bristol, UK, 1990.
[15.22] Kröger, N., Deutzmann, R., Sumper, M., Polycationic peptides from diatom biosilica that direct silica nanosphere formation. *Science*, 286, 1129–1132, 1999.
[15.23] Kröger, N., Lorenz, S., Brunner, E., Sumper, M., Self-assembly of highly phosphorylated silaffins and their function in biosilica morphogenesis. *Science*, 298, 584–586, 2002.
[15.24] Poulsen, N. and Kröger, N., Silica morphogenesis by alternative processing of silaffins in the diatom *Thalassiosira pseudonana*. *J. Biol. Chem.*, 279, 42993–42999, 2004.
[15.25] Martin-Jézéquel, V. and Lopez, P.J., Silicon – a central metabolite for diatom growth and morphogenesis. *Prog. Mol. Subcell. Biol.*, 33, 99–124, 2003.
[15.26] Martin-Jézéquel, V., Hildebrand, M., Brzezinski, M.A., Silicon metabolism in diatoms: Implications for growth. *J. Phycol.*, 36, 821–840, 2000.
[15.27] Gröger, C., Sumper, M., Brunner, E., Silicon uptake and metabolism of the marine diatom *Thalassiosira pseudonana*: Solid-state ^{29}Si NMR and fluorescence microscopic studies. *J. Struct. Biol.*, 161, 55–63, 2008.
[15.28] Sumper, M. and Brunner, E., Silica biomineralisation in diatoms: The model organism *Thalassiosira pseudonana*. *ChemBioChem*, 9, 1187–1194, 2008.
[15.29] Sjoberg, S., Silica in aqueous environments. *J. Non-Cryst. Solids*, 196, 51–57, 1996.
[15.30] Iler, R.K., *The chemistry of silica: Solubility, polymerization, colloid and surface properties, and biochemistry*, Wiley, New York, 1979.
[15.31] Kinrade, S.D., Gillson, A.-M.E., Knight, C.T.G., Silicon-29 NMR evidence of a transient hexavalent silicon complex in the diatom Navicula pelliculosa. *J. Chem. Soc., Dalton Trans.*, 307–309, 2002.
[15.32] Azam, F., Hemmingsen, B.B., Volcani, B.E., Role of silicon in diatom metabolism V. silicic acid transport and metabolism in the heterotrophic diatom *Nitzschia alba*. *Arch. Microbiol.*, 97, 103–114, 1974.
[15.33] Brunner, E., Gröger, C., Lutz, K., Richthammer, P., Spinde, K., Sumper, M., Analytical studies of silica biomineralization: towards an understanding of silica processing by diatoms. *Appl. Microbiol. Biotechnol.*, 84, 607–616, 2009.
[15.34] Annenkov, V.V., Danilovtseva, E.N., Pal'shin, V.A., Aseyev, V.O., Petrov, A.K., Kozlov, A.S., Patwardhan, S.V., Perry, C.C., Poly(vinyl amine)-silica composite nanoparticles: Models of the silicic acid cytoplasmic pool and as a silica precursor for composite materials formation. *Biomacromolecules*, 12, 1772–1780, 2011.

[15.35] Belton, D.J., Deschaume, O., Patwardhan, S.V., Perry, C.C., A solution study of silica condensation and speciation with relevance to *in vitro* investigations of biosilicification. *J. Phys. Chem. B*, 114, 9947–9955, 2010.
[15.36] Coradin, T., Durupthy, O., Livage, J., Interactions of amino-containing peptides with sodium silicate and colloidal silica: A biomimetic approach of silicification. *Langmuir*, 18, 2331–2336, 2002.
[15.37] Demadis, K.D., Recent developments in controlling silica and magnesium silicate in industrial water systems, in: *Science and Technology of Industrial Water Treatment*, Z. Amjad, (Ed.), pp. 179–203, CRC Press, London, Chapter 10, 2010.
[15.38] Chen, C.Y., Zhang, Y., Jiao, L.N., Bai, N., Zhang, S.G., Xia, M.Z., Study on the mechanism of the interactions between the dendritic macromolecule PAMAM and silicic acids in neutral solution. *Silicon*, 10, 1497–1502, 2018.
[15.39] Overton, P., Danilovtseva, E., Karjalainen, E., Karesoja, M., Annenkov, V., HeikkiTenhu, H., Aseyev, V., Water-dispersible silica-polyelectrolyte nanocomposites prepared via acid-triggered polycondensation of silicic acid and directed by polycations. *Polymers*, 8, 96, 2016.
[15.40] Icopni, G.A., Brantley, S.L., Heavey, P.J., Kinetics of silica oligomerization and nanocolloid formation as a function of pH and ionic strength at 25°C. *Geochim. Cosmochim. Acta*, 69, 293–303, 2005.
[15.41] Coradin, T. and Livage, J., Effect of some amino acids and peptides on silicic acid polymerization. *Colloids Surf. B: Biointerfaces*, 21, 329–336, 2001.
[15.42] Binauld, S. and Stenzel, M.H., Acid-degradable polymers for drug delivery: a decade of innovation. *Chem. Commun.*, 49, 2082–2102, 2013.
[15.43] Zhao, Z.-G., Adsorption of phenylalanine from aqueous solution onto active carbon and silica gel. *Chin. J. Chem.*, 10, 325–330, 1992.
[15.44] Soppimath, K.S., Aminabhavi, T.M., Kulkarni, A.R., Rudzinski, W.E., Biodegradable polymeric nanoparticles as drug delivery devices. *J. Controlled Release*, 70, 1–20, 2001.
[15.45] Kennedy, J.H., HPLC Purification of pergolide using silica gel. *Org. Process Res. Dev.*, 1, 68–71, 1997.
[15.46] Ehrlich, H., Demadis, K.D., Koutsoukos, P.G., Pokrovsky, O., Modern views on desilicification: Biosilica and abiotic silica dissolution in natural and artificial environments. *Chem. Rev.*, 110, 4656–4689, 2010.
[15.47] Coradin, T., Eglin, D., Livage, J., The silicomolybdic acid spectrophotometric method and its application to silicate/biopolymer interaction studies. *Spectroscopy*, 18, 567–576, 2004.
[15.48] Alexander, G.B., The preparation of monosilicic acid. *J. Am. Chem. Soc.*, 75, 5655–2888, 1953.
[15.49] Eaton, A.D., Clesceri, L.S., Rice, E.W., Greenberg, A.E., Franson, M.H., Silicon, in: *Standard Methods for Examination of Water and Wastewater*, American Public Health Association, Denver, Colorado, USA, 2005.
[15.50] Truesdale, V.W. and Smith, C.J., The spectrophotometric characteristics of aqueous solutions of α- and β-molybdosilicic acids. *Analyst*, 797–805, 1975.
[15.51] Truesdale, V.W., Smith, P.J., Smith, C.J., Kinetics of α- and β-molybdosilicic acid formation. *Analyst*, 897–918, 1979.
[15.52] Annenkov, V.V., Danilovtseva, E.N., Pal'shin, V.A., Verkhozina, O.N., Zelinskiy, S.N., Krishnan, U.M., Silicic acid condensation under the influence of water-soluble polymers: From biology to new materials. *RSC-Adv.*, 7, 20995–21027, 2017.
[15.53] Annenkov, V.V., Danilovtseva, E.N., Filina, E.A., Likhoshway, Y.V., Interaction of silicic acid with poly(1-vinylimidazole). *J. Polym. Sci.: Part A: Polym. Chem.*, 44, 820–827, 2006.

[15.54] Neofotistou, E. and Demadis, K.D., Silica scale inhibition by polyaminoamide STARBURST® dendrimers. *Colloids Surf. A: Physicochem. Eng. Asp.*, 242, 213–216, 2004.
[15.55] Demadis, K.D., A structure/function study of polyaminoamide dendrimers as silica scale growth inhibitors. *J. Chem. Technol. Biotechnol.*, 80, 630–640, 2005.
[15.56] Stathoulopoulou, A. and Demadis, K.D., Enhancement of silicate solubility by use of "green" additives: linking green chemistry and chemical water treatment. *Desalination*, 224, 223–230, 2008.
[15.57] Demadis, K.D. and Stathoulopoulou, A., Novel, multifunctional, environmentally friendly additives for effective control of inorganic foulants in Industrial Water and Process Applications. *Mater. Perform.*, 45, 1, 40–44, 2006.
[15.58] Demadis, K.D. and Stathoulopoulou, A., Solubility enhancement of silicate with polyamine/polyammonium cationic macromolecules: Relevance to silica-laden process waters. *Ind. Eng. Chem. Res.*, 45, 4436–4440, 2006.
[15.59] Ketsetzi, A., Stathoulopoulou, A., Demadis, K.D., Being "green" in chemical water treatment technologies: Issues, challenges and developments. *Desalination*, 223, 487–493, 2008.
[15.60] Demadis, K.D., Tsistraki, A., Popa, A., Ilia, G., Visa, A., Promiscuous stabilisation behavior of silicic acid by cationic macromolecules: The case of phosphonium-grafted dicationic ethylene oxide bolaamphiphiles. *RSC-Adv.*, 2, 631–641, 2012.
[15.61] Preari, M., Spinde, K., Lazic, J., Brunner, E., Demadis, K.D., Bioinspired insights into silicic acid stabilization mechanisms: The dominant role of polyethylene glycol-induced hydrogen bonding. *J. Am. Chem. Soc.*, 136, 4236–4244, 2014.
[15.62] Demadis, K.D., Neofotistou, E., Mavredaki, E., Tsiknakis, M., Sarigiannidou, E.-M., Katarachia, S.D., Inorganic foulants in membrane systems: Chemical control strategies and the contribution of "green chemistry". *Desalination*, 179, 281–295, 2005.
[15.63] Demadis, K.D. and Preari, M., "Green" scale inhibitors in water treatment processes: The case of silica scale inhibition. *Desalin. Water Treat.*, 55, 749–755, 2015.
[15.64] Spinde, K., Pachis, K., Antonakaki, I., Paasch, S., Brunner, E., Demadis, K.D., The influence of polyamines and related macromolecules on silicic acid polycondensation: Relevance to "soluble silicon pools"? *Chem. Mater.*, 23, 4676–4687, 2011.
[15.65] Papathanasiou, K.E., Vassaki, M., Spinthaki, A., Alatzoglou, F.-E.G., Tripodianos, E., Turhanen, P., Demadis, K.D., Phosphorus chemistry: From small molecules to polymers to pharmaceutical and industrial applications. *Pure Appl. Chem.*, 91, 421–441, 2019.
[15.66. Spinthaki, A., Skordalou, G., Stathoulopoulou, A., Demadis, K.D., Modified macromolecules in the prevention of silica scale. *Pure Appl. Chem.*, 88, 1037–1047, 2016.
[15.67] Demadis, K.D., Ketsetzi, A., Pachis, K., Ramos, V.M., Inhibitory effects of multicomponent, phosphonate-grafted, zwitter-ionic chitosan biomacromolecules on silicic acid condensation. *Biomacromolecules*, 9, 3288–3293, 2008.
[15.68] Demadis, K.D., Pachis, K., Ketsetzi, A., Stathoulopoulou, A., Bioinspired control of colloidal silica *in vitro* by dual polymeric assemblies of zwitterionic phosphomethylated chitosan and polycations or polyanions. *Adv. Colloid Interface Sci.*, 151, 33–48, 2009.
[15.69] Spinthaki, A., Zerfass, C., Paulsen, H., Hobe, S., Demadis, K.D., Pleiotropic role of recombinant silaffin-like cationic polypeptide P5S3: Peptide-induced silicic acid stabilization, silica formation and inhibition of silica dissolution. *ChemistrySelect*, 2, 6–17, 2017.
[15.70] Demadis, K.D., Brückner, S., Brunner, E., Paasch, S., Antonakaki, I., Casolaro, M., The intimate role of imidazole in the stabilization of silicic acid by a pH-responsive, histidine-grafted polyampholyte. *Chem. Mater.*, 27, 6827–6836, 2015.
[15.71] Mavredaki, E., Neofotistou, E., Demadis, K.D., Inhibition and dissolution as dual mitigation approaches for colloidal silica (SiO_2) fouling and deposition in process water systems: Functional synergies. *Ind. Eng. Chem. Res.*, 44, 7019–7026, 2005.

[15.72] Demadis, K.D. and Neofotistou, E., Synergistic effects of combinations of cationic polyaminoamide dendrimers/anionic polyelectrolytes on amorphous silica formation: A bioinspired approach. *Chem. Mater.*, 19, 581–587, 2007.
[15.73] Neofotistou, E. and Demadis, K.D., Cationic polymeric chemical inhibitors and multifunctional blends for the control of silica scale in process waters. *Int. J. Corros. Scale Inhib.*, 3, 28–34, 2014.
[15.74] Spinthaki, A., Stathoulopoulou, A., Demadis, K.D., Profound "turn-off" effects of anionic polymers on the inhibitory activity of cationic polyallylamine in the prevention of silica scale. *Int. J. Corros. Scale Inhib.*, 4, 85–95, 2015.
[15.75] Spinthaki, A., Stathoulopoulou, A., Demadis, K.D., The interplay between cationic polyethyleneimine and anionic polyelectrolytes for the control of silica scale formation in process waters. *Int. J. Corros. Scale Inhib.*, 4, 125–138, 2015.
[15.76] Spinthaki, A. and Demadis, K.D., Chemical methods for scaling control, in: *Corrosion and Fouling Control in Desalination Industry*, V.S. Saji, A.A. Sorour, A.A. Meroufel (Eds.), pp. 307–342, Springer Nature, Cham, Switzerland, 2020, Chapter 15.

16

The Mesopores of Raphid Pennate Diatoms: Toward Natural Controllable Anisotropic Mesoporous Silica Microparticles

Mohamed M. Ghobara[1*], Richard Gordon[2,3†] and Louisa Reissig[1‡]

[1]*Department of Physics, Freie Universität Berlin, Berlin, Germany*
[2]*Gulf Specimen Marine Laboratory, Panacea, FL, United States*
[3]*C.S. Mott Center for Human Growth & Development, Department of Obstetrics & Gynecology, Wayne State University, Detroit, MI, United States*

Abstract

In recent years, mesoporous materials have gained increasing attention through their potential and demonstrated applicability in a range of technologies. While to date many commercially used examples are obtained through synthetic routes, nature can provide us with promising alternative porous materials. Through biomineralization, some microorganisms synthesize remarkable, well-engineered, inorganic frameworks with patterns down to the nanometer scale that might benefit them and us. Among those microorganisms, diatoms construct genetically controlled inorganic (siliceous) microstructures, called frustules, with distinct macro- and mesoporosity of high application potential. The implementation of such frustules in applications created a new rapidly developing field called "Diatom nanotechnology". In this chapter, we focus on raphid pennate diatoms that build siliceous perforate pore occlusions, which have pores in the mesoscale with distinct similarities to mesoporous synthetic silica micro- and nanoparticles, with the aim of combining knowledge on biological mesoporous materials with that on mesoporous synthetic silica, for comparison and future inspiration. The literature demonstrates that the synthetic control of structural parameters of mesoporous synthetic silica, such as their shape, size, mesoporous pattern, surface area, and pore volume, could also be achieved in living diatoms by altering physicochemical parameters, by prospective genetic engineering, or by post-harvest modifications.

***Keywords*:** Natural mesoporous silica, synthetic mesoporous silica, raphid pennate diatom, hymenate pore occlusion, diatom nanotechnology

**Corresponding author*: mohamedghobara@rocketmail.com
†*Corresponding author*: dickgordoncan@xplornet.com
‡*Corresponding author*: Louisa_Reissig@gmx.de

Vadim Annenkov, Joseph Seckback and Richard Gordon (eds.) Diatom Morphogenesis, (383–410) © 2022 Scrivener Publishing LLC

16.1 Introduction

Porosity is defined as a measure of empty spaces within a material and has become an important parameter in materials characterization and their potential applications. The IUPAC has generally classified pores according to their size as follows: micropores with sizes smaller than 2 nm, mesopores with sizes between 2 and 50 nm, and macropores with sizes larger than 50 nm [16.97]. Different materials of various pore sizes, shapes, and patterns have been demonstrated to possess unique functions making them potential target materials for advanced applications [16.16, 16.80, 16.133]. Compared with ordinary bulk materials, they stand out due to their vast surface area per volume as well as their characteristic ability to interact with ions, atoms, and molecules, thus making them perfect materials for applications in filtration, separation, sensing, catalysis, or even drug delivery. However, the controlled introduction of especially the smaller pores into the target materials currently faces high cost, complicated procedures and large energy consumption, as well as the undesirable presence of traces of toxic residues in the to-be-utilized samples that make it necessary to find alternative, more suitable synthetic routes to easily commercialize such potential materials [16.24, 16.115].

Porous synthetic silicas, especially mesoporous silica [16.153], have been commercially used in many applications [16.80]. The synthesis, chemistry, and applications of different silicas [16.149, 16.150], including mesoporous silica [16.153, 16.80] have been discussed extensively in a range of highly cited publications. As a green and cheap alternative, attention is increasingly directed toward natural systems, where a large number of examples of structures with different porosity scales can be found, since most cellular molecular mechanisms take place at the nano- or microscale. A large number of living organisms have porous inorganic internal or external skeletal supports, with porosities variable in size and pattern [16.19]. Within these organisms, a few classes exhibit macro- and mesoporous microstructures made out of silica (Table 16.1), making them to some extent comparable to synthetic macro- or mesoporous silica, depending on their pore size (see Section 16.4).

Among these biomaterials, the most interesting range of porosity scales, ranging from pore diameters of 2 μm down to 3 nm (see Figure 16.1 as an example), can be found in the siliceous frustules of diatoms [16.27, 16.95]. Therefore, macropores (Figure 16.1a), mesopores (Figure 16.1b), and possibly micropores are present in diatom frustules. However, the presence of micropores has not been yet confirmed [16.122]. Among the different diatom taxa, the finest porosity (in the mesoscale between 10 and 3 nm) is usually observed in raphid pennate frustules within a very delicate layer that is called the hymenate layer or hymenate pore occlusion (Figure 16.1b), resembling a porous protective skin covering larger periodically arranged macropores in the frustules (see Section 16.2).

Based on the recent understanding of the diatom frustule's ultrastructure, porosity and its unique properties, a new field emerged [16.31], now called diatom nanotechnology, with increasing numbers of articles, applications, and patents published in the last decade [16.27, 16.30, 16.34, 16.67]. There is a large potential in a number of established applications to use both fresh cultivated or fossil diatoms (diatomaceous earth or diatomite, see review [16.28]), with cleaned frustules.

Table 16.1 Major classes of microorganisms with porous siliceous structures. Example 6 is added for completion and comparison, despite of the porous walls being made of calcite. The pore diameters, if not mentioned in a reference, were measured via ImageJ software based on SEMs.

Class of microorganism	Cell wall composition	Main porosity scale range (nm)	Example references
1. Bacillariophyceae (diatoms)	Silica cell wall (dimensions range from 1.5 μm up to 5 mm)	3–2,000	[16.27]
2. Synurophyceae	Silica scales form a protective layer (dimensions of such scales range from 1.1 μm to 5.8 μm)	Base plate pores 15–135	[16.47, 16.48]
3. Silicoflagellata	Life cycle has a stage with a siliceous structure surrounding the cell	6,000–15,000	[16.74]
4. Radiolaria	Siliceous cell skeleton with different porosity scales	1,000–65,000	[16.53]
5. Heliozoa	Some species of heliozoa have siliceous scales, some very similar to the scales of Synurophyceae	15–400	[16.129]
6. *Prymnesiophyceae (Coccolithophyceae)*	*Calcareous plates (coccoliths) form a protective layer around the cell, some with large slits and others with quasi-regular pores*	*30–500*	[16.15]

This chapter aims to combine the knowledge and concepts derived for mesoporous silica by materials scientists and for diatom porous silica by diatomists to encourage the scientific community to more efficiently utilize such natural porous silica in applications, replacing or substituting conventional synthetic materials, and to look beyond the traditionally used few centric and pennate diatom species, which could have mesoporous silica but with comparably large pore sizes (from 30 to 50 nm), such as the two centrics: *Coscinodiscus* sp. and *Thalassiosira eccentrica* by Losic *et al.* [16.68], to, e.g., the group of raphid pennate diatoms featuring hymenate pore occlusions that have generally smaller pore sizes ranging from 15 down to 3 nm with the majority below 10 nm [16.13]. Furthermore, it aims to provide the reader with a brief overview on our abilities to control the dimensions and periodicities of the pores and frustules of diatoms by altering physicochemical parameters, as well as current trends toward their control through genetic engineering, analogous to the chemist's ability to control the characteristics of synthetic mesoporous silica.

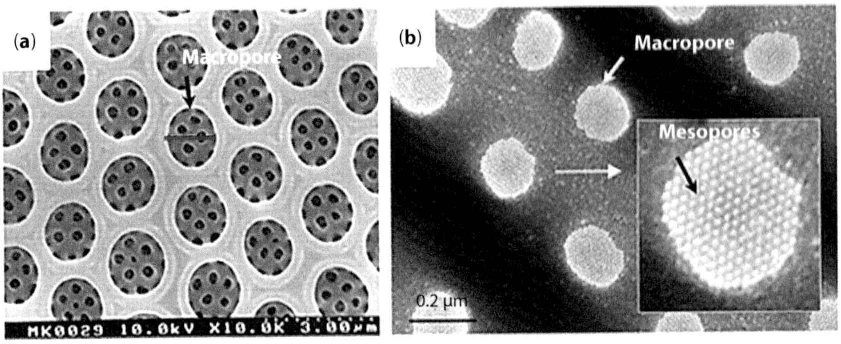

Figure 16.1 (a) SEM illustrates the muliscalar porosity appearing in valve view of the centric diatom *Coscinodiscus wailesii*, reproduced from Kucki and Fuhrmann [16.135] under a Creative Commons Attribution 4.0 International License. The macropores are called foramen and have a mean diameter of ≈ 1.4 µm (red line). The layer of smaller macropores below them has diameters down to 140 nm. (b) TEM of part of the pennate diatom *Pseudo-nitzschia pungens* var. *pungens* illustrates the circular areolae (macropores) with hymenate pore occlusions, reproduced from [16.136] with the kind permission of John Wiley & Sons. The mesopores (the smallest) have a mean diameter of ≈ 6 nm. Scale bars, 3 µm (white dotted line) and 0.2 µm = 200 nm, respectively. Inset magnified by a factor of 2.6.

16.2 Morphology and Very Fine Ultrastructure of Diatom Frustules

Diatoms are known for their amorphous siliceous cell walls, the diatom frustules, with dimensions extending from 1.5 µm [16.12] up to 5 mm in length or diameter [16.106]. Characteristic of diatoms and specific for each diatom species are not only the overall geometry of the frustule but also the structural parameters of its multiscalar multilayer porosity [16.95]. Figure 16.2a depicts the various parts of the two main classes of distinct frustule symmetry: centric and pennate frustules. Each frustule is composed of two theca, the epitheca and the hypotheca, each of which contains a valve. The two valves are connected via one or more girdle bands (as illustrated in Figures 16.2c and d). The individual frustule parts are in general porous, and these pores are usually arranged in rows, called striae (Figure 16.2a), which usually have a distinct species-specific distribution and symmetry within the frustule [16.95]. In most cases, this porosity indicates the underneath presence of 3D structures called areolae. The areolae have diameters ranging mainly from 200 nm to more than 2 µm [16.95] and are a result from the intercrossing of rib systems during the biosynthesis of the valve [16.95]. These areolae have different 3D designs including lineolate (with elongated pores such as *Navicula* spp.), punctate (a round or oval pore of simple structure similar to a nano- or microchannel in the silica wall such as *Gomphoneis* spp.) (Figure 16.4b), or loculate (Figures 16.3d and 16.4a). Only loculated areolae have a chamber, and they occur mainly in centrics but also in a number of pennates such as *Donkinia*, *Psammodictyon*, and *Pleurosigma* [16.95]. It can be seen in Figures 16.3d, 16.4a, and 16.5 that such an areola is not just a channel, such as shown for a punctate areola in Figure 16.4b, but rather a hexagonal, spherical, oval or conical-shaped chamber with an inner space, pore occlusion and a foramen.

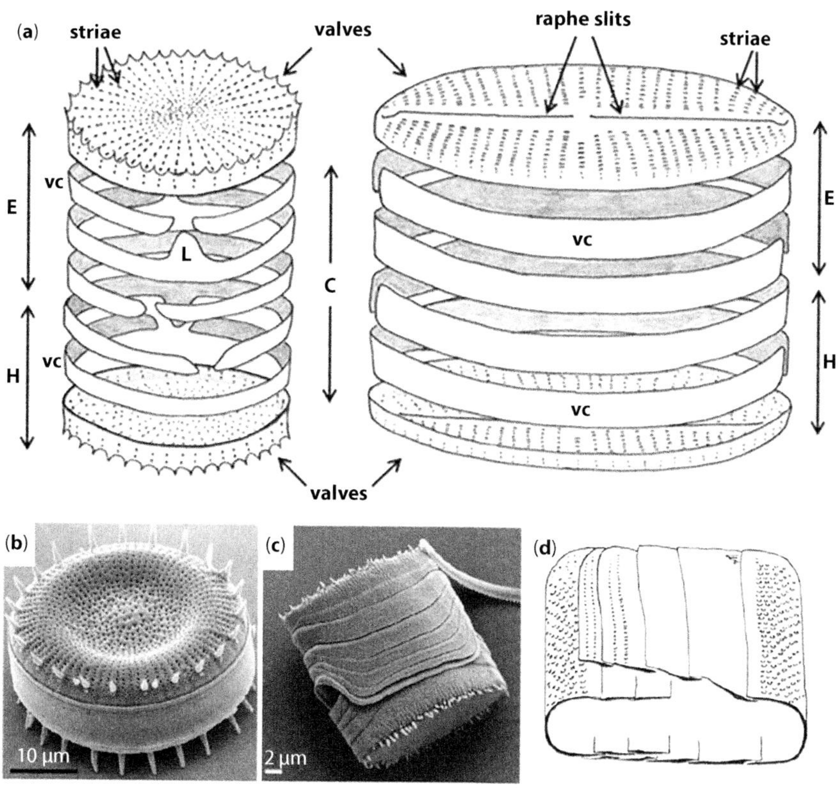

Figure 16.2 (a) Diagram showing the frustule structure of a centric frustule on the left and a raphid pennate frustule on the right. E, older epitheca; H, younger hypotheca; C, girdle bands; L, tongue-like extension; and VC, valvocopula. Reproduced from [16.137] under the Creative Commons Attribution 4.0 License. (b) SEM of the valve view of a complete frustule of *Stephanodiscus* sp. (c) SEM of the girdle view of *Diatoma* sp. showing the epitheca at the top overlapping the hypotheca at the bottom. (d) Schematic diagram of a cross section in a complete pennate frustule illustrating the girdle band relations with the epi- and hypovalve. (b), (c), and (d) are reproduced from [16.138] with kind permission of Springer.

The areola is usually occluded, either from the internal (Figures 16.3d and 16.5) (also see Figure 16.3 in [16.68]) or the external side (Figure 16.4) (see Figure 16.2 in [16.68]), with a fine membranous siliceous layer with its own finer porosity [16.68, 16.95]. Figures 16.3 to 16.5 give a visualization of this layer and its structural relationship with the overall valve's architecture. This fine membranous layer is also known as a "pore occlusion" with mesopores generally averaging from 3 to 50 nm in diameter [16.13, 16.127]. In some studies, the alternative term "velum" has been used. For a *Plagiotropis tristis* cell, the thickness of the mesoporous layer has been estimated to be around 110 nm from the TEM (transmission electron micrograph) of a cross section through the structure of a cell featuring a very clear valve [16.76].

It has been found that diatom mesopore distribution patterns, shapes, and average sizes are morphogenetic characteristics, which differ from taxon to taxon [16.71, 16.95, 16.127, 16.128]. For instance, species of the same genus, *Pseudo-nitzschia*, have different pore occlusions with distinct pore shapes, sizes, and distribution pattern [16.114].

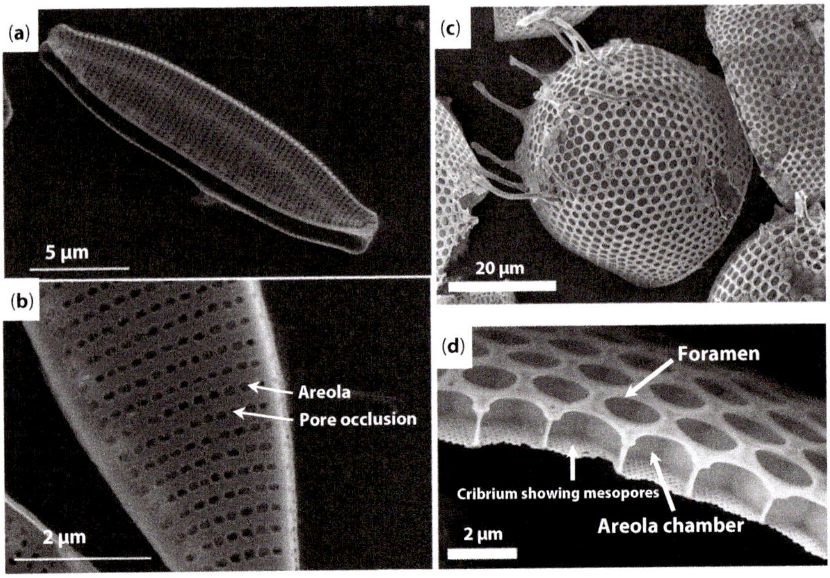

Figure 16.3 (a) SEM of the external valve face of the raphid pennate diatom *Nitzschia palea*. (b) SEM of a close-up on the external face of a *Nitzschia palea* valve showing small areolae with hymenate pore occlusion covering the areolae entrance. (c) SEM of the external valve face of the centric diatom *Stephanopyxis turris*. (d) SEM of a cross section in a single valve of *Stephanopyxis turris* shows areolae with external foramen layer and internal cribrae layer with mesopores. (a) and (b) were obtained by MG, while (c) and (d) are reproduced from [16.139] under Creative Commons Attribution 3.0 Unported License.

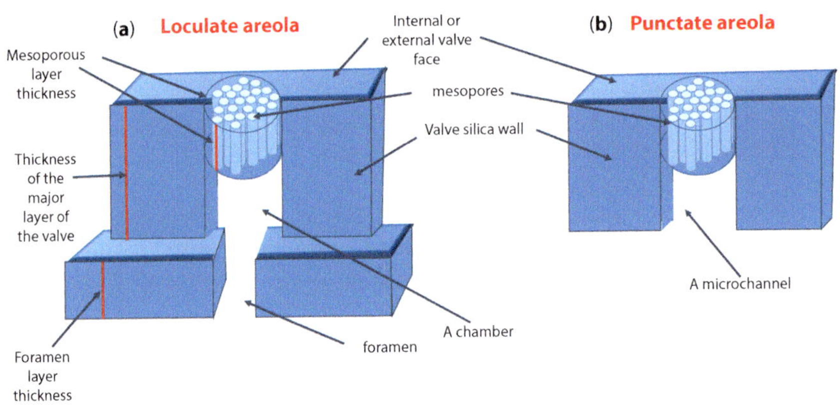

Figure 16.4 A schematic diagram representing the loculate areolae (a) and punctate areolae (b) as occur in pennates with pore occlusion layer.

It should be noted that in living cells, the frustule is usually associated with organic components as for example found in the organic envelope covering it [16.5, 16.116]. Therefore, sample processing is necessary to remove organic matter surrounding the frustules in order to clearly visualize their silica structures [16.126]. In some cases, the fine details, such as the mesopores of pore occlusions, are, however, lost during diatom frustule preparation [16.119] due to extended use of high temperature or strong oxidizers. Only through gentle

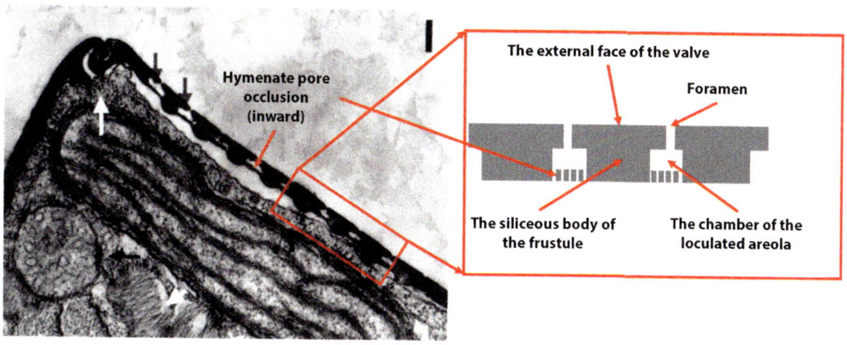

Figure 16.5 A transmission electron micrograph (TEM) showing a cross section through a cell of *Plagiotropis tristis*. The loculate structure of the areolae appeared with hymenate pore occlusion (inward) and slit-like opening (outward). The cross section of the valve inside the red rectangle is represented with the diagram at the right. The thickness of the pore occlusion layer is around 110 nm. Scale bar = 1 μm. Adapted from [16.76] with the kind permission of Taylor & Francis.

handling of the frustules during cleaning procedures, as would be the case when using, e.g., the hot hydrogen peroxide method [16.126], can such fine mesoporous silica structures be preserved, further studied, and utilized in applications.

Furthermore, only since the development of high-resolution imaging instruments, such as the scanning electron microscope (SEM) and the transmission electron microscope (TEM), has the fine artwork of diatoms with structures below 200 nm, down to 3 nm, been revealed, leading, among other things, also to significant updates to their taxonomy [16.71, 16.95]. To date, SEM is considering the best tool for diatom taxonomists as it well-describes the surface morphology (see for example Figure 16.3) [16.95]. Combined with complimentary information obtained by TEM or atomic force microscopy (AFM) the finest ultrastructure within the valves, which could be in the order of 1 nm, could now be observed [16.13, 16.68]. Recently, SEM and AFM have also been integrated with digital holography microscopy (DHM) in order to represent the nanometric details detected by SEM and AFM in a 3D model that can be used for further studies and simulations [16.89].

There are surprisingly few published works concerning diatom pore occlusions and their mesopores [16.13, 16.71, 16.93, 16.94, 16.95]. According to Cox [16.13], pore occlusions of diatoms can be classified into (i) finely perforate occlusions (ricae = hymenes), (ii) coarsely perforate occlusions (two types: cribrae and rotae), and (iii) flap-like occlusions (volae) (Figure 16.6).

The fine hymen is a very thin and delicate perforate layer of silica (for example, with thickness in the range of 110 nm as in Figure 16.5) covering the whole areolae openings from the internal (e.g., many raphid pennates displayed in Round *et al.* [16.95]) or the external face (e.g., *Cymbellonitzschia* [16.95]) of the valve with mesopores, the latter's diameter of no more than 15 nm (usually even less than 10 nm) (Figures 16.1b, 16.6e and f, 16.9b). Such pore occlusions have only been found in raphid pennates [16.13]. In contrast, the coarsely perforate occlusions (cribrae and rotae) are mainly found in centrics, with comparably large mesopore diameters (usually exceeding 30 nm, and thus, visible by SEM, while the mesopores of the hymenes require a higher resolution SEM or TEM) (Figures 16.6a–c). In the case of rotae, a bar can be found spanning across the pore, usually with circular elements (Figure 16.6c). A few examples of the occurrence of cribrae have been reported in the

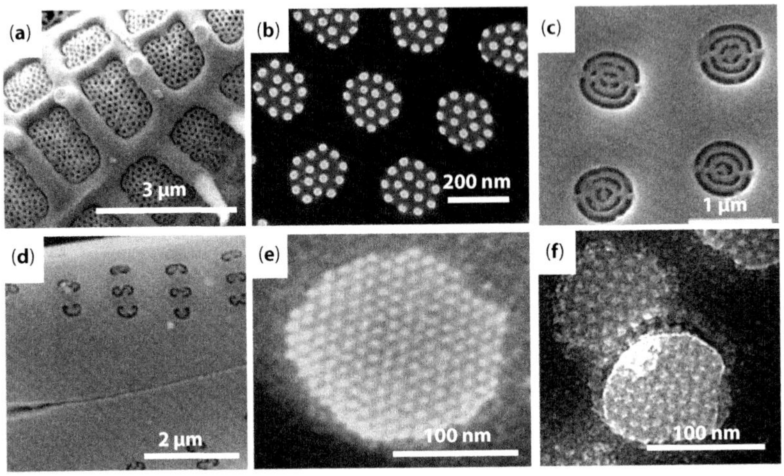

Figure 16.6 Scanning and transmission electron micrographs showing different types of pore occlusions that cover the areolae of diatom frustules. (a) SEM of the external face of a valve of pennate diatom *Neofragilaria anomala* with imperfectly hexagonally arranged cribrae, reproduced from [16.140] under a Creative Commons Attributive license. (b) TEM of a valve of a centric species *Coscinodiscus* sp. with centric cribrae, reproduced from Wang *et al.* [16.141] with the kind permission of Springer. (c) SEM of the external view of pennate diatom *Rhaphoneis amphiceros* valve with mature rotae, reproduced from [16.142] with the kind permission of John Wiley and Sons. (d) SEM of the external view of the center of pennate diatom *Gomphonema demersum* valve with flap-like occlusions, reproduced from Reichardt [16.143] under a Creative Commons Attributive license. (e) TEM of hymenate pore occlusions, with small hexagonally arranged mesopores, distributed in the valve of pennate diatom *Pseudo-nitzschia pungens* var. *pungens*, reproduced from [16.136] the kind permission of John Wiley & Sons. (f) TEM of pennate diatom *Nitzshcia palea* illustrates hymenate pore occlusion with small quasi-circular mesopores.

raphid taxa, but to date, no examples of rotae. Flap-like occlusions are the simplest type of pore occlusion and resemble a flap of (frequently unperforated) silica extending from one side of the pore, covering most of it, and thus, leaving only a small curved slit. A variety of flap-like occlusions has been found in raphid diatoms, with the most apparent form seen in *Gomphonema* spp. (Figure 16.6d, also see [16.13]).

In this chapter, we will mainly focus on the hymenate type of pore occlusion, found in raphid pennates, because of its interesting small mesoporosity, which is very similar to a type of synthetic mesoporous silica, in which the small mesopores are beneficial for certain applications, such as drug encapsulation for oral drug administration. Moreover, the frustules featuring hymenate pore occlusions are very delicate and small in size compared to other diatom species, which may give rise to materials of potentially high surface area with suitable properties of controllable loading, encapsulation, and molecular sieving for applications such as the fabrication of sensors, catalysts, and optoelectronic components.

Three geometric arrangements of pore occlusions have been reported: hexagonal, centric, and regularly scattered [16.71]. The latter are known as "spacing patterns" in the morphogenesis of other organisms [16.6, 16.18, 16.23, 16.44, 16.50, 16.52, 16.56, 16.81, 16.91, 16.104, 16.113]. Clear examples of hymenate pore occlusions have been found in many species of the following raphid pennate genera [16.13, 16.95]: *Achnanthidium, Amphipleura, Amphora, Anomoeoneis, Bacillaria, Berkeleya, Brachysira, Campylopyxis, Cavinula, Climaconeis, Cocconeis, Cosmioneis, Craticula, Cymbellonitzschia, Denticula,*

Diadesmis, Diploneis, Donkinia, Entomoneis, Eucocconeis, Fallacia, Fragilariopsis, Frustulia, Gyrosigma, Hantzschia, Haslea, Luticula, Navicula, Neidium, Nitzschia, Parlibellus, Pinnularia, Plagiotropis, Pleurosigma, Proschkinia, Psammodictyon, Pseudogomphonema, Rhoicosphenia, Stauroneis, Staurophora, Scolioneis, Sellaphora, Seminavis, Toxonidea, Trachyneis, Thalassiophysa, and *Tryblionella*.

16.3 Synthetic Mesoporous Silica

As already mentioned, since the discovery of the potential of mesoporous materials for novel applications, a range of examples have been reported for different materials such as titania [16.121], zirconia [16.111], and silica [16.80]. The invention and fabrication of synthetic mesoporous silica—distinct from ordinary porous silica by their pore size, surface area, and the high order of pores—started about five decades ago [16.80]. Since then, mesoporous silicas of various particle shapes have been synthesized with spherical and hollow spherical shapes constituting the majority of the reported examples [16.37, 16.38, 16.132] due to the simplicity of their preparation methods. Nevertheless, during the last two decades, other non-spherical mesoporous silica particles were successfully developed, including semispherical and discoid [16.124], platelet and sheet [16.37], spiral [16.118], rod-shaped and hollow rod-shaped [16.37], nano-ellipsoid and hollow nano-ellipsoid [16.38], micro-ellipsoid [16.105], cubic to parallelepiped shape [16.79], hollow cubic [16.37], coralline shaped [16.64], and even irregular shapes [16.99] (see Figure 16.7). The size of synthesized mesoporous silica particles ranges from 30 nm [16.70] to 237 μm [16.49], reported pore sizes range from 1.5 nm [16.80] to 25 nm [16.85], and the BET surface area ranges from 323.85 m^2/g [16.99] to 1500 m^2/g [16.124]. The pore's symmetry could be hexagonal, cubic

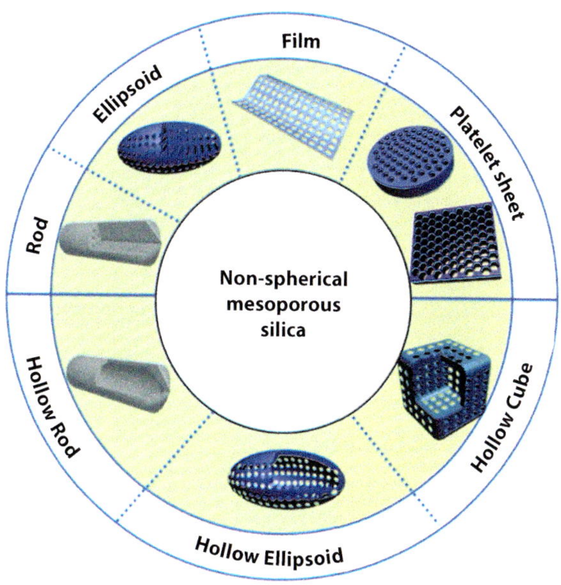

Figure 16.7 Some examples of non-spherical mesoporous silica structures. Adapted from [16.37] with the kind permission of the Royal Society of Chemistry.

or lamellar, with a degree of crystallinity ranging from amorphous, as in the Santa Barbara amorphous type (SBA-15), to crystalline, as in the Mobil crystalline materials (MCM-41). All of these parameters are highly dependent on the fabrication method and the parameters used [16.80]. The alteration of reaction conditions and starting precursors, as well as the choice of surfactant type during synthesis, has been shown to lead to different degrees of crystallinity as well as the particles' and pores' structural parameters, silica wall thickness, and pore symmetry [16.37, 16.80].

The applications of mesoporous silicas of a range of shapes, sizes, and pore volumes were investigated with focus on drug delivery systems [16.14, 16.72], catalysis [16.22, 16.77], separation, molecular sieving, and sensing applications [16.102, 16.131]. Moreover, they have been successfully utilized as targeted carriers for a range of insoluble drugs, proteins, enzymes, DNA, as well as RNA [16.80] (as illustrated in Figure 16.8). Studies revealed the main advantages of such mesoporous silica compared to the bulk material to be their large surface area, controllable pore size, and arguable biocompatibility [16.3].

The particle size of mesoporous silica might affect its potential for a given application [16.3]. For instance, it has been claimed that mesoporous silica microparticles (MSMs) (with sizes in the order of one micrometer) enhanced the cytotoxicity of anticancer platinum drugs compared to equivalent mesoporous silica nanoparticles (MSNs) [16.112]. Moreover, MSNs are

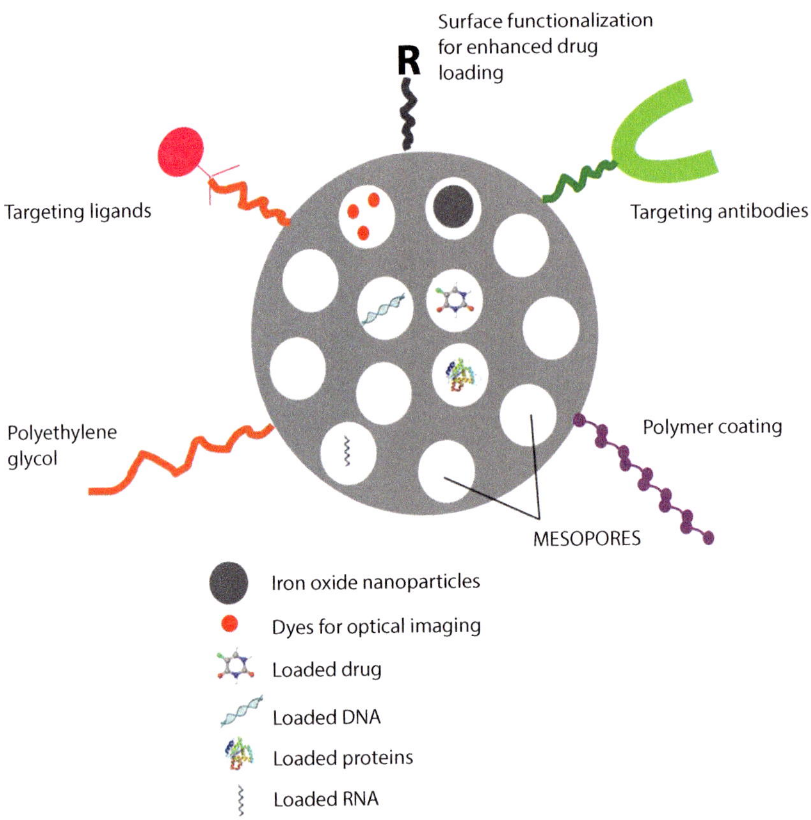

Figure 16.8 Different applications of mesoporous silica. From [16.80] under a Creative Commons Attribution License (CC-BY 4.0).

comparably difficult to retrieve from the environment (once released). Thus, as an alternative in certain applications, attention was directed toward MSMs, which are currently successfully used in drug delivery formulations including those for oral administration [16.40, 16.41, 16.65, 16.66, 16.88, 16.112], as well as in catalysis or sensing applications [16.131].

16.4 The Potential of Raphid Pennates' Mesoporous Bio-Silica, Similarities, and Dissimilarities Compared With Synthetic MSM/Ns

16.4.1 The Current Potential of Diatom Porous Silica in Applications

Diatom biosilica is not only natural, non-toxic and authorized by the FDA (USA Food and Drug Administration) but also biocompatible, highly porous, of relatively high surface area, chemically inert, commercially available in considerable amounts (especially diatomite) and relatively cheap [16.8, 16.115]. It can also be easily modified using well-established and controllable surface chemistry [16.8, 16.115]. That is why diatom frustules, especially diatomite, have already been used extensively, either modified or unmodified, in filtration, separation, catalysis, target delivery [16.28, 16.58], liquid chromatography [16.151], and sensing applications [16.54, 16.55, 16.109].

However, diatomite samples have distinct disadvantages due to their heterogeneity as they consist of a highly fragmented mixture of biosilica [16.58], and sometimes, samples are composed of several species. The use of such a mixture of frustules with different designs and ultrastructures are not possible in applications that require homogeneity. Therefore, using freshly cultivated biosilica could be more promising due to its higher purity, expected specific surface area, and homogeneity comparing with diatomaceous earth [16.152]. One disadvantage here lies, however, in limitations in obtaining fresh samples from many species as it requires expertise and optimized conditions [16.119]. Despite that, large-scale production of a few diatom species has been started, with potential production output exceeding 120 MT of cell dry weight per hectare per year [16.125], demonstrating the potential in becoming commercially competitive to both diatomite and synthetic porous silica samples, especially when considering the cost of synthetic silica production. Thus, more efforts should be placed on increasing the cultivation yield of a greater number of species. Bubble farming might guarantee little or no contamination at lower cost than open raceways or photobioreactors [16.35].

Moreover, it should be noted that the cell size reduces at each generation during asexual reproduction [16.95], and the signals for triggering sexual reproduction are still unknown for most species [16.78]. Therefore, more research is required to optimize cultivation [16.7, 16.75, 16.82, 16.98, 16.120, 16.125] and silica harvesting processes, and to engineer sexual/asexual diatom growth cycles [16.87] to optimize the production of freshly cultivated biosilica.

16.4.2 Why Should We Be Interested in the Mesoporous Silica of Raphid Pennate Frustules if the Frustules of Other Species With Larger Pores Work?

A good example would be the fossil frustules of the *Aulacoseira* spp. which have shown great potential as alternative drug carriers [16.4, 16.69], despite being usually macroporous.

For instance, their areolae diameters range typically from 300 to 500 nm [16.69], sometimes with a preserved finer mesh of pore occlusions with 50- to 100-nm pores [16.83], but which are still in the macroporosity range. As another example, *Coscinodiscus* and *Thalassiosira* spp. have been successfully used in a number of different applications, and their pore occlusions are in the mesoscale range but still relatively large, such as the examples studied by Losic *et al.* [16.68] with diameters around 40 nm and in De Stefano *et al.* [16.17] with diameters ranging from 30 to 40 nm.

As can easily be understood, there are no absolute criteria for the best pore size, but it always depends on the nature of the application. Thus, for some applications, small mesopores would be desirable, while for certain applications only a specific size range is advantageous. For instance, the insulin loading capacity was found to be different between SBA15 (pore size of 6.06 nm) and MCM-41 (pore size of 2.8 nm), with the latter having no absorption ability for insulin, which has a molecular size around 2.7 nm, while the insulin loading capacity of SBA-15 was 15.1% [16.107]. Insulin can also fit in pores which are larger than 6.06 nm, but it will not be protected from proteolytic degradation in the stomach, if used in oral drug formulations. Therefore, using mesoporous raphid pennate frustules (with smaller pore sizes, between 10 and 3 nm) might increase the conformational stability of specific adsorbed or chemically immobilized molecules, and simultaneously ensure their protection [16.20]. As another example that explaining why pore size could matter, the blue photoluminescence of *Nitzschia frustulum* frustule has increased dramatically with the presence of mesopores of 5 nm size [16.134], a property that can be used in the fabrication of photoluminescence-based sensors [16.147, 16.148]. Furthermore, frustules with such mesopores could thus be beneficial for a wide range of applications including the fabrication of different types of sensors, such as Surface-Enhanced Raman Spectroscopy (SERS)–based sensors [16.51, 16.90], as the interactions of biomolecules and their receptors within the mesopores might increase the stability, selectivity and accuracy of the results.

Thus, when selecting diatom frustules with optimized properties, through searching through the vast number of species, or through our ability to modify the species specific parameters (see Section 16.5), one can predict an even higher potential for use in selected applications. Consequently, including raphid pennate frustules, which have smaller mesopores, as target materials, might be a suitable alternative for use in established or novel applications especially size selective protein separations, immobilization, or encapsulation within pores for medical purposes [16.39, 16.46, 16.59], SERS-based sensors [16.51, 16.90], and photoluminescence-based sensors [16.147, 16.148].

16.4.3 Similarities and Dissimilarities Compared With Synthetic MSM/Ns

It has been previously suggested that some diatom species with microsized frustules featuring mesopores [16.13, 16.67], especially raphid pennates [16.13], exhibit similarities to synthetic MSM/Ns [16.1] (Figure 16.9). There are four major points for comparison between synthetic MSM/Ns and diatom mesoporous silica:

- The particle shape and size;
- The homogeneity of mesoporosity over the entire surface;
- The surface area; and
- The ability to control the mesopore size range (see Section 16.5)

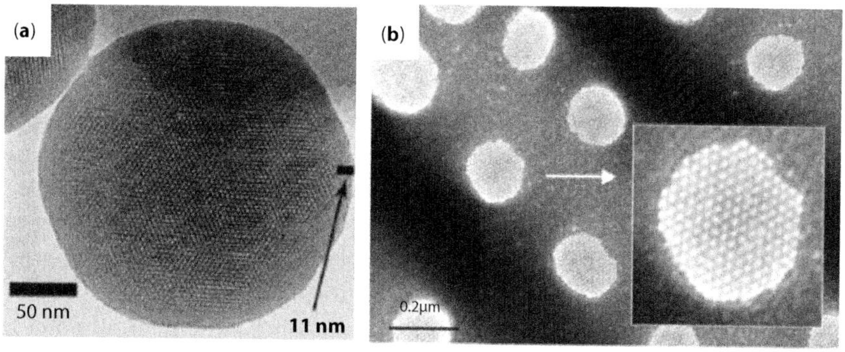

Figure 16.9 Two TEMs for a comparison between (a) synthetic MSMs with diameter 245 nm, reproduced from Slowing *et al.* [16.144] with kind permission of John Wiley & Sons, and (b) the hymenate pore occlusions (diameter around 200 nm), with very fine porosity, distributed in the valve of *Pseudo-nitzschia pungens* var. *pungens*, reproduced from [16.136] with the kind permission of John Wiley & Sons.

The synthetic MSM/Ns could be fabricated in various shapes (Figure 16.7) and sizes as mentioned before in Section 16.3, while raphid pennate frustules mostly have a hollow micro-ellipsoid shape, composed of two valves. The dimensions of most pennate frustules range from 1.5 μm to 600 μm in length [16.27] depending on the species. Thus, they can be compared with intermediate size or large MSMs.

One of the main differences lies in the coverage of the mesoporosity over the entire surface. While MSM/Ns display a continuous mesoporosity, diatom frustules have discrete patches of hymenate layers, with only a fraction of the surface being covered with mesopores (Figure 16.9). The thickness of the hymenate layer is species-specific, and it might be variable according to the culture conditions. Thus, raphid pennate frustules are anisotropic mesoporous microparticles with a relatively small specific surface area. However, for an adequate comparison, the nature of the silica itself needs to be taken into account, which is, in the case of frustules, composed of aggregated silica nanoparticles. This nanoparticle nature adds to the surface area, through increased roughness, as well as probable minute spaces between pores that can reach down to 2 nm (thus being themselves micropores). Furthermore, although diatom frustules do not have a homogeneous 3D structure, the accessible inner space of areolae could be of potential use (Figure 16.4) and increases the surface area to which materials can be attached [16.58].

In the literature, reported surface areas of diatom frustules have reached values as high as 189 m^2/g for raw diatomite and 222 m^2/g for diatomite leached in 5M HCl solutions for 1 hour at 75°C [16.101]. A specific surface area of 211 m^2/g has been measured for ferrihydrite-modified diatomite [16.130]. Although these values usually do not reach commonly reported values for synthetic MSMs of similar size (e.g., 935 m^2/g [16.41]), fresh diatom frustules could arguably reach higher values through approaches discussed in Section 16.5.

Furthermore, when comparing the surface area of centric diatom *Coscinodiscus asteromphalus* frustules estimated by nitrogen adsorption to be 89 m^2/g, to the surface area of pennate diatom *Navicula pelliculosa* frustules reported as 123 m^2/g [16.63, 16.152], it can be seen that raphid pennate diatom frustules may be more promising due to their small size of their frustules and their small mesopores. This is supported by the recently reported surface area of biosilica retrieved from freshly cultivated *Amphora* sp., a raphid pennate diatom, estimated by nitrogen adsorption to be 332 m^2/g [16.146].

It was previously mentioned in Section 16.3 that chemists have the ability to modify the structural parameters of MSM/Ns, especially the size and density of their mesopores. In the case of diatoms, it is still not clear to what extent we can modify their mesopores. But recent results suggest that we may be able to modify the pore size as well as other frustule parameters sufficiently during cultivation (see Section 16.5). Furthermore, it was also reported that it is possible to incorporate into freshly cultivated diatom frustules or their pores functional materials [16.96, 16.110] or proteins [16.58] *in vivo* during cultivation.

It can also be expected that the selective functionalization of the internal pores in MSM/Ns that assist protein loading within the pores could also be extended to diatom frustules. The same can be said for recent advances in the selective functionalization of larger-pored (>7 nm) mesoporous microparticles, in addition to available selective functionalization techniques that are limited to smaller pores (<5.5 nm). This could lead to changed chemical properties, which in turn could potentially lead to their usability in different applications [16.103]. Another interesting aspect when using diatom frustules might be their specific interaction with the adsorbed molecules and ions with characteristic hydrodynamic properties within their pores [16.92] that might open new pathways for *in vitro* immobilization of protein molecules or nanoparticles specifically within the pores for a range of applications [16.58].

In conclusion, we suggest that fresh cultivated raphid pennate frustules might promise comparable or even superior properties when compared with synthetic MSMs in certain applications.

16.5 Our Ability to Control the Diatom Frustule's Ultrastructure

A major question often asked by physicists and chemists, when discussing the use of diatoms in material science, is centered around our ability to control the pore sizes, shapes, patterns, volumes, and, with this, the surface area of diatom frustules. This is crucial when assessing the real potential of siliceous frustules in becoming alternative sources for the currently used synthetic MSMs in which almost all of the structural parameters can be controlled [16.80].

To answer this question, we should first consider the diatom morphogenesis and biomineralization process, which may not follow the same mechanism for each structural size scale [16.27]. It should be noted that the understanding of diatom morphogenesis is still incomplete [16.2]. Overall patterns have been approximated by diffusion limited precipitation simulations [16.33], sometimes aided by microtubules [16.84], or immiscible fluids models [16.60]. The large variety of patterns must have, of course, ultimately genetic correlates, perhaps through modifying physical and chemical parameters constraining silica precipitation. Some studies have found species-specific protein templates [16.21, 16.86], but these "prepatterns" [16.33] also require explanation. A top down explanation for diatom morphogenesis has been proposed [16.42], which contradicts expectations of molecular reductionism, but may indeed prevail, as possibly in multicellular organisms [16.36]. Thus, the frustule's ultrastructure results from as yet unknown interactions between the genotype and the physicochemical parameters inside the silicalemma [16.33]. For example, given a nascent pore, diffusion limited precipitation simulations do an excellent job of predicting the mesoporous structure of pore occlusions [16.127], as had previously been shown for patterns inside analogous centric diatom midrings [16.33]. Furthermore, the conditions inside the silicalemma can be altered by conditions in the environment [16.110]. Thus, multiple approaches to manipulating the diatom frustule ultrastructure are potentially available:

1. Altering physicochemical parameters of the medium surrounding living diatom cells [16.110];
2. Post-harvest modifications [16.110];
3. Genetic engineering [16.43, 16.57];
4. Deliberate selective breeding of diatoms [16.9, 16.10, 16.11] for particular pattern characteristics with culling of diatoms that do not meet desired patterns, a process that could be automated [16.29].

Biomimicry of diatom frustule synthesis should also be mentioned as a potential future strategy for the alternative controlled and green synthesis of MSM/Ns [16.145]. For this, a number of organic molecules [including silaffins, silacidins, cingulins, silicanins, long-chain polyamines (LCPAs), and polysaccharides] that have been found to be connected to the biomineralization process in diatoms have been successfully isolated, leading the way toward the mimicry of the synthesis of diatom frustules [16.2, 16.86, 16.145]. However, so far, this approach has been hampered by low yield and comparably low structural order. Nevertheless, with the joint efforts of a number of research groups, evident in rising numbers of publications, this approach can be predicted to become valuable in future years.

In all this discussion, one should, however, not forget that any alterations which might be achieved will have to be within the tolerance range of a given taxon, which limits not only the methods which can be used (e.g., to which extent the medium can be changed) but also the achievable results. Thus, we believe that in a first step a suitable species has to be selected to which small amendments can be induced. Nevertheless, we do not yet know how far we can push diatoms to change their morphology [16.29].

16.5.1 Physicochemical Parameters Alteration Approach

There is no doubt about the fast response of diatom growth to the physicochemical parameters of the surrounding medium [16.26, 16.100]. Thus, one can expect the alteration of physicochemical parameters, within the tolerance range of a given taxon, to have an effect on the ultrastructure of diatom frustules. This approach, once optimized, could finally lead to the control of a range of structural parameters of the frustules.

In 1999, it was for instance already suggested that salinity could affect frustule pore size [16.84] and probably other morphological characters. This might be, because NaCl can penetrate into and crystallize within the silicalemma [16.32]. A decade later, Vrieling *et al.* [16.123] and Leterme *et al.* [16.61] demonstrated that lowered salinity levels in the growth medium decrease the pore size of three diatom frustules including two centrics and one pennate (*Cocconeis placentula*). In addition, Leterme *et al.* [16.62] found that at elevated salinity levels the pore size of *Cocconeis placentula* and *Cocconeis pinnata* increased. However, no consistent picture was found when considering the changes in pore surface area with salinity level, which increased in the case of *Cocconeis pinnata* but decreased in the case of *Cocconeis placentula* with increasing salinity. These results might indicate a species-specific dependence of pore shape, size, and surface area on the salinity of the growth medium, which would have to be studied for each chosen target species in more detail. Moreover, Marchand *et al.* [16.73] reported a progressive transition from a "fusiform morphotype" to an "oval morphotype" of *Phaeodactylum tricornutum* through decreases in the salinity levels in the culture.

Besides salinity, the spatial arrangement of pores can be manipulated by silica limitation [16.33], but changes in pore dimension or shape were not examined. Recent results on the cultivation of isolated *Nitzschia palea* from a single source suggest that the concentration of nutrients also affects frustule dimensions and ultrastructure including the mesopore size within the pore occlusions [16.25].

Su *et al.* [16.110] reviewed the main abiotic factors during growth [temperature, salinity, pH, heavy metals, light, $Si(OH)_4$, and $Ge(OH)_4$] that could affect the diatom's ultrastructure, and could therefore be used as a starting guide for the approach of tailoring the frustule's parameters through changes in physiological condition. A complete and expansive study of the optimization of this approach is one of the top priorities for diatomists with the expectation of expanding diatom use in nanotechnology.

16.5.2 Genetic Engineering Approach

As an alternative strategy, gene editing could be valuable for tailoring the frustule's ultrastructure, which has been the focus of increasing research in recent years [16.45, 16.57, 16.117]. The results described in Tesson *et al.* [16.117] show the first considerable attempt of gene manipulation that really affected the frustule's ultrastructure. Figure 16.10 depicts

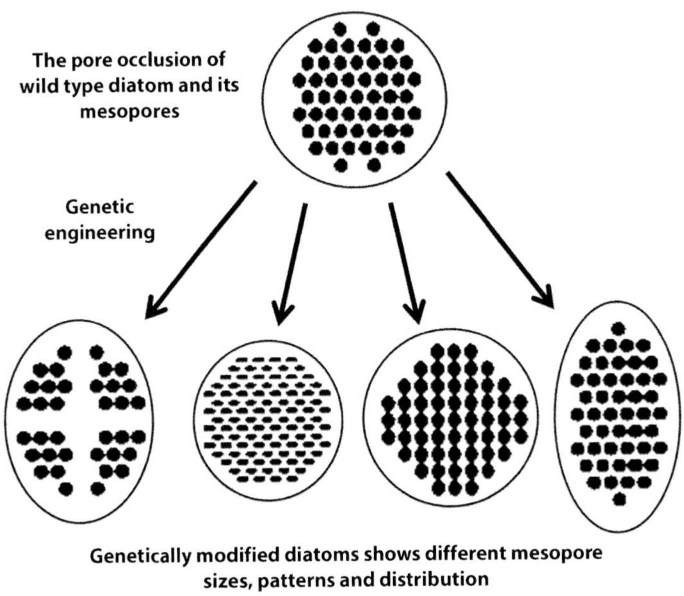

Figure 16.10 A schematic diagram of the concept of engineering diatom ultrastructure. The black circles and ellipses represent pore occlusion within a single areola, and black dots and other shapes indicate its mesopores. By generating genetically modified diatom strains that contain tailored alterations in components of the cellular silica forming machinery (e.g., knockout of certain protein and expression of mutant protein), it may be possible to alter the mesopore sizes patterns and distribution. Modified from [16.57] with the kind permission of Elsevier.

the main idea of controlling morphological characters of pores, especially mesopores, via genetic engineering.

Despite all the advances in understanding diatom biomineralization, the accurate understanding of the mechanics for the 3D assembly of condensed silica to form the overall diatom frustule 3D structure still has considerable gaps. Furthermore, for most diatom species, we still lack sufficiently promising genetic engineering methods often already hampered by the missing knowledge of their genome. Thus, while promising, this approach will require a considerable amount of additional research to become applicable.

16.6 Conclusion

Raphid pennate diatom frustules could be one of the most promising alternative natural sources for MSMs with highly competitive surface area especially when considering recent advances in our potential ability to control their structural parameters. In future, the area of desired mesopore sizes might be increased and the desired thickness of the siliceous membranous layer of pore occlusion might be tailored. More attention should be paid by diatomists and material scientists in order to obtain the best taxa, before trying optimizing it, for their selected application in diatom nanotechnology.

Acknowledgment

This work was supported by the Deutsche Forschungsgemeinschaft (DFG) (grant number: RE-3109\2-1; project number: 422038271).

References

[16.1] Abo-Shady, A.M., Zalat, A.A., Al-Ashkar, E.A., Ghobara, M.M., Nanoporous silica of some Egyptian diatom frustules as a promising natural material. *Nanosci. Nanotechnol.-Asia*, 9, 3, 414–425, 2019.

[16.2] Annenkov, V.V., Danilovtseva, E.N., Gordon, R., Steps of silicic acid transformation to silica frustules: Main hypotheses and discoveries, in: *Diatom Morphogenesis [DIMO, Volume 3 in the series: Diatoms: Biology & Applications*, V. Annenkov, J. Seckbach, R. Gordon (Eds.), Wiley-Scrivener, Beverly, MA, USA Submitted, 2021.

[16.3] Asefa, T. and Tao, Z., Biocompatibility of mesoporous silica nanoparticles. *Chem. Res. Toxicol.*, 25, 11, 2265–2284, 2012.

[16.4] Bariana, M., Aw, M.S., Kurkuri, M., Losic, D., Tuning drug loading and release properties of diatom silica microparticles by surface modifications. *Int. J. Pharm.*, 443, 1, 230–241, 2013.

[16.5] Beninger, P.G. and Decottignies, P., What makes diatoms attractive for suspensivores? The organic casing and associated organic molecules of *Coscinodiscus perforatus* are quality cues for the bivalve *Pecten maximus*. *J. Plankton Res.*, 27, 11–17, 2005.

[16.6] Bonner, J.T. and Hoffman, M.E., Evidence for a substance responsible for the spacing pattern of aggregation and fruiting in the cellular slime molds. *J. Embryol. Exp. Morphol.*, 11, 571–589, 1963.

[16.7] Chew, K. W., Chia, S. R., Show, P. L., Yap, Y. J., Ling, T. C., Chang, J.-S., Effects of water culture medium, cultivation systems and growth modes for microalgae cultivation: A review. *Journal of the Taiwan Institute of Chemical Engineers*, 91, 332–344, 2018.

[16.8] Chandrasekaran, S. and Voelcker, N.H., Chapter 7: The potential of modified diatom frustules for solar energy conversion, in: *Diatom Nanotechnology: Progress and Emerging Applications*, The Royal Society of Chemistry, United Kingdom (UK), pp. 150–174, 2018.

[16.9] Chepurnov, V.A., Chaerle, P., Mann, D.G., How to breed diatoms: Examination of two species with contrasting reproductive biology, in: *The Science of Algal Fuels: Phycology, Geology, Biophotonics, Genomics and Nanotechnology*, R. Gordon and J. Seckbach (Eds.), pp. 323–340, Springer, Dordrecht, 2012.

[16.10] Chepurnov, V.A., Chaerle, P., Roef, L., Van Meirhaeghe, A., Vanhoutte, K., Classical breeding in diatoms: Scientific background and practical perspectives, in: *Diatom World*, vol. 19, Seckbach, J.K.J.P. (Ed.), pp. 167–194, 2011.

[16.11] Chepurnov, V.A. and Vanhoutte, K., *Diatom breeding: Scientific background and practical perspectives*, SBAE Industries, Oostmoer, Belgium, 2009.

[16.12] Collier, A. and Murphy, A., Very small diatoms: Preliminary notes and description of *Chaetoceros galvestonensis*. *Science*, 136, 3518, 780–&, 1962.

[16.13] Cox, E.J., Pore occlusions in raphid diatoms-a reassessment of their structure and terminology, with particular reference to members of the Cymbellales. *Diatom*, 20, 33–46, 2004.

[16.14] Croissant, J.G., Fatieiev, Y., Almalik, A., Khashab, N.M., Mesoporous silica and organosilica nanoparticles: Physical chemistry, biosafety, delivery strategies, and biomedical applications. *Adv. Healthcare Mater.*, 7, 4, 1700831, 2018.

[16.15] Cros, L. and Fortuño, J.M., Atlas of Northwestern Mediterranean Coccolithophores. *Sci. Mar.*, 66, S1, 1–182, 2002.

[16.16] Davis, M.E., Ordered porous materials for emerging applications. *Nature*, 417, 6891, 813–821, 2002.

[16.17] De Stefano, L., Rendina, I., De Stefano, M., Bismuto, A., Maddalena, P., Marine diatoms as optical chemical sensors. *Appl. Phys. Lett.*, 87, 23, 233902, 2005.

[16.18] Deblandre, G.A., Wettstein, D.A., Koyano-Nakagawa, N., Kintner, C., A two-step mechanism generates the spacing pattern of the ciliated cells in the skin of *Xenopus* embryos. *Development*, 126, 21, 4715–4728, 1999.

[16.19] Diab, M. and Mokari, T., Bioinspired hierarchical porous structures for engineering advanced functional inorganic materials. *Adv. Mater.*, 30, 41, 1706349, 1–9, 2018.

[16.20] Dordick, J.S. and Freeman, A., Biocatalysis as a discovery tool: From nanoscale to high-throughput and beyond. *Curr. Opin. Biotechnol.*, 17, 6, 559–561, 2006.

[16.21] Ehrlich, H. and Witkowski, A., Biomineralization in diatoms: The organic templates, in: *Evolution of Lightweight Structures: Analyses and Technical Applications*, C. Hamm (Ed.), pp. 39–58, Springer Netherlands, Dordrecht, 2015.

[16.22] Eslami, M., Dekamin, M.G., Motlagh, L., Maleki, A., MCM-41 mesoporous silica: a highly efficient and recoverable catalyst for rapid synthesis of α-aminonitriles and imines. *Green Chem. Lett. Rev.*, 11, 1, 36–46, 2018.

[16.23] Geisler, M., Nadeau, J., Sack, F.D., Oriented asymmetric divisions that generate the stomatal spacing pattern in *Arabidopsis* are disrupted by the *too many mouths* mutation. *Plant Cell*, 12, 11, 2075–2086, 2000.

[16.24] Gerardin, C., Reboul, J., Bonne, M., Lebeau, B., Ecodesign of ordered mesoporous silica materials. *Chem. Soc. Rev.*, 42, 9, 4217–4255, 2013.

[16.25] Ghobara, M., *Studies on the frustules of some diatom species and its applications in nanotechnology*, Master thesis, Faculty of Science, Tanta University, Tanta, 2016.

[16.26] Ghobara, M. and Salem, Z.E., Spatiotemporal fluctuations in phytoplankton communities and their potential indications for the pollution status of the irrigation and drainage water in the Middle Nile Delta Area, Egypt, in: *Conventional Water Resources and Agriculture in Egypt*, A.M. Negm (Ed.), pp. 317–345, Springer International Publishing, Cham, 2019.

[16.27] Ghobara, M.M., Mazumder, N., Vinayak, V., Reissig, L., Gebeshuber, I.C., Tiffany, M.A., Gordon, R., On light and diatoms: A photonics and photobiology review [Chapter 7], in: *Diatoms: Fundamentals & Applications [DIFA, Volume 1 in the series: Diatoms: Biology & Applications, series editors: Richard Gordon & Joseph Seckbach]*, J. Seckbach and R. Gordon (Eds.), pp. 129–190, Wiley-Scrivener, Beverly, MA, USA, 2019.

[16.28] Ghobara, M.M. and Mousa, A.M., Diatomite in use: Nature, modification, commercial applications, and prospective trends, in: *Diatoms: Fundamentals & Applications [DIFA, Volume 1 in the series: Diatoms: Biology & Applications, series editors: Richard Gordon & Joseph Seckbach]*, J. Seckbach and R. Gordon (Eds.), pp. 471–510, Wiley-Scrivener, Beverly, MA, USA, 2019.

[16.29] Gordon, R., Computer controlled evolution of diatoms: design for a compustat. *Nova Hedwig. Beih.*, 112, 215–220, 1996.

[16.30] Gordon, R., Diatoms and nanotechnology: Early history and imagined future as seen through patents, in: *The Diatoms: Applications for the Environmental and Earth Sciences*, 2nd, J.P. Smol and E.F. Stoermer (Eds.), pp. 585–602, Cambridge University Press, Cambridge, 2010.

[16.31] Gordon, R. and Aguda, B.D., Diatom morphogensis-natural fractal fabrication of a complex microstructure, in: *Proceedings of the Annual International Conference of the Ieee Engineering in Medicine and Biology Society*, Pts 1–4, pp. 273–274, 1988.

[16.32] Gordon, R. and Brodland, G.W., On square holes in pennate diatoms. *Diatom Res.*, 5, 2, 409–413, 1990.

[16.33] Gordon, R. and Drum, R.W., The chemical basis of diatom morphogenesis. *Int. Rev. Cytol. - Survey Cell Biol.*, 150, 243–372, 1994.

[16.34] Gordon, R., Losic, D., Tiffany, M.A., Nagy, S.S., Sterrenburg, F.A.S., The Glass Menagerie: Diatoms for novel applications in nanotechnology. *Trends Biotechnol.*, 27, 2, 116–127, 2009.

[16.35] Gordon, R., Merz, C.R., Gurke, S., Schoefs, B., Bubble farming: Scalable microcosms for diatom biofuel and the next Green Revolution [Chapter 22], in: *Diatoms: Fundamentals & Applications [DIFA, Volume 1 in the series: Diatoms: Biology & Applications, series editors: Richard Gordon & Joseph Seckbach]*, J. Seckbach and R. Gordon (Eds.), pp. 583–654, Wiley-Scrivener, Beverly, MA, USA, 2019.

[16.36] Gordon, R. and Stone, R., Cybernetic embryo, in: *Biocommunication*, World Scientific Press, pp. 111–164, London, UK, 2016.

[16.37] Hao, N., Li, L., Tang, F., Shape matters when engineering mesoporous silica-based nanomedicines. *Biomater. Sci.*, 4, 4, 575–591, 2016.

[16.38] Hao, N., Nie, Y., Tadimety, A., Closson, A.B., Zhang, J.X.J., Microfluidics-mediated self-template synthesis of anisotropic hollow ellipsoidal mesoporous silica nanomaterials. *Mater. Res. Lett.*, 5, 8, 584–590, 2017.

[16.39] Hartmann, M. and Jung, D., Biocatalysis with enzymes immobilized on mesoporous hosts: the status quo and future trends. *J. Mater. Chem.*, 20, 5, 844–857, 2010.

[16.40] He, Q., Zhang, Z., Gao, Y., Shi, J., Li, Y., Intracellular localization and cytotoxicity of spherical mesoporous silica nano- and microparticles. *Small*, 5, 23, 2722–2729, 2009.

[16.41] Heikkila, T., Santos, H.A., Kumar, N., Murzin, D.Y., Salonen, J., Laaksonen, T., Peltonen, L., Hirvonen, J., Lehto, V.P., Cytotoxicity study of ordered mesoporous silica MCM-41 and SBA-15 microparticles on Caco-2 cells. *Eur. J. Pharm. Biopharm.*, 74, 3, 483–494, 2010.

[16.42] Hildebrand, M. and Lerch, S.J.L., Diatom silica biomineralization: Parallel development of approaches and understanding. *Semin. Cell Dev. Biol.*, 46, 27–35, 2015.

[16.43] Hildebrand, M., Lerch, S.J.L., Shrestha, R.P., Understanding diatom cell wall silicification: Moving forward. *Front. Mar. Sci.*, 5, 125, 1–19, 2018.

[16.44] Holder, N. and Glade, R., Skin glands in the axolotl: The creation and maintenance of a spacing pattern. *J. Embryol. Exp. Morphol.*, 79, 97–112, 1984.

[16.45] Huang, W. and Daboussi, F., Genetic and metabolic engineering in diatoms. *Philos. Trans. R. Soc. B: Biol. Sci.*, 372, 1728, 20160411, 2017.

[16.46] Hudson, S., Cooney, J., Magner, E., Proteins in mesoporous silicates. *Angew. Chem. Int. Ed.*, 47, 45, 8582–8594, 2008.

[16.47] Jo, B.Y. and Kim, H.S., Newly recorded species of the genus *Synura* (Synurophyceae) from Korea. *J. Ecol. Environ.*, 41, 1, 1, 2017.

[16.48] Jo, B.Y., Kim, J.I., Škaloud, P., Siver, P.A., Shin, W., Multigene phylogeny of *Synura* (Synurophyceae) and descriptions of four new species based on morphological and DNA evidence. *Eur. J. Phycol.*, 51, 4, 413–430, 2016.

[16.49] Ju, M., Li, Y., Yu, L., Wang, C., Zhang, L., Two-phase diffusion technique for the preparation of ultramacroporous/mesoporous silica microspheres via interface hydrolysis, diffusion, and gelation of TEOS. *Langmuir*, 34, 5, 2046–2056, 2018.

[16.50] Kagan, M.L. and Sachs, T., Development of immature stomata: Evidence for epigenetic selection of a spacing pattern. *Dev. Biol.*, 146, 1, 100–105, 1991.

[16.51] Kamińska, A., Winkler, K., Kowalska, A., Witkowska, E., Szymborski, T., Janeczek, A., Waluk, J., SERS-based immunoassay in a microfluidic system for the multiplexed recognition of interleukins from blood plasma: Towards picogram detection. *Sci. Rep.*, 7, 1, 10656, 2017.

[16.52] Kim, J.Y., Cho, S.W., Song, W.C., Lee, M.J., Cai, J.L., Ohk, S.H., Song, H.K., Degan, A., Jung, H.S., Formation of spacing pattern and morphogenesis of chick feather buds is regulated by cytoskeletal structures. *Differentiation*, 73, 5, 240–248, 2005.

[16.53] Kling, S.A., Chapter 9 - Radiolaria, in: *Introduction to Marine Micropaleontology*, Second Edition, B.U. Haq and A. Boersma (Eds.), pp. 203–244, Elsevier Science B.V., Amsterdam, 1998.

[16.54] Kong, X., Squire, K., Wang, A.X., Microfluidic Diatomite Analytical Devices for Ultra-Sensitive Detection of Hazardous Chemicals. *Conference on Lasers and Electro-Optics*, Optical Society of America, San Jose, California, p. STh3J.5, 2018.

[16.55] Kong, X.M., Chong, X.Y., Squire, K., Wang, A.X., Microfluidic diatomite analytical devices for illicit drug sensing with ppb-Level sensitivity. *Sens. Actuators B-Chem.*, 259, 587–595, 2018.

[16.56] Koto, A., Kuranaga, E., Miura, M., Programmed cell death controls spacing pattern of *Drosophila* sensory bristles. *Differentiation*, 80, S48–S49, 2010.

[16.57] Kröger, N., Prescribing diatom morphology: Toward genetic engineering of biological nanomaterials. *Curr. Opin. Chem. Biol.*, 11, 6, 662–669, 2007.

[16.58] Kröger, N., Dubey, N.C., Kumari, E., Chapter 6: Immobilization of proteins on diatom biosilica, in: *Diatom Nanotechnology: Progress and Emerging Applications*, pp. 126–149, The Royal Society of Chemistry, United Kingdom (UK), 2018.

[16.59] Lee, C.-H., Lin, T.-S., Mou, C.-Y., Mesoporous materials for encapsulating enzymes. *Nano Today*, 4, 2, 165–179, 2009.

[16.60] Lenoci, L. and Camp, P.J., Diatom structures templated by phase-separated fluids. *Langmuir*, 24, 1, 217–223, 2008.

[16.61] Leterme, S., Ellis, A., Mitchell, J., Buscot, M.-J., Pollet, T., Schapira, M., Seuront, L., Morphological flexibility of *Cocconeis placentula* (Bacillariophyceae) nanostructure to changing salinity levels. *J. Phycol.*, 46, 715, 2010.

[16.62] Leterme, S.C., Prime, E., Mitchell, J., Brown, M.H., Ellis, A.V., Diatom adaptability to environmental change: a case study of two *Cocconeis* species from high-salinity areas. *Diatom Res.*, 28, 1, 29–35, 2013.

[16.63] Lewin, J.C., The dissolution of silica from diatom walls. *Geochim. Cosmochim. Acta*, 21, 3, 182–198, 1961.

[16.64] Li, Y., Song, F., Cheng, L., Qian, J., Chen, Q., Functionalized large-pore mesoporous silica microparticles for gefitinib and doxorubicin codelivery. *Mater. (Basel)*, 12, 5, 766, 1–19, 2019.

[16.65] Li-hong, W., Xin, C., Hui, X., Li-li, Z., Jing, H., Mei-juan, Z., Jie, L., Yi, L., Jin-wen, L., Wei, Z., Gang, C., A novel strategy to design sustained-release poorly water-soluble drug mesoporous silica microparticles based on supercritical fluid technique. *Int. J. Pharm.*, 454, 1, 135–142, 2013.

[16.66] Limnell, T., Heikkilä, T., Santos, H.A., Sistonen, S., Hellstén, S., Laaksonen, T., Peltonen, L., Kumar, N., Murzin, D.Y., Louhi-Kultanen, M., Salonen, J., Hirvonen, J., Lehto, V.-P., Physicochemical stability of high indomethacin payload ordered mesoporous silica MCM-41 and SBA-15 microparticles. *Int. J. Pharm.*, 416, 1, 242–251, 2011.

[16.67] Losic, D., *Diatom Nanotechnology: Progress and Emerging Applications*, The Royal Society of Chemistry, Cambridge, UK, 2018.

[16.68] Losic, D., Rosengarten, G., Mitchell, J., Voelcker, N., Pore architecture of diatom frustules: Potential nanostructured membranes for molecular and particle separations. *J. Nanosci. Nanotechnol.*, 6, 982–989, 2006.

[16.69] Losic, D., Yu, Y., Aw, M.S., Simovic, S., Thierry, B., Addai-Mensah, J., Surface functionalisation of diatoms with dopamine modified iron-oxide nanoparticles: toward magnetically guided drug microcarriers with biologically derived morphologies. *Chem. Commun.*, 46, 34, 6323–6325, 2010.

[16.70] Lu, F., Wu, S.-H., Hung, Y., Mou, C.-Y., Size effect on cell uptake in well-suspended, uniform mesoporous silica nanoparticles. *Small*, 5, 12, 1408–1413, 2009.

[16.71] Mann, D., Sieves and flaps: siliceous minutiae in the pores of raphid diatoms, in: *Proceedings of the 6th Symposium on Recent and Fossil Diatoms,* R. Ross (Ed.), pp. 279 -300, Koeltz Botanical Books, Koenigstein, Germany, 1981.

[16.72] Manzano, M. and Vallet-Regí, M., Mesoporous silica nanoparticles in nanomedicine applications. *J. Mater. Sci.: Mater. Med.*, 29, 5, 65, 2018.

[16.73] Marchand, J., Moreau, B., Ghobara, M., Solymosi, K., Schoefs, B., Le changement de forme chez les diatomées: un phénomène spectaculaire [poster], in: *36ème Colloque de l'Association des Diatomistes de Langue Française (ADLaF)*, 2017, https://www.researchgate.net/publication/320014019_Le_changement_de_forme_chez_les_diatomees_un_phenomene_spectaculaire.

[16.74] McCartney, K., Witkowski, J., Jordan, R.W., Daugbjerg, N., Malinverno, E., van Wezel, R., Kano, H., Abe, K., Scott, F., Schweizer, M., Young, J.R., Hallegraeff, G.M., Shiozawa, A., Fine structure of silicoflagellate double skeletons. *Mar. Micropaleontol.*, 113, 10–19, 2014.

[16.75] Merz, C.R. and Main, K.L., Microalgae (diatom) production—The aquaculture and biofuel nexus, in: *Oceans-St. John's, 2014*, pp. 1–10, IEEE, Newfoundland, Canada, 2014.

[16.76] Mikkelsen, D.M., Ellegaard, M., Lundholm, N., Cytoplasmic and frustular ultrastructure and morphology of *Plagiotropis tristis* sp. nov. (Bacillariophyceae). *Diatom Res.*, 20, 1, 97–113, 2005.

[16.77] Miletto, I., Paul, G., Chapman, S., Gatti, G., Marchese, L., Raja, R., Gianotti, E., Mesoporous silica scaffolds as precursor to drive the formation of hierarchical SAPO-34 with tunable acid properties. *Chemistry (Weinheim an der Bergstrasse, Germany)*, 23, 41, 9952–9961, 2017.

[16.78] Mitchell, J.G., Chapter 1: Whence is the diversity of diatom frustules derived? in: *A book chapter In Diatom Nanotechnology: Progress and Emerging Applications*, D. Losic (Ed.), pp. 1–13, The Royal Society of Chemistry, Cambridge, UK, 2018.

[16.79] Nanaki, S., Siafaka, P.I., Zachariadou, D., Nerantzaki, M., Giliopoulos, D.J., Triantafyllidis, K.S., Kostoglou, M., Nikolakaki, E., Bikiaris, D.N., PLGA/SBA-15 mesoporous silica composite microparticles loaded with paclitaxel for local chemotherapy. *Eur. J. Pharm. Sci.*, 99, 32–44, 2017.

[16.80] Narayan, R., Nayak, U.Y., Raichur, A.M., Garg, S., Mesoporous silica nanoparticles: A comprehensive review on synthesis and recent advances. *Pharmaceutics*, 118, 10, 3, 1–49, 2018.

[16.81] Nardi, J.B. and Magee-Adams, S.M., Formation of scale spacing patterns in a moth wing I. Epithelial feet may mediate cell rearrangement. *Dev. Biol.*, 116, 2, 278–290, 1986.

[16.82] Orefice, I., Musella, M., Smerilli, A., Sansone, C., Chandrasekaran, R., Corato, F., Brunet, C., Role of nutrient concentrations and water movement on diatom's productivity in culture. *Sci. Rep.*, 9, 1479, 2019.

[16.83] Pan, Z., Lerch, S.J.L., Xu, L., Li, X., Chuang, Y.-J., Howe, J.Y., Mahurin, S.M., Dai, S., Hildebrand, M., Electronically transparent graphene replicas of diatoms: a new technique for the investigation of frustule morphology. *Sci. Rep.*, 4, 1, 6117, 2014.

[16.84] Parkinson, J. and Gordon, R., Beyond micromachining: the potential of diatoms. *Trends Biotechnol.*, 17, 5, 190–196, 1999.

[16.85] Pavel, I.-A., Girardon, M., El Hajj, S., Parant, S., Amadei, F., Kaufmann, S., Tanaka, M., Fierro, V., Celzard, A., Canilho, N., Pasc, A., Lipid-coated mesoporous silica microparticles for the controlled delivery of β-galactosidase into intestines. *J. Mater. Chem. B*, 6, 35, 5633–5639, 2018.

[16.86] Pawolski, D., Heintze, C., Mey, I., Steinem, C., Kröger, N., Reconstituting the formation of hierarchically porous silica patterns using diatom biomolecules. *J. Struct. Biol.*, 204, 1, 64–74, 2018.

[16.87] Poulíčková, A. and Mann, D.G., Diatom sexual reproduction and life cycles, in: *Diatoms: Fundamentals & Applications [DIFA, Volume 1 in the series: Diatoms: Biology & Applications, series editors: Richard Gordon & Joseph Seckbach]*, J. Seckbach and R. Gordon (Eds.), pp. 243–270, Wiley-Scrivener, Beverly, MA, USA, 2019.

[16.88] Prokopowicz, M., Czarnobaj, K., Szewczyk, A., Sawicki, W., Preparation and *in vitro* characterisation of bioactive mesoporous silica microparticles for drug delivery applications. *Mater. Sci. Eng.: C*, 60, 7–18, 2016.

[16.89] Rea, I., Dardano, P., Ferrara, M.A., de Stefano, L., Chapter 5: Micro- and nano-optical devices from diatom nanostructures: Light control by mother nature, in: *Diatom Nanotechnology: Progress and Emerging Applications*, D. Losic (Ed.), pp. 111–125, The Royal Society of Chemistry, Cambridge, UK, 2018.

[16.90] Ren, F., Campbell, J., Wang, X., Rorrer, G.L., Wang, A.X., Enhancing surface plasmon resonances of metallic nanoparticles by diatom biosilica. *Opt. Express*, 21, 13, 15308–15313, 2013.

[16.91] Renaud, O. and Simpson, P., Movement of bristle precursors contributes to the spacing pattern in *Drosophila*. *Mech. Dev.*, 119, 2, 201–211, 2002. #Pii s0925-4773(02)00381-7.

[16.92] Rosengarten, G. and Herringer, J.W., Chapter 2: Interactions of diatoms with their fluid environment, in: *Diatom Nanotechnology: Progress and Emerging Applications*, pp. 14–54, The Royal Society of Chemistry, Cambridge, UK, 2018.

[16.93] Ross, R., Cox, E., Karayeva, N.I., Mann, D., Paddock, T.B.B., Simonsen, R., Sims, P., An amended terminology for the siliceous components of the Diatom cell. *Nova Hedwig. Beih.*, 64, 513–533, 1979.

[16.94] Ross, R. and Sims, P.A., The fine structure of the frustule in centric diatoms: A suggested terminology. *Br. Phycol. J.*, 7, 2, 139–163, 1972.

[16.95] Round, F.E., Crawford, R.M., Mann, D.G., *Diatoms: Biology and Morphology of the Genera*, Cambridge University Press, Cambridge, UK, 1990.
[16.96] Rorrer, G.L. CHAPTER 4. Functionalization of Frustules From Diatom Cell Culture for Optoelectronic Properties. In: *Progress and Emerging Applications,* D. Losic (Ed.), 79–110, The Royal Society of Chemistry, UK, 2018.
[16.97] Rouquerol, J., Avnir, D., Fairbridge, C.W., Everett, D.H., Haynes, J.H., Pernicone, N., Ramsay, J.D.F., Sing, K.S.W., Unger, K.K., Recommendations for the characterization of porous solids. *Pure Appl. Chem.*, 66, 8, 1739–1758, 1994.
[16.98] Sabu, S., Singh, I.S.B., Joseph, V., Optimisation of critical medium components and culture conditions for enhanced biomass and lipid production in the oleaginous diatom *Navicula phyllepta*: a statistical approach. *Environ. Sci. Pollut. Res.*, 24, 34, 26763–26777, 2017.
[16.99] Sahare, P., Ayala, M., Vazquez-Duhalt, R., Pal, U., Loni, A., Canham, L.T., Osorio, I., Agarwal, V., Enhancement of peroxidase stability against oxidative self-inactivation by co-immobilization with a redox-active protein in mesoporous silicon and silica microparticles. *Nanoscale Res. Lett.*, 11, 1, 417, 2016.
[16.100] Salem, Z., Ghobara, M., El Nahrawy, A.A., Spatio-temporal evaluation of the surface water quality in the middle Nile Delta using Palmer's algal pollution index. *Egypt. J. Basic Appl. Sci.*, 4, 3, 219–226, 2017.
[16.101] Şan, O., Gören, R., Özgür, C., Purification of diatomite powder by acid leaching for use in fabrication of porous ceramics. *Int. J. Miner. Process.*, 93, 1, 6–10, 2009.
[16.102] Sancenón, F., Pascual, L., Oroval, M., Aznar, E., Martínez-Máñez, R., Gated silica mesoporous materials in sensing applications. *ChemistryOpen*, 4, 4, 418–437, 2015.
[16.103] Schlipf, D.M., *Biomolecule localization and surface engineering within size tunable nanoporous silica particles,* University of Kentucky, Chemical and Materials Engineering Department, USA, 2015.
[16.104] Schnittger, A., Folkers, U., Schwab, B., Jurgens, G., Hülskamp, M., Generation of a spacing pattern: The role of TRIPTYCHON in trichome patterning in *Arabidopsis. Plant Cell*, 11, 6, 1105–1116, 1999.
[16.105] Shen, S., Gu, T., Mao, D., Xiao, X., Yuan, P., Yu, M., Xia, L., Ji, Q., Meng, L., Song, W., Yu, C., Lu, G., Synthesis of nonspherical mesoporous silica ellipsoids with tunable aspect ratios for magnetic assisted assembly and gene delivery. *Chem. Mater.*, 24, 1, 230–235, 2012.
[16.106] Smetacek, V., The giant diatom dump. *Nature*, 406, 6796, 574–575, 2000.
[16.107] Zakeri Siavashani, A., Haghbin Nazarpak, M., Fayyazbakhsh, F. *et al.*, Effect of amino-functionalization on insulin delivery and cell viability for two types of silica mesoporous structures. *J. Mater. Sci.*, 51, 10897–10909, 2016.
[16.108] Spaulding, S.A., Lubinski, D.J., Potapova, M., Diatoms of the United States: Glossary, USA, 2016, http://westerndiatoms.colorado.edu/glossary.
[16.109] Squire, K., Kong, X.M., LeDuff, P., Rorrer, G.L., Wang, A.X., Photonic crystal enhanced fluorescence immunoassay on diatom biosilica. *J. Biophotonics*, 11, 10, e201800009, 2018.
[16.110] Su, Y., Lundholm, N., Ellegaard, M., Effects of abiotic factors on the nanostructure of diatom frustules—ranges and variability. *Appl. Microbiol. Biotechnol.*, 102, 14, 5889–5899, 2018.
[16.111] Tang, S., Huang, X., Chen, X., Zheng, N., Hollow mesoporous zirconia nanocapsules for drug delivery. *Adv. Funct. Mater.*, 20, 15, 2442–2447, 2010.
[16.112] Tao, Z., Toms, B., Goodisman, J., Asefa, T., Mesoporous silica microparticles enhance the cytotoxicity of anticancer platinum drugs. *ACS Nano*, 4, 2, 789–794, 2010.
[16.113] Taylor, A.B. and Fallon, J.R., Dendrites contain a spacing pattern. *J. Neurosci.*, 26, 4, 1154–1163, 2006.

[16.114] Teng Sing, T., Leaw Chui, P., Lim Hong, C., Lim Po., T., The genus *Pseudo-nitzschia* (Bacillariophyceae) in Malaysia, including new records and a key to species inferred from morphology-based phylogeny. *Bot. Mar.*, 56, 4, 375–398, 2013.

[16.115] Terracciano, M., De Stefano, L., Rea, I., Diatoms: Green nanotechnology for biosilica-based drug delivery systems. *Pharmaceutics*, 10, 4, 242, 1–15, 2018.

[16.116] Tesson, B. and Hildebrand, M., Characterization and localization of insoluble organic matrices associated with diatom cell walls: Insight into their roles during cell wall formation. *PLoS One*, 8, 4, e61675, 2013.

[16.117] Tesson, B., Lerch, S.J.L., Hildebrand, M., Characterization of a new protein family associated with the silica deposition vesicle membrane enables genetic manipulation of diatom silica. *Sci. Rep.*, 7, 1, 13457, 2017.

[16.118] Travaglini, L. and De Cola, L., Morphology control of mesoporous silica particles using bile acids as cosurfactants. *Chem. Mater.*, 30, 12, 4168–4175, 2018.

[16.119] Vinayak, V., Joshi, K.B., Gordon, R., Schoefs, B., CHAPTER 3 Nanoengineering of Diatom Surfaces for Emerging Applications, in: *Diatom Nanotechnology: Progress and Emerging Applications*, pp. 55–78, The Royal Society of Chemistry, Cambridge, UK, 2018.

[16.120] Vitug, L.V.D. and Baldia, S.F., Enhancement of some culture conditions for optimizing growth and lipid production in the diatom *Nitzschia palea*. *Acta Manila.*, 62, 25–34, 2014.

[16.121] Vivero-Escoto, J.L., Chiang, Y.-D., Wu, K., Yamauchi, Y., Recent progress in mesoporous titania materials: adjusting morphology for innovative applications. *Sci. Technol. Adv. Mater.*, 13, 1, 013003, 2012.

[16.122] Vrieling, E.G., Beelen, T.P.M., Sun, Q., Hazelaar, S., van Santen, R.A., Gieskes, W.W.C., Ultrasmall, small, and wide angle X-ray scattering analysis of diatom biosilica: interspecific differences in fractal properties. *J. Mater. Chem.*, 14, 13, 1970–1975, 2004.

[16.123] Vrieling, E.G., Sun, Q., Tian, M., Kooyman, P.J., Gieskes, W.W.C., van Santen, R.A., Sommerdijk, N.A.J.M., Salinity-dependent diatom biosilicification implies an important role of external ionic strength. *Proc. Natl. Acad. Sci. U. S. A.*, 104, 25, 10441–10446, 2007.

[16.124] Waldron, K., Wu, W.D., Wu, Z., Liu, W., Selomulya, C., Zhao, D., Chen, X.D., Formation of monodisperse mesoporous silica microparticles via spray-drying. *J. Colloid Interface Sci.*, 418, 225–233, 2014.

[16.125] Wang, J.K. and Seibert, M., Prospects for commercial production of diatoms. *Biotechnol. Biofuels*, 10, 1–3, 2017.

[16.126] Wang, Y., Cai, J., Jiang, Y., Jiang, X., Zhang, D., Preparation of biosilica structures from frustules of diatoms and their applications: current state and perspectives. *Appl. Microbiol. Biotechnol.*, 97, 2, 453–460, 2013.

[16.127] Willis, L., Cox, E.J., Duke, T., A simple probabilistic model of submicroscopic diatom morphogenesis. *J. R. Soc. Interface*, 10, 83, 20130067, 2013.

[16.128] Willis, L., Page, K., Broomhead, D., Cox, E., Discrete free-boundary reaction-diffusion model of diatom pore occlusions. *Plant Ecol. Evol.*, 143, 297–306, 2010.

[16.129] Wujek, D., Freshwater silica-scaled heterotrophic Protista: Heliozoa, Thaumatomonad Flagellates, Amoebae, and Bicosoecids, from the Lake Itasca Region, Minnesota. *J. Minn. Acad. Sci.*, 79, 1–13, 2015.

[16.130] Xiong, W. and Peng, J., Development and characterization of ferrihydrite-modified diatomite as a phosphorus adsorbent. *Water Res.*, 42, 19, 4869–4877, 2008.

[16.131] Xu, P., Yu, H., Li, X., Functionalized mesoporous silica for microgravimetric sensing of trace chemical vapors. *Anal. Chem.*, 83, 9, 3448–3454, 2011.

[16.132] Zhao, X., Chen, Z., Liu, Y., Huang, Q., Zhang, H., Ji, W., Ren, J., Li, J., Zhao, Y., Silk fibroin microparticles with hollow mesoporous silica nanocarriers encapsulation for abdominal wall repair. *Adv. Healthcare Mater.*, 7, 21, 1801005, 2018.

[16.133] Zheng, H.Q., Gao, F.F., Valtchev, V., Nanosized inorganic porous materials: fabrication, modification and application. *J. Mater. Chem. A*, 4, 43, 16756–16770, 2016.
[16.134] Qin, T., Gutu, T., Jiao, J., Chang, C.-H., Rorrer, G.L., Photoluminescence of Silica Nanostructures from Bioreactor Culture of Marine Diatom *Nitzschia frustulum*. *J. Nanosci. Nanotechnol.*, *8*, 5, 2392–2398, 2008.
[16.135] Kucki, M. and Fuhrmann-Lieker, T., Staining diatoms with rhodamine dyes: control of emission colour in photonic biocomposites. *J. R. Soc. Interface*, 9, 727–733, 2012.
[16.136] Lim, H., Leaw, C., Su, S., Teng, S., Usup, G., Mohammad-Noor, N., Lundholm, N., Kotaki, Y., Lim., P., Morphology and molecular characterization of Pseudo-nitzschia (Bacillariophyceae) from Malaysian Borneo, including the new species *Pseudo-nitzschia circumpora* sp. nov. *J. Phycol.*, 48, 1232–1247, 2012.
[16.137] Cox, E.J., Diatom identification in the face of changing species concepts and evidence of phenotypic plasticity. *J. Micropalaeontol.*, 33, 111–120, 2014.
[16.138] Mann, D.G., Crawford, R.M., Round, F.E., Bacillariophyta, in: *Handbook of the Protists*, J.M. Archibald, A.G.B. Simpson, C.H. Slamovits (Eds.), pp. 205–266, Springer, Cham, 2017.
[16.139] Jantschke, A., Fischer, C., Hensel, R., Braun, H.G., Brunner., E., Directed assembly of nanoparticles to isolated diatom valves using the non-wetting characteristics after pyrolysis. *Nanoscale*, 6, 11637–11645, 2014.
[16.140] Li, C.L., Ashworth, M.P., Witkowski, A., Dąbek, P., Medlin, L.K., Kooistra, W.H.C.F., Sato, S., Zgłobicka, I., Kurzydłowski, K.J., Theriot, E.C., Sabir, J.S.M., Khiyami, M.A., Mutwakil, M.H.Z., Sabir, M.J., Alharbi, N.S., Hajarah, N.H., Qing, S., Jansen., R.K., New insights into Plagiogrammaceae (Bacillariophyta) based on multigene phylogenies and morphological characteristics with the description of a new genus and three new species. *PLoS One*, 10, 2015, e0139300, 2015.
[16.141] Wang, Y., Zhang, D., Pan, J., Cai., J., Key factors influencing the optical detection of biomolecules by their evaporative assembly on diatom frustules. *J. Mater. Sci.*, 47, 6315–6325, 2012.
[16.142] Sato, S., Watanabe, T., Nagumo, T., Tanaka., J., Valve morphogenesis in an araphid diatom *Rhaphoneis amphiceros* (Rhaphoneidaceae, Bacillariophyta). *Phycological Res.*, 59, 236–243, 2011.
[16.143] Reichardt, E., New and recently described *gomphonema* species (Bacillariophyceae) from Siberia. *Fottea*, 9, 289–297, 2009.
[16.144] Slowing, I., Trewyn, B., Giri, S., Lin., V., Mesoporous silica nanoparticles for drug delivery and biosensing applications. *Adv. Funct. Mater.*, 17, 1225–1236, 2007.
[16.145] Patwardhan, S.V., Clarson, S.J., Perry, C.C., On the role(s) of additives in bioinspired silicification. *Chem. Commun.*, 2005, 9, 1113, 2005.
[16.146] Sasirekha, R., Sheena, T.S., Anitha, R., Santhanam, P., Kulandaivel, J., Characterizations and analysis of genus *Amphora* diatom frustules: A promising biomaterial. *Bioinspired Biomimetic Nanobiomater.*, 8, 4, 224–230, 2019.
[16.147] Stefano, L.D., Rendina, I., Stefano, M.D., Bismuto, A., Maddalena, P., Marine diatoms as optical chemical sensors. *Appl. Phys. Lett.*, 87, 23, 233902, 2005.
[16.148] Rea, I. and Stefano, L.D., Recent advances on diatom-based biosensors. *Sensors*, 19, 23, 5208, 2019.
[16.149] Brinker, C.J. and Scherer, G.W., *Sol-Gel Science: The Physics and Chemistry of Sol-Gel Processing*, Academic Press, New York, 1990.
[16.150] Iler, R.K., *The Chemistry of Silica: Solubility, Polymerization, Colloid and Surface Properties and Biochemistry of Silica*, p. 177, Wiley, New York, 1979.
[16.151] Annenkov, V.V., Gorshkov, A.G., Zelinskii, S.N. *et al.*, Macroporous liquid-chromatography matrices based on siliceous valves of diatoms. *Dokl. Chem.*, 432, 175–177, 2010.

[16.152]Vrieling, E.G., Beelen, T.P., Santen, R.A., Gieskes, W.W., Diatom silicon biomineralization as an inspirational source of new approaches to silica production. Progress in Industrial Microbiology Marine Bioprocess Engineering. *Proceedings of an International Symposium Organized under Auspices of The Working Party on Applied Biocatalysis of the Eurpean Federation of Biotechnology and The European Society for Marine Biotechnology*, pp. 39–51, 1999.
[16.153]Tang, F., Li, L., Chen, D., Mesoporous Silica Nanoparticles: Synthesis, Biocompatibility and Drug Delivery. *Adv. Mater.*, 24, 12, 1504–1534, 2012.

Glossary

Areola: perforation, or pore, in the diatom valve. Usually, many are grouped to form striae [16.108]. Plural is areolae

Centric diatom: the group of radial symmetric diatoms [16.108].

Pore occlusion: a delicate mesoporous silica layer covering the opening of an areola [16.13]. Velum is another term for pore occlusion. plural is vela.

Cribrum: a type of pore occlusion. Visible only using a scanning electron microscope. Plural form is cribrae [16.108].

Epitheca: see Theca.

Foramen: the large opening in one wall of a loculate areola. Located on the external or internal wall of the areola, depending on species, opposite the velum [16.108].

Frustule: the siliceous parts of the diatom cell wall [16.108].

Girdle: The series of linking, siliceous bands associated with a valve. Each band is called a girdle band [16.108].

Hymen: a kind of pore occlusion with very fine mesopores found mainly in raphid pennates. Plural is hymenes [16.13].

Hymenate: see hymen.

Hypotheca: see Theca.

Lineola: is an elongated areola in the apical direction. The plural is lineolae [16.108].

Macropore: pore with a diameter scale larger than 50 nm [16.97].

Mesopore: pore with a diameter scale ranging from 2 nm to 50 nm [16.97].

Mesoporous: a material with mesopores.

Micropore: pore with a diameter scale smaller than 2 nm [16.97].

MSM: mesoporous silica microparticles.

MSN: mesoporous silica nanoparticles.

Pennate: the group of generally bilaterally symmetric diatoms. This is a heterogeneous group that includes araphid, monoraphid, and biraphid taxa [16.108].

Raphe slit: a fissure through the valve face of raphid pennate diatoms that living cells usually use for gliding over surfaces [16.108].

Raphe: One or two slits, or fissures, through the valve face of raphid pennate diatoms. In some taxa (monoraphid), there is a raphe only on a single valve, while others (biraphid) have a raphe on both valves. Raphe position may be (1) axial, along the apical axis; (2) eccentric, along one margin; or (3) circumferential, around the whole margin of the valve [16.108].

Raphid diatoms: pennate diatoms that have one or two raphe slits [16.95].

Rica: the finest type of pore occlusion. See Hymen. Plural is ricae [16.13].

Rota: a type of pore occlusion, plural is Rotae [16.13].

Silicalemma: is a membrane inside the dividing living cells surrounding the vesicle where the silica deposition and new valve morphogenesis occurs.

Stria: a line of areolae or pores on the diatom surface. Plural is striae [16.108].

Theca: a theca is composed of a valve and its associated cingulum. From Greek, theca means sheath or case (plural is thecae). The larger theca of the frustule is called the epitheca and it overlaps the smaller theca, which is called the hypotheca [16.108].

Valve: a siliceous part of the diatom frustule that lies at each end of the frustule. The two valves are linked by the girdle [16.108].

Valvocopula: the girdle band in direct contact with the valve. Accessory structures such as septa may arise from the valvocopula [16.108].

Velum: see Pore occlusion.

Vola: a type of pore occlusions in which the pore is closed by a flap of silica, attached to the wall of the pore by a fairly broad base and extending most of the way across it, leaving only a curved slit. Plural is volae [16.13].

Index

Also of Interest

Check out these published and forthcoming related titles from Scrivener Publishing

The Mathematical Biology of Diatoms
Edited by Janice L. Pappas and Richard Gordon
Forthcoming 2022. ISBN 978-1-119-74985-1

Diatoms Morphogenesis
Edited by Vadim V. Annenkov, Joseph Seckbach and Richard Gordon
Published 2021. ISBN 978-1-119-48795-1

Diatoms Motility
Edited by Stanley A. Cohn, Kalina M. Manoylov and Richard Gordon
Published 2021. ISBN 978-1-119-52635-3

Diatoms
Fundamentals and Applications
Edited by Joseph Seckbach and Richard Gordon
Published 2019. ISBN 978-1-119-37021-5

www.scrivenerpublishing.com

Printed and bound by CPI Group (UK) Ltd, Croydon, CR0 4YY
14/09/2026
14962012-0001